ISBN 978-0-265-39634-6
PIBN 10026242

FB &c Ltd, Dalton House, 60 Windsor Avenue, London, SW19 2RR.
Company number 08720141. Registered in England and Wales.

BACTERIOLOGY

GENERAL, PATHOLOGICAL

AND

INTESTINAL

BY

ARTHUR ISAAC KENDALL, B.S., Ph.D., Dr.P.H.

PROFESSOR OF BACTERIOLOGY IN THE NORTHWESTERN UNIVERSITY MEDICAL SCHOOL, CHICAGO, ILLINOIS

ILLUSTRATED WITH 98 ENGRAVINGS AND 9 PLATES

LEA & FEBIGER
PHILADELPHIA AND NEW YORK

TO

THEOBALD SMITH, M.D., LL.D.

PREFACE.

"In the study of the microscopic forms known as bacteria we have what might be fitly called the focal points of the various branches of biological science. Though their investigation may require careful morphological researches, yet the unmistakable monotony of form combined with a considerable variation of physiological activity has compelled the bacteriologist to pay much attention to means by which such physiological variations may be more or less accurately registered in order that they may serve as a supplementary basis for classification. Again, with unicellular organisms the manifestations of cell activity become the most important phenomena for study. These manifestations bring together the fields of physiology and chemistry and make bacteriology in one sense a branch of physiological chemistry."[1]

"There is no ulterior interest in the study of bacteria as such, which is a strong impulse in many other departments of biological science. It is what bacteria do, rather than what they are, that commands attention, since our interest centers in the host rather than the parasite."[2]

The development of bacteriology has followed very closely the gradual improvement of the optical parts of the compound microscope, and to a lesser degree, the perfection of other instruments of precision on the one hand, and the production of anilin dyes and a great expansion of the fields of organic and physical chemistry on the other hand. Naturally the greatest advances in bacteriology have been made along the lines of morphology, staining and diagnosis, because the application of the microscope, anilin dyes, and the preparation and use of cultural media to bacterial problems is relatively simple and direct. The final chapters of bacteriology, in which the problems of immunology are of paramount interest, will be intimately associated with an unfolding of the chemistry of cellular activity, as Theobald Smith has so clearly pointed out in the opening paragraphs of this discussion.

[1] Theobald Smith. The Fermentation Tube, Wilder Quarter Century Book, 1893, p. 187.

[2] Theobald Smith. Some Problems in the Life History of Pathogenic Microörganisms, Amer. Med., 1904, viii, 711.

The chemistry of bacterial activity is not thoroughly studied at the present time and many of its problems must await the development of new methods of chemistry and physics, as well as a refinement of existing methods. Nevertheless, sufficient information exists to warrant its presentation in concrete form, partly to emphasize its deficiencies, chiefly to indicate its relation to the biology of the bacteria, which are potentially "living chemical reagents," as Professor Folin has so aptly termed them.

In the last analysis, the interest and importance of bacteria centers around "what they do rather than what they are," and the elucidation of this aspect of bacteriology lies largely within the field of biochemistry.

The relation of the chemistry of bacterial nutrition to the study of intestinal bacteriology in health and disease is self-evident; some of the more general aspects of this subject are briefly set forth in the chapter relating to intestinal bacteria.

It is with great pleasure that the writer acknowledges his indebtedness to his colleagues in the Northwestern Univeristy Medical School for many valuable suggestions, to Doctors Noguchi and Amoss, of the Rockefeller Institute, for the privilege of using the original plates illustrating the Treponemata and Poliomyelitis, and to Mrs. N. M. Frain for the line drawings in the text. Finally, the writer would acknowledge his deep obligation to Miss Bertha J. Schwarz, Secretary of the Department of Bacteriology, for her invaluable assistance in the preparation of the manuscript and in reading the proof of the book.

A. I. K.

CHICAGO, 1916.

CONTENTS.

SECTION I.—GENERAL BACTERIOLOGY.

INTRODUCTION.—THE DEVELOPMENT AND SCOPE OF BACTERIOLOGY.

CHAPTER I.

THE MORPHOLOGY OF BACTERIA.

CHAPTER II.

THE PHYSIOLOGY OF BACTERIA AND THE EFFECT OF ENVIRONMENTAL INFLUENCES.

CHAPTER III.

THE CHEMISTRY OF BACTERIA.

CHAPTER IV.

BACTERIAL METABOLISM.

CHAPTER V.

SAPROPHYTISM, PARASITISM AND PATHOGENISM.

CHAPTER VI.

INFECTION AND IMMUNITY.

CHAPTER VII.

CHAPTER VIII.

ANTIGENS AND THE TECHNIQUE OF SERUM REACTIONS.

CHAPTER IX.

BACTERIOLOGICAL TECHNIQUE.

CHAPTER X.

BACTERIOLOGICAL EXAMINATION OF MATERIAL FROM PATIENT AND CADAVER.

CHAPTER XI.

PRACTICAL STERILIZATION, ANTISEPSIS AND DISINFECTION.

SECTION II.—PATHOGENIC BACTERIA.

CHAPTER XII.

CHAPTER XIII.

CHAPTER XIV.

CHAPTER XV.

CHAPTER XVI.

CHAPTER XVII.

CHAPTER XVIII.

CHAPTER XIX.

CHAPTER XX.

CHAPTER XXI.

CHAPTER XXII.

Hemoglobinophilic Bacilli.

CHAPTER XXIII.

Tubercle Bacillus Group.

SECTION I.

GENERAL BACTERIOLOGY.

INTRODUCTION—THE DEVELOPMENT AND SCOPE OF BACTERIOLOGY.

Bacteriology is that branch of Natural Science which treats of the structure, functions and chemistry of bacteria. Bacteria are intimately related to many fields of human activity, therefore bacteriology is inseparably associated with a number of the arts and sciences. In those branches of science which treat of the diseases of plants, of animals and of man, bacteria enter into complex reciprocal relations with their hosts as parasites or pathogens, relations which are neither purely bacterial, animal nor vegetal in their limitation. A new science, Immunology, is rapidly developing which is concerned chiefly with the elucidation of these relationships between host and parasite.

Bacteria are the smallest in size and simplest in structure of known visible living organisms. They are rigid unicellular organisms devoid of chlorophyll or other photodynamic pigment; they possess no morphologically demonstrable nucleus and reproduce by simple transverse fission, the resulting individuals being of approximately equal size.

Bacteria are ubiquitous in their distribution; they are found in all climates in association with animal and vegetable life. Some thrive at temperatures but slightly above the freezing point of water; the majority flourish between 15° and 40° Centigrade; some even develop in thermal springs at a temperature of 70° Centigrade. Free or atmospheric oxygen is essential for most types of bacteria, but to a few it is actually a poison.

Bacteria are ordinarily classed as plants, but they exhibit several prominent characteristics which suggest a relationship with the lowest animals. The most important of these is the absence of photodynamic pigment (chlorophyll), which implies an analytical or destructive function in the economy of Nature.

The great majority of bacteria are saprophytic, living upon dead organic matter, which they transform into simple compounds suitable for plant use. These bacteria are Nature's analysts. Some are parasitic on living plants and animals; a few are progressively pathogenic for man and animals. It is this last group, few in numbers, but formidable in that their activities are in partial opposition to those of man and animals, that has given to bacteria all the notoriety which they possess.

Anton von Leeuwenhoek, a Dutch spectacle maker, appears to have been the first to see bacteria: in 1675, with lenses of his own grinding, he examined various putrescent fluids, drops of water, scrapings from his teeth, and his own diarrheal discharges. He says in his writings, collected and edited by Robert Hooke,[1] "With great astonishment I observed everywhere through the material which I was examining, animalcules of the most minute size, which moved themselves about very energetically." It is possible to recognize cocci, bacilli and spirilla in his drawings, and it is almost certain that he actually observed motility among his organisms. The learned monk, Athanasius Kircher, observed and described "minute living worms" as early as 1659, but his optical equipment was inferior to that of von Leeuwenhoek and it is doubtful if he actually saw bacteria.

Improvements in the microscope opened a new world for investigation and speculations concerning the doctrine of the Spontaneous Generation of Life led to numerous experiments of increasing refinement that finally resulted in the brilliant researches of Pasteur, and Tyndall, who showed by numerous ingenious and carefully executed experiments that the phenomena in putrescible fluids erroneously interpreted as spontaneous generation did not take place when proper precautions in manipulation were observed. About 1835 achromatic lenses for the microscope reached a state of perfection compatible with the examination of minute objects and the microscope was almost immediately applied to the study of various morbid processes, with remarkable success. Bassi (1837) discovered a fungus which caused a contagious disease of silk worms known as muscardine; Cagniard de Latour and Schwann observed and described the yeast plant in liquids undergoing alcoholic fermentation.

[1] Collected Memoirs of Anton v. Leeuwenhoek, Royal Society of London, 1675, 1683.

Ehrenberg (1838) began his classification of animalcules and in his group of Vibrionia described several "species" of organisms, as follows:

1. Bacterium—rigid and filamentous organisms.
2. Vibrio—flexuous and filamentous organisms.
3. Spirillum—rigid spiral filamentous organisms.
4. Spirocheta—flexuous spiral filamentous organisms.

This classification, which contains terms widely used in bacterial nomenclature today, was followed in 1872 by the important contributions of Cohn upon "Bacteria," the starting-point of modern bacterial classification.

The diseases of man naturally attracted much attention and in 1839 Schoenlein examined the crusts of that disease of the scalp known as favus with the microscope and found the mycelia of the fungus now known in his honor as Achorion schoenleinii.

The extensive studies of Pasteur upon yeasts and the "diseases" of beer and wine, upon the diseases of the silk worm (pébrine and flacherie), upon furunculosis and puerperal sepsis,[1] upon anthrax and anthrax immunization (attenuated viruses) chicken cholera, and somewhat later, rabies laid broad foundations for the development of the science of bacteriology.

Among the most important technical discoveries which have contributed to the development of bacteriology are: The improvement in the achromatic lens (about 1835) and the perfection of the substage condenser (Abbé); the use of cotton for air filters in flasks and test-tubes by Schroeder and von Dusch (1854), the sterilization of culture media by heat (Pasteur, Tyndall, Koch and others), the introduction of anilin dyes as staining reagents by Weigert and Ehrlich (1877), and finally, the use of solid culture media and the plate method for pure cultures by Koch in 1881.

Sir Joseph Lister (1867) published an epoch-making contribution entitled, "On the Antiseptic Principle of the Practice of Surgery," in which is clearly set forth the importance of bacteria in surgery and the principles of surgical asepsis that have revolutionized this branch of medicine.

About 1878 Koch isolated the anthrax bacillus in pure culture from the blood of infected animals, grew the organisms for several generations in the clear aqueous humor of the eye of the ox, and then reinjected the organisms into experimental animals and reproduced the disease. For the first time a specific microbe was clearly and convincingly

[1] *Compt. rend. Acad. d. Sci.*, 1880, xc, 1033.

shown to be the etiological factor of a bacterial disease. Koch also found that the anthrax bacillus formed spores.

From this time bacteriology developed with amazing rapidity. In 1882 Koch startled the world with the announcement of the discovery of the tubercle bacillus; and in rapid succession, typhoid, diphtheria, cholera, tetanus and other well-known pathogenic bacteria were isolated and studied in pure culture.

In 1882 Metchnikoff published the first of his highly important contributions to immunity and phagocytosis, and a decade later von Behring and Kitasato announced the discovery of diphtheria antitoxin.

The last three decades have not only witnessed the rise and development of those most brilliant chapters of medicine, infection and immunity; but sanitation, agriculture, many industries and other fields of human activity have benefited largely by the development of bacteriology.

In medicine the diagnosis of bacterial disease has reached a high degree of precision, and bacteriological diagnosis is an important branch of medical science. The most important problem for the future is to create a system of Bacterial Therapeutics of equal efficiency.

CHAPTER I.

THE MORPHOLOGY OF BACTERIA.

A. NORMAL FORMS: COCCI, BACILLI, SPIRILLA.

The normal forms of the true bacteria are very simple, and are included in three fundamental types: the sphere (coccus, plural cocci), the straight rod (bacillus, plural bacilli), and the curved rod (spirillum, plural spirilla). There is in addition a group of organisms intermediate between the true bacteria and the molds, which is characterized by a filamentous type of growth. The members comprising this group of filamentous organisms are commonly known as the higher bacteria or Chlamydobacteriaceæ. An organism belonging to one of these groups always reproduces its kind under normal conditions; that is, a coccus always reproduces a coccus, a bacillus always reproduces a bacillus, and a spirillum always reproduces a spirillum.

Cocci.—A single coccus is typically spherical, although those organisms in which division is taking place may be temporarily somewhat elongated in one diameter, thus appearing oval in outline at this stage of their development. They may even resemble very short bacilli in extreme instances. The habitual occurrence of cocci in pairs, frequently with their proximate surfaces flattened, is a noteworthy morphological characteristic of certain members of this group. They are referred to as diplococci. The flattening of the proximated surfaces may be associated with an elongation of the axes of the organisms parallel to the plane of apposition, which leads to "coffee bean" shaped diplococci, exemplified in the meningococcus and gonococcus, or to an elongation of the axes perpendicular to the plane of apposition, in which event the organisms are "lance-shaped" diplococci, as for example the pneumococcus.

Bacilli.—Bacilli are rod-shaped, cylindrical organisms in which a longer and a shorter dimension may be recognized. They are typically circular in cross-section. When division is taking place the shorter bacilli may be temporarily oval or even circular in outline. The dimensions of bacilli vary considerably: some are habitually long, some are short, some are thick, some are thin. The ends may be convex, less commonly flat or even concave. A few bacilli are not typically isodiametric, but appear in outline as club-shaped, spindle-shaped, or even more or less conical (cuneate) rods. Less commonly, slightly curved rods are met with; the curvature takes place along the longer dimension.

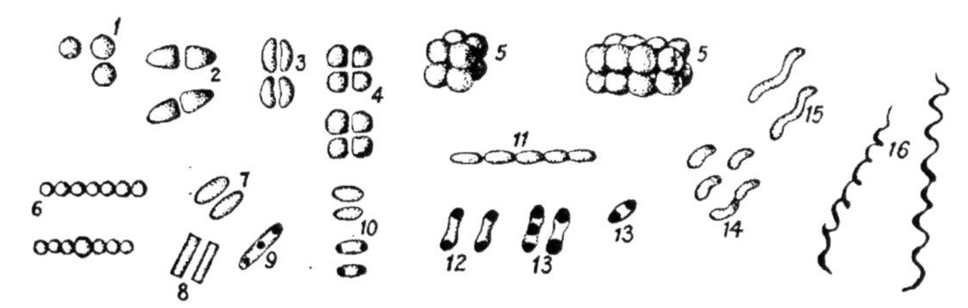

FIG. 1.—The normal types of bacteria. 1–6, cocci; 7–13, bacilli; 14–16, spirilla; 1, micrococcus; 2 and 3, diplococci; 4, tetracoccus; 5, sarcina; 6, streptococcus (the lower chain includes an arthrospore); 7 and 8, bacilli; 9, 10, 12, and 13, bacilli with various granules; 11, streptobacillus; 14, vibrio; 15, spirillum; 16, Spirocheta trepo-nema.

Spirilla.—Spiral bacteria, like the bacilli, exhibit a longer and a shorter dimension; unlike the bacilli, the longer axis is curved in three planes of space. The curvature may be slight, less than a complete turn, in which event the organism is "comma-shaped" when viewed under the microscope; it may be a series of open curves, giving the organism a sinuous outline; or it may be very much curved, so that the organism resembles a somewhat closely coiled spring in outline. As a rule, the curvature is symmetrical and uniform in each instance.

The cocci, through almost imperceptible morphological gradations, merge into the bacilli, and the bacilli, through the slightly curved forms, merge into the spirilla. Even in the spirilla slight differences in curvature are usually discernible. Thus, a culture of the cholera vibrio may contain many straight, uncurved organisms in addition to the slightly curved rods which are the characteristic morphologic forms. There are a few bacteria in which the morphology is still a subject of controversy. For example, Micrococcus melitensis is called Bacillus melitensis by some observers. The vast majority of bacteria, however, are easily referable to their proper morphological type by simple inspection under the microscope.

B. ABNORMAL FORMS: VARIATION, DEGENERATION AND INVOLUTION, PLEIOMORPHISM AND BRANCHING.

Variation.—The composition of the medium in which bacteria are growing, the age of the culture, and to a limited degree even the temperature of incubation influence somewhat the average size of bacteria. Given constant conditions, however, bacteria growing in a favorable environment exhibit constancy of form and size, although a few organisms in every culture are somewhat larger or smaller than their fellows, appearing as occasional giants or dwarfs. These occasional giant and dwarf forms represent normal variations in size from the average or mean.

Degeneration and Involution.—Bacteria growing in an unfavorable environment, brought about by the accumulation of waste products, by undue changes in reaction resulting in excessive acidity or alkalinity, by the presence of harmful chemicals, or by specific antagonistic substances, may gradually assume atypical shapes, probably the direct result of these harmful influences. These atypical organisms may exhibit little or no resemblance to the normal organism, either in form or size; they may or may not develop into normal organisms when they are placed again in a favorable environment. If the change is a morphological one, the atypical organisms are designated involution forms: thus, plague bacilli grown in nutrient agar containing 3 per cent. common salt appear as swollen, balloon-like bodies, notably unlike the typical short rod-shaped bacillus. If, on the contrary, the organisms permanently lose some morphological or chemical characteristic, they are spoken of as degeneration forms. Thus, anthrax bacilli heated for several hours at 43° to 44° C. lose their ability to form mature spores.

Pleiomorphism.—By pleiomorphism is meant a permanent or semipermanent change in the normal form of the organism. A pleiomorphic organism would be one which might at one time resemble a bacillus, again a coccus, or even a spirillum, depending upon the age and growth of the organism or the fitness of the culture medium. This phenomenon is rarely or never met with among the pathogenic bacteria.

Branching.—Among the individual organisms comprising a culture in artificial media of tubercle, diphtheria or glanders bacilli, and to a lesser extent of other bacilli, a certain number appear as definitely branched rods: the typical organism in each instance does not exhibit branching. Branching has also been demonstrated in the spirilla.[1] Bacillus bifidus appears habitually as a rod-shaped organism with bifurcated ends in artificial media, although it is an unbranched bacillus in its normal habitat, the intestinal tract of nurslings. Occasionally, bacteria, as the tubercle bacillus, may exhibit branching in the animal body as well as in cultures, although less commonly.

The cause of this branching is unknown, and at least two theories have been advanced in explanation of it: each theory has a certain amount of evidence in its favor. One theory assumes that branching is the result of unfavorable environmental conditions, and it has been shown that old broth cultures of diphtheria bacilli contain branched organisms; young cultures contain few or no branched forms. The assumption is that old cultures contain harmful products of metabolism which cause the diphtheria bacillus to assume branched forms. The second theory asserts that the appearance of branched forms among bacteria demonstrates a relationship between them and higher organisms, which are habitually branched. Bacteria, according to this theory, exhibit branching as a part of their normal development.

Branching does not necessarily take place under conditions which would appear to be unfavorable or partially inimical to their growth, and, on the other hand, it may be observed occasionally when environmental conditions should be favorable for development. It appears to be reasonable to assume that branching may be a normal developmental process in the life history of the organism, although the phylogenetic significance of branching is as yet undetermined.

C. SIZE AND WEIGHT OF BACTERIA.

Size.—The unit of measurement for microscopic objects is the micron (μ), which is 0.001 of a millimeter, or approximately $\frac{1}{25000}$ of an inch, in length. Bacteria are the smallest known living organisms which have been seen with the microscope. Measured with this unit, they exhibit considerable differences in size. The average sized pus-

[1] Reichenbach, Centralbl. f. Bakteriol., 1901, xxix ,553.

producing coccus is 0.8 micron in diameter; Micrococcus melitensis, the smallest of the Coccaceæ, varies in diameter from 0.3 to 0.5 micron. The largest known bacillus, B. bütschlii[1] is 3 to 6 microns in diameter and from 40 to 60 microns in length. The smallest known bacillus, B. influenzæ, is but 0.2 by 0.5 micron in diameter; an average sized bacillus would measure about 2 microns in length and 1 micron in diameter. Spirillum colossum[2] is from 2.5 to 3.5 microns in diameter. The cholera vibrio is about 2.5 microns long and 1 micron in diameter. There are certain living viruses of unknown morphology, so-called ultramicroscopic or filtrable viruses, which are either somewhat smaller than any known bacteria or more plastic. Viruses belonging to this group derive their name from the fact that they retain their viability even after passage through the pores of standard, unglazed porcelain filters, which will hold back even the smallest bacteria.

Weight of Bacterial Cell.—The weight of a bacterial cell is dependent upon its size and its specific gravity. According to Rubner,[3] the specific gravity of common bacteria varies between 1.038 and 1.065.[4] B. coli is an average sized cylindrical rod (bacillus), measuring 1 micron in diameter and 2 microns in length. The volume of a cylinder is the product of the diameter squared, multiplied by 0.7854, multiplied by the length of the cylinder. The volume of a single colon bacillus consequently would be $(0.001)^2 \times 0.7854 \times 0.002$, or 0.00000000157 c.mm. The weight of a single colon bacillus would be the volume multiplied by the specific gravity, which is approximately 1.040 or 0.00000000163 mg.; that is to say, sixteen hundred million colon bacilli would weight approximately one milligram. For purposes of comparison it may be stated that a single red blood corpuscle (human) weighs about 0.00008 mg., about fifty thousand times the weight of a single colon bacillus.

D. STRUCTURE AND CONSTITUENTS OF THE BACTERIAL CELL.

The typical bacterial cell consists essentially of protoplasmic cell substance, endoplasm, enclosed by a rigid cell membrane, ectoplasm.

1. **Cell Membrane.—Ectoplasm.**—Bacteria appear to possess a special external boundary layer, cell membrane, or ectoplasm, which is

[1] Schaudinn, Arch. f. Protistenk., 1902, i, 306.

[2] Errera, Recueil de l'Instit. botanique (Université de Bruxelles), 1901, v, 347.

[3] Arch. f. Hyg., 1903, xlvi, 41; 1890, xi, 385.

[4] Stigell (*Cent. f. Bakt.*, 1908, xlv, 487) finds that the specific gravity of the same organism varies somewhat with the medium in which it is grown. The specific gravity of ordinary bacteria varies commonly between 1.120 and 1.35, older cultures being as a rule of less specific gravity than younger cultures of the same kind.

rigid and maintains the shape of the organism. Generally speaking, this cell membrane is intermediate in character between that characteristic of animal and of plant cells respectively, being somewhat more developed than the former, less highly specialized as a rule than the latter. Some authorities consider the cell membrane of bacteria to be merely a concentrated external layer of endoplasm.

The thickness of the cell membrane varies among different varieties of bacteria, and it appears to be somewhat thinner in young organisms of a given variety than in the older individuals of the same kind. Ordinarily it is not seen, and special stains are required to demonstrate it clearly. In certain spore-forming bacteria, however, the cell membrane is occasionally seen after the spore has matured within the cell, as a thin, feebly staining shadow, outlining the original contour of the organism. Bacteria which plasmolyze easily also show the cell wall clearly after the cell contents have shrunken away from it.

Capsule.—A considerable number of bacteria are surrounded by mucin-like envelopes, particularly when they are observed in the animal body or grown in albuminous fluids. This envelope or capsule frequently disappears when the organisms are grown in ordinary media. This has led to the theory that a capsule represents an hypertrophy of the ectoplasm. The significance of capsules is still a matter of controversy. Two principal theories have been advanced to explain the significance of capsules: according to one theory, bacterial capsules are purely degenerative phenomena; the more widely accepted theory, which has much evidence in its favor, maintains that capsule formation is closely related to the virulence of the organisms.[1] The demonstration of capsules may be an important factor in the identification of certain bacteria, for example, the pneumococcus.

Zoöglea.—A very few bacteria exhibit a slimy intracellular substance which causes cohesion between considerable numbers of bacterial cells. This intracellular substance, zoöglea, is colored lightly by ordinary staining methods. It is not found in any of the pathogenic bacteria.

2. **Cell Substance.**—**Cytoplasm.**—The cytoplasm or endoplasm of living bacteria (particularly in young cultures) is usually a clear, colorless, highly refractile, homogeneous appearing substance, although at times various granules may be seen within it. Vacuoles also are met with, usually in older bacteria. The cytoplasm usually stains readily with basic anilin dyes. A few bacteria, notably B. viride and

[1] Eisenberg, Centrabl. f. Bakteriol., 1908, xlv, 148.

B. chlorinum, contain a yellowish pigment in the cytoplasm suggesting chlorophyll, and the so-called purple bacteria similarly possess a purple colored pigment, bacteriopurpurin.

Nucleus.—The occurrence of a demonstrable morphological nucleus in bacteria is by no means definitely settled: the typical bacterial cell can not be separated chromoscopically into a nucleus and cytoplasm. Those who have thoroughly studied the question by staining methods, notably Nakanishi,[1] believe that the whole bacterial cell, as it is ordinarily seen, is potentially a nucleus surrounded by a very thin film of cytoplasm. Others believe the nucleus substance is distributed throughout the cell in very finely divided granules: Zettnow[2] is the champion of the latter theory. He believes that the bacterial cell, as it is viewed following the usual staining processes, is endoplasm in which the nuclear substance is finely divided and uniformly distributed. Some observers deny that a nucleus exists at all. Chemical analyses show beyond doubt that bacteria contain a relatively high percentage of substances usually regarded as essentially of nuclear origin. It is quite certain, therefore, that, although there may be no morphologic nucleus demonstrable by ordinary staining methods, nuclear material is present in abundance in the organism.

Metachromatic Granules.—Certain types of bacteria, notably members of the diphtheria and hemorrhagic septicemia groups, exhibit one or more highly refractile granules in an otherwise homogeneous endoplasm when they are examined unstained with the higher powers of the microscope. These granules are few in number in the diphtheria bacillus group and are distributed somewhat irregularly throughout the cell, one or more granules usually being greater in diameter than the cell itself, thus giving the rod a swollen appearance. In the hemorrhagic septicemia group these granules are arranged symmetrically, one at each end of the organism, polar granules. Such granules are called Ernst-Babes or metachromatic granules. They color differently from the rest of the cell when they are stained with methylene blue, appearing as mahogany-red spots in the deep blue endoplasm. They retain the stain rather tenaciously. Many theories have been advanced to explain their significance, but nothing definite is known about them, except that these granules appear to differ widely in chemical composition. Some are colored brown with iodine, suggesting that they may be related to glycogen.[3] Some stain black with osmic acid, suggesting

[1] Centralbl. f. Bakteriol., 1901, xxx, 97, 145, 193, 225.
[2] Ztschr. f. Hyg., 1899, xxx, 1; Festschr. z. 60 Geburtstage von R. Koch, 1903, p. 383.
[3] A. Meyer, Flora, 1899, lxxxvi, 428.

that they may be fatty or lipoidal in composition, while others are probably complex phosphorus-containing compounds.[1] Not all of these varieties of granules are met with in the same organism.[2] Among the higher bacteria granules of sulphur or of iron are demonstrable respectively in the sulphur and the iron bacteria.

Flagella.—All minute particles suspended in water or other fluids of low viscosity are in constant motion. This motion, which is irregular and tremulous, was first described by Brown:[3] it is variously termed Brownian movement, pedesis, or molecular movement. Brownian movement may be rapid or slow, extensive or circumscribed, depending upon the nature of the particles and the composition and temperature of the fluid in which they are suspended. This is not true motility, even though each individual particle moves independently of the other particles in an irregular orbit, for the particles as a whole do not permanently change their relative positions. Dead bacteria and many

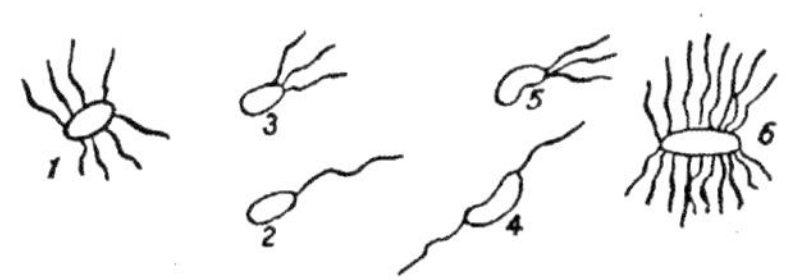

FIG. 2.—Flagella. 1 and 6, peritrichic flagella; 2 and 4, monotrichic flagella; 3 and 5, lophotrichic flagella.

living bacteria, notably the cocci, exhibit the Brownian movement. Many bacilli and spirilla, on the contrary, possess the power of independent motility, that is, they can progressively and permanently change their relative positions in space. Motile bacteria are provided with one or more long, delicate, contractile filaments—flagella—which are probably the organs of locomotion. These flagella cannot be demonstrated on living bacteria, except possibly by dark-ground illumination, and ordinary staining reactions usually fail to reveal them. Special staining methods show them clearly. They appear to arise from the cell membrane.[4] Their arrangement and number is varied among bacteria in general, but relatively constant for a particular variety of bacterium: they are thinner as a rule on younger bacterial cells, thicker on older organisms.[5] A cholera vibrio has a single

[1] Grimme, Centralbl. f. Bakteriol., 1904, xxxvi, 952.
[2] For literature see Marx and Woithe, Centralbl. f. Bakteriol., 1900, xxviii, 1, 33, 65, 97; Krompecher, ibid., 1901, xxx, 385, 425; Gauss, ibid., 1902, xxxi, 92.
[3] Edinburgh Phil. Jour., 1828, v, 358; 1830, viii, 41.
[4] Schaudinn, Arch. f. Protistenk., 1903, i, 421.
[5] De Grandi, Centralbl. f. Bakteriol., 1903, xxxiv, 97.

flagellum at one or both ends of the organism; in the typhoid bacillus they are distributed around the sides of the organism but do not occur at the ends.

Spores.—*Endospores.*—Many bacteria die when their environment becomes unsuited for further growth. Death may result from the presence of inimical substances, the absence of essential foods, or the intervention of unsuitable physical conditions. Death is manifested by a cessation of chemical interchange between the bacterial cell and its environment. There is a group of bacteria, however, usually of saprophytic origin, which is able to survive even prolonged exposure to unfavorable environmental conditions by passing into a latent stage during which chemical interchange with the environment is at an extremely low ebb. This latent stage or hibernation has been known to last for more than two decades in certain instances, and yet the organisms have resumed their original luxuriant growth when placed

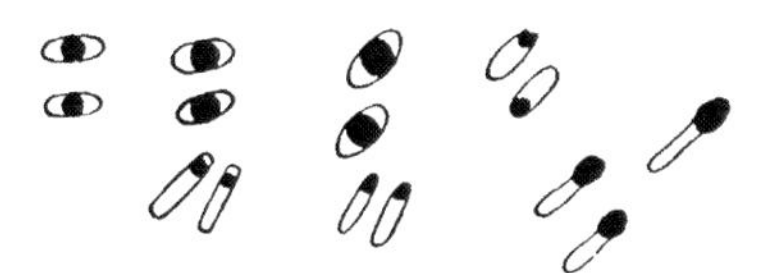

Fig. 3.—Types of bacterial spores.

under favorable conditions. The bacteria which exhibit this latent state produce within their substance highly refractile, spherical or oval bodies called spores. Spores are not found in very young, actively growing cultures, as a rule. Spore formation is ushered in by a clouding of the endoplasm of the bacterial cell, which gradually becomes granular. The granules coalesce, eventually appearing as the mature spore which is surrounded by a dense membrane, frequently exhibiting a double contour when stained by dilute carbol fuchsin.[1] The spore membrane (ectoplasm) is relatively impermeable to heat and disinfectants and confers the resistance to physical agents which spores exhibit upon them. But one spore is formed in an individual bacterium, except under most unusual conditions. It is to be emphasized, consequently, that spore formation is not a reproductive process. The mature spore may form in the center of the bacterium, at, or near one end. The spore may be round or oval, and greater or lesser in diameter than the parent cell. If the spore is greater in diameter it distends the cell membrane, producing a spindle-shaped organism if the spore is

[1] Meyer, A., *Practicum der botanischen Bakterienkunde*, Jena, 1903.

in the center of the rod: if the spore is at one end, a drumstick-shaped organism results. Usually the size and position of the spore is fairly constant in a given type of bacteria. Spore formation is most common among the anaërobes, fairly common among the saprophytic bacteria, practically absent in the pathogenic bacteria, and practically never takes place spontaneously in the human or animal body. The spiral organisms rarely produce spores, and, with the exception of Sarcina pulmonum, spore formation is practically never observed in the cocci.

It has never been satisfactorily determined whether spore formation is a regular definite stage in the life history of bacteria which produce them or whether spores are produced rather under the stress of unfavorable evironmental conditions.

Germination of Spores.—When bacterial spores are placed in an environment favorable to the vegetative activity of the cell, they germinate: the dense membrane which constitutes the ectoplasm of the spore softens, usually at the pole or the equator, and the vegetative rod emerges, at first as a small bud, then rapidly assumes the typical size and shape of the fully mature cell. The development of the anthrax bacillus from the spore is usually in the line of the longer axis, polar germination: B. subtilis, on the contrary, usually emerges at right angles to the larger axis of the spore, equatorial germination. Many spores are circular in outline, and in such cases the relation of the developing vegetative cell to the axis of the spore is unknown. Frequently the remnants of the spore membrane remain attached to one end of the newly formed vegetative cell, appearing as a cap, as it were. Some spores do not appear to rupture as germination takes place—the newly forming organism appears to absorb the entire spore and its ectoplasm, incorporating the entire structure by solution in the vegetative cell.

Arthrospores.—Certain organisms belonging to the coccal group, more particularly the streptococci, exhibit from time to time cells which are decidedly larger than their fellows. These cells are more highly refractile, they usually possess a granular cytoplasm, and frequently stain somewhat irregularly. They have been designated by Hueppe as arthrospores. These arthrospores appear to have no unusual resisting powers, and they are in no sense to be regarded as true spores. It is very probable that they are involution forms.

E. REPRODUCTION AND CELL DIVISION.

Bacteria are structurally the simplest known organisms which maintain an independent existence: all their vital functions are exhibited in a single asexual cell devoid of a morphologically definable nucleus. The absence of sexual characters and of a morphologic nucleus makes bacterial reproduction mechanically a simple process, and doubtless the rapid sequence of generations observed in various bacteria depends in part upon this simplicity of structure.

Reproduction takes place in the following manner: A bacterial cell placed in a favorable environment increases in size until it reaches a maximum which is relatively constant for each variety; then a slight equatorial constriction occurs, which deepens until a distinct septum is produced by invagination, which divides the original cell into two morphologically complete, fully mature individuals of approximately equal size. It is obvious that this septum consists ordinarily of at least two layers, since one layer is required to complete each of the dividing individuals. Successive generations may be produced at intervals which may be as frequent as every fifteen minutes in the more rapidly growing types. Septation usually takes place deliberately; that is to say, the septum forms relatively slowly. Diptheria bacilli and possibly related bacteria divide somewhat differently; the parental cell appears to be under tension when the septum becomes visible, and the daughter cells spring apart suddenly when septation is completed. So forcible is this separation that the daughter cells lie at an angle with each other: Nakanishi[1] has observed that the septum in this group of organisms frequently forms at a metachromatic granule. Septation in the Bacillaceæ and Spirillaceæ normally takes place at right angles to the long axis of the organism, and midway between the ends, thus effecting the separation into two individuals with the minimal expenditure of material; in the Coccaceæ, which are usually isodiametric, no economy of material in septation is apparent, and no known force determines the initial plane of septation: subsequent fission may be definitely related to the initial plane. Noguchi[2] has brought forward striking evidence and photographic illustrations in favor of the view that the Spirocheta (Treponemata) may reproduce by longitudinal fission rather than by transverse fission. If this view be generally adopted, it would contradict the "minimal requirement

[1] Centralbl. f. Bakteriol., 1900, xxvii, 641.
[2] Jour. Exper. Med., 1912, xv, 201.

theory," which assumes that transverse fission, the more economical process both with respect to amount of material and expenditure of energy, holds universally for bacteria, as has previously been maintained.

F. CELL GROUPING.

In Bacilli and Spirilla, where septation typically occurs at right angles to the long axis of the organism, it is obvious that no geometrical arrangement of cells is possible other than the formation of chains of rods or of spirals if the individual organisms remain adherent. The cocci, on the other hand, are spherical and have no longer or shorter axis, consequently a definite sequence of septation in one, two or three planes of space can give rise to (1) chains of cocci, if the plane of septation is always in one plane of space; (2) groups of four cocci, if septation takes place alternately in two planes of space; or (3) in packets of cocci, if septation is alternate in three planes of space. Many cocci do not exhibit a definite sequence of planes of septation.

G. CLASSIFICATION OF BACTERIA.

Relation to Higher Plants.—The position of Bacteria in the Plant Kingdom is indicated in the following table:

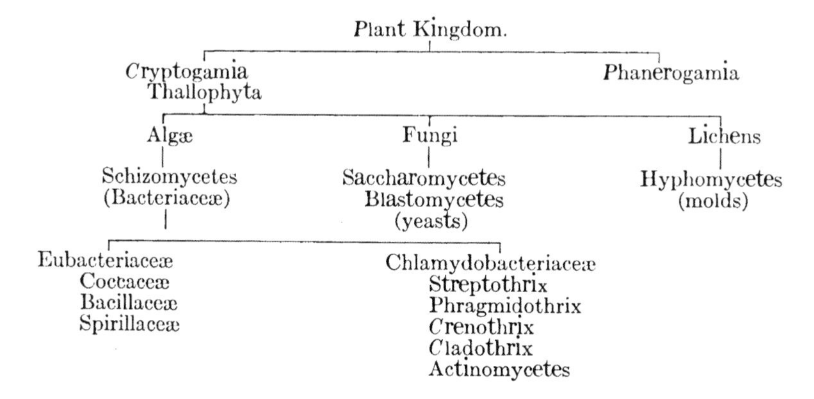

A complete natural classification of bacteria is impossible at the present time. The monotony of form observed in this group of organisms merely suffices to classify them into three great divisions: Cocci, Bacilli, and Spirilla. Further subdivision into groups which are potentially families, genera and species is accomplished by arranging them according to their physiological and chemical activities. Even this artificial procedure is unsatisfactory, for bacteriological diagnosis is a subject which has developed under the stress of practical needs, and as bacteria play a part in many fields of activity, it has inevitably followed that the criteria whereby they are recognized vary greatly according to the art or science in which they are contemplated. Even the same species may be identified by wholly different characteristics. Notwithstanding the difficulties which surround the grouping of bacteria, Migula[1] has worked out a system of classification based upon purely morphological characteristics, which effects a primary separation of bacteria into smaller subdivisions, which is moderately satisfactory so far as it goes, and it is the one commonly adopted.

With certain additions it is as follows:

THE TRUE BACTERIA: EUBACTERIACEÆ.

1. *Coccaceæ.* Cells in the free state spherical.
   - (*a*) Micrococcus. Cells spherical. No definite sequence of planes of septation.
   - (*b*) Diplococcus. Organisms habitually occur in pairs.
   - (*c*) Streptococcus. Plane of septation parallel. Form longer or shorter chains.
   - (*d*) Tetracoccus. Planes of septation alternate, and at right angles in two planes of space. Form groups of four or tetrads.
   - (*e*) Sarcina. Planes of septation alternate, at right angles, in three planes of space. Form packets.
   - (*f*) Planococcus. Motile cocci, provided with flagella.
   - (*g*) Planosarcina. Motile sarcina, provided with flagella.
2. *Bacillaceæ.* Cells elongated and cylindrical; straight.
   - (*a*) Bacterium. Non-motile. No flagella.
   - (*b*) Bacillus. Cells motile. Peritrichic flagellation.
   - (*c*) Pseudomonas. Cells motile. Polar flagellation. Single flagellum or tufts of flagella at one or both poles of the organism.
3. *Spirillaceæ.* Cells elongated and cylindrical; spirally twisted about the long axis.
   - (*a*) Spirasoma. Cells rigid and slightly curved; without flagella.
   - (*b*) Microspira. Cells rigid and slightly curved; with one, rarely several, polar flagella.
   - (*c*) Spirillum. Cells rigid, loosely coiled; with tuft of polar flagella.
   - (*d*) Spirocheta. Cells flexuous, closely coiled; flagellation unknown.

THE HIGHER BACTERIA.

4. *Chlamydobacteriaceæ.* Cells enclosed in a sheath.
   - (*a*) Streptothrix. Cell division always in one plane.
   - (*b*) Phragmidothrix. Cell division in three planes of space; very delicate sheath.
   - (*c*) Crenothrix. Cell division in three planes of space; sheath well developed.
   - (*d*) Cladothrix. Cells more or less branched.
5. *Beggiatoaceæ* (Thiothrix). Cells contain sulphur granules.

[1] System d. Bakterien, Jena, 1907.

H. MUTATION: CONSTANCY OF TYPES.[1]

True mutation or discontinuous variation is rarely observed among bacteria, although a few instances are on record which have been subjected to satisfactory scrutiny. Mutation must be carefully differentiated from the loss of one or more characteristics of bacteria during cultivation; the loss or suppression of one or more characteristics is fairly commonly observed among bacteria. Pigment production, and proteolytic activity—as for example the ability to liquefy gelatin—are frequently lost to cultures of bacteria during prolonged cultivation, but these properties may be regained when the organisms are placed once more in a suitable environment. Similarly, strains of fermenting bacteria may temporarily, or even permanently, become unable to decompose certain carbohydrates. Change in virulence, or loss of virulence is rather commonly noticed among pathogenic bacteria grown outside the animal body. It is even possible to so parasitize organisms by prolonged cultivation upon one medium that they will develop not at all, or slowly at best, on other media. Thus, a strain of B. proteus has been grown continuously upon agar with frequent transfers for four years, and the organism will no longer grow in broth. Similarly, B. bulgaricus is an obligate milk parasite. Exposure to unfavorable environmental conditions may also suppress important characters: Pasteur's celebrated experiment of growing anthrax bacilli at 43° C. for some hours and establishing an asporeless variety is a familiar example. The suppression of characters as outlined above is frequently important as the starting point for new adjustments between pathogenic bacteria and their hosts.

Turning to the production of disease in man, it is certain that at least some organisms produce the same reaction today they did years ago: tuberculosis appears to be the same disease today it was centuries ago, as is evidenced by the lesions found in Egyptian mummies. Clinically, the observations of Hippocrates would be a fair exposition of the phenomena seen in tuberculous patients at the present time. Leprosy also appears to be the same entity now it was during the middle ages, although the geographical distribution is much more restricted. With respect to more acute diseases, which require more careful examination to differentiate them, the evidence is less certain, although typhoid bacilli do not appear to have changed since they were first isolated

[1] Eisenberg, Ueber Mutationen bei Bakterien und anderen Mikroörganismen in Ergebnisse d. Immunitätsforsch. experimentellen Therapie, Bakteriologie und Hygiene, Berlin, 1914, pp. 28–142, for summary.

by Gaffky three decades ago. It appears to be reasonably certain from what is known of bacteria and the manifestations of disease they induce that mutation is an infrequent phenomenon: attenuation and the partial suppresion of characteristics, on the contrary, appear to be quite common. The available evidence indicates that bacterial types are stabile under natural conditions: there is no definite evidence in favor of the view that bacteria change slowly or abruptly either in their morphology or in the changes they induce in their environment in the sense that entirely new, unrelated types are developed *de novo* from preëxisting types. This does not preclude the possibility that such changes have taken place in the past, rather that such changes, if they have taken place, have not been definitely established.

CHAPTER II.

GENERAL PHYSIOLOGY OF BACTERIA—THE EFFECT OF ENVIRONMENT ON BACTERIA.

A. RATE OF REPRODUCTION.

One of the striking characteristics of the Bacteriaceæ is their rapidity of reproduction. Among the most actively growing types of bacteria, as, for example, the cholera vibrio, successive generations may appear at intervals as frequent as every fifteen minutes when the environmental conditions are most favorable: that is to say, ninety-six generations are theoretically possible in twenty-four hours. If this rate of reproduction could be maintained for three days, the progeny of a single organism would occupy a space not less than that of the combined waters of the earth. Fortunately, nature imposes many restraints which limit the numbers of bacteria. The rapid accumulation of waste products, the exhaustion of nutrient material, and the enormous death rate in culture media even after a comparatively few hours' growth, together with other factors restrict development to such a degree that the actual number of living descendants of bacteria in cultures or in nature falls far short of the theoretical number. Many bacteria develop more slowy than this, however. They may require hours or even days to arrive at maturity. The tubercle bacillus, for example, grows comparatively slowy in artificial media (where such observations are of necessity made), and the frequency of septation, even in the most rapidly growing bacteria, is greatly affected by environmental factors.

Generally speaking, when nutritional conditions are favorable, the rate of reproduction is influenced by temperature, growth being most rapid when the temperature is optimum for the organism, less rapid when the temperature exceeds or falls below this point.

B. MOTILITY: RATE OF MOTION.

The rhythmic contractions of the flagella, with which practically all motile bacteria are provided, drive the organisms through fluid media in which they may be suspended, some slowly, some rapidly. Not all bacteria even in the same culture exhibit motility. The character of the motion may be direct, serpentine, oscillatory, or irregular. Rarely, the flagella appear to produce local currents in the medium which immediately surrounds the organism. Various environmental factors incite or inhibit motility. Chemotactic substances may attract bacteria, thus in a sense directing their line of movement. Other substances, as protoplasmic poisons, paralyze bacterial movements. Oxygen appears to increase the motility of aërobic bacteria, and it inhibits motility in the anaërobes. Generally speaking, in favorable media motility increases with the rise in temperature to the optimum. If this temperature is exceeded, even by a very few degrees, motion ceases.

The rate at which bacteria progress through a fluid is a variable one, although with a given organism under favorable conditions it appears to be fairly constant. It must be remembered that the apparent rate of motion observed under the microscope is increased proportionately to the increase in magnification. Lehmann and Fried[1] have measured the average speed of certain bacteria in fluid media in millimeters per second. They find that of the cholera vibrio to be 0.03, typhoid bacillus 0.018, B. subtilis 0.01, B. megatherium 0.0075. If a man traveled at a rate of speed in proportion to his size as great as that of the cholera vibrio, he would average more than a mile a minute.

C. SPORULATION: GERMINATION OF SPORES.

Many saprophytic bacteria form within themselves spores which appear apparently under the stimulus of the stress of conditions unfavorable for the continued vegetative growth of the organism. Sporulation, in other words, appears to be a specialized mechanism for the

[1] Arch. f. Hyg., 1903, xlvi, 314.

perpetuation of the organism during periods of environmental unfitness. Whether spore formation is a definite phase in the life-history of spore-forming bacteria is not definitely settled. Sporulation is rarely observed when the temperature of the environment falls much below 15° C., although considerable latitude is observed among the spore-forming bacteria in this respect. Spores are rarely, if ever, produced within the tissues of the animal body: if the tissues are exposed to the air, however, particularly postmortem, spore formation may take place. No bacteria progressively pathogenic for man are known to form spores.

The unusual resistance of mature spores to desiccation, to exposure to dry and moist heat, and to disinfectants may be due either to their low content of water, for spores contain less than half of the water contained in the normal vegetative cell, to the relatively thick

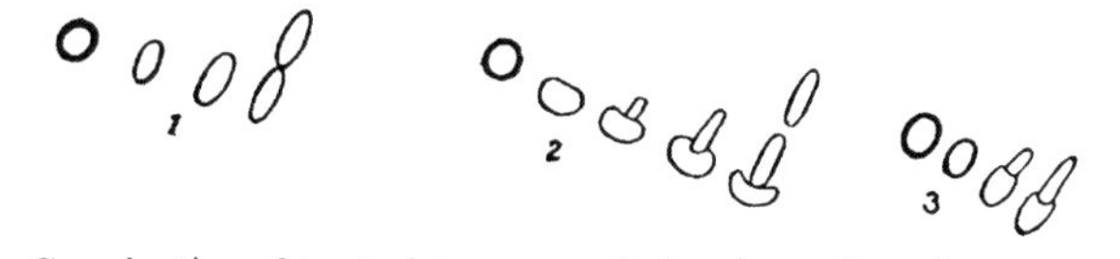

FIG. 4.—Germination of bacterial spores. 1, by absorption of spore membrane; 2, equatorial germination; 3, polar germination.

refractile spore membrane, or to unusual concentrations of fatty and lipoidal substances. Experiments by Lewith[1] would suggest that the relative desiccation of the contents of spores as compared with the moisture content of the vegetative organism would be the most plausible explanation of their resistance to heat without apparent injury. He found that egg albumen (dried) suspended in 5 per cent. of water coagulated at 145° C.; suspended in 18 per cent. of water, coagulation took place at 90° C.; with 25 per cent. of water, at 80° C.; and in a considerable volume of water (amount not stated) coagulation occurred when the temperature reached 56° C.

The resistance of spores to physical conditions varies somewhat according to the organism in which they are formed. Generally speaking, however, several minutes' exposure to the temperature of boiling water (100° C.) may fail to kill them. Dry heat is less effective than moist heat, for an exposure of 160° C. for one and one-half hours is required to certainly sterilize glassware containing spores. Ten to fifteen pounds live steam pressure for fifteen minutes is required

[1] Arch. f. exp. Path. u. Pharmakol., 1890, xxvi.

to effect sterilization of liquids and organic matter in general. Direct sunlight will kill spores after days of exposure.

Germination of bacterial spores takes place when they are placed in a suitable nutritive environment in which the temperature, moisture and oxygen relations are favorable. The vegetative cell breaks through the spore membrane apparently after the latter has lost its refractility, and reproduction by fission proceeds anew, and persists until environmental conditions again lead to sporulation.

D. LONGEVITY.

The duration of life in the individual non-spore-forming bacterium is unknown, but it is greatest apparently when the organism is quiescent or nearly so. This condition is realized most commonly when bacteria are exposed to temperatures slightly above freezing in a dark place. This question has been studied recently under unusual conditions. A mastodon was discovered in Siberia which had been uncovered by an unusual recession of the ice. This animal was found to be practically intact, and cultures made with proper precautions from the center of the proboscis contained bacteria indistinguishable from Sarcina lutea and other well known air organisms.[1] If these cultures are authentic, a most unexpected instance of bacterial longevity has been unearthed, for this animal has undoubtedly been frozen for hundreds of years.

Spores have been dried and kept in a cool dry place for more than two decades, and yet developed with their usual luxuriance when placed in a favorable environment. Dried anthrax spores thus retain not only their viability but their virulence unimpaired for years. Practically, the average duration of life among bacteria is comparatively brief.

E. MOISTURE AND DESICCATION.

Bacteria normally contain at least 80 per cent. of moisture in their substance, and they develop typically only in media containing considerable amounts of moisture. Bacteria do not vegetate normally in desiccated media, but many varieties resist drying for considerable periods. Advantage is taken of the restriction of bacterial development in the absence of suitable amounts of moisture in various processes of drying meats and other foodstuffs; desiccated foods will keep for weeks under the proper conditions. Bacterial spores pro-

[1] Russian Academy of Science, 1911–1912.

tected from direct sunlight are extremely resistant to drying, but they develop with characteristic vigor when environmental conditions become suitable. Even non-sporogenic bacteria may develop after days or weeks of desiccation. Many pathogenic bacteria are eliminated from the body enveloped in albuminous material, as in sputum. These organisms thus protected may resist drying for many days, provided they are not exposed to direct light. The following table indicates the relative viability of various bacteria pathogenic for man to air drying.[1]

1. Gonococcus, few hours.
2. Cholera vibrio, few hours to two days.
3. Plague bacillus, one to eight days.
4. Diphtheria bacillus, twenty to thirty days.
5. Streptococcus pyogenes, fourteen to thirty-six days.
6. Pneumococcus, nineteen to fifty-five days.
7. Staphylococcus pyogenes, fifty-five to one hundred days.
8. Typhoid bacillus, up to seventy days.
9. Tubercle bacillus, two to three months.

F. OXYGEN: AËROBIOSIS AND ANAËROBIOSIS.

Oxygen, either in the free state or combined, is essential to the growth of all known bacteria. The majority of bacteria grow best in the presence of free (atmospheric) oxygen, although the percentage of this gas necessary to support bacterial life may be considerably less than that occurring normally in the air. Some bacteria appear to be wholly dependent upon free oxygen, and they are called obligate aërobes. A small group of bacteria, on the contrary, grow only in the absence of free oxygen, and more than minimal concentrations of this gas are actually poisonous to them. Those bacteria which grow only in the absence of free oxygen are called obligate anaërobes. The vast majority of bacteria are facultative with respect to their oxygen requirement, growing best in the presence of atmospheric oxygen but able to develop either in the presence of small amounts of free oxygen, as in the tissue of the body and certain parts of the intestinal tract, or they are able to obtain their oxygen from chemical compounds, as certain simple sugars, if free oxygen is not available. These organisms are called facultative anaërobes. The maximum tolerance of bacteria for oxygen varies very considerably, as the following table indicates:

Oxygen content of the air is taken as 100 per cent.

[1] Fischer, Vorlesungen über Bakterien 1903, II Aufl., 110.

MAXIMUM OXYGEN TOLERANCE.

	Atmospheric oxygen, Per cent.
B. (clostridium) butyricus	1.35
B. chauvei	5.00
B. edematis maligni	3.25
Purple bacteria (Molisch) . . . about	90.00
Thiosulphate bacteria (Nathansson)	400.00
B. prodigiosus	3000.00

G. TEMPERATURE.

1. **General.**—The extreme temperature limits of bacterial growth are very slightly above 0° C. to 80° C. inclusive. Some bacteria, notably those found in the Arctic regions, appear to develop even at 0° C.; others, chiefly those found in soil, feces, and certain thermal springs, grow even at 80° C., a degree of heat considerably above that at which the protoplasm of most animals and plants is coagulated. The vast majority of bacteria, however, develop best within a range of temperature from 15° C. as a minimum to 40–43° C. as a maximum. All bacteria exhibit three cardinal thermic points: a minimum temperature, below which growth ceases; an optimum temperature, at which growth is most luxuriant and rapid; and a maximum temperature, above which growth ceases, and the organisms die. Fischer[1] has classified bacteria according to their thermic relations as follows:

	Minimum.	Optimum.	Maximum.	
1. Psychrophilic bacteria	0	15–20	30	Many water bacteria.
2. Mesophilic bacteria	15–25	37	43	Pathogenic bacteria and others.
3. Thermophilic bacteria	25–45	50–55	85	Spore-forming bacteria from soil, feces, and thermal springs.

Bacteria which are progressively pathogenic for man and warm-blooded animals develop within a much narrower range of temperature than the saprophytic bacteria which are found chiefly in nature, as the following table, also taken from Fischer,[2] indicates:

	Minimum.	Optimum.	Maximum.	Difference between minimum and maximum.
B. phosphorescens	0	20	38	38
B. fluorescens	5	20–25	38	33
B. subtilis	6	30	50	44
Vibrio choleræ	10	37	40	30
B. anthracis	12	37	45	33
B. diphtheriæ	18	33–37	45	27
Mic. gonorrheæ	25	37	39	14
B. tuberculosis	30	37	42	12
B. thermophilus	40	60	80	40

[1] Vorlesungen über Bakterien, 1903, II Aufl.
[2] Loc. cit., 106.

The saprophytic bacteria, as for example B. subtilis, which develop through a relatively wide range of temperature are also called *Eurythermic* bacteria. The pathogenic bacteria, as for example the tubercle bacillus, which exhibit but little latitude in this respect, are called *Stenothermic* bacteria.

2. **Cold.**—All bacteria grow best and most rapidly in an environment which is maintained at the optimum temperature for the organism. If this temperature is lowered even a few degrees, the rate of reproduction is proportionately reduced. As the temperature approaches 0° C., there is complete or nearly complete cessation of growth with a corresponding complete or nearly complete restriction of chemical interchange between the organism and its environment. The viability, and in the pathogenic bacteria the virulence, is not seriously impaired even by exposure to these low temperatures for considerable periods of time. Practical advantage is taken of this restriction of bacterial development by cold in the preservation of food by refrigeration or by cold storage, and also for the preservation of laboratory cultures of many non-spore-forming bacteria by placing them in the ice-box at 5–10° C. So resistant are bacteria to low temperatures that they may be actually frozen solid and kept in this state for days and even weeks without killing all the individuals of the cultures. Alternate freezing and thawing is much more disastrous to them than simple freezing. Thus, typhoid bacilli may be suspended in water and exposed to a freezing mixture of ice and salt at −18° C. for several weeks without killing all the organisms, although the majority of them are killed within a few hours. At the end of a week fully 90 per cent. are dead; over 95 per cent. succumb by the end of four weeks' continual freezing; but from four to six months' continuous freezing is required to kill all of the typhoid bacilli. The survivors appear to be no more resistant to subsequent freezing than similar organisms which have not been frozen. It is a noteworthy fact that bacteria suspended in colloidal substances, as egg albumen, are much more resistant to freezing than similar organisms frozen in water. Alternate freezing and thawing in colloids is much less disastrous to bacteria, in other words, than the same freezing in aqueous solutions. It is probable that the mechanical factor of crystallization which takes place when water is frozen actually crushes many of the bacteria, thus accounting, in part at least, for the greater death rate in aqueous solutions than that observed in colloids. When bacteria are once frozen, further lowering of the temperature has surprisingly little influence upon the death rate. Typhoid and colon bacilli will survive freezing, in moderate

numbers at least, in liquid air (−176° C.) or even liquid hydrogen (−252° C.) for several hours, and develop vigorously when they are again placed in a suitable environment at the optimum temperature.

3. **Heat.**—Bacteria are distinctly injured by exposure to even slight increases of temperature above that optimum for their growth, although there are considerable differences met with among different kinds of organisms in this respect. Generally speaking, the saprophytic bacteria exhibit greater latitude than the pathogenic bacteria. If the maximum temperature of growth be exceeded by even a very few degrees, the death of the organisms follows rather promptly. The greater the degree of heat, the shorter the time required to kill them. Therefore, the thermal death point of bacteria, that temperature at which specific organisms die, is dependent not only upon the actual temperature to which they are exposed, but also to the length of time of exposure. A standard exposure of ten minutes has been proposed, so that the thermal death point of the bacterium may be defined as the lowest temperature to which it must be exposed for ten minutes under constant conditions to ensure the sterility of the culture. The determination of the thermal death point is influenced by many factors besides the kind of organism under observation and the temperature. Older cultures are usually less resistant than younger cultures of the same kind. The reaction of the medium (acids particularly decrease thermal resistance), the presence of extraneous substances as mucin and other non-conductors of heat, all play a part. Certain modifications in the characteristics of bacteria are observed when they are exposed for several hours at the maximum temperature of growth or a degree or two above this point. For example, anthrax bacilli, which habitually form spores, lose this property when they are exposed to 44° C. for several hours.

Dry Heat, Moist Heat—Dry heat is less effective in killing bacteria than moist heat. This is shown by the high temperature to which glassware and other apparatus must be exposed in order to kill spores, a temperature of 160° C. for one and one-half hours being required to ensure sterility. Moist heat, which is best obtained by dry steam under pressure, will kill even the most resistant spores in fifteen minutes at fifteen pounds pressure.

H. HEAT PRODUCTION.

The energy liberated by bacteria during the decomposition of organic substances by bacterial growth is partly utilized by them for their anabolic requirements. A larger part, however, is dissipated as heat. The heat generated in actively growing cultures of bacteria can be detected with sensitive thermometers, provided losses due to radiation and evaporation are guarded against. The heat production is not great as a rule, although in certain fermentations it may rise as high as 12–15° above the uninoculated controls. The decomposition of protein and protein derivatives (putrefaction) usually gives rise to less heat than the decomposition of carbohydrates (fermentation) under the same conditions.

I. LIGHT AND ELECTRICITY.

The vast majority of plants possess a photodynamic pigment, chlorophyll. This pigment can synthesize inorganic substances, as CO_2 and water, together with nitrates, into complex organic compounds through the energy of the sun's rays acting upon it. Plants possessed of this pigment, therefore, are the synthetic agents of nature. Usually this pigment is green; it may, however, be brown or red, the latter pigment being characteristic of certain algæ. A group of the higher bacteria, the Rhodobacteriaceæ, possess a photodynamic pigment, bacteriopurpurin, which appears to be analogous to chlorophyll of the green plants. These sulphur bacteria prefer light and move toward it.[1] The action of sunlight on this bacteriopurpurin enables them to decompose CO_2 and to utilize the oxygen thus obtained to oxidize H_2S.

All other known bacteria have no photodynamic pigment. Light is not a source of energy to them, and they are distinctly harmed by it; they grow best in darkness. Direct daylight kills them rapidly, and even prolonged exposure to diffuse light may be fatal. Bacteria are more rapidly killed by exposure to the sun's rays in June, July and August[2] than exposure of the same time in November, December and January. Expressed differently, many bacteria which are killed after an exposure of from one to two hours' direct sunlight in summer require an exposure of from two to three hours in winter to accomplish the same result.

Of the spectral rays, the red and infra-red rays, aside from the heating effect, are without noteworthy action on bacteria. The blue, violet, and ultraviolet rays, on the contrary, are distinctly bactericidal.

[1] Yost, *Plant Physiology*, 223.
[2] In the Northern Hemisphere.

These rays are chemodynamic and it is very probable that the death of bacteria exposed to them in organic media results from the formation of H_2O_2 or other germicidal substances from the substrate. Bacteria are also killed in non-decomposable media when they are exposed to the ultraviolet rays. It should be remembered that one of the most important characteristics of ultra spectral emanations is their very short wave length. Glass is opaque to them where quartz is transparent.

Electricity.—It is difficult to differentiate sharply between purely electrical effects and chemical changes which are induced in media of various kinds by the action of electric currents. Generally speaking, strong electrical currents sterilize media in which bacteria are growing, but it is by no means certain that the electric current *per se* is the important factor. Zeit[1] has made a careful, extensive and accurate study of the action of various kinds of electric currents on bacterial growth, and his conclusions are as follows:

"1. A continuous current of 260 to 320 milliamperes passed through bouillon cultures kills bacteria of low thermal death points, in ten minutes by the production of heat—98.5° C. The antiseptics produced by electrolysis during this time are not sufficient to prevent growth of even non-spore-bearing bacteria. The effect is a purely physical one.

"2. A continuous current of 48 milliamperes passed through bouillon cultures for from two to three hours does not kill even non-resistant forms of bacteria. The temperature produced by such a current does not rise above 37° C. and the electrolytic products are antiseptic but not germicidal.

"3. A continuous current of 100 milliamperes passed through bouillon cultures for seventy-five minutes kills all non-resistant forms of bacteria even if the temperature is artificially kept below 37° C. The effect is due to the formation of germicidal electrolytic products in the culture. Anthrax spores are killed in two hours. Subtilis spores were still alive after the current was passed for three hours.

"4. A continuous current passed through bouillon cultures of bacteria produces a strongly acid reaction at the positive pole, due to the liberation of chlorin which combines with oxygen to form hypochlorous acid. The strongly alkaline reaction of the bouillon culture at the negative pole is due to the formation of sodium hydroxid and the liberation of hydrogen in gas bubbles. With a current of 100 milliamperes for two hours it required 8.82 milligrams of H_2SO_4 to neutralize 1 c.c. of the

[1] Jour. Am. Med. Assn., November, 1901.

culture fluid at the negative pole, and all the most resistant forms of bacteria were destroyed at the positive pole, including anthrax and subtilis spores. At the negative pole anthrax spores were killed also, but subtilis spores remained alive for four hours.

"5. The continuous current alone, by means of DuBois-Reymond's method of non-polarizing electrodes and exclusion of chemical effects by ions in Kruger's sense, is neither bactericidal nor antiseptic. The apparent antiseptic effect on suspensions of bacteria is due to electric osmose. The continuous electric current has no bactericidal nor antiseptic properties, but can destroy bacteria only by its physical effects—heat—or chemical effects, the production of bactericidal substances by electrolysis.

"6. A magnetic field, either within a helix of wire or between the poles of a powerful electro-magnet, has no antiseptic or bactericidal effects whatever.

"7. Alternating currents of a three-inch Ruhmkorff coil passed through bouillon cultures for ten hours favor growth and pigment production.

"8. High frequency, high potential currents—Tesla currents—have neither antiseptic nor bactericidal properties when passed around a bacterial suspension within a solenoid. When exposed to the brush discharges, ozone is produced and kills the bacteria.

"9. Bouillon and hydrocele-fluid cultures in test-tubes of non-resistant forms of bacteria could not be killed by Röntgen rays after forty-eight hours' exposure at a distance of 20 mm. from the tube.

"10. Suspensions of bacteria in agar plates and exposed for four hours to the rays, according to Rieder's plan, were not killed.

"11. Tubercular sputum exposed to the Röntgen rays for six hours at a distance of 20 mm. from the tube, caused acute miliary tuberculosis of all the guinea-pigs inoculated with it.

"12. Röntgen rays have no direct bactericidal properties. The clinical results must be explained by other factors, possibly the production of ozone, hypochlorous acid, extensive necrosis of the deeper layers of the skin, and phagocytosis."

J. GRAVITY, OSMOTIC PRESSURE, AGITATION, CHEMOTAXIS.

1. **Gravity.**—The majority of bacteria suspended in liquids are not killed even by four hours' exposure to direct pressure of from 2000 to 3000 atmospheres (one atmosphere of pressure is equal to approxi-

mately 15 pounds to the square inch, or one kilogram per square centimeter of surface). Bacteria are weakened, however, by these great pressures, as is evidenced by a diminution in virulence, decreased pigment production, and the partial or complete inability to multiply. It is a curious fact that motile bacteria may retain their motility after an exposure of several hours to 2000 atmospheres from the pressure liquids, even although their powers of reproduction are quite lost.

Liquids are practically non-compressible, consequently direct pressure does not affect the volume of the liquid in which bacteria are suspended, nor does this pressure affect the amount of gas dissolved in the liquid. If, however, bacteria are exposed in liquids to gas pressure in the place of direct pressure, the germicidal action of the gas plays the prominent part in the final result. The amount of gas dissolved in the liquid increases with increase of pressure, consequently feebly germicidal gases may become powerfully germicidal as the pressure is increased. Thus, bacteria suspended in water overlaid by CO_2, which is feebly germicidal at ordinary pressures, are rapidly killed if the pressure is gradually increased; that is, CO_2 under these conditions becomes strongly bactericidal. According to Certes,[1] 600 atmospheres pressure of an inert gas, as nitrogen, will not kill anthrax bacilli.

Diminished Pressure.—Diminished pressure, aside from lowering the oxygen tension to a point below that necessary for the growth of aërobic bacteria, does not interfere seriously with bacterial growth.

2. **Osmotic Pressure.**—The boundary layer, ectoplasm, of every bacterial cell reacts like a semi-permeable or osmotic membrane. Through this membrane must pass all the elements necessary to the nutrition of the organism. A normal bacterial cell always tends to maintain a greater concentration of solutes within its substance than exists in the surrounding medium; hence the pressure from within upon the cell membrane is somewhat greater than the pressure from without upon the cell membrane, and the cell is consequently in a state of continual turgor. The osmotic pressure exerted by dissolved substances varies very greatly. Those of high molecular weight, as albuminoses or peptones, exert little or no osmotic pressure. Crystalloids, on the contrary, may exert very considerable pressure. Thus, a 30 per cent. solution of dextrose exerts a pressure of about 22 atmospheres. A bacterial cell placed in such a solution is under a great strain. If bacteria which are in a state of equilibrium with reference

[1] Compt. rend. Acad. de sc., 1884, 99, 385.

to the osmotic pressure of a solution are suddenly introduced into media containing a greater concentration of solutes, the contents of the cell diminish somewhat in amount, due to the rapid withdrawal of water leaving the rigid cell membrane visible. This shrinkage of the cell contents is spoken of as plasmolysis.[1] This shrinkage of the cell contents would indicate that the cell membrane is differentially more rapidly permeable to water than to crystalloids. All bacteria are not plasmolyzed when they are suddenly introduced into hypertonic solutions, and some organisms exhibit the phenomenon of plasmolysis to a much greater extent than others. Plasmolysis does not necessarily result in the death of the organism. It appears to be a fact that older bacteria are frequently more readily plasmolyzed than younger individuals of the same kind. The observations of Nicolle and Auclaire[2] would indicate that bacteria which retain the Gram stain are less readily plasmolyzed than Gram-negative bacteria. Whether Gram-positive bacteria which have become Gram-negative due to prolonged cultivation in artificial media invariably follow the same rule is not known.

If bacteria are gradually subjected to solutions of greater or lesser osmotic pressure, they usually accommodate themselves to these changes without visible effect. If bacteria are introduced abruptly into solutions of low osmotic pressure or distilled water, water rapidly passes through the cell membrane of the bacteria faster than the solutes within the cell can pass out, thus rapidly increasing the intracellular pressure until frequently the cell membrane ruptures, permitting the escape of some of the cell contents. This phenomenon is called plasmoptysis.[3] Most bacteria do not plasmoptyze readily, and it is problematical how much importance should be attached to either plasmolysis or plasmoptysis in practical bacteriology.

3. **Agitation.**—Bacteria grow best in quiet surroundings, although a slight amount of agitation is usually harmless and may be even beneficial if it tends to dislodge waste products from the immediate surroundings of sedimented organisms. Rapid agitation frequently retards the multiplication of bacteria in fluid cultures, and Meltzer[4] and Horvath[5] have shown that violent shaking gradually kills bacteria; not, however, by rupturing the cell membrane. The organisms undergo

[1] Fischer, loc. cit., p. 23.
[2] Ann. de l'Inst. Pasteur, 1909, xxiii, 547.
[3] Fischer, loc. cit., p. 48.
[4] Ztschr. f. Biol., 1894, xxx.
[5] Pflüger's Arch., 1887, xvii.

a gradual disintegration, and the injurious effects observed are said by these observers to be not purely mechanical.

4. **Chemotaxis.**—Bacteria respond to various chemical stimuli. Substances which can be used by them for nutritional purposes, as various constituents of laboratory media, appear to attract bacteria. Harmful substances, as acids or alkalis, may act in the reverse manner. Oxygen is a powerful chemotactic agent for many aërobic bacteria, while many anaërobes are repelled by it. The mutual chemotactic relations of bacteria and leukocytes, and the well-defined tendency of certain invasive bacteria to localize in definite tissues or organs of the animal body are interesting fields for speculation. Nothing conclusive is known about these relations.

K. ENZYMES, TOXINS, PTOMAINS.

Enzymes.—The phenomena of chemical interchange between bacteria and their environment indicate that enzyme activity plays an important part in bacterial metabolism.

Enzymes may be defined as substances of unknown composition produced by living cells which incite specific chemical reactions without permanently combining with the products of reaction. A small amount of enzyme acting under favorable conditions will cause a relatively extensive transformation of substance without itself being used up or inactivated. There is, however, a limit to the amount of transformation which a given amount of enzyme can accomplish, for the accumulation of reaction products tends to restrict enzyme action; the removal of reaction products appears to extend enzyme action somewhat. All bacterial cells appear to produce or to possess enzymes, probably several, which may be divided somewhat arbitrarily into two classes, the extracellular or exo-enzymes, and the intracellular or endo-enzymes.

Exo-enzymes.—Exo-enzymes are those which are excreted from the organism and appear as soluble, filterable and, frequently, diffusible enzymes, which may be obtained in an active state from filtrates of cultures of bacteria. Their diffusion from the bacterial cell and their filterability suggests that they may be relatively simple in molecular aggregation. Their function is essentially a "preparatory" one, for they transform potential nutritional substances, as proteins, carbohydrates or fats, to simpler compounds which are assimilable by the bacteria. It is very probable that the exo-enzymes work uneconomically in the sense that they transform more material than the organisms require: this phenomenon is exhibited in the extensive liquefaction of gelatin by proteolytic bacteria, as B. proteus. The organism which elaborates such an exo-enzyme probably derives but little energy from its activity, and; conversely, probably expends comparatively little energy in the elaboration and secretion of the exo-enzyme.

Endo-enzymes.—Comparatively little is known of the endo-enzymes: it is generally believed that they are comparatively non-diffusible, at least in an active state, and that they are non- or but slightly filterable. This suggests that they are relatively complex in their molecular aggregation. Their function is probably to act upon the nutrient substances which the cell has assimilated, partly to liberate energy from them, and partly to participate in the organization of the cell constituents. These endo-enzymes work economically in contradistinction to the exo-enzymes in the sense that the substrate is apparently changed by them in proportion to the requirements of the cell. Endo-enzymes may be obtained from bacterial cells when the latter disintegrate, provided the rupture of the cells is not accomplished by violent chemical means. Probably the phenomena of autolysis which many bacteria exhibit when they are placed in an environment free from food may be due, in part at least, to the autodigestion of the organisms by their endo-enzymes.

Classification of Enzymes.—Enzymes are usually classified according to the substrate they act upon: thus, proteolytic enzymes, or proteases, split proteins or protein derivatives into simpler compounds; carbohydrolytic enzymes split starches or polysaccharides into simpler carbohydrates; fat-splitting ferments, lipases, split fats into glycerin and fatty acids. The above enzymes are hydrolytic in character, that is, they effect cleavage of protein or carbohydrate or fat or of glucosides by splitting the molecule into simpler molecules which simultaneously take up hydrogen and oxygen in the proportions to form water, thus:

1.

$$\underset{\text{Glycyl-glycine.}}{CH_2NH_2CO\text{-}NHCH_2COOH} + H_2O\ (+\ \text{enzyme}) = \underset{\text{Glycine.}}{CH_2NH_2COOH} + \underset{\text{Glycine.}}{CH_2NH_2COOH}.$$

2.

$$\underset{\text{Lactose.}}{C_{12}H_{22}O_{11}} + H_2O\ (+\ \text{lactase}) = \underset{\text{Dextrose.}}{C_6H_{12}O_6} + \underset{\text{Galactose.}}{C_6H_{12}O_6}.$$

3.

$$\underset{\text{Triacetin.}}{\begin{array}{l} CH_2O\text{-}CO\text{-}CH_3 \\ | \\ CHO\text{-}CO\text{-}CH_3 \\ | \\ CH_2O\text{-}CO\text{-}CH_3 \end{array}} + 3\ H_2O\ (+\ \text{lipase}) = \underset{\text{Glycerin.}}{\begin{array}{l} CH_2OH \\ | \\ CHOH \\ | \\ CH_2OH \end{array}} + \underset{\text{Acetic acid.}}{3\ CH_3COOH}$$

The question of specificity of action of bacterial enzymes is not definitely settled. There is some evidence in favor of the view that exo-proteolytic enzymes produced by various bacteria act upon a variety of proteins: thus, the cholera vibrio produces a soluble proteolytic enzyme which will digest casein, coagulated blood serum, egg albumen, fibrin and gelatin. Other organisms, as the staphylococcus, produce an exo-enzyme which will hydrolyze casein, coagulated blood serum and gelatin: its action upon other proteins is not definitely established. The important question—are the products of hydrolysis of the *same* protein by proteolytic enzymes from *different* bacteria the same—is not definitely settled; it is probable, however, that the products differ. This suggests that the proteolytic enzymes of bacteria are not mere "catalyzers" which accelerate reactions in relatively unstable substances that would take place spontaneously but much more slowly; these enzymes (proteolytic enzymes) may not only incite reaction, they may guide it, as it were, along lines of cleavage which would not be followed in the absence of this enzyme. The carbohydrate—and the fat-splitting enzymes have much less latitude in splitting the carbohydrates and fats respectively than the proteolytic enzymes, for these substances are less complex in structure and composition than the proteins and protein derivatives.

Fuhrmann[1] has classified enzymes of bacterial origin into four types as follows:

A. Schizases (Hydrolytic) Cleavage Enzymes.
  1. Proteases, protein-splitting enzymes. Pepsin, Trypsin (Lysins, Coagulases).
  2. Carbohydrate-splitting enzymes. Amylase, Cellulase, Pectinase, Gelase, Invertase, Lactase.
  3. Glucoside-splitting enzymes. Emulsin (Synaptase).
  4. Fat-splitting enzymes. Lipases (esterases).

B. Oxidizing Enzymes.
  Tyrosinase, Acetic bacteria, Oxydase.

C. Reducing Enzymes.
  Reductases.

D. Fermentation Enzymes.
  Zymase, Urease, Lactic acid enzyme.

The bacteriolysins are of particular importance in bacteriology: of the bacteriolysins, those which liberate unchanged hemoglobin from red blood cells (hemolysins) and those which digest hemoglobin (hemodigestins[2]) are intermediary in their general properties between enzymes and toxins, if indeed there is any tangible distinction between

[1] Vorlesungen über Bakterienenzyme, Jena, 1907.
[2] Van Loghem, Centralbl. f. Bakteriol., 1912–1913, lxvii, 410.

them. Vaughan[1] has studied both enzymes and toxins extensively, and has summarized admirably the points of resemblance between exo-enzymes and exo-toxins as follows:

"1. Both are destroyed by heat.[2]

"2. They act in very dilute solution.

"3. When repeatedly injected into animals in non-fatal doses they cause the body cells to elaborate antibodies which neutralize the toxin (or the enzyme) both *in vivo* and *in vitro*.

"4· In the development of their effects a period of incubation is required.

"5· It has been shown (by Abderhalden) by optical methods that they have a cleavage effect upon proteins—they split complex proteins into simpler bodies; in other words, they have a proteolytic action.

"6· They are specific in two senses: (*a*) they are specific according to the cell which produces them; (*b*) they are specific in the antibody elaborated in the animal body after repeated injections of non-fatal doses."

Bacterial toxins are usually classified as exo- or soluble (extra-cellular) toxins, and endo- (intracellular) toxins. The former are soluble and diffuse out from the bacterial cell into the surrounding medium. Very few bacteria produce exo-toxins: the best known are those of the diphtheria, tetanus, and botulismus bacilli. To these specific antitoxins are known. Endo-toxins are non-diffusible and are locked up in the bacterial cell; they are liberated only when the cell disintegrates. No specific antitoxin has been produced for an endo-toxin.

Ptomains.—Ptomains are soluble, basic, nitrogen-containing substances formed from proteins or protein derivatives by the action of microörganisms. They are non-specific, relatively poor in oxygen content, and probably simpler in composition than either exo- or endo-toxins. No antibodies have been produced against them. Some are poisonous, many are not.

L. PIGMENTS.

With the exception of bacteriopurpurin, which occurs in the sulphur bacteria and is supposed to be photodynamic and, therefore, somewhat analogous to the chlorophyll of the higher plants, the significance of

[1] *Protein Split Products*, Lea & Febiger, Philadelphia and New York, 1913.

[2] Although they are somewhat more resistant to heat than the cells which produce them.

pigment formation, which is a striking cultural characteristic of many bacteria, is wholly unknown. The pigment they produce does not protect them against strong light, and achromogenic strains may be cultivated from the chromogenic varieties without apparent loss in the cultural or chemical characters of the organisms. It is very probable that these pigments are chiefly waste products of metabolic origin.

Pigments are produced in darkness and sunlight rapidly destroys many of them. Oxygen is not necessary for their production, for the non-colored leukobase is the form in which the pigment is excreted by bacteria, but oxygen is necessary for the development of color from this leukobase.

Pigment-producing bacteria may be grouped into four classes:

1. Bacteria producing photodynamic pigment. Certain sulphur bacteria which produce bacteriopurpurin.

2. Phosphorogenic bacteria which produce a luminous substance somewhat analogous to that of glow-worms. These organisms are chiefly marine forms, as B. phosphorescens.

3. Fluorogenic bacteria which produce a pigment soluble in water and culture media; this usually exhibits complementary colors as it is viewed by reflected and transverse light respectively.

4. Chromogenic bacteria. The pigment produced is usually insoluble in water and soluble in organic solvents. The color varies according to the organism producing it. The more common colors are red, orange, yellow, green, blue, violet, brown, and black pigment. These colored pigments are usually referred to as lipochromes because of their solubility in organic solvents and their general relationship to fats. Many of them give well-defined and constant absorption when they are viewed spectroscopically in solutions.[1]

M. SYMBIOSIS, ANTIBIOSIS AND COMMENSALISM.

The biological relations of bacteria are of the greatest importance in the economy of nature and in the production of disease. Bacteria do not grow in pure culture in nature, although they may do so in the tissues of man or animals, as disease-producing bacteria (pathogenic bacteria). In nature, where the reduction of dead complex organic material to mineralized salts is the striking function of bacteria, the successive steps in the degradation of organic matter are carried on by different kinds of microbes. The various steps appear to vary

[1] Sullivan, Jour. Med. Research, 1905, xiv, 109.

somewhat, but the process is on the whole an orderly and definite one. The association of various kinds of bacteria in this process, where each succeeding kind profits by the activities of the preceding kind, is a symbiotic one; that is, the several types of organisms mutually profit by their combined activities.

It frequently happens that the products of symbiotic activity may be greater than the sum of the products of the separate activities of the organisms.[1] On the contrary, many instances are known in which one kind of organism by its activity actually crowds out a preëxisting organism, as for example, the lactic acid bacteria which sour milk. They produce sufficient lactic acid from the fermentation of the lactose to kill the proteolytic forms. This substitution of one type of organism by another is known as antibiosis: the latter organism profits wholly at the expense of the first organism.

It not infrequently happens that one type of bacterium profits by the activity of another type of organism without benefiting the former in return. If two types of bacteria are concerned, the process is known as metabiosis; if the bacterium is living on a host, the relationship is spoken of as parasitism.

N. MEDIA—COMPOSITION AND REACTION.

Most bacteria grow best in a medium containing a large percentage of moisture in which diffusible proteins or protein derivatives are present as sources of nitrogen: these substances are better adapted to the dietary needs of the majority of bacteria than are ammonium salts or even simple amino acids. A very few bacteria (nitrifying bacteria) cannot grow in media containing organic nitrogen compounds: a few strictly pathogenic bacteria appear to require nitrogen as it exists in the highly complex tissues of man or animals for their growth. Many bacteria can utilize carbohydrates for their carbon, hydrogen, and oxygen requirements. Some bacteria appear to be able to utilize fats for their carbon requirement.

A neutral or feebly alkaline reaction is best adapted to the development of the vast majority of bacteria; a few types develop best in a medium which is distinctly acid—the aciduric bacteria.[2] Mineral acids are germicidal; organic acids may be utilized by bacteria for foods.

[1] Kendall, Jour. Am. Med. Assn., 1911, lvi, 1084.
[2] Kendall, Jour. Med. Research, 1910, xxii, 153.

O. GROWTH OF BACTERIA IN THE ANIMAL BODY.

The vast majority of bacteria do not grow in the tissues of the body, although a small number of organisms, the parasitic bacteria, live habitually on the surface of the body or on mucous membranes, usually without producing noticeable effects. A small, formidable group of bacteria, the progressively pathogenic bacteria, actually invade the tissues; they may produce within the host inhibition of function or anatomical changes incompatible with health.

CHAPTER III.

THE CHEMISTRY OF BACTERIA. THE EFFECT OF BACTERIA ON THEIR ENVIRONMENT.

A. GENERAL CHEMISTRY OF BACTERIA.

The practical significance of bacteria is summed up in the nature and extent of the chemical changes which they induce in their environment, the result of their multiplication and vegetative activity. These changes are essentially analytical, for the function of bacteria in nature is to transform dead organic matter from complex unstable combinations of carbon, hydrogen, nitrogen, and oxygen, which are worthless in the economy of nature, to fully mineralized, stable inorganic compounds of these elements, which may be resynthesized by plants.

A small but formidable group of bacteria, chiefly those pathogenic for plants, animals and man, act directly upon the living plant or animal organism, producing changes in them which may be temporarily incompatible with their well-being, and not infrequently lead to their death and eventually to their mineralization. The pathogenic bacteria, therefore, are also analytical in their activities and do not differ essentially in this respect from the saprophytic types.

It is necessary to consider briefly the method of the interchange of material between the vegetable and animal kingdoms in order to understand the full significance of bacterial action in the economy of nature. All animals require preformed organic compounds for their sustenance. They are unable to build up these compounds of which their tissues are composed from chemical elements or from simple inorganic salts. They are, therefore, dependent directly or indirectly upon the synthetic activities of green plants for their foodstuffs. The green plants by virtue of the chlorophyll contained within their leaves and stems possess the power of combining CO_2, water and nitrogenous salts under the influence of sunlight directly into the highly complex proteins and carbohydrates essential for animal food. These products of the synthetic activity of the plants are utilized by the animal kingdom for food; directly by the herbivora, indirectly by the carnivora. These substances are either broken down within the digestive tract of the animal body and reconstructed to form the tissues and supply energy to the animal, or eliminated as excreta. The excreta of animals are not sufficiently simple in composition, as a rule, to be used directly by plants, and the tissues of dead animals and plants are of little value in their complex state for plant foods. Further cleavage, both of the excreta of animals and the dead bodies of plants and animals, is necessary to make the elements contained within them utilizable by plants, and this cleavage is brought about by bacterial activity. Various saprophytic bacteria act successively upon these complex organic compounds, changing them, chiefly by hydrolytic cleavage, into stable, fully mineralized salts, which are directly utilizable in this state by the chlorophyll-bearing plants. There is, therefore, a constant rotation of the various elements which enter into the composition of animal and plant tissues between the plant and animal kingdoms respectively by means of an anabolic or constructive process in the one (plants), and a catabolic or destructive process in the other (animals). The cycle as outlined, however, is not a continuous one, for there are important gaps in the process of cleavage and in the process of synthesis which if left unbridged by the bacteria would eventually arrest all vital activity both of plants and animals, and all life would then inevitably cease on this planet. These gaps between the animal and vegetable kingdoms are filled by the analytical activity of bacteria.

A small group of bacteria, on the other hand, is also important from the synthetical point of view. A certain amount of nitrogen is lost in the animal and vegetable kingdoms by various natural agencies, and this supply of nitrogen must be made good from sources which are not directly available either to plants or to animals. Approximately 80 per cent. of the atmosphere is made up of nitrogen, and a certain group of bacteria, "the nitrogen-fixation" bacteria so-called, which are found chiefly on the nodules or roots of leguminous plants, are able to draw upon this great reservoir of atmospheric nitrogen and synthesize it into nitrogen-containing compounds which plants can utilize directly.

Another type of bacterial activity of importance is the oxidation of ammonia, the final step in the degradation of protein, into nitrites and nitrates. This is carried on by the nitrifying bacteria of the soil. Contrary to the generally accepted idea, therefore, the activities of the majority of bacteria are not in opposition to the activities of man, animals, and plants; bacteria are indispensable agents in the economy of nature.

B. CHEMICAL COMPOSITION OF BACTERIA.

1. **Elementary Composition.**—Bacteria normally contain the same elements in their substance that the higher plants and animals contain, viz., carbon, nitrogen, hydrogen, oxygen and phosphorus, together with smaller amounts of sodium, chlorine, sulphur, potassium, calcium, magnesium, and traces of iron.

2. **Chemical Constitution.**—The elements carbon, hydrogen, nitrogen and oxygen, and to a certain extent phosphorus, and perhaps sulphur are united to form proteins, nucleoproteins, carbohydrates, and fats. The inorganic substance of bacteria is made up of the other elements mentioned above in variable proportions. Of these elements, carbon, hydrogen, nitrogen, oxygen and phosphorus are the most important.[1]

TABLES ILLUSTRATING THE CHEMICAL COMPOSITION OF BACTERIA.

1. Percentage of the Elements in Ash-free "Mycoprotein."[2]

C per cent.	H per cent.	N per cent.
52.1–52.6	7.3–7.38	14.5–14.9

2. Percentage Composition with Respect to Organic and Inorganic Constituents.

	Putrefactive bacteria.[3]	Bacillus prodigiosus.[4]	Tubercle bacilli.[5]
Water	83.42	85.45	85.00
Protein	13.96	10.33	8.50
Extractive	1.00	0.70	4.00
Ash	0.78	1.75	1.40
Residue	0.84	1.77	1.10

[1] Certain acid-fast bacteria can be grown in media containing theoretically but five elements: carbon, hydrogen, nitrogen, oxygen, and phosphorus. Löwenstein, Centralbl. f. Bakteriol., Original, 1913, lxviii, 591. Wherry, Centralbl. f. Bakteriol., 1913, lxx, 115. Kendall, Day and Walker, Jour. Inf. Dis., 1914, xv, 428.

[2] Kruse, Allgemein. Microbiol., p. 62.

[3] Nencki and Scheffer, Ueber die chemische Zusammensetzung der Fäulnisbakterien, Beitr. z. Biol. d. Spaltpilze. Nencki, Leipzig, 1880, Jour. f. prakt. Chemie, N. F., xix, u. xx.

[4] Kappes, Analyze d. Massen Kulturen einiger Spaltpilze u. d. Soorhefe, Leipzig, Diss., 1889.

[5] Ruppel, Die Proteine, 1900, Heft 4, Beitr. z. exp. Therapie., Ztschr. f. physiol. Chemie, xxvi.

COMPOSITION OF BACTERIA.[1]

	Water, per cent.	In per cent. dry residue N	Acetone extract.	$CHCl_3$ extract.[3]	Phosphorus, per cent. in fat.[4]
Glanders	76.5	10.5	11.7	8.6	2.5
Chicken cholera	79.3	10.8	7.5	6.3	2.4
Cholera	73.4	9.8	8.7	6.8	2.4
Dysentery (Shiga)	78.2	8.9	12.8	10.6	1.6
Proteus vulgaris	80.0	10.7	10.9	7.1	1.6
Typhoid	78.9	8.3	15.4	10.6	1.2
Anthrax[2]	81.7	9.2	6.3	1.5	0.9
Pseudotuberculosis	78.8	10.4	15.6	10.3	0.8
B. pneumoniæ	85.5	10.4	15.4	10.3	0.8
B. coli	73.3	8.3	15 2	11.8	0.8
B. prodigiosus	78.0	10.5	9.0	6.6	0.5
B. psittacosis	78.0	9.5	11.1	7.0	0.5
B. diphtheriæ	84.5	..	7.0	5.2	0.2
B. pyocyaneus	75.0	9.8	15.8	10.7	0.2

It will be seen that from 75 to 86 per cent. of the bacterial cell is water. The remainder of the cell consists chiefly of protein, carbohydrate-like bodies, extractives (fats, fatty acids, waxes and lipoids), and inorganic salts. Of these, the nitrogenous substances vary greatly in amount, depending upon the composition of the medium in which the organisms are grown. The extractives (fats, waxes, lipoids, and fatty acids) are most prominent in the tubercle bacillus and the acid-fast group. Some extractives, however, are found in all bacteria, they being greater in amount on a medium containing carbohydrate and protein than on one containing protein alone. The chemical determination of the extractives is very unsatisfactory, partly because of the difficulty in breaking up the cell sufficiently to facilitate the entrance of the solvent.

3. **Chemical Composition of Bacteria.**—The percentages of the elements and various constituents of bacteria, as indicated in the above tables, is at best only approximate. Other factors very markedly influence the composition of the organisms.

Of these, the age of the culture, the temperature at which it is grown, and the composition of the medium in which the organisms are grown are the most important. Generally speaking, young cultures appear to contain rather more dry residue than older cultures, and bacteria grown at 37° C. contain more dry residue than those grown at 20° C.[5] The inorganic constituents of the broth influence

[1] Nicolle and Alilaire, Ann. l'Inst. Past., 1909, xxiii, 547.
[2] Asporeless. [3] From acetone extract. [4] From $CHCl_3$ extract.
[5] The decrease in dry residue observed in old cultures is partly attributable to autolysis of bacteria; this is usually observed earlier in cultures maintained at 37° C. than in corresponding cultures kept at 20° C. Growth is more rapid at this higher temperature, and recessive changes due partly to the accumulation of waste products are seen earlier.

the composition of bacteria markedly. Cramer[1] has found that the percentage composition of the ash of the cholera vibrio varies within very considerable limits as the organism is grown under different conditions. The following table indicates in a general way the influence of these factors:

	1 per cent. soda bouillon. Regular broth. 1 per cent. NaOH.	Phosphate bouillon. Regular broth. 4 per cent. Na phosphate.	NaCl bouillon. Regular broth. 3 per cent. NaCl.
Ash content of bacteria in dry substance	9.30	22.30	25.90
Ash content of moist mass	1.34	2.75	3.73
Ash content of medium in moist mass	1.25	2.50	4.12
Phosphoric acid in bacterial ash	28.70	34.80	10.90
Phosphoric acid in media ash	7.90	39.80	2.10
Chlorine in bacterial ash	16.90	7.97	40.70
Chlorine in media ash	23.00	11.40	49.20

The phosphorus content of the medium in these experiments, as shown in the above table, was varied almost twenty times, but in the bacterial organisms it varied scarcely three times. The variation in chlorine content was somewhat greater.

Even as important an element as nitrogen is subject to rather wide variations in bacteria, as Cramer[2] and Lyons[3] have shown. The following tables summarize Cramer's and Lyons's results. They were obtained by growing certain bacteria mentioned specifically below on a medium consisting fundamentally of 1.5 per cent. agar, to which were added various substances, as indicated in the tables, respectively Media A, B, and C. The general procedure was to grow the bacteria at 37° C. for several days, to wash them off with salt solution, to free them from adherent media by centrifugalization and washing, to dry the washed organisms *in vacuo* to constant weight, and to analyze the dry residue for extractives and ash.

CRAMER.

Medium.	Nitrogen substance.			Ether-alcohol extractives.			Ash.		
	A	B	C	A	B	C	A	B	C
Organism.									
Pfeiffer bacillus	66.6	70.0	53.7	17.7	14.63	24.0	12.56	9.10	9.13
Bacillus H-28[4]	73.1	79.6	59.0	16.9	17.83	18.4	11.42	7.79	9.20
Pneumonia bacillus	71.7	79.8	63.6	10.3	11.40	22.7	13.94	10.36	7.88
Rhinoscleroma bacillus	68.4	76.2	62.1	11.1	9.06	20.0	13.45	9.33	9.44

[1] Quoted by Kruse, Allgemeine Mikrobiologie, p. 88.
[2] Arch. f. Hyg., 1893, 151.
[3] Ibid., 1897, xxiii, 30.
[4] From water.

LYONS.

Medium.	Nitrogen-containing substance.			Ether extractives.			Alcohol extractives.			Ash.		
	A	B	C	A	B	C	A	B	C	A	B	C
Organism.												
Pfeiffer bacillus . .	62.75	58.88	45.88	1.68	3.50	2.67	12.17	17.30	29.60	7.16	2.79	3.09
Bacillus No. 28[1] . .	71.81	59.12	46.25	3.32	3.84	2.84	11.39	15.19	22.78	6.51	3.66	4.18
"Thread bacillus." .	61.06	44.31	33.25	1.74	2.24	1.87	18.40	21.80	27.50	8.09	4.50	3.02

Medium A agar, 1.5 per cent.; peptone, 1 per cent.
Medium B agar, 1.5 per cent.; peptone, 5 per cent.
Medium *C* agar, 1.5 per cent.; peptone 1 per cent.; dextrose, 5 per cent.

It will be seen that the nitrogen content of the bacteria grown in a medium containing nitrogen plus carbohydrate is almost 25 per cent. less than the nitrogen content in the same organisms grown in the *same* nitrogen medium but with *no* carbohydrate. The nitrogen content is greatest in the carbohydrate-free medium, the extractives are greater in the carbohydrate-containing medium. This decrease in the nitrogen content in pathological bacteria grown in sugar media may be of considerable importance, particularly in the preparation of vaccines and other antigens. Nothing is known definitely of the distribution of nitrogen in bacteria, but this reduction of 25 per cent. in the nitrogen content may well influence somewhat the immunizing value of vaccines.

C. COMPOSITION OF THE MORPHOLOGICAL COMPONENTS OF THE BACTERIAL CELL.

1. **Cell Membrane.**—Typical cells of higher plants contain cellulose, and bacteria were formerly differentiated sharply from the plant kingdom because cellulose could not be found in them. Later observations would suggest that cellulose or substances chemically closely related to it are demonstrable in certain bacteria. Dreyfuss[2] appears to have identified cellulose in bacteria from pus and in B. subtilis; Hammerschlag[3] claims to have isolated cellulose from tubercle bacilli, Dzierzgowski and Rekowski[4] appear to have found cellulose in diphtheria bacilli; more recently Tamura[5] has demonstrated a hemi-cellulose

[1] From water.
[2] Ztschr. f. phys. Chemie, 1893, xviii, 375.
[3] Sitzber. Akad. Wiss., Wien, xiii, 12.
[4] Arch. Soc. Biol., St. Petersburg, 1892.
[5] Ztschr. f. phys. Chem., 1914, lxxxix, 289.

in the same organism. So that the ability of at least certain bacteria to elaborate cellulose can hardly be doubted.

Emmerling[1] identified chitin in Bacterium xylinum, and Irvanoff[2] gives the following percentage composition of the cell membranes of B. pyocyaneus, B. megatherium and B. anthracis: C, 46 per cent.; H, 6.7–7 per cent.; N, 8.4–8.8 per cent.; which is empirically very similar to chitin. Chitin is chemically a polymer of glucoseamine, CH_2OH.-$(CHOH)_3.CHNH_2.CHO$, which in turn is an amino hexose very similar to dextrose, except that it has an amino group adjacent to the aldehyde group. Chitins are typically animal in origin, and are rarely, if ever, found in typical plants, hence the distribution between cellulose and chitin in bacteria is important as suggesting relationships to the vegetable or animal kingdoms.

Many bacteria stain brown with iodin, and the assumption is that the cell membrane of such organisms, or the cell substance contains substances similar to glycogen. According to Arthur Meyer,[3] many bacteria color blue with very small amounts of iodin; brown or red-brown with an excess of iodin; indicating that there is a very small amount of starch and a relatively large amount of glycogen or amylodextrin in the substance. Similar observations have been made by Heinze[4] and Levene,[5] who have isolated a substance from tubercle bacilli which reacts chemically like glycogen.

2. **Capsule.**—The capsules of the capsule-forming bacteria contain considerable amounts of a mucinous substance apparently a glycoprotein. Cultures of bacteria which do not ordinarily exhibit capsules occasionally produce spontaneously viscid, mucinous substances in artificial media; thus, strains of rabbit septicemia bacilli and glanders bacilli may become viscid after repeated transfers.[6] Broth cultures of tubercle bacilli may similarly become mucinous.[7] Rettger's observations[8] make it very probable that these viscid substances are true mucins.

3. **Cytoplasm.**—The cytoplasm of bacteria consists chiefly of the bacterial protein, which appears to be specific in character for any

[1] Berichte d. chem. Gesell., 1899, 541.
[2] Hofmeister's Beiträge, 1902, i, 524.
[3] Flora, 1899.
[4] Centralbl. f. Bakteriol., 2te Abt., 1903, xii; 1904, xiv.
[5] Jour. Med. Research, 1901, vi, 135.
[6] Theobald Smith, Transactions of First Annual Meeting of National Association for the Study and Prevention of Tuberculosis.
[7] Weleminsky, Berl. klin. Wchnschr., 1912, xlix, 1320; Kendall, Walker and Day, Jour. Infec. Dis., 1914, No. 11.
[8] Jour. Med. Research, 1903, x, 101.

given organism, together with enzymes and at least minimal quantities of all the products of its metabolism.

Regarding the nature of the protein substance in bacteria, but little is known, although 50–80 per cent. of the dried substance of the bacterial cell consists of protein and protein derivatives. Conspicuous among these protein derivatives are the nuclein constituents, nucleins, nucleoproteins, and nucleic acids; they occur constantly in bacteria and apparently the greater part of the protein of the bacterial cell consists of these nuclear constituents. Nucleins and nucleoproteins have been isolated from many bacteria: from B. subtilis by Van de Velde;[1] from the plague bacillus by Lustig and Galeotti;[2] from the typhoid bacillus by Paladino-Blandini;[3] from the tubercle bacillus by Von Ruck[4] and Ruppel;[5] from the diphtheria bacillus by Aronson;[6] and Carapelle[7] has identified a glyco-nucleo-protein in B. prodigiosus.

Numerous observations indicate that nuclein bases (xanthin bases) are found in bacterial cells; thus, Lustig and Galeotti[8] identified xanthin in plague bacilli. Nashimura[9] obtained xanthin bases in the dried residue of a water bacillus in the following amounts: xanthin 0.07 per cent.; guanin, 0.14 per cent.; adenin, 0.08 per cent. No hypoxanthin was found.

The amino-acids of bacterial protein have not been thoroughly studied. The variable nitrogen content even of the same organism as it is grown in different media and under different conditions would suggest that quantitative determinations of nitrogenous substances would be somewhat unsatisfactory. Qualitatively, so far as available data show, many amino-acids found in protein of higher animals and plants have been isolated or identified in bacterial cells. These amino-acids appear to differ in amount in different organisms, and several have not been isolated at all up to the present time. Vaughan, Wheeler, and Leach[10] conclude that the bacterial substance contains carbohydrates, nuclein bodies and polymers of mono- and diamino-acids. They are glyco-nucleo-proteins. Kruse[11] and Vaughan[12] have arrived at

[1] Ztschr. f. phys. Chem., viii.
[2] Deutsch. med. Wchnschr., 1897, 225.
[3] Baumgarten's Jahresberichte, 1901, 228, ref.
[4] Prophylactic Immunization against Tuberculosis, Report No. 1, Asheville, 1912, 3.
[5] Ztschr. f. phys. Chem., 1898, xxvi.
[6] Arch. f. Kinderheilkunde, vol. xxx.
[7] Centralbl. f. Bakteriol., 1907, xliv, 440.
[8] Loc. cit.
[9] Arch. f. Hyg., xviii, 325.
[10] Tr. Assn. Am. Phys., 1902, p. 243.
[11] Allgemeine Microbiologie, p. 65.
[12] Protein Split Products, p. 437.

the same conclusion. The analysis of one hundred grams of dried tubercle bacilli by Ruppel[1] indicates the importance of the nucleins in bacterial proteins.

	Grams.
Nucleic acid (tuberculinic acid)	8.5
Nucleoprotamin	25.5
Nucleoproteid	23.0
Albuminoids (keratin, etc.)	8.3
Fat and wax	26.5
Ash	9.2

Carbohydrates.—Glycogen or some similar carbohydrate, which is readily detected by the mahogany color it gives with iodine, is found in many bacteria, as has been stated previously, but it is extremely difficult to decide definitely whether it is limited exclusively to the cell membrane or scattered somewhat diffusely through the cytoplasm as well.

Fats and Fatty Derivatives.—Fats, fatty acids, lipoids and waxes, which may be demonstrated by staining bacteria with Sudan III, Scharlach R, and osmic acid, occur in variable amounts in the tubercle bacillus and other acid-fast bacilli. The amount of these extractives may be very great in the acid-fast group, varying from 26 to 40 per cent. of the total dry residue. Considerable discussion has centred around the distribution of these substances, many authorities claiming that the fats and waxes are contained in the cell wall of the organism, while others maintain that these substances are scattered throughout the cell substance as well. In the acid-fast bacilli it is probable that these fats are both intra- and extracellular, for analyses show that a certain amount of them can be extracted from intact bacilli, while still more can be extracted when the organisms are broken up. The following table from Kresling[2] illustrates the distribution of the fatty substance of the tubercle bacillus:

I. CONTENTS OF THE DRIED TUBERCLE BACILLI IN THE PREPARATION OF TUBERCULIN.

	Per cent.
Moisture (dried at 100°–110° *C.*)	3.9375
Moisture (dried in desiccator)	3.08
Ash	2.55
Nitrogen	8.575
Nitrogen-containing substances (albumin) reckoned by multiplying the amount of N by the factor 6.25 (the N of lecithin and other substances soluble in chloroform, benzol, ether, and alcohol were not reckoned)	53.59
Fatty substances in medium after the first four determinations	38.95
Other N-free substances, reckoned as the difference	0.9725

[1] Loc. cit.

[2] Centralbl. f. Bakteriol., 1901, xxx, 909.

II. FATTY SUBSTANCE OBTAINED BY EXTRACTION WITH CHLOROFORM, POSSESSES THE FOLLOWING CHARACTERISTICS:

Melting point	46° C.
Acid number	23.08
Reichert-Meissl number	2.007
Hehner number	74.236
Saponification number	60.70
Ether number	36.62
Iodine number (according to Hübl)	9.92

III. THE FATTY SUBSTANCE OBTAINED BY EXTRACTION WITH CHLOROFORM CONTAINS:

	Per cent.
Free fatty acids	14.38
Neutral fats and esters of fatty acids	77.25
Alcohols separated from the fatty acid esters (with melting point 43.5–44° *C.*)	39.10
Lecithin	0.16
Cholesterin	Not determined
Substances directly soluble in water	0.73
Substances soluble in water which are formed by the complete saponification of the fatty substances	25.764

Inorganic Constituents.—The most conspicuous inorganic element found in the ash of bacteria is phosphorus, and the content of phosphorus, recovered as phosphoric acid, frequently reaches as high as half the total ash weight. It is probable that a considerable part of this phosphorus is combined with nucleic acid to form nucleo-protein.

4. **Spores.**—The chemical composition of spores is not well determined, but the generally accepted theory is that they contain relatively less water and consequently a greater proportion of proteins and ash. Reinke[1] has suggested that the sporoplasm is an anhydride of the cytoplasm of the vegetative cell. Sporulation implies that relatively considerable amounts of water must be taken up by the spore substance in order to regain the proportion of this substance found in the parent organism.

D. FOOD RELATIONSHIPS OF BACTERIA.

1. **General.**—Food is any substance which a living organism may utilize, either by making it a part of its living material or as a source of energy. Food which is suitable for utilization by any organism must contain all the elements necessary for the building up and maintenance of that organism. Analyses of bacterial cells, which have been given in preceding tables, show them to be made up of the same elements as those of the higher plants and animals; viz., carbon, hydrogen, oxygen, nitrogen, and phosphorus, together with smaller amounts

[1] Quoted by Kruse, Allgem. Mikrobiol., p. 57.

of sodium, potassium, sulphur, calcium, and magnesium. Foods to be fully suitable for bacterial needs, therefore, should contain these elements. It should be stated, however, that the food requirements of bacteria vary within wide limits, but the above statements are generally applicable.

2. **Sources of Food.**—(*a*) **Nitrogen.**—The nature of the compounds in which nitrogen must be presented to bacteria as food varies greatly among the different groups. The nodule bacteria found in the nodules on the roots of many leguminous plants actually utilize atmospheric nitrogen: nitrifying bacteria found chiefly in the soil derive their nitrogen requirement chiefly from mineral salts which are oxidized through their activities to nitrites and eventually to nitrates. From this very simple source of nitrogen these bacteria are able to synthesize the complex nitrogen-containing proteins of their bodies.

The majority of bacteria, including not only the saprophytic organisms but most of those pathogenic for man, animals, and plants as well, thrive in media in which nitrogen is presented to them as peptones, albumoses, or even certain amino-acids; in other words, upon the products of protein digestion. The more strictly pathogenic organisms, as the gonococcus, may require nitrogen in the form of highly specific tissue proteins. Generally speaking, animal protein or its derivatives is more easily utilized by bacteria than protein of vegetable origin.

(*b*) **Carbon.**—The simplest carbon compound which occurs naturally, CO_2, cannot be used by bacteria, except certain nitrifying bacteria, as a source of energy, for it is already fully oxidized. The carbon of proteins and their derivatives, of carbohydrates, and of fats, on the contrary, is readily utilizable by most bacteria. As a rule, hydrocarbons of the aliphatic series are not attacked by the microörganisms, but compounds containing oxygen as well as carbon and hydrogen are better adapted for microbial food. Organic acids, as acetic acid, aspartic, tartaric, and many oxyacids are utilizable by some bacteria. The simpler alcohols can be used, but by very few bacteria. The complex alcohols, like glycerin and mannite, on the other hand, are available food materials for many.

The best nitrogen-free food compounds for microörganisms are the carbohydrates, particularly those containing six and twelve carbon atoms, the hexoses and bioses respectively. Carbohydrates containing four, five, or any number of carbon atoms not a multiple of three are usually not readily attacked by bacteria. Starches and cellulose are not generally utilizable, although certain types of organisms, notably those found in the intestinal tracts of herbivora, appear to decompose them very readily.

(*c*) **Hydrogen.**—Hydrogen is readily obtained by microörganisms from organic compounds containing available carbon, nitrogen, and hydrogen, but not apparently from water.

(*d*) **Oxygen.**—Oxygen is indispensable to the life of all living organisms as a source of energy and for structural purposes. A few bacteria, the obligately aërobic bacteria, can live only in the presence of free oxygen; another small group, the obligately anaërobic bacteria, live either in the absence of free oxygen or at best in the presence of minimal amounts of it; more than minimal amounts of free oxygen act as specific poisons to them. The majority of bacteria are facultative with respect to their oxygen requirements; that is, they can either live in the presence of free oxygen or derive their oxygen needs from organic compounds, usually the carbohydrates or proteins.

(*e*) **Inorganic Salts.**—Inorganic salts are used by bacteria almost wholly for structural purposes. The requirement for mineral compounds is very little, for these substances do not on the average make up more than 7 to 10 per cent. of the solid matter of the bacterial cell. The essential elements and the percentage of them found in the ash of certain bacteria have been referred to previously, and it was stated that the amount of inorganic salts found in the bodies of the bacteria bore a rather direct relationship to the salt concentration of the media. Of the inorganic elements, phosphorus is the most important, for it makes up nearly 50 per cent. of the ash. Phosphorous in contradistinction to any other inorganic salt is absolutely indispensable to bacterial growth. It is combined organically in nucleo-proteins, glyconucleo-proteins, and nucleic acids, which form the greater part of the protein of the bacterial cell.

CHAPTER IV.

BACTERIAL METABOLISM.

I. GENERAL BACTERIAL METABOLISM.

Two distinct phases may be recognized in the life-history of a bacterial cell; an anabolic or constructive phase, during which the cell becomes morphologically complete; and a catabolic, vegetative, or fuel phase, in which the mature organism reacts chemically upon its environment to provide the energy (fuel) necessary for the maintenance of the cell. Chronologically, the anabolic phase precedes the catabolic phase; that is to say, the bacterial cell must be morphologically complete before it can bring about its characteristic energy transformations; practically the two phases overlap somewhat.

The actual amount of material required for the anabolic phase of the bacterial cell is very small, for the actual weight of the average bacterium is but 0.000,000,0016 of a milligram, approximately (see page 25). The structural phase is practically ended, aside from the replacement of comparatively slight losses of substance incidental to the elaboration of soluble enzymes or to additional requirements for the formation of structural elements, such as capsules, when the organism is morphologically complete. The waste incidental to the utilization of material for purely anabolic needs is likewise very small in amount, and the total environmental change attributable to the purely constructive phase of bacterial metabolism is slight andordinarily disregarded.[1]

[1] Kendall, Jour. Med. Res., 1911, N. S., xx, 140.

The amount of material required for the catabolic (vegetative or fuel) phase of the bacterial cell, on the contrary, is relatively large. The energy requirement of cellular organisms varies rather with the area of their surface than according to their actual volume; consequently, very minute organisms, as bacteria, in which the surface is relatively very great in comparison with their size, would require much more material for energy purposes than for structural purposes. For example, the total surface area of a million average-sized cocci (each 1 micron in diameter) would be approximately 3.1416 sq. mm.; the weight of these organisms, assuming the specific gravity to be 1.030 (which is reasonably accurate), would be about 0.00054 mg. The combined surface of all the cocci in an actively growing broth culture of such organisms would be very considerable. It must be remembered, however, that these figures do not carry any specific basis for the measurement of bacterial activity in terms of chemical or physical phenomena; they merely express in a very general manner the physical basis for the apparent disproportion observed between the size of bacteria and the amount of change they induce in their environment.

The energy phase commences theoretically when the cell is morphologically complete, and it is a continuous process which ends only with the death of the cell. It may be reduced to a minimum when the cell enters upon a latent state of existence, as in spore formation; it is greatest when the organism is growing in a favorable medium at the optimum temperature, and it is restricted proportionately when environmental conditions become unfavorable.

The life-history of a culture in which innumerable bacteria are growing can not be sharply divided into the anabolic and catabolic phases. During the first few hours after inoculation, however, the anabolic aspect predominates; later the catabolic aspect predominates. Thus, colon bacilli inoculated into dextrose broth fermentation tubes do not produce gas in visible amounts during the first few hours of incubation, although the medium gradually becomes turbid, due to the rapid multiplication of bacteria. Somewhat later gas formation is observed, and it then proceeds with considerable rapidity. The production of gas is indicative of a period of great vegetative activity in which large numbers of mature colon bacilli utilize the dextrose for their energy requirements. Still later the production of gas ceases, the activities of the organisms diminish, and the culture finally dies out as waste products accumulate in sufficient amounts.

Those bacteria habitually pathogenic for man induce less striking physical and chemical changes in their environment, as a rule, than do the saprophytic types, as Theobald Smith[1] showed long ago. Thus, typhoid bacilli are relatively inert culturally; they form no gas in sugar media, no indol, and do not liquefy gelatin; on the contrary, B. coli and even more strikingly B. proteus are characterized by striking cultural changes; B. coli produces deep-seated changes in protein, resulting in the production of indol; it produces gas from sugar media, but it does not liquefy gelatin. B. proteus behaves much like B. coli in sugar media, but liquefies gelatin as well. These marked changes in the composition of the medium, namely, the production of indol from protein, the production of gas from sugar, and the liquefaction of gelatin, are all phenomena associated with the vegetative or fuel phase of bacteria.

II. THE NATURE OF BACTERIAL METABOLISM.

Chemically considered, the anabolic phase of bacterial activity is one characterized by the synthesis of relatively simple substances, chiefly nitrogen-containing, into the complex specific bacterial protoplasm through a series of synthetic reactions among which reductions and condensations appear to be the more prominent. It is very probable that many of these condensation reactions are hydrogenic in nature; that is, two simpler molecules are united into one molecule of greater complexity through the removal of hydrogen and oxygen from them in the proportions to form water.

As simple illustrations: the formation of lactose from a molecule each of dextrose and galactose,

$$\underset{\text{Dextrose.}}{C_6H_{12}O_6} + \underset{\text{Galactose.}}{C_6H_{12}O_6} = \underset{\text{Lactose.}}{C_{12}H_{22}O_{11}} + H_2O$$

the formation of a polypeptid, glycyl-glycin, from two molecules of glycocoll,[2]

$$\underset{\text{Glycocoll.}}{NH_2.CH_2.COOH} + \underset{\text{Glycocoll.}}{H.NH.CH_2.COOH} = \underset{\text{Glycyl-glycin.}}{NH_2.CH_2.CO.NH.CH_2.COOH} + H_2O$$

and the formation of the glyceride of a fatty acid from glycerin and acetic acid may be cited,

$$\begin{array}{lll} CH_2.OH + HOOC.CH_3 = & CH_2.O.O.CH_3 & \\ \vert & \vert & \\ CH.OH + HOOC.CH_3 = & CH.O.O.CH_3 & + 3\ H_2O \\ \vert & \vert & \\ CH_2.OH + HOOC.CH_3 & CH_2O.O.CH_3 & \\ \text{Glycerin.} \quad \text{Acetic acid.} & \text{Triacetin.} & \end{array}$$

[1] Fermentation Tube, Wilder Quarter Century Book, 1893, p. 219. (See also Kendall, Day and Walker, Jour. Am. Chem. Assn., 1913, xxxv, 1201–1249, for analytical data.)
[2] Fischer, Ber. d. deutsch. chem. Gesell, 1906, xxxix, 530.

The catabolic phase is essentially analytic; it is characterized chemically by a series of reactions in which the cleavage of more complex compounds to simpler ones with their simultaneous or subsequent oxidation, involving the liberation of energy, is a noteworthy feature. The catabolic phase is chiefly a series of oxidations of carbon and hydrogen. (For illustrative catabolic reactions see infra, pp. 73, 76.)

III. NITROGEN METABOLISM.

Bacteria, like all known living things, contain nitrogen in their substance, and nitrogen in some form is absolutely indispensable for the building up of their structure. Nitrogen, in other words, is an absolutely essential element in the constructive phase of the bacterial cell. The form in which nitrogen must be presented to bacteria in order to be utilizable by them varies with the kind of organism. The nitrogen-fixing bacteria found on the roots of leguminous plants can utilize the nitrogen of the atmosphere; some nitrifying bacteria can utilize the nitrogen of ammonium salts. (These two groups of organisms appear to be the only ones which can oxidize nitrogen.) Many bacteria can obtain their nitrogen from amino-acids. The majority of bacteria pathogenic for man and the higher animals are somewhat more exacting in this respect and require more highly organized nitrogen, as peptones and proteoses, while a small group of obligately human pathogenic bacteria, as the gonococcus, grows only in media containing nitrogen as it exists in the highly specialized protein of human origin, at least during their first growth outside the human body on artificial media.

The vegetative phase of bacterial metabolism is essentially a series of oxidations of carbon and hydrogen; nitrogen can not be oxidized by the great majority of bacteria, and consequently it appears to yield little or no energy to them. When nitrogen-containing compounds as amino-acids, peptones, albumoses, or proteins are utilized for the energy requirements of these organisms, the nitrogen (amino nitrogen) is usually eliminated from the amino-acid complex incidental to the oxidation of the carbon and hydrogen; the nitrogen thus eliminated appears in soluble form in the culture medium as ammonia. This process is true deaminization. Nitrates and even nitrites may be sources of energy to many bacteria, usually, however, because of their valuable oxygen content. To summarize, bacteria must have available nitrogen for their structural needs, but nitrogen, except for the nitrogen-fixing and nitrifying bacteria, is not as a rule a source of energy to them, because the great majority of bacteria can not oxidize it.

IV. CARBON METABOLISM.

Carbon is an important structural element for bacteria, and it is equally indispensable as a source of energy, for the oxidation of carbon is an important feature of the catabolic activity of the majority of microörganisms. The reduced form in which this element is present in amino-acids and other protein derivatives appears to be particularly adapted for structural purposes; for fuel purposes it is less available, possibly because of the necessity of introducing free oxygen into the carbon complex to provide the requisite energy for the vegetative activities of bacteria, as well as the additional amount of work required to eliminate the nitrogen of the amino-acid molecule (deaminization). It is generally stated that bacteria with relatively few exceptions fail to grow with their customary vigor in sugar-free media from which free (atmospheric) oxygen is excluded; the relative absence of available oxygen in such compounds would explain this phenomenon, in part at least.

The carbohydrate molecule, which contains no nitrogen and in which the carbon is already partially oxidized, can be utilized for fuel purposes by most bacteria with less expenditure of energy for its preparation than can be the case with most amino-acids, peptones, or proteins; for this reason it is very probable that utilizable carbohydrate is acted upon by many bacteria in preference to protein carbon. In this sense utilizable carbohydrate protects or shields protein or protein derivatives from bacterial attack for their fuel requirements; it does not protect protein from bacterial breakdown to supply their structural requirements, however.

The net result of this selective protective action of carbohydrates for protein is important because the amount of material required to provide energy for the bacterial cell far exceeds the amount of material required to build up the bacterial cell. The chemical transformations incidental to the anabolic phase of bacterial metabolism are insignificant in amount and ordinarily not noticeable; on the contrary, the chemical transformations associated with the catabolic phase of bacterial metabolism are relatively very considerable in amount; and the nature and extent of those chemical reactions which are associated with the transformation of material for energy are important not only for the identification of bacteria, they collectively comprise the important specific function of bacteria.

V. QUALITATIVE CATABOLIC REACTIONS OF BACTERIA.

The chemical changes observed in cultures of ordinary bacteria are chiefly those associated with the breakdown of organic substances for energy—they are reactions of the catabolic phase of bacterial metabolism. It should be again emphasized that the energy reactions—the catabolic reactions—are those which are most profoundly influenced by the composition of the nutritive substrate upon which the organisms are grown.

A. **Reactions of Bacteria in Media Containing Only Nitrogenous Substances (Proteins or Protein Derivatives) Which are Utilized for the Energy Requirements of Bacteria.**—Proteins are composed of amino-acids, of which some seventeen are recognized. Bacteria which decompose protein appear to act upon these amino-acids in the last analysis, and several types of reaction are recognized at the present time. Each kind of organism utilizes protein or protein derivatives somewhat differently and characteristically, but in general one or more of the following types of reactions are involved either successively or simultaneously in the catabolism of these substances. The reactions follow:[1]

1. $R.CH_2.CHNH_2.COOH + H_2 = R.CH_2.CH_2.COOH + NH_3$. Reductive deaminization of amino-acid to fatty acid with the same number of carbon atoms.

2. $R.CH_2.CHNH_2.COOH + H_2O = R.CH_2.CHOH.COOH + NH_3$. Hydrolytic deaminization of amino-acid to oxy-acid with the same number of carbon atoms.

3. $R.CH_2.CHNH_2.COOH + O_2 = R.CH_2.CO.COOH + NH_3$. Oxidative deaminization of amino-acid to keto-acid with same number of carbon atoms.

4. $R.CH_2.CHNH_2.COOH \rightarrow R.CH_2.CH_2.HN_2 + CO_2$. Carboxylic decomposition of amino-acid to amine with one less carbon atom.

5. $R.CH_2.CH_2.COOH \rightarrow R.CH_2.CH_3. + CO_2$. Carboxylic decomposition of fatty acid.

6. $R.CH_2.CH_2.COOH + 3O = CH_2.COOH + CO_2 + H_2O$. Carboxylic decomposition with the formation of a fatty acid with one less carbon atom.

A few illustrations will indicate the nature of these changes in amino-acids with the production of certain substances of clinical interest:

1. Formation of indol from tryptophan. Indol is a substance pro-

[1] See Kruse, Allgem. Mikrobiol., 505–536, for literature.

duced in the intestinal tract from tryptophan (an amino-acid found in protein), chiefly by B. coli and B. proteus. The reactions through which tryptophan is changed to indol by these organisms are as follows.[1]

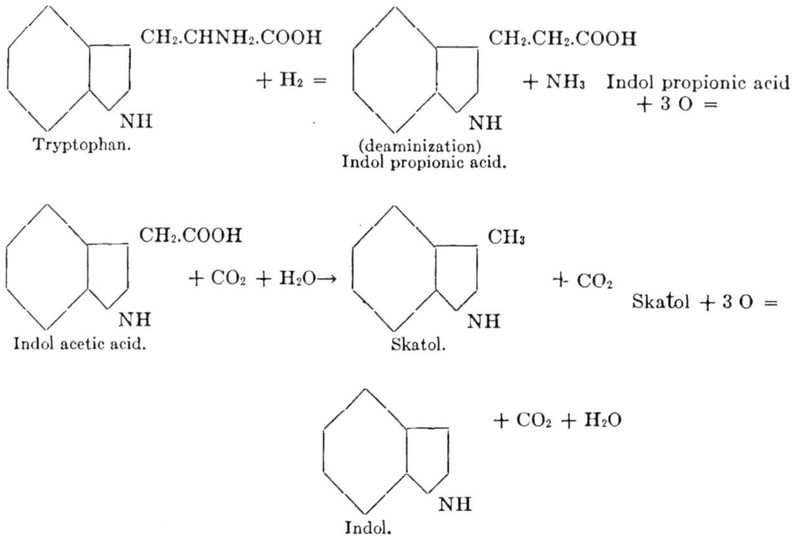

$CH_2.CHNH_2.COOH$ NH Tryptophan. $+ H_2 =$ $CH_2.CH_2.COOH$ NH (deaminization) Indol propionic acid. $+ NH_3$ Indol propionic acid $+ 3\ O =$

$CH_2.COOH$ NH Indol acetic acid. $+ CO_2 + H_2O \rightarrow$ CH_3 NH Skatol. $+ CO_2$ Skatol $+ 3\ O =$

NH Indol. $+ CO_2 + H_2O$

Indol contains little or no energy for most bacteria, and it is left as such in the culture medium or the intestinal tract. Indol is frequently absorbed from the intestinal tract, but it has little or no energy for the human body—it is oxidized in the liver to indoxyl

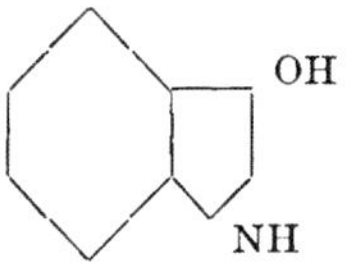

OH NH and is excreted and appears in the urine as indican O – S(=O)(=O) – ONa NH

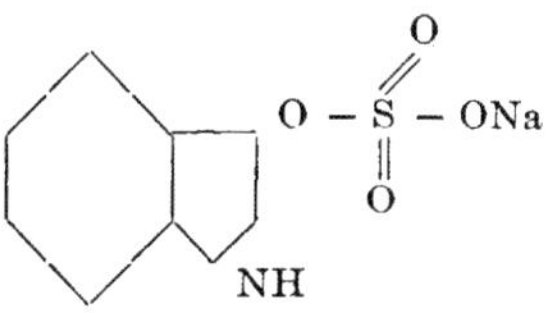

B. coli, B. proteus, and other organisms which "form indol" utilize the alanin radical of the tryptophan molecule (alpha amino propionic acid) for energy, first eliminating the nitrogen (deaminization), then oxidizing the carbon. The indol radical which is left is not a source of energy; it can not be oxidized by these organisms, consequently it remains as such in culture media or is absorbed from the intestinal tract.

[1] Nencki, Sitzungsber. Wien. Akad., 1898, II Abt., xcviii, 412.

2. Production of phenolic bodies from tyrosine.

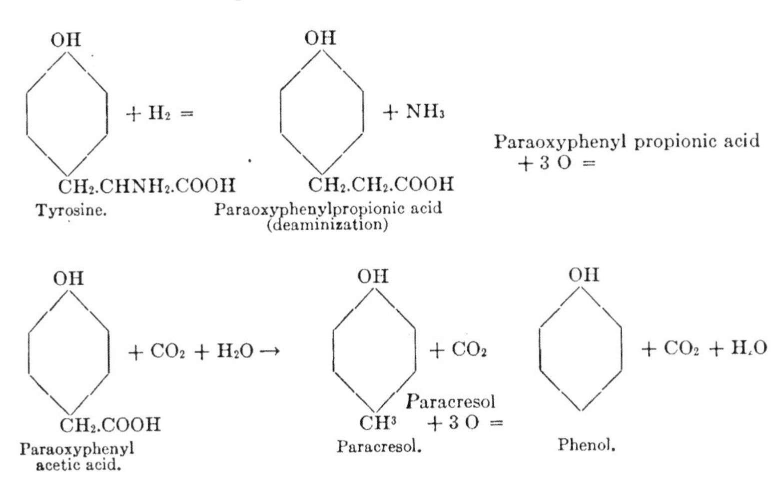

Phenol is not oxidizable by bacteria, hence it remains as such unchanged in the culture media. Phenol (or cresol) may be absorbed from the intestinal tract, but it appears eventually in the urine as an ethereal sulphate, precisely as indol appears in the urine as indican. Indol and phenolic bodies are not found in cultures containing utilizable carbohydrate—the bacteria which produce indol and phenols from tryptophane and tyrosine, respectively, can obtain their requisite energy far more directly and economically from the sugar than from the nitrogen-containing amino acid. Doubtless the same general principle applies to the formation of these aromatic substances in the intestinal tract.

3. Formation of amines from amino-acids by bacterial action.

(*a*) Cadaverin from lysine.[1]

$$\underset{NH_2}{CH_2}.CH_2.CH_2.CH_2.\underset{NH_2}{CH}.COOH \rightarrow \underset{NH_2}{CH_2}.CH_2.CH_2.CH_2.\underset{NH_2}{CH_2} + CO_2$$

Lysine. Cadaverin.

(*b*) Putrescin from ornithin.[2]

$$\underset{NH_2}{CH_2}.CH_2.CH_2.\underset{NH_2}{CH}.COOH \qquad \underset{NH_2}{CH_2}.CH_2.CH_2.\underset{NH_2}{CH_2} + CO_2$$

Ornithin. Putrescin.

[1] Ladenburg, Ztschr. f. phys. Chem., 1886, xix, 780.

[2] Ellinger, Ztschr. f. phys. Chem., 1902, xxix, 334; Ber. d. deut. chem. Gesell., 1889, xxxi, 3183; ibid., 1900, xxxii, 3542.

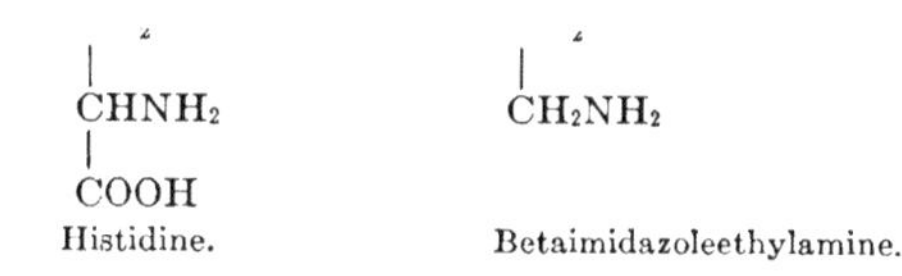

Histidine. Betaimidazoleethylamine.

According to Vaughan,[1] betaimidazoleethylamine is possibly the active poisonous principle of the protein molecule. Recent investigations would suggest that its liberation in the intestinal tract as the result of bacterial decomposition of protein there and its absorption into the body may be associated with symptoms of considerable severity. The substance is not formed as a product of bacterial metabolism in media containing utilizable carbohydrates.

B. **Reactions of Bacteria in Media Containing Both Utilizable Nitrogenous Substances (Protein and Their Derivatives) and Carbohydrates.**—Carbohydrates contain no nitrogen; consequently pure carbohydrate solutions are not complete foods for bacteria, they are important chiefly as sources of energy to them. Generally speaking, carbohydrates containing two, four, five, seven or eight carbon atoms are not readily fermentable by bacteria. Those containing six carbon atoms, particularly dextrose, are most readily utilizable, those with three carbon atoms, generally speaking, somewhat less so. Bioses, containing twelve carbon atoms, and starches appear to be hydrolyzed to sugars containing six carbon atoms before they are finally oxidized.

The final utilization of sugars for energy by bacteria varies according to the type of organism; the following qualitative reactions are illustrative of some of the general types of decomposition usually met with. It must be remembered that the exact quantitative utilization of carbohydrates by bacteria and the nature and composition of many of the intermediary products formed from them are still uncertain.

1. $\underset{\text{Lactose.}}{C_{12}H_{22}O_{11}} + H_2O = \underset{\text{Dextrose.}}{C_6H_{12}O_6} + \underset{\text{Galactose.}}{C_6H_{12}O_6}$.

Hydrolytic cleavage of a biose to two molecules of hexose sugar.

2. $C_6H_{12}O_6 = 3CH_3COOH$. Pure acetic acid fermentation.
3. $C_6H_{12}O_6 = 2CH_3CHOHCOOH$. Pure lactic acid fermentation.

[1] *Protein Split Products.*

4. $C_6H_{12}O_6 = CH_3.CH_2.CH_2.COOH + 2CO_2 + 2H_2$. Pure butyric acid fermentation.

5. $C_6H_{12}O_6 = 2CH_3CHOH + 2CO_2$. Pure alcoholic fermentation.

6. $2C_6H_{12}O_6 + H_2O = 2CH_3.CH_2OH.COOH + CH_3.COOH + C_2H_5OH + 2CO_2 + 2 + 2H_2$. The type of fermentation produced by B. coli in dextrose broth.[1]

The sugars containing six carbon atoms appear to be somewhat more utilizable than their corresponding alcohols: thus, the Shiga bacillus (B. dysenteriæ) can not ferment mannite; it can, however, readily ferment dextrose. This would suggest that the aldehyde group—CHO—is somewhat more readily attacked than the alcohol group—CH_2OH—, for mannite has no aldehyde group and dextrose has an aldehyde group. The alcohols in general appear to be less readily acted upon by bacteria than are the corresponding aldehydes or even organic acids, provided the latter are not too greatly dissociable.

The products of fermentation of higher alcohols, as mannite, by bacteria are somewhat different from those of the corresponding sugars (aldoses). The chief points of difference, according to our present knowledge, consist principally in the production of more alcohol when the higher alcohols are utilized than when the corresponding aldoses are concerned. This has been worked out satisfactorily for certain bacteria, notably the colon and the typhoid bacilli, by Harden.[2] It is not definitely known for many other organisms. The gas-forming bacteria, as a rule, produce more gas and more alcohol from the alcohols of the C_6 series than from their corresponding aldoses. This gas formation appears to result from the decomposition of formic acid by the activity of a specific enzyme, formiase, according to the equation $HCOOH = CO_2 + H_2O$.[3] Thus, B. coli and related gas-forming bacteria, according to this theory, produce the ferment, formiase, while B. typhosus, which also produces formic acid from the decomposition of dextrose, does not possess this ferment and consequently, forms no gas in sugar solutions. Formic acid is, therefore, somewhat prominently represented among the decomposition products of carbohydrates by the typhoid bacillus, while formic acid is either not present or present in small amounts in corresponding cultures of colon bacilli.[4]

The qualitative changes produced in fats and lipoidal substances by bacteria are not well known.

[1] Kruse, Allgemeine Mikrobiologie, p. 294.
[2] Jour. Hygiene, 1905.
[3] Franzen and Stuppuhn, Ztschr. f. physiol. Chem., 1912, lxxvii, 129.
[4] Clark, Science, November 7, 1913.

VI. QUALITATIVE INFLUENCE OF UTILIZABLE CARBOHYDRATES UPON THE ELABORATION OF PROTEOLYTIC ENZYMES.

Certain bacteria, as for example B. proteus, characteristically produce proteolytic enzymes which rapidly dissolve gelatin by hydrolytic cleavage. These enzymes are exo-enzymes; that is, they may be obtained sterile and free from bacteria simply by passing gelatin liquefied by their action through sterile unglazed procelain filters. Although the bacteria which elaborated the enzymes are removed by this filtration, the sterile filtrate still contains the active enzyme which will liquefy considerable amounts of gelatin. The function of these enzymes is to prepare the gelatin for assimilation by the proteus bacillus: the gelatin is broken down by enzyme action to gelatin peptone or even to polypeptids. The proteus bacillus does not produce soluble gelatin-splitting enzymes in gelatin containing utilizable carbohydrate, although sugar-free gelatin contains them in considerable amounts. These gelatinases, however, will liquefy sugar-gelatin quite as readily as sugar-free gelatin, indicating that the enzyme itself is not inactivated by the sugar, at least in the amount usually employed, 1 per cent. The same phenomenon is observed in cultures of the cholera vibrio and many other bacteria which liquefy sugar-free gelatin. Extensive investigations by Auerbach,[1] and by Kendall, Day and Walker[2] have shown that the gelatinase, which, as has been noted, is produced only in sugar-free gelatin, although it liquefies sterile sugar gelatin, prepares protein for utilization by these bacteria for purely catabolic purposes; if the organisms have access to utilizable carbohydrate the enzyme is not produced by them, because they utilize the sugar, not the protein, under these conditions as the source of their energy. These observations indicate how fundamentally the metabolism of bacteria is influenced by the nature and composition of the substrate upon which they are grown.

VII. QUANTITATIVE MEASURE OF BACTERIAL METABOLISM.

It is possible to measure the nitrogen metabolism of bacteria under varying conditions with a very considerable degree of accuracy in spite of the minute amounts of products involved. Such measurements are not only indicative of the nature and degree of the decom-

[1] Arch. f. Hyg., 1897, xxxi, 311.
[2] Jour. Am. Chem. Assn., 1914, xxxvi, 1962.

position of purely nitrogenous substances by bacteria; they furnish quantitative evidence of the extent of the utilization of carbohydrates by bacteria in preference to nitrogenous substances for fuel (catabolic) purposes; that is to say, such measurements evaluate the nitrogen metabolism of bacteria in purely protein solutions, and their nitrogen metabolism in media containing both protein and utilizable carbohydrate.

Such determinations have been made for a large series of bacteria by Kendall and Farmer,[1] and Kendall, Day and Walker.[2] The general method followed is to measure the amount of ammonia (deaminization) which appears in fluid cultures of bacteria under various conditions of growth. The following table shows, respectively, the change in reaction (to neutral red as an indicator in terms of $\frac{N}{1}$ acid or alkali per 100 c.c. media) and the increase in ammonia (milligrams per 100 c.c. media), as certain bacteria are grown for ten days in plain and dextrose broth respectively. The broths are identical in initial composition and reaction, except that the "dextrose broth" contains in addition to the ingredients of the "plain broth" 1 per cent. of chemically pure dextrose. All other conditions are exactly parallel. The results are averages of several strains of the same organism in various lots of media. It will be seen that B. alcaligenes, for example, which ferments no sugars, produces an alkaline reaction (indicated as "—" in the table) both in plain and dextrose broth: the amounts of ammonia in both media are nearly the same.

All the organisms which ferment dextrose produce less ammonia in the dextrose medium than in the corresponding sugar-free medium, although the numbers of living bacteria were found to be greater in the former than the latter. The small amount of ammonia in the dextrose broth appears to be largely the nitrogenous waste incidental to the utilization of protein for structural purposes: the relatively large amount of ammonia observed in the corresponding sugar-free broths is the combined "structural waste" and the "deaminization" incidental to the utilization of protein for their energy requirement. The progressively pathogenic bacteria, as the diphtheria, typhoid and dysentery bacilli, produce much less ammonia in sugar-free media than do the same organisms in various lots of media.[3] (Kendall, Day and Walker.)

[1] Jour. Biol. Chem., 1912, xii, 13, 215, 219, 465; xiii, 63. Methods given here.
[2] Jour. Am. Chem. Soc., 1913, xxxv, 1201–1249.
[3] Ibid.

Ten-day observations.	Sugar-free broth. Reaction.[1]	NH.3	Sugar broth. Reaction.[1]	NH.3
B. alcaligenes	−1.25	+3.50	−1.15	+5.30
B. dysenteriæ (Shiga)	−0.30	+4.20	+2.80	0.00
B. dysenteriæ (Flexner)	−0.25	+3.10	+2.45	0.00
B. typhosus	−0.45	+5.40	+3.30	+0.60
B. diphtheriæ	−0.50	+3.10	+2.80	+1.05
B. of hemorrhagic septicemia	−0.20	+4.70	+2.25	+0.35
B. paratyphosus alpha and beta	−0.10	+7.50	+3.90	+1.20
B. icteroides	−0.10	+4.20	+3.80	+2.10
B. of hog cholera avirulent	−1.25	+16.45	+3.70	+1.05
B. of hog cholera virulent	−0.75	+8.40	+2.65	+1.05
B. of fowl cholera	−1.00	+13.65	+3.35	+0.70
B. of Morgan	−1.33	+29.50	+3.90	+29.66[2]
B. coli	−1.00	+24.40	+4.90	+0.35
B. cloacæ	−1.20	+39.20	−0.30	+36.40[2]
B. proteus	−1.98	+58.40	+3.55	+1.40
Sp. choleræ	−1.45	+62.80	+2.00	+0.70
Sp. of Finkler and Prior	−1.00	+27.30	+1.50	+0.70
Sp. of Metchnikoff	−4.30	+41.30	+2.70	+0.70
B. pyocyaneus	−1.85	+30.30	−1.33	+41.50
Streptococcus	+0.70	+1.40	+5.00	+0.70
Staphylococcus	−0.75	+38.70	+3.75	+0.70
Mic. tetragenus	+1.00	+2.10	+3.00	+0.70
Mic. melitensis	−0.10	+6.30	+3.50	+0.70

VIII. SIGNIFICANCE OF BACTERIAL METABOLISM, WITH SPECIAL REFERENCE TO THE SPARING ACTION OF UTILIZABLE CARBOHYDRATE FOR PROTEIN.

Considerable emphasis has been placed upon the sparing action of utilizable carbohydrate for protein in the preceding pages. It now remains to summarize the salient features of this aspect of bacteriology and to indicate briefly by means of a few illustrations precisely how a comprehension of the principles underlying bacterial metabolism may be made use of in controlling, or at least influencing the action of these microörganisms upon their environment. The examples selected are chosen rather with a view of indicating the extreme range of the subject than for completeness along any limited line of investigation.

1. **The Composition of Bacteria.**—Experiments quoted previously (page 60) show very clearly that the percentage of composition of the bacterial cell varies according to the medium in which it is grown. Particularly striking is the difference in nitrogen content when the same bacterium is grown in media of the same nitrogenous composition and reaction with and without the addition of utilizable carbohydrate.

[1] Neutral red, − = alkaline reaction, + = acid reaction.

[2] These organisms can utilize 1 per cent. of dextrose without forming enough acid to inhibit their growth; after the dextrose is used up they attack the protein for their fuel needs—hence the ammonia production in a medium containing utilizable sugar. During the initial period when sugar is present, the ammonia value is very little, and the reaction is acid.

2. **The Recognition of Bacteria.**—The recognition of many kinds of bacteria, as for example members of the intestinal group, depends upon the reactions these organisms induce in various sugars. Thus, B. alcaligenes ferments no sugars; B. dysenteriæ ferments dextrose with the production of acid; B. proteus ferments dextrose and saccharose with the evolution of gas and the production of acid; B. coli ferments dextrose and lactose with the evolution of gas and the production of acid; B. coli coagulates milk, while B. proteus characteristically peptonizes it. All of these reactions are explained perfectly upon the theory that utilizable carbohydrate protects protein from bacterial breakdown. Thus, B. alcaligenes does not utilize any carbohydrate; as is well known, it is carnivorous. B. dysenteriæ can utilize dextrose, and consequently it produces acid in a medium containing both protein derivatives and this sugar: similarly, B. proteus and B. coli ferment dextrose and in addition a specific biose. B. proteus, however, does not ferment lactose, hence it attacks the protein of milk; while B. coli, which does ferment lactose, produces an acid coagulation in milk: the acid resulting from the fermentation of the milk sugar (lactose) protects the proteins of the milk. In each instance the organisms attack the utilizable carbohydrate whenever it is present, in preference to the protein for their energy requirements. If bacteria did not habitually utilize carbohydrate in preference to protein for their fuel needs, these fermentation reactions would be of no value whatsoever as diagnostic tests for these various microörganisms.

3. Certain bacteria, notably B. proteus, produce active, soluble (extracellular) enzymes when grown in sugar-free gelatin, that bring about an energetic liquefaction of this medium, which becomes alkaline in reaction. If the organisms are grown in dextrose gelatin no liquefaction takes place; the bacilli produce CO_2 and H_2 as well as acid in dextrose gelatin, using the sugar in preference to the protein for their energy needs. The liquefied gelatin containing the soluble gelatinase may be sterilized by passage through a Berkefeld filter, thus removing all bacteria. The filtrate will liquefy sterile plain or sterile dextrose gelatin, thus proving that the soluble enzyme, which prepares gelatin for assimilation by proteus bacilli (and which is only produced in a carbohydrate-free medium), acts specifically on the protein irrespective of other substances which may be present. In this instance the presence of utilizable sugar in cultures of living proteus bacilli protects the protein (gelatin in the instance cited)

from bacterial attack, and inasmuch as proteus bacilli prepare gelatin for assimilation through the action of a proteolytic ferment, the ferment is not elaborated by them under these conditions. A precisely similar restriction of the development of gelatin-liquefying ferments by utilizable sugars occurs in cultures of cholera vibrios and other bacteria which habitually liquefy this medium. In each instance the same explanation holds true.

4. Diphtheria bacilli do not produce their characteristic powerful extracellular toxin in the presence of utilizable carbohydrate—dextrose—as Theobald Smith[1] showed several years ago. The toxin is only formed in sugar-free media. In this case again the dextrose shields the protein of culture media from attack by the diphtheria bacillus, and consequently prevents the formation of toxin which is apparently a true excretion produced incidental to the utilization of protein for energy by these organisms. Similarly, tetanus and Shiga bacilli fail to produce toxin in the presence of utilizable carbohydrates.

5. Colon and proteus bacilli produce considerable amounts of indol in sugar-free media, but no indol in the same media to which utilizable sugar has been added. Here again the carbohydrate is attacked by these organisms in preference to the protein. The following table summarizes briefly the salient features of the above discussion:

Chemical composition of bacteria.	Sugar-free media.[2] Nitrogen substance, per cent.	Dextrose media.[2] Nitrogen substance, per cent.
1. Pfeiffer bacillus	70.0	53.7
Pneumo bacillus	79.8	63.6
Rhinoscleroma bacillus	76.2	62.1

	Sugar-free media.	Sugar media.
2. Diphtheria bacillus	Powerful extracellular toxin of which on the average 0.005 c.c. kills guinea-pigs.	No toxin produced; several cubic centimeters medium fails to kill guinea-pigs.
3. B. tetani	Powerful extracellular toxin produced.	No toxin produced.
B. dysenteriæ (Shiga)	Toxin present.	No toxin present.
4. B. proteus	Soluble, extracellular gelatinase formed.	No gelatinase formed.
Sp. choleræ	Soluble, extracellular gelatinase formed.	No gelatinase formed.
5. B. coli, B. proteus:		
Odor	Foul.	None.
Reaction	Strongly alkaline.	Strongly acid.
Products	H_2S, indol, phenols, ammonia, etc.	H_2, CO_2, lactic acid.

[1] Tr. Assn. Am. Phys., 1896.

[2] Nitrogenous constituents and reaction precisely the same in both sugar-free and sugar-containing media. The only difference is that the dextrose medium contains 1 per cent. of dextrose in addition. The organisms studied have, therefore, a choice between protein and sugar for catabolic purposes.

IX. FERMENTATION AND PUTREFACTION.

The terms "fermentation" and "putrefaction" have been confused and even used synonymously in bacteriological, chemical and even legal nomenclature, but they represent essentially distinct and generic types of bacterial activity. They indicate, or should indicate respectively, microbic decomposition of two quite distinct types of organic compounds, the carbohydrates and closely related nitrogen-free compounds, on the one hand (fermentation), and nitrogenous organic substances on the other hand, putrefaction. There are substances intermediate in character between carbohydrates and proteins, or fats and nitrogen-containing compounds in which it would be difficult to predict *a priori* which term would be correct —glucose amine is such a substance. Glucose amine is an amino-aldose, containing both nitrogen and carbohydrate groupings. Such instances, however, are uncommon and do not militate against the correctness of the general theory that fermentation and putrefaction are distinct processes.[1]

Fischer[2] has defined fermentation in the broad sense it should be used in bacteriology, essentially in the following terms: "Fermentation is the biochemical decomposition of nitrogen-free compounds, chiefly carbohydrates, by the action of microörganisms." Similarly, putrefaction is defined as "The biochemical decomposition of nitrogenous organic compounds by the action of microörganisms."

Fermentation and putrefaction are probably enzyme phenomena.

Transposing the sparing action of utilizable carbohydrate for protein, which has been repeatedly emphasized in the preceding pages, it may be stated that in the catabolic phase of bacterial metabolism "fermentation takes precedence over putrefaction,"[3] meaning by that that bacteria which can utilize carbohydrate derive their energy requirements from the utilizable carbohydrate when they are growing in media containing both carbohydrate and protein. The results of this sparing action of utilizable carbohydrate for protein have been indicated in the preceding pages, sections V–VIII, inclusive.

[1] Kendall, Jour. Med. Research, 1911, N. S., xx, 140–144.
[2] Vorlesungen über Bakterien, 1903, II Aufl., 206.
[3] Kendall, Jour. Med. Research, 1911, N. S., xx, 140–144.

CHAPTER V.

SAPROPHYTISM, PARASITISM, AND PATHOGENISM.

I. DEFINITIONS AND LIMITS.

The most conspicuous and important function of bacteria in the economy of Nature is to maintain a continuity between the Animal and Vegetable Kingdoms by restoring in utilizable form to the Plant World the elements contained in the complex organic compounds which comprise the dead bodies of plants, animals and their products. Bacteria dissipate much of the energy accumulated in these dead bodies and oxidize the elements contained in them to inorganic, fully mineralized salts. These salts are resynthesized by the chlorophyll-bearing plants through the energy of sunlight to carbohydrates, proteins and fats, and in these complex combinations the elements are again available for animal food.

The bacteria which live upon this dead organic matter, and whose function it is to effect its degradation and ultimate mineralization, are called saprophytic bacteria. They are specifically the most numerous, chemically the most active, and economically the most important members of the phylum Bacteriaceæ. They are rarely pathogenic, that is, they rarely initiate disease in man or the lower animals. Whenever they are found associated with morbid processes their presence is usually to be explained on the ground that they are secondary invaders.

A smaller group of bacteria are *parasitic,* that is, they exist upon the bodies of living plants, animals or men. Many of them are rarely met with in Nature far removed from their respective hosts. Their activities are not usually in opposition to those of their host and their presence is therefore unnoticed. They may become invasive, however, whenever the natural barriers, which ordinarily suffice to keep them out, are impaired.

From the parasitic bacteria there has been gradually evolved a small but formidable group of organisms, the *pathogenic* bacteria, whose activities are in partial opposition to those of their host. The pathogenic bacteria, like the *parasitic* bacteria, require a living host, but they differ from the parasitic forms in that they actually invade their hosts and induce progressive disease from host to host.

There are no sharply definable limits between these three groups of bacteria, the saprophytic, parasitic, and pathogenic; the latter appear to have arisen from the former by a process of evolution. Certain general modifications in the general types of chemical activity manifested by these groups are discernible, however, which are partly the result and partly the cause of their change in environment as they have passed from a saprophytic to a parasitic existence. Prominent among these modifications and activities is a gradual decrease in the intensity with which the parasitic and pathogenic bacteria act upon their environment.

The essential function of the saprophytic bacteria in Nature is to effect a rapid, deep-seated degradation of organic matter to simple compounds; these organisms decompose a relatively large amount of substance in a relatively short time. They are chemically active and many of them form highly resistant spores which enable them to survive prolonged periods of environmental vicissitude. The habitually parasitic bacteria, on the other hand, which exist upon the bodies of living animals, and the progressively pathogenic bacteria which develop within the tissues of animals are not subjected to extremes of temperature and food supply; they rarely or never form spores. The chemical activity of these organisms is usually much less pronounced than that of the saprophytic bacteria.[1] Indeed, intense chemical activity would be incompatible with their continued parasitic existence, for the damage to their host would be insupportable. The parasitic and pathogenic bacteria do not, for example, produce widespread liquefaction of the tissues, even when large numbers of them are actually growing in the body of the host. The growth of invasive organisms in the animal body is characterized by subtle changes in the composition of the tissues of the host and the development of these reciprocal reactions between host and parasite, which collectively are included in the newly developed science of Immunology.

It would appear, therefore, that in their evolution toward parasitism, those bacteria which could thrive without producing deep-seated and rapid degradation of proteins, that is to say, whose metabolism approached more closely the intracellular metabolism of their host, would be the more adaptable to a parasitic existence, and this is in accord with what is known of the chemistry of these organisms. Their metabolism approaches rather closely that of their host.

II. THE CYCLE OF PARASITISM.

The cycle of parasitism for bacteria whose life cycle is such that but a limited excursion outside their host is possible for them—and this appears to be the case for the majority of organisms parasitic on man—consists of three separate and well-defined stages, as Theobald Smith[2] has so clearly pointed out. They must—first—reach an appropriate host; secondly—multiply at least temporarily thereon, and thirdly—escape to other suitable hosts. Each phase of this parasitic existence must be exactly fulfilled, otherwise the cycle is broken and that particular strain dies out. It is not surprising, therefore, that the bacteria habitually parasitic for man are found variously upon the surface of the body—in the upper respiratory tract, the gastro-intestinal tract, or upon the mucous surfaces which are in direct communication with the exterior. Escape from the body of the host to other hosts is readily accomplished from these positions.[3]

Under special conditions, parasitic bacteria may actually invade the body of the host and become, therefore, temporarily pathogenic.

[1] Theobald Smith, Am. Med., October 22, 1904, viii; Kendall, Boston Med. and Surg. Jour., 1913, clxix, 749.
[2] Theobald Smith, loc. cit.
[3] Theobald Smith, loc. cit.

Such an invasion is usually subsequent to a preëxisting disease or to local weakening of the tissues which under normal conditions suffice to exclude these organisms. The disease produced by parasitic organisms is usually non-specific in character and sporadic in distribution, and ordinarily it does not attain epidemic proportions. The bacteria which have penetrated into the tissues of the host are locked up there, as it were, and their descendants cannot escape to other hosts, at least, in numbers sufficient to perpetuate the invasive strain, for these organisms have not perfected their pathogenic cycle. Parasitic organisms, in other words, are "opportunists," as Theobald Smith has admirably called them, rarely initiating disease, but usually able to penetrate the body as secondary or terminal invaders. The colon bacillus, for example, is an habitual parasite in the gastro-intestinal tract of man and many animals. Under certain conditions it may become invasive, causing cystitis, appendicitis, peritonitis, or other inflammatory lesions, but it does not ordinarily become progressively pathogenic for successive hosts, producing epidemics of cystitis, appendicitis or peritonitis. The staphylococcus is a common inhabitant of the skin of healthy man. When the continuity of the epidermis is destroyed, the organism may become invasive, causing furuncles, osteomyelitis, or endocarditis. The pneumococcus is found in the respiratory tract of many normal men, particularly in large cities, where it exists as an "opportunist," ordinarily producing no harmful effects, but frequently becoming invasive and producing a variety of lesions when the general resistance of the host is lowered.[1] These parasitic bacteria have not perfected their mechanism of entry into the tissues of the host, and of escape from the tissues to the exterior, consequently those strains which accidentally become invasive are locked up in the body and, as a rule, either are overwhelmed by their host or perish with it. They are imperfectly pathogenic, in other words.

III. THE CYCLE OF PATHOGENISM.

Habitually pathogenic bacteria—those organisms which produce progressive, specific disease from host to host—actually invade the living bodies of animals or man. This invasion may be *direct*, in which event the microörganisms actually enter the tissues or body fluids

[1] Recent studies by Cole and his associates indicate that the ordinary "mouth" pneumococcus differs serologically from the strains found in the saliva of pneumonia cases. It is not improbable that similar serological differences may be demonstrated in the group of the streptococci.

and multiply there, or it may be *indirect*, in which instance their soluble toxins alone are absorbed by the host. The cycle of pathogenism, therefore, is more complex than the cycle of parasitism; it necessitates lodgement of the invading microbe on the body of the host, the location and penetration of the necessary portal of entry (which involves an initial skirmish between the organism and the non-specific natural defences of the host), growth within the tissues of the host in the presence of opposition there, escape from the tissues to the surfaces of the host or to some channel in communication with the exterior and, finally, the transmission of the organism, directly or indirectly, to other suitable hosts. If the organism cannot force an entrance to the tissues of the host, that is, if the natural defences of the host suffice to keep out the prospective invader, the latter usually perishes and no infection takes place; if the organism does penetrate the tissues of the body, the invasion and growth of the microörganism leads to disturbances of structure, function or composition of the host, which are abnormal and inimical to his well-being. The production of disease, therefore, depends ordinarily upon the ability of the microörganism to multiply in the tissues or the body fluids of the host; bacteria which cannot force an entrance into the tissues of the host, multiply there and escape to the exterior and eventually to other susceptible hosts do not produce *progressive* disease.

The *nature* and *extent* of the disease produced depends upon several factors: (1) the kind of microörganism; (2) the number of micro-organisms; (3) their ability to locate and force an entrance to the tissues of the body (their virulence, in other words); (4) the location and extent of their multiplication in the tissues of the host; (5) the response of the tissues of the host to this invasion, and (6) the nature and extent of the secondary, specific defense of the host in response to the invasion.

The contagiousness of a disease depends upon the ability of the invading organisms to escape from their host in sufficient numbers to infect new hosts and to survive environmental vicissitudes until new hosts are reached. A few examples will indicate the principal variants of the pathogenic cycle commonly met with among progressively pathogenic bacteria.

The tubercle bacillus ordinarily gains entrance to the host through the air passages. The organisms pass through the alveoli of the lungs, set up infection there, and gradually are shut off from communication with the exterior through the formation of the tubercle. After a longer or shorter time, these tubercles eventually break down, typically into the air passages—and discharge there large numbers of tubercle bacilli. These are coughed up by the patient and are eliminated from the body, usually in enormous numbers, by droplets and in the sputum. Pulmonary tuberculosis is typically a chronic, focal disease. The perpetuation of the tubercle bacillus is assured through their elimination from the diseased body in enormous numbers through long periods of time, their ability to resist desiccation, and the relative directness with which they reach other hosts.

The typhoid bacillus gains entrance to the body through the mouth and the intestinal tract. The organisms penetrate the intestinal mucosa, develop in the internal organs, particularly the spleen, and after a rather definite excursion in the tissues of the body, enter into the intestinal tract again, either through ulcers or the gall-bladder or, occasionally, they appear in the urine. They are eliminated from the body in great numbers, either with the feces, or less commonly, the urine, and they gain access immediately to other subjects through direct contact or more or less indirectly through water or food, in sufficient numbers to set up infection in at least some of them.

The gonococcus is transmitted directly by contact. Occasionally the infection may be somewhat less direct, involving the conjunctiva.

The plague bacillus may be transmitted from host to host, either directly in the case of pneumonic plague, where great numbers of plague bacilli are coughed up from the lungs of one patient and transmitted through inhalation to other patients, or somewhat more indirectly, as is the case in bubonic plague. Bubonic plague appears to be a true septicemia; the plague bacilli circulate, at least temporarily, in the blood, and they are removed from the blood of one patient and transmitted to another patient (either man or rat) through the agency of the flea, which acts potentially as an hypodermic syringe, as it were, in this instance. Plague bacilli are locked up in the tissues of the host and were it not for the agency of a suctorial insect, as the flea, bubonic plague would almost certainly disappear, because the organisms have not perfected for themselves any mechanism of escape from one host to the other.

IV. DISTRIBUTION OF PARASITIC AND PATHOGENIC BACTERIA IN NATURE.

It has been shown in previous sections that comparatively few, if indeed any, of those bacteria habitually parasitic or pathogenic for man are found in Nature far removed from rather intimate association with their hosts. This is in accordance with the fact that few, if any, of these organisms are provided with spores which would enable them to survive exposure to long periods of conditions unfavorable to their growth. It is true, however, that some, at least, of these organisms, as for example, the typhoid bacillus, can survive for longer or shorter periods of time in the soil, particularly if it be frozen, or in water, for days or even weeks. There is little evidence that these bacteria multiply extensively outside the body; on the contrary, they tend to die off rather rapidly. In any event, their existence depends upon their reaching a suitable host again within a comparatively brief period.

There are a few spore-forming bacteria which occasionally infect man when associated conditions are favorable for them. Of these the bacillus of lockjaw, B. tetani; of botulism, B. botulinus; the gas bacillus, B. aërogenes capsulatus; and the anthrax bacillus are well-known. These organisms are not habitual parasites, however; they are "saprophytic opportunists." That is, they could in all probability exist if man were eliminated from their environment.

V. HOW PARASITIC AND PATHOGENIC BACTERIA REACH MAN.

A. **The Occurrence of Parasitic Bacteria upon the Bodies of Healthy Men and Animals.**—The continual exposure of the skin of man to his environment makes it almost inevitable that microbes shall collect there. It is quite probable, however, that the large number of microorganisms which reach the skin are not only non-pathogenic, they are not even habitually parasitic. Most of them are found there only transiently. Certain organisms, however, occur among these adventitious microbes, which appear to be habitual parasites, and many of these bacteria, under certain conditions, produce disease. Of these, Staphylococcus aureus and albus and Streptococcus pyogenes are almost invariably present not only on the skin, but on the exposed mucous membranes, particularly those of the nose and throat. The influenza bacillus, diphtheria bacillus, the pneumococcus, and even the tubercle bacillus, meningococcus and other organisms may also be occasionally found, particularly in the nose and throat of healthy men. The occurrence of these organisms is readily explained; the secretions of the nose and throat, as well as that of the skin are excellent culture media for these organisms, which collect at these sites and grow upon the various secretions and desquamated cells.

The majority of these organisms, however, particularly the coccal forms, as the staphylococcus, streptococcus and pneumococcus are to be regarded as "opportunists"; they do not of themselves initiate disease, as a rule. They are to be regarded rather, as Theobald Smith has called them, "organisms of the diseased state," because of their invasion of the body secondary to other, intercurrent diseases. Even the tubercle bacillus and the diphtheria bacillus, particularly the latter, have been found in the mouths of men who apparently have had neither tuberculosis nor diphtheria, yet these organisms appear to be virulent when tested in the usual manner and presumably might be able to incite disease whenever conditions favor their entrance to the tissues of the body. Theoretically at least, people who harbor these organisms are potential sources of danger to others. Even the internal organs of healthy individuals may contain parasitic bacteria without harm, although these organisms naturally are not present in large numbers. Tubercle bacilli have been found occasionally in lymph glands in normal man and in cattle. Intestinal bacteria also occur not infrequently in the apparently healthy tissues of the body. In rare instances, B. coli may be present in the urinary bladder without causing noteworthy symptoms.

B. **How Pathogenic Bacteria Reach the Body.**—The manner in which bacteria of the "opportunist" type reach the body has been considered above. It is now necessary to consider the manner in which bacteria which cause progressive disease from man to man reach the body.

1. **Air-borne Infection.**—Bacteria which cause progressive disease, particularly of the respiratory tract, are discharged from the diseased body principally through the mouth and nose and find lodgment in the environment of the patient through the medium of the air, from whence they settle upon various substances, as food, clothing, and walls and floors of rooms. These bacteria probably do not proliferate to any extent outside of the body, but they resist drying and may remain fully virulent for considerable periods of time and potentially able to infect a certain proportion of those individuals who may be exposed to them.

These air-borne infections are transmitted in at least two rather distinct ways: (*a*) by dust, and (*b*) by droplet infection.

(*a*) Organisms which are transmissible through dust must first of all be able to survive considerable periods of drying. The larger particles of dust to which bacteria may become attached soon settle from the air, but smaller particles may remain suspended for some time, depending on the velocity of air currents and the nature, size and shape of the particles. Dusting and sweeping in rooms naturally stir up particles which have settled from the air, and even larger particles may be resuspended in this way. Tuberculosis has frequently been suspected to have been transmitted through the inhalation of infected dust particles, that is, particles of dust which have dried tubercle bacilli adhering to them. Careful investigation has shown that houses in which careless consumptive patients have lived have been responsible for the transmission of tuberculosis. The ward-room of a battle ship is known to have become infected with tubercle bacilli early in its career and at least two successive details of officers contracted tuberculosis in this place. Guinea-pigs exposed on the floor of these so-called tuberculous rooms are quite frequently successfully infected with the tubercle bacillus.

The extent to which dust dissemination is a factor in transmitting disease, however, is not at all definitely known. It must be emphasized that the transmission of disease through dust is not necessarily a very direct one, because the inciting organisms may pass a very considerable period of time in dust before they reach a favorable host. In this sense, transmission of disease by dust is a relatively latent one.

(*b*) *Droplet Infection.*—Flügge[1] and his pupils were the first to demonstrate that minute droplets of spray may be eliminated from the mouth during talking, sneezing and coughing. These droplets are frequently carried through the air for some distance, even as much as ten meters in a quiet room. Usually the more minute particles remain suspended in the air for some time. The possibility of droplet infection has been definitely proven in the following manner: Agar plates containing sodium carbonate are placed at various heights and distances from the experimenter, who places in his mouth a solution of phenolphthalein and then talks in a natural manner, expelling droplets containing phenolphthalein during his speech. This dye is transmitted with the droplets until they reach the agar plates, where bright red spots are produced which are very readily observed. In like manner, cultures of B. prodigiosus placed in the mouth will infect agar plates at similar distances.

The transmission of disease by droplet infection may be, and frequently is, a very direct one. Bacteria which are air-borne or borne by droplets may remain alive for several weeks in indirect sunlight,

[1] Ztschr. f. Hyg., 1897, xxv, 179.

but all of them are readily killed if they are exposed to direct sunlight. The virus of whooping-cough, mumps, measles, influenza, cerebrospinal meningitis, pneumonic plague, tuberculosis, the exanthemata, the diphtheria bacillus, and possibly the pneumococcus may be spread in this manner. Air-borne infections probably rarely take place in the open air where the sunlight is strong. This does not apply to droplet infections where one individual coughs, talks or sneezes directly into the face of another. Air-borne infections, particularly droplet infections, are potentially common where overcrowding occurs, as in tenements, public gatherings, railway trains, schools, and factories.

2. **Soil-borne Infections.**—Those bacteria which are occasionally pathogenic for man and produce sporadic disease in man, and whose habitat is the soil, are for the most part spore-forming organisms. They commonly enter the body through wounds. Of these the bacillus of tetanus, malignant edema, symptomatic anthrax, of anthrax, and the gas bacillus are the best known but, with the exception of the latter, they are not habitually human parasites. Of those bacteria which are habitually pathogenic for man, typhoid, cholera, paratyphoid and probably dysentery may be soil-borne, but ordinarily infection with these organisms does not take place through the soil.

3. **Water-borne Infection.**—The viruses of excrementitious diseases—typhoid, paratyphoid, dysentery, and cholera—are not infrequently transmitted from man to man through contaminated water. Feces containing these organisms get into water supplies, reach man again, incite disease in man, again escape in the feces and reënter water courses, thus being recirculated. The cycle may be somewhat more complex, as for example, when typhoid dejecta are thrown upon the ground and are eventually washed directly into water supplies and thus reach man again.

4. **Food-borne Infection.**—A considerable number of pathogenic bacteria may reach man through food, although food which is infected is usually rendered so through the handling of it by man. Milk is probably the most common food thus to be infected and it is particularly dangerous for two reasons. In the first place, its opacity makes it difficult to distinguish foreign substances which may be in it; and again, it contains all the elements which are necessary for the food of man and incidentally for the majority of bacteria. Scarlet fever, diphtheria, tuberculosis both human and bovine, Malta fever, epidemic sore throat or tonsillitis, typhoid, dysentery, foot-and-mouth disease, many diarrheas of children, milk sickness, and the organisms

of cholera infantum, and, rarely, Asiatic cholera as well, all may be transmitted from milk.

Shell-fish, particularly oysters, have been known to transmit enteric diseases. This has been due, in the past, largely to their exposure in the estuaries of rivers where sewage flowed freely over them. Typhoid bacilli enter the mantle cavity of the shell-fish, remain alive there and enter the digestive tract in a viable state when the shell-fish are consumed in an uncooked condition.

Meats, particularly from beef and swine, have been known to transmit paratyphoid fever, botulismus (sausage poisoning) and meat-poisoning as well. There is, in addition, a group of cases with somewhat insidious symptoms, which are probably due to the consumption of food, particularly meat, which has been decomposed by saprophytic bacteria.

5. **Animal Carriers.**—The microbic diseases which are transmissible to man from animals and from man to man by animals are varied in character. They comprise protozoan and bacterial infections and the so-called "filterable viruses." Of these diseases, comparatively few are common to man and animals. Microörganisms may be transmitted to man by animal carriers in at least four distinct ways:

(*a*) By direct contact.
(*b*) By indirect transfer.
(*c*) By mechanical transfer.
(*d*) By intermediary hosts.

(*a*) *Direct Contact.*—The transfer of glanders from the horse to man, of anthrax from cattle and sheep to man and of hydrophobia from dogs to man represents direct transfer of the virus from the sick animal to the well man. Other diseases are thus transmitted, but the examples given suffice for illustration.

(*b*) *Indirect Transfer.*—Insects are common carriers in the indirect transmission of the virus of disease from man to man. Flies are known to have carried typhoid bacilli from typhoid dejecta to milk or other food, which in turn has been consumed by man, resulting in infection. The same insect, doubtless, when conditions are favorable, can and does carry other enteric bacteria—paratyphoid, dysentery and even cholera organisms. It is very probable that other insects also participate in the indirect transmission of bacteria. Acute conjunctivitis, particularly that form which is prevalent in Egypt, is supposed to be spread in this manner.

(*c*) *Mechanical Transfer.*—Suctorial insects are known to transmit the viruses of certain diseases which circulate in the blood stream of animals to man, incidental to feeding. Thus, the flea transmits the plague bacillus from rat to man, from man to man, and possibly from man to the rat. The louse similarly spreads the virus of typhus from man to man. In the instances cited the insect is probably not a true intermediary host, for the virus does not necessarily multiply in the insect, nor does the virus undergo any essential transformation, so far as is known, in the insect. Nevertheless, the transmission of the viruses of these diseases—bubonic plague and septicemia, for example, depends upon the agency of suctorial insects for their passage from host to host. Other insects also transmit disease, but the evidence in a majority of instances is somewhat less definite than the cases cited.

(*d*) *Intermediary Hosts.*—Certain insects, notably mosquitoes, transmit disease from man to man only after the virus has passed an extracorporeal cycle in the extrinsic host—the mosquito in this instance. Thus, Anopheles transmits malaria from man to man and *Stegomyia fasciata*, or as it is now called, *Aedes calopus*, transmits in similar manner, the virus of yellow fever. Transmission in these cases is through the female insect and a definite interval (latent period) must elapse between the time of biting the patient and the time when the mosquito becomes infective to the non-immune host.

6. **Human Carrriers.**—Individuals who are apparently healthy occasionally harbor within their bodies (in free communication with the exterior, however, either through the respiratory tract, the gastro-intestinal tract, the urinary tract or the skin) bacteria which are capable of inciting disease in others. Such individuals are known as bacillus carriers; frequently they eliminate these pathogenic bacteria in large numbers.

The bacillus carrier may or may not give a history indicating recovery from an infection of the specific organism which he "carries." Bacillus carriers may be temporary carriers, in which event they harbor the pathogenic bacteria for but a few weeks; or they may become habitual carriers, in which case the organism may be excreted for considerable periods of time, even years. The excretion may be constant or intermittent.

The typhoid bacillus is a common organism to be thus carried. It appears to localize eventually in the gall-bladder or the bile ducts, less commonly in the urinary bladder, and it may appear occasionally in large numbers in the feces or urine of the carrier. Women are more commonly found to be typhoid carriers than men. Similarly, paratyphoid, dysentery and cholera organisms may be excreted in the feces through long periods of time, rarely or never, however, in the urine.

Slowly progressing focal diseases, as pulmonary tuberculosis are, in a sense, spread by carriers, for the patient may survive for years, excreting daily large numbers of tubercle bacilli. The line of demarcation, in other words, between the human bacillus carrier and the patient in whom a focal disease is chronic for long periods of time is not sharply circumscribed.

7. **Contact Infection.**—The direct transmission of bacteria from man to man is well exemplified in the venereal diseases, gonorrhea and syphilis, which are usually transmitted by direct contact. Diseases of the respiratory tract, as tuberculosis, diphtheria and whooping-cough, may be transmitted directly from patient to patient by kissing, swapping chewing gum, by eating utensils, etc. Soiled fingers may transmit the typhoid bacillus from a typhoid patient to other individuals. All of the excrementitious diseases may be spread in a similar manner, under certain conditions.

8. **Germinal and Prenatal Infection.**—True germinal infection implies that a disease-producing microörganism is carried by the ovum or spermatozoa and incorporated in the embryo prior to its development. This method of transmission is not definitely worked out, although it has been claimed that syphilis may be thus transmitted by the male to the ovum *in utero*, the mother remaining uninfected by the disease.

In prenatal infections the organisms must pass the placental barrier. This implies that the fetus becomes infected directly from the maternal blood stream, or by continuity of growth of the organisms through the placenta. The placental form of infection is not conceded by all observers, but it is reasonably certain that congenital syphilis may be contracted thus. Smallpox, measles, dysentery, various pyogenic infections, and, rarely, pneumonia are occasionally said to be prenatally transmitted to the fetus. With respect to tuberculosis, there is difference of opinion. A very few cases are on record in which prenatal infection seems almost certainly to have taken place, for the newborn infant exhibited lesions which were so far advanced that no other explanation than prenatal infection suffices to explain them.

C. **Portal of Entry; Atria of Invasion.**—The bacteria which cause infection in the human body may be provisionally divided into two great groups: those of exogenous origin, which are not habitual parasites of man; and those of endogenous origin, which are habitual parasites of man.

The great majority of specific microbic diseases (in contradistinction to non-specific inflammations) are incited by bacteria of exogenous origin. These organisms must enter the host directly through their respective appropriate atria to produce characteristic disease. For example, the typhoid bacillus only causes typhoid fever when the organism is swallowed and enters the body through the intestinal tract. Infection of a skin-wound with typhoid bacilli will not result in typhoid fever. Similarly, cholera vibrios do not produce the disease cholera unless they enter the body through the gastro-intestinal tract, although if cholera vibrios are introduced through the skin in experimental animals they tend to migrate toward the intestinal tract, thus suggesting a special affinity for the intestinal tissues. Pathogenic bacteria of exogenous origin produce in general, progressive specific disease from man to man. Bacteria of endogenous origin, on the other hand—those which occur habitually as "opportunists" on the surface of the body or on mucous membranes opening to the exterior —ordinarily exist as harmless parasites. They may, however, and occasionally do, become invasive, inciting local or generalized inflammatory reactions as a rule, rather than well-defined clinical syndromes which are frequently so characteristic of infections with exogenous pathogenic bacteria. The bacteria of the "opportunist" type do not ordinarily gain entrance to the tissues of the body through sharply-circumscribed atria and the disease they produce is usually not epidemic in character.

1. **Skin and Adnexa**—The intact skin is a natural barrier which protects the underlying tissues of the body from bacterial invasion. Its free exposure to the environment suggests that a great variety of organisms find lodgment upon it from time to time; a majority of these organisms are harmless, and probably transient saprophytes which come and go irregularly. The moisture and excretions, however, appear to favor the limited development of a few types of bacteria, mainly those of the coccal group, which occur with sufficient regularity to be regarded provisionally as habitually parasitic bacteria. Of these, the pyogenic cocci are usually the most numerous; they exist as "opportunists" on the surface of the skin or penetrate into hair follicles and the ducts of the cutaneous glands, ordinarily, however, without becoming invasive so long as the continuity of the skin is maintained. Abrasions and cuts furnish a portal of entry to the subcutaneous tissues, in which these parasitic bacteria frequently set up inflammatory reactions. Friction may actually force them through

the intact skin. The plague bacillus and certain types of staphylococci are said to pass through the skin occasionally in this manner.

Streptococci and staphylococci are the more common habitual bacterial parasites found on the skin. Staphylococcus epidermidis albus (Welch), a variant of Staphylococcus pyogenes albus, is a particularly common factor in the causation of the troublesome, but relatively benign stitch abscesses which frequently develop where sutures are introduced through the skin.

The damaged skin is the usual portal of entry for spore-forming bacteria as well as the cocci mentioned above. Spores of the bacilli of tetanus, anthrax, symptomatic anthrax, malignant edema and the "gas bacillus," (B. aërogenes capsulatus, Welch) may pass to the underlying tissues through abrasions of the skin and cause either localized infections or widely distributed lesions. Even so insignificant an abrasion as an insect bite may furnish the necessary atrium for infection. The umbilicus of the newborn furnishes a portal of entry for certain bacteria: particularly severe is the infection of the stump of the umbilicus with B. tetani, causing that very fatal "tetanus neonatorum" which has been so common in the tropics in the past. Contused wounds and compound fractures are particularly dangerous; the inflamed tissues furnish anaërobic conditions particularly favoring the growth of anaërobic bacteria, as the tetanus and gas bacilli. Clean-cut wounds are usually less liable to infection with anaërobic bacteria. The free flow of blood with its bactericidal properties washes out many bacteria, inhibits the growth of residual microbes, and by virtue of the clot which soon seals the wound prevents the entrance of other organisms.

The sebaceous secretions, particularly of the axilla and external genitalia, are good culture-media for certain acid-fast bacteria, particularly B. smegmatis. The cerumen of the external ear is frequently infected with Micrococcus cereus flavus, and the puncture of the tympanic membrane may lead to direct infection of the middle ear from the outside, with this or other organisms. Infection of the middle ear may also take place directly through the Eustachian tube. The blood and lymph may also deposit bacteria in the middle ear.

The conjunctiva, by virtue of its very exposed position, must receive bacteria upon it very frequently. Its polished surface and the mechanical cleansing by the flow of tears (which do not possess germicidal properties) usually suffice to remove adventitious bacteria and to prevent bacterial development under ordinary conditions. The conjunctival sac, which receives the washings from the conjunctiva, is probably the recipient of many bacteria; of these B. xerosis occurs with sufficient regularity in the conjunctival sac to be regarded as a normal inhabitant. The pneumococcus is also found there. These organisms are "opportunists," occasionally causing severe acute conjunctivitis, although usually they are benign. Certain bacteria affect the conjunctiva fairly readily. Among these organisms, the gonococcus is particularly troublesome, causing a most severe inflammation. Ophthalmia neonatorum, a gonorrheal infection of the conjunctivæ of the newborn of infected mothers, has been in the past a most common cause of blindness. It has been claimed that the meningococcus may occasionally pass from the eye through the tear duct to the nasal cavity, and from there to the meninges.

Subcutaneous Tissue.—Many bacteria, particularly exogenous pathogenic bacteria, do not develop in the subcutaneous tissues, as for example, the majority of those organisms which induce specific progressive disease from man to man such as typhoid and cholera organisms. On the other hand, many of those bacteria which are habitually parasitic on the skin may produce infections of the subcutaneous tissues which vary in severity from mild inflammations to severe cellulitis. The staphylococci and streptococci are among the more important of this type.

Tonsils.—The crypts of the tonsils afford mechanical protection to bacteria which gain access to them and the secretions and tissue undoubtedly provide the necessary nutritive elements, consequently it is not surprising to find many types of bacteria in them. Staphylococci are almost invariably present and streptococci, particularly non-hemolytic varieties, are very common. The tonsils, which are in very direct communication with the lymphatic system, are important atria of invasion, particularly for streptococci, and many cases of low-grade infections of the body appear to have originated from the passage of bacteria through the tonsils to the tissues of the body. The extent to which the normal tonsils destroy bacteria—their value in the non-specific initial defense of the body against bacterial invasion in other words—is not clearly established. Generally speaking, however, the tonsils appear to bear the brunt of attack in certain diseases and they are of undoubted importance in shielding the body from invasion through the lymphatic tract by directly holding back these bacteria. The promiscuous removal of tonsils, particularly in the young, has no justification from available knowledge. The removal of diseased tonsils is quite a different matter.

Salivary Glands.—The salivary glands of the mouth are sometimes invaded by bacteria.

Nasal Cavity.—Large numbers of bacteria, indeed practically all known bacteria may at one time or another gain access to the nose through the inhalation of air containing dust, by droplet infection, from the tear ducts, and in other ways. The air which is inhaled is freed from bacteria before it enters the trachea, largely during its tortuous passage over the turbinates; the moist surface of the nasal mucosa effectively arrests the progress of bacteria, which adhere to it. The constant secretion of mucus encloses many of these organisms, which are removed mechanically with the mucus. There is no evidence that the nasal secretions are germicidal. The permanent nasal flora is very limited, however. The pseudodiphtheria bacillus is very frequently found there and pneumococci, streptococci and staphylococci are relatively common. The true diphtheria bacillus is found in the nasal cavity of about 1 per cent. of healthy individuals.

Lungs.—The expired air in quiet, normal breathing is sterile: also, the inhaled air is practically sterile before it reaches the bronchi, for the moist tortuous passages of the nasal cavity mechanically retain bacteria; the same mechanism prevents the expulsion of bacteria during exhalation, unless the breath is expelled forcibly either through the nose or mouth. Bacteria leave the nose or mouth in expired air only when the expiration is forcible enough to eject finely divided droplets from the mouth or nose respectively.

The lungs are protected from bacterial invasion not only by the tortuous nasal air passages, but by the ciliated epithelium which covers the surface of the mucosa of the bronchi and bronchioles. The rhythmic contractions of these cilia carry upward and outward those bacteria which may have penetrated so deeply into the respiratory passages. Inhibition of the activity of these cilia by cold or other environmental conditions may be a potent factor in the establishment of infection in the respiratory tract. Occasionally bacteria succeed in reaching the terminal bronchioles and alveoli of the lungs: they are normally removed by the phagocytic activity of leukocytes (microphages) or of certain fixed tissue cells (macrophages). In spite of these barriers, however, the lungs occasionally become infected. The pneumococcus and tubercle bacillus are the most common primary invaders of the lungs. Streptococci are more frequently secondary invaders, although many primary lobular pneumonias are caused by this organism.

2. **Mucous Membranes.**—The moist surface of mucous membranes makes them excellent culture media for many bacteria which can grow at the temperature of the body. The physiological secretions which bathe these membranes, with the exception of the stomach, are usually without germicidal properties; at best, their antiseptic properties are weak. The removal of bacteria from such surfaces is probably for the most part mechanical. The secretion of mucus, which has been shown to enclose bacteria, may be an important factor in their elimination.

Mouth.—The mouth is a most important portal of entry for the great majority of bacteria, both pathogenic and non-pathogenic, which are associated with man. All of the intestinal bacteria, harmful or benign, many of the bacteria which are associated with morbid processes of the respiratory tract, and several which induce specific lesions of the brain and spinal cord enter through this atrium. A great majority of viruses which infect the respiratory tract and the cerebrospinal axis also leave the body through the mouth or nose.

The normal flora of the mouth is quite varied,[1] including not only bacteria which are ordinarily regarded as harmless, but also organisms which occasionally or frequently incite disease. Thus, from 20 to 40 per cent. of healthy individuals living in large cities harbor typical and apparently virulent pneumococci in their mouths;[2] about 2 per cent. of school children harbor typical diphtheria bacilli in their mouths.[3] Rarely, tubercle bacilli have been detected in the mouths of apparently normal individuals.

It is worthy of note that an occasional abscess in the cervical region may contain spiral organisms; frequently a careful examination will reveal a sinus connecting the abscess with the mouth, perhaps originating at the base of a carious tooth. Dental caries is usually regarded as a bacteriological process. The removal of bacteria from the teeth and gums can not be satisfactorily accomplished by antiseptic mouth washes and the saliva possesses no germicidal properties. Bacteria are removed from the teeth mechanically by friction and are transported from the mouth to the stomach during the processes of mastication and deglutition. The oral flora is most numerous before

[1] For full literature and descriptions see Miller, Die Mikroörganismen der Mundhöhle, Leipzig, 1892, and Goadby, Mycology of the Mouth, 1903.

[2] Recent observations by Cole and his associates indicate that the ordinary "mouth" pneumococci differ in their serological reactions from pneumococci isolated directly from pneumonia lesions.

[3] Moss, Guthrie and Gelien have found a much larger proportion of diphtheria bacillus carriers during a period when diphtheria was epidemic.

eating and almost absent immediately after eating a hearty meal. Tubercle bacilli are swallowed thus and many of them eventually appear in the feces.

Stomach.—The acidity of the stomach during gastric digestion, by virtue of the free hydrochloric acid of the gastric juice, is a potent factor in the destruction of bacteria which reach the stomach both from the mouth and the respiratory tract. Mineral acids are much more powerful germicides than organic acids. The normal stomach, therefore, is quite free from inflammations or irritations attributable to the activity of bacteria. Many bacteria, however, run the gauntlet of the stomach successfully, especially when the stomach is empty (when the concentration of hydrochloric acid is very low) and pass into the intestinal tract, where the conditions are much more favorable for their growth. The passage of bacteria through the stomach probably takes place either very early in gastric digestion, when the hydrochloric acid is not at its "digestive concentration" (about 0.2 per cent.), or after gastric digestion has ceased. When water or other fluids are drunk, which do not call forth gastric juice, bacteria doubtless pass through the stomach unharmed, and it is probable that organisms included mechanically within food particles may occasionally escape the action of the gastric acidity.

Certain aciduric bacteria[1] and even yeasts which are tolerant of acid may be found occasionally in the normal stomach, but rarely or never pathogenic bacteria. Abnormally, particularly when the hydrochloric acid is deficient, many bacteria are found in the stomach contents. Obstruction of the pylorus tends to increase the number of bacteria in the stomach by promoting stasis of food. This condition is particularly common in carcinoma of the pylorus. The Oppler-Boas bacillus, sometimes called B. geniculatus, one of the aciduric bacteria, is so frequently found in this pathological condition it was at one time supposed to be an accessory factor; it is now known to have no relationship to gastric carcinoma. B. geniculatus is also found very commonly in cases of achlorhydria. Sarcina ventriculi is also found in similar conditions.

The gastric acidity will destroy the toxins of B. diphtheriæ and B. tetani; the toxin of B. botulinus is not inactivated by the gastric juice. The toxins of the paratyphoid group of bacteria also appear to be resistant to gastric digestion.

[1] Kendall, Jour. Med. Research, 1910, N. S., xviii, 153.

Intestines.—The abundant intestinal contents, which vary somewhat in composition and reaction at different levels, provide conditions which make the intestinal tract a very efficient combined incubator and culture medium. Many kinds of bacteria may theoretically find conditions well adapted to their rapid development there and it is not surprising to find that bacterial proliferation is greater both in nature and extent in the intestinal tract than in any other known medium. It has been conservatively estimated that the average daily fecal excretion of bacteria in a healthy adult on a normal diet is expressed by the truly enormous number, 33×10^{12}. About 47 per cent. of the nitrogen of the feces is contained in the bodies of these bacteria which, when dried, weigh nearly 0.5 gram.

The upper level of the intestinal tract, particularly the duodenum, is relatively free from bacteria during interdigestive periods. The duodenal bacterial population increases rapidly when food enters this section of the alimentary canal and decreases when the food passes to lower levels. The numbers of bacteria increase very greatly where stasis of food becomes more marked and in the cecum and large intestines generally there are continually present enormous numbers of bacteria.[1]

The types of bacteria found in the intestinal tract are influenced markedly by the nature of the food of the host and by the ability of the organisms themselves to change their metabolism to meet variations in the composition of this food. Those bacteria which can best meet alternations in diet of the host are the ones which naturally persist. The bacteria contained in the food itself may also play a prominent part in determining the nature of the organisms which are found in the intestinal tract. The colon bacillus is particularly labile in meeting dietary alternations in the intestines and this organism constitutes about 80 per cent. of the bacteria which can be isolated from the feces of the adult.

At birth the intestinal tract is sterile and the embryonal feces, the meconium, which is passed during the first eighteen hours after birth, is sterile. Following this period of sterility there is a period lasting about three days on the average, in which various adventitious organisms are met with in the dejecta. The normal nursling flora begins to appear by the end of the third day, following the ingestion of breast milk. The dominant organism of this nursling flora is ordinarially an obligate anaërobe, Bacillus bifidus, which is one of the

[1] Kendall, Jour. Med. Research, 1911, xxv, 126–130.

best known examples of obligately fermentative organisms. It does not thrive on a purely protein diet but requires carbohydrate, which is normally supplied by the breast milk. Breast milk, it will be remembered, contains on the average about 7 per cent. of lactose, 3 per cent. fat and but 1.5 per cent. protein. The proportion of carbohydrates to protein in the diet decreases as the infant becomes older and the diet becomes more liberal, and this decrease in the percentage of carbohydrate is associated with a diminution in the number of the obligately fermentative bacteria, particularly of Bacillus bifidus, and their gradual replacement by organisms which can thrive well on a diet containing variable proportions of carbohydrate and protein.[1]

Bacillus coli is a most labile organism with respect to its ability to develop in the carbohydrate and protein constituents of the intestinal contents at the ileocecal region and lower levels; this organism is represented to the extent of fully 80 per cent. in the feces of healthy men. Smaller numbers of other bacteria, as Micrococcus ovalis, Bacillus acidophilus, B. proteus, B. mesentericus, B. aërogenes capsulatus and many other varieties are found transiently or semi-permanently in the intestinal contents. Exogenic bacteria occasionally invade the tissues of the body through the intestinal mucosa. Thus typhoid, paratyphoid and dysentery bacilli and cholera vibrios may produce severe infections. The tubercle bacillus may pass through the apparently intact intestinal wall without leaving any evidence of its passage. It is supposed that this organism penetrates the intact mucosa and enters lymphatic channels suspended in fats and eventually proliferates in deeper tissues.

3. **Genito-Urinary System.**—*Vagina.*—The vagina has an acid reaction and it harbors very few bacteria, but immediately afterchildbirth the reaction may become temporarily alkaline. The bacillus of Döderlein, however, occurs so commonly, that it may be provisionally regarded as a normal inhabitant and a few strains of aciduric cocci are not infrequently detected in cultures from the fundus of the vagina. The Gonococcus and Treponema pallidum are the more common pathogenic organisms whose portal of entry is the vagina.

Uterus.—The normal uterus is sterile and the acid reaction of the vagina and the closure of the cervix uteri tends to maintain sterility under normal conditions. During menstruation and childbirth the mechanical defenses of the uterus are impaired. The organ itself appears to possess no specialized powers of resistance to infection.

[1] A more detailed discussion on intestinal bacteria and their significance will be found in Chapter xxx.

Urethra.—The urethra in health is practically free from bacteria. The flow of urine mechanically frees it from bacteria. The external orifice of the urethra, however, frequently contains an acid-fast organism, Bacillus smegmatis, which can be differentiated from the tubercle bacilli only by animal inoculation, and, very frequently, Bacillus coli. The gonococcus and Treponema pallidum may invade the tissues through the urethra.

Urinary Bladder and Ureter.—The slightly alkaline reaction of the urine affords a good culture medium for many bacteria and infection of the bladder by B. coli, B. proteus, B. typhosus and other microorganisms is by no means uncommon. It is probable that infection occurs much more frequently through the blood or lymph than through an ascending infection from the urethra. B. proteus appears to grow with great luxuriance in the urinary bladder and a typical cystitis may be readily incited in dogs by injecting virulent cultures of the organism directly into the bladder. Occasionally a descending infection from an inflamed kidney may result in cystitis: whether a true ascending infection through the ureter to the kidney takes place is not definitely proven.

Kidneys.—The kidneys are normally free from bacteria, but infection of one or both kidneys through the blood stream is a well-established phenomenon. A variety of organisms may thus infect the kidney. The cocci of suppuration frequently incite acute nephritis and tubercle bacilli induce chronic infection. Theoretically, any invasive organism which enters the blood stream may localize in the kidney and establish metastatic foci there. The organ is susceptible to specific bacterial toxins as well as to the bacteria themselves.

D. **Where Bacteria Multiply in the Body.**—Practically no organ or part of the body, except such structures as the nails, are free from invasion with one or another kind of organism. The obvious complexity of the subject makes it difficult or even impossible to present in concrete form, a statement which shall indicate specifically the types of organisms which incite infection in association with the particular organs or tissues where they become localized. It is important in this connection, however, to remember that a great majority of progressively pathogenic bacteria exhibit rather marked affinities for special tissues, and that they invade the tissues through definite atria. The organisms which are habitually parasitic, on the contrary —the "opportunists"—as Theobald Smith has so clearly pointed out, are less exacting in this respect, as a rule, and they may invade the tissues whenever the natural barriers—skin, mucous membranes, and so on—weaken and become vulnerable.

The following table indicates the more common and important bacteria, parasitic or pathogenic, which may invade the tissues, and the organs where they tend to localize and develop.

SKIN:
- Staphylococcus and streptococcus groups.
- Acid-fast group: tubercle bacilli, lepra bacilli, smegma bacilli.
- Anaërobic group: tetanus, gas bacillus.
- Anthrax.
- "Bottle" bacillus (spore of Melassez).

NOSE, THROAT AND ADNEXA:
- Staphylococcus group.
- Streptococcus and pneumococcus group.
- Diphtheria and pseudodiphtheria group.
- Influenza and pertussis group.
- Pneumobacillus, rhinoscleroma and ozena group.
- Bacillus fusiformis and spirillum group.
- Meningococcus and catarrhalis group.
- Acid-fast group—chiefly tubercle bacilli and leprosy.
- Blastomycetes and hyphomycetes.
- Virus of poliomyelitis and unknown viruses, mumps, etc.
  (Organisms of dental caries and pyorrhea not included above.)

EYE AND EAR:
- Streptococcus and pneumococcus group.
- Staphylococcus group.
- Diphtheria and pseudodiphtheria group.
- Influenza group.
- Koch-Weeks and Morax-Axenfeld group.
- Gonococcus.
- Proteus group.
- Pyocyaneus group.

LUNGS:
- Streptococcus and pneumococcus group.
- Pneumobacillus group.
- Acid-fast group: tubercle bacillus.
- Influenza and pertussis group.
- Plague bacillus, anthrax bacillus and B. psittacosis.
- Colon and typhoid group.
- Actinomyces and hyphomycetes.

PELVIC ORGANS:
- Streptococcus and staphylococcus group.
- Gonococcus and Treponema pallidum.
- Tubercle bacillus and smegma bacillus.
- Micrococcus melitensis.

SEROUS FLUIDS:
1. Cerebrospinal fluid:
   (a) Fluid usually clear: tubercle bacillus and Treponema pallidum. Virus of poliomyelitis.
   (b) Fluid turbid: Pneumococcus, streptococcus, meningococcus, B. influenzæ, B. typhosus, B. coli.
2. Pleural and pericardial fluids:
   (a) Fluid usually clear: tubercle bacillus.
   (b) Fluid turbid as a rule: pneumococcus, streptococcus, B. influenzæ, pneumobacillus group, Bacillus typhosus, staphylococcus.
3. Peritoneal fluid:
   - Streptococcus group.
   - Coli and typhoid group.
   - Tubercle bacillus (?).

BLOOD:
- Streptococcus and pneumococcus group.
- Staphylococcus group.
- Typhoid, paratyphoid and dysentery group.
- B. coli.
- Recurrent fever and treponemata.
- B. pestis.
- Certain filterable viruses: Yellow fever, poliomyelitis (?).
- Tubercle bacillus (occasionally).

INTESTINAL CONTENTS, FECES:
- B. bifidus and B. acidophilus group (chiefly in infants).
- B. coli, B. lactis aërogenes, proteus and cloacæ group.
- Alcaligenes, paratyphoid, typhoid and dysentery group.
- Streptococcus and Micrococcus ovalis groups.
- Mucosus capsulatus group.
- Spore-forming group: Aërobic—B. mesentericus, B. subtilis, B. anthracis.
  Anaërobic—B. aërogenes capsulatus, B. tetani, B. botulinus.
- Acid-fast group: tubercle bacilli, bovine and human; grass bacilli.
- Spiral group: Vibrio choleræ, Sp. of Finkler and Prior.

E. **Where and How Bacteria Escape from the Body.**—It appears from foregoing considerations that those microörganisms which are progressively pathogenic for man habitually invade the tissues through atria characteristic for each microbe. Their escape from the tissues through appropriate channels in direct communication with the outside is equally important. Bacteria of the "opportunist" type frequently perish within the tissues because they lack a perfected mechanism of escape to the outside. Progressively pathogenic bacteria leave the body through two principal avenues—the mouth and nose, and the feces. Less commonly, certain types may pass to the outside in the urine. The skin is not a very important factor in the elimination of pathogenic bacteria. The paths of pathogenic bacteria from the tissues to the outside are varied, but very constant for each special organism and the discussion of this phase of their activity is reserved for the Section on Specific Organisms.

VI. BALANCED PATHOGENISM; EPIDEMIOLOGY.

It has been helpful, for clearness and discussion, to distinguish rather sharply between parasitic and pathogenic bacteria and in a majority of specific instances such a differentiation can be readily established. There is no hard and fast line of demarcation, however, between organisms of the "opportunist" type and those progressively pathogenic, for it is undoubtedly true that some "opportunists" may exhibit epidemic tendencies for limited periods if a combination of conditions arise which favor the distribution of the organisms and either increase the invasive powers of the microbe or decrease the resistance of the host. The limited spread of such bacteria is far more frequently attributable to unusually direct transfer of organisms by a common vehicle through a series of susceptible hosts than to the escape of the microbes from one host to another. Thus, milk-borne epidemics of septic sore throat may be extensive and involve many patients, but secondary transfer from man to man is relatively uncommon. These bacteria have not, as a rule, perfected their mechanism of escape from the tissues of one host to those of another. The epidemics are usually of brief duration and it is probable that the surviving microbes return to their original parasitic state.

Of far greater importance is a probable tendency of many progressively pathogenic bacteria to act more and more on the defensive; to gradually disembarrass themselves, on the one hand, of the offensive weapons which originally conferred upon their possessors the ability to invade their host, and, on the other hand, to perfect whatever defensive weapons they may have possessed the rudiments of.[1] Such a change, as Theobald Smith has pointed out, would be difficult to detect, because an elimination of the more aggressive type and its gradual replacement by a strain in which the defensive elements were more prominently represented would require years for its accomplishment. Such a change in the activities of the microörganisms would probably be accompanied by reciprocal activities of the host, so that eventually a strain of microörganisms would be evolved which had reached a state of relative equilibrium with the host. Unusually virulent strains of microbes would tend to perish with their hosts, and unusually susceptible hosts would tend to perish with their invaders. A mutual adjustment of virulence and resistance between the surviving hosts and microbes would lead eventually to one of three conditions:

1. Gradual extinction of the microörganism;
2. The gradual assumption of a parasitic or "opportunist" existence, or
3. A more perfect pathogenism in which the mechanism of invasion, multiplication within the tissues and escape to other hosts is accomplished without acute damage to the host.

It might well happen that the introduction of such "balanced" strains into new fields would lead to temporary disaster, as for example, the highly fatal epidemic of measles when this virus first gained a foothold in the South Sea Islands.

[1] Theobald Smith (Some Problems in the Life History of Pathogenic Microörganisms, Am. Med., 1904, viii, 711) clearly stated and discussed this hypothesis over a decade ago, and it is surprising how little cognizance has been taken of it.

Theobald Smith[1] has mentioned the diphtheria bacillus as an organism which possibly exhibits a tendency toward a parasitic existence. The toxin of the diphtheria bacillus is not a poison specific for man; many animals, as the horse and guinea-pig, are very susceptible to it. Yet the diphtheria bacillus is almost obligately a human pathogen. The ever-increasing occurrence of avirulent, non-toxin producing strains which are otherwise perfectly typical, and the frequent occurrence of individuals whose serum contains small amounts of natural antitoxin might be interpreted as an indication that strains of this organism are becoming gradually accustomed to a purely parasitic existence in the upper respiratory tract of man on the one hand, and that man has acquired some specific resistance to the microbe on the other hand.

The tubercle bacillus (typus humanus) is an excellent example of an exquisitely balanced pathogenic microörganism. Its metabolism is not markedly different from that of the host and the typical disease excited by it is focal, chronic, and slow-going. Years may elapse before the host finally succumbs. The development of the organisms within the tissues of the host does not appear to lead to the formation of substances which arouse the latent offensive and defensive mechanism of the host to acute antagonism. During this long period the tubercle bacilli establish communication with the outside and, in a majority of cases, countless myriads of bacilli escape from the host before death removes him as a source of infection. Occasionally tubercle bacilli become widely disseminated in the body, causing rapidly fatal, generalized miliary tuberculosis. These organisms perish with their host.

It is well known that the virulence of bacteria, many kinds at least, can be increased decidedly by passage from animal to animal by providing an artificial portal of entry and of exit from animal to animal. This is accomplished by injecting the organisms into a first animal and reinjecting them, at brief intervals, into other animals. In such instances there is a direct continuity of growth from animal to animal, greater than is met with in naturally occurring infections. It is worthy of note that bacteria of the "opportunist" type are, generally speaking, more successfully exalted in virulence under these conditions than the progressively pathogenic forms.

There is yet another feature of Pathogenism which is worthy of note. From time to time almost any bacterial disease, for example,

[1] Theobald Smith, loc. cit.

typhoid, plague or influenza, may leap suddenly to epidemic proportions, spread rapidly and then subside again, to be succeeded by sporadic cases which gradually diminish in numbers and in severity. The bacteria causing these outbreaks appear to acquire somehow and somewhere, an unusual degree of invasiveness and they spread rapidly, especially in thickly settled areas, and as rapidly lose their unusual activities and subside to what appears to be their usual level of virulence. It is very probable that those strains of pathogenic bacteria in general, which suddenly acquire unusual virulence are short-lived, partly because their hosts perish before the microbes can escape to new hosts. Not infrequently, these or similar epidemics are preceded by mild, atypical disease, which may not be specifically recognized, and during this initial period the bacteria may be quite widely distributed.[1]

[1] Kendall, Boston Med. and Surg. Jour., 1915, clxxii, 851.

CHAPTER VI.

IMMUNITY AND INFECTION.

GENERAL PHENOMENA OF IMMUNITY.

IT has long been recognized that man and animals exhibit refractoriness to infection with specific bacteria or other microörganisms which cause serious epizoötics in closely related animals. Man is, as a rule, quite free from the epizoötic diseases of animals domesticated by him, and the domestic animals are usually not infected with the organisms which incite progressive disease in man. Thus, man does not contract chicken cholera and domestic animals do not become infected with the typhoid bacillus. Furthermore, closely related animal species may exhibit striking differences in susceptibility to the same disease; for example, field mice are readily infected with the glanders bacillus, but house mice are quite resistant to infection with this organism, and ordinary sheep readily succumb to anthrax although Algerian sheep are practically immune to infection with the anthrax bacillus.

This inherent or congenital resistance or refractoriness to infection with a specific microörganism, when general among the individuals of a species or group of animals or of man is termed natural or inherited immunity. It is not necessarily absolute; lowering the natural resistance of the individual may render him susceptible to infection. Thus hunger, experimental (phloridzin) diabetes, fatigue produced by prolonged exercise in treadmills, and excessive chilling by the removal of hair have been shown to decrease resistance in experimental animals.

It is also a matter of common observation that the uniform exposure of a number of theoretically susceptible individuals of the same species to a virus does not lead to uniform infection; a certain small number usually perish, a larger proportion become mildly or severely ill and recover. The greatest number are not especially affected, as a rule. Those individuals who escape infection in one epidemic may succumb to infection during a subsequent epidemic of the same disease. This phenomenon of individual variation in susceptibility is well exemplified in water- and milk-borne epidemics of typhoid fever where typhoid bacilli are widely distributed in a water or milk supply. A small number become infected, but the greater number escape the disease. The incidence of scarlet fever, of diphtheria or of other infectious diseases among the members of the same family frequently illustrates this same phenomenon. This resistance to infection exhibited by certain individuals of a susceptible species is termed inherited immunity.

Susceptible individuals who survive a naturally acquired or artificially induced infection—as smallpox, measles, typhoid fever or vaccinia—are frequently resistant or refractory to subsequent infection with the same virus. They have developed a resistance to specific infection, they have acquired immunity, in other words. This type of immunity, which results from actual infection, is termed active, acquired immunity. It is the outcome of a successful struggle between the host and the invading microbe during which the former, through cellular activity, produces or increases antibodies specifically inimical to the latter. The immunity which is thus laboriously produced is frequently fairly persistent. It is more commonly observed following invasion by exogenous, progressively pathogenic bacteria than infection with endogenous microörganisms of the "opportunist" type. Indeed, infection with the latter not infrequently results in increased susceptibility to subsequent infection with the same species of microbe. Thus, recovery from one attack of typhoid fever usually confers lasting immunity upon the individual; one attack of lobar pneumonia, on the other hand, appears to predispose the individual to subsequent infection with the pneumococcus.

The injection of specific immune substances or antibodies into susceptible individuals may confer upon them transient or temporary immunity to the specific infection; the host is a passive recipient of antibodies in such instances. These alien antibodies, however, soon diminish in potency or disappear, leaving the susceptibility of the individual to infection at its original level. Immunity induced by the injection of specific antibodies is termed passive acquired immunity. The transitory immunity to diphtheria or tetanus following the injection of diphtheria or tetanus antitoxin is an example of passive acquired immunity.

Immunity may be localized or general in the same individual, and different individuals frequently exhibit varying degrees of resistance or susceptibility to the same virus.

I. CLASSIFICATION OF IMMUNITY.

Both immunity and susceptibility are relative; there is probably neither absolute immunity nor complete susceptibility to any infection. There is furthermore, no hard and sharp line of demarcation between the various types of immunity; nevertheless, it is convenient to assemble the prominent manifestations of immunity into several types or classes.

A. **Natural or Inherited Immunity.**—The inherited power of resisting specific infection manifested by a large proportion of the individuals comprising a family, genus or species is termed inherited or natural immunity. It may be:

1. **Racial.**—Observed in specific families, genera or species of the animal kingdom, or

2. **Individual.**—Observed in individuals of the same species. Individual natural immunity may also be sexual—observed in males or females of the same species.

B. **Acquired or Induced Immunity.**—The resistance or non-susceptibility to infection following naturally acquired or artificially induced specific diseases, or resistance passively brought about by the introduction of specific protective substances is termed acquired or induced immunity.

1. **Active Immunity.**—(*a*) *Natural.*—Following naturally acquired disease, as for example, immunity following recovery from smallpox or typhoid fever.

(*b*) *Artificial.*—Brought about by the introduction of attenuated or killed viruses, vaccines or toxic products of bacteria into a susceptible host. The toxic products of bacteria may be either those

8

excreted during life, or products arising from their disintegration. Immunity to smallpox following vaccination and immunity to typhoid fever following the injection of killed cultures of typhoid bacilli are familiar examples of this type of immunity.

There is usually a period of increased susceptibility to infection immediately after the introduction of the virus or its products, in artificially acquired immunity. This period of susceptibility is followed by an increase in resistance to the virus. If the process of immunization is repeated several times, the initial level of resistance to infection may be raised very materially. Thus, prophylactic vaccination with killed typhoid bacilli (anti-typhoid vaccination) increases the resistance of the recipient of the vaccine to typhoid infection to such a degree that his chances of acquiring the disease are greatly lessened. It is also probable that in the event of infection of the protected individual with the typhoid bacillus, both the duration and severity of the attack will be diminished.

2. **Passive Immunity.**—(*a*) *Antibody Immunity.*—Introduction into the host of specific products of immunity (antibodies) as diphtheria antitoxin.

(*b*) *Chemotherapy.*—The use of chemicals for preventing or modifying infection.

Passive immunity is induced by the injection of antibodies into the host, which have been developed in another animal. The recipient of these antibodies is protected only so long as they remain in the body. The immunity, however, is effective almost immediately after injection; there is no latent period.

3. **Mixed, Active and Passive Immunity.**—Mixed artificially acquired immunity is induced by the simultaneous injection of specific antibodies and the weakened or attenuated virus; resistance to infection is usually increased at once (passive immunity), while at the same time the host begins to react to the virus and to produce antibodies thereto (artificially acquired immunity).

The factors which predispose the host to or protect him from invasion by microörganisms are usually varied and complex. Relatively simple explanations of the mechanism involved suffice to account for the phenomenon in specific instances, however. For example, frogs and hens are not naturally susceptible to infection with the anthrax bacillus, whose optimum temperature of growth is 37° C., yet infection could take place if the body temperature of either animal were brought to this level, as Pasteur showed nearly two decades ago. A change in environment may predispose to infection; the carnivora in their native state are quite resistant to infection with the tubercle bacillus, whereas in captivity they may succumb readily. Similarly, man placed in bad hygienic surroundings appears to be distinctly more vulnerable to many infectious diseases than he is when his environment is more sanitary. Unhygienic conditions, however, are relatively complex in their reactions on man, for the attendant evils of overcrowding, underfeeding and increased exposure to infection undoubtedly play a part.

Heredity also appears to be an important factor in determining the average severity of infection in certain types of endemic disease. Measles is a common and usually fairly mild disease of childhood among civilized people. Among aboriginal populations, as those of the South Sea Islands where the inhabitants had not been exposed to measles previous to the advent of Europeans, the introduction of the virus has resulted in a veritable plague during which large numbers of the people died. This phenomenon of hereditary acquired tolerance for specific endemic disease may conceivably be even more specific; for example, strains of a given organism might produce mild disease in areas where it has been endemic for generations and yet be rapidly fatal for alien populations who may have in turn become partly tolerant for other strains of the same organism. If such prove to be the case, unrestricted emigration may lead to temporary disturbances in the balance between specific microörganisms on the one hand and hosts on the other—a feature which Theobald Smith called attention to many years ago.[1]

Racial differences in susceptibility are occasionally met with even in the same species. Negroes and Indians are more susceptible to infection with tubercle bacillus than the Caucasian race. The Jews appear to be somewhat more resistant to infection with the tubercle bacillus than the other branches of the Caucasian race.

II. INFECTION.

Pathogenic bacteria which reach the host do not necessarily incite disease; they may be, and undoubtedly are, frequently overcome by the body without inducing symptoms. This initial resistance to infection involves an initial struggle between host and microörganism which brings into play non-specific lines of defense of the macroörganism

[1] Theobald Smith, Tr. Assn. Am. Phys., 1893.

consisting collectively of the skin, mucous membranes of the respiratory and gastro-intestinal tracts and other intact barriers discussed in the preceding chapter. If this initial line of defense holds, the host overcomes the prospective invader and the latter frequently perishes. Repeated microbic assaults may be successful if the first fails. On the other hand, if the microbe prevails and penetrates the initial line of defense, invasion of the tissues of the host occurs and the microorganism encounters a second line of defense which is made up of two rather distinct factors—cellular and humoral. The cellular defense of the host resides in the leukocytes which circulate in the the body fluids and in certain fixed tissue cells in the lungs, lymph-spaces and glands, the Kupfer cells of the liver, as well as large cells which appear in serous cavities. These cells engulf and destroy certain types of invading microörganisms. The humoral defense resides in the natural, non-specific power of the blood and lymph to destroy limited numbers of microörganisms or to so interfere with their nutrition or other functions as to prevent their development within the body. The humoral defense is frequently effective against bacteria which do not succumb to the cellular defense of the body, and *vice versa.*

It is recognized that certain environmental factors predispose to infection. Thus, extreme climates, excessive humidity, or exposure to unhygienic conditions, bad air, poor or insufficient food, lack of exercise or fatigue may react upon the individual in ways not definitely understood and reduce his resistance to microbic invasion on the one hand, and his ability to rally his specific, anti-microbic mechanism on the other hand. Intracurrent disease frequently weakens the initial lines of defense, permitting bacteria of the "opportunist" type to become invasive. Thus, furunculosis frequently is a complication of diabetes, pneumonia not uncommonly terminates a case of tuberculosis. Renal and cardiac disease may weaken the normal barriers of the body, permitting a variety of infections with endogenous bacteria.

It is a well-attested fact that certain occupations or professions cause or promote pathological conditions which predispose to infection. Prominent among these is participation in arts or industries which involve exposure to poisonous or irritating dust or fumes. The incidence of tuberculosis among those frequently exposed to organic or inorganic dust is a striking example of the relation of occupation to infection.

When an invading microörganism has reached a suitable atrium of the body, overcome the initial defense of the host at that point, and has successfully resisted the normal humoral or cellular opposition of the host, a new phase of the struggle becomes prominent, during which the host gradually develops a specific attack upon the invader, bringing into action latent forces which constitute the third and last defense of the body. The invader also may change its weapons to some degree to meet the antimicrobic activity of the host and the result of the struggle may be complete recovery from infection, chronic disease, the bacillus carrier state, or death of the host.

The initial and secondary defensive powers of the host, therefore, are both cellular and humoral in character. The intact skin and mucous membranes of the gastro-intestinal, respiratory and genito-urinary tracts are important initial non-specific lines of defense. The phagocytic activity of leukocytes and certain fixed tissue cells, and the natural, normal bactericidal substances of the blood and lymph, which bathe the initial line of defense, are important adjuvants in maintaining the integrity of these initial barriers to infection. In certain infections the humoral factors are the more important, while in others the cellular mechanism is conspicuous.

The defensive mechanism against the same bacterium may be different in one or another animal. For example, dogs and rats are relatively immune to infection with the anthrax bacillus. The immunity observed in the dog appears to be due to phagocytic activity of leukocytes which engulf and destroy anthrax bacilli which may have gained entrance to the body.[1] The rat, on the contrary, enjoys immunity not because its leukocytes engulf and destroy anthrax bacilli; the blood of the rat possesses soluble, non-specific bactericidal substances which destroy anthrax bacilli. Frequently both the cellular and humoral elements are engaged either simultaneously or successively as the struggle between host and invading organism proceeds.

III. THEORIES OF IMMUNITY.

Two distinct explanations have been advanced to account for the mechanism of immunity as it is observed during the course of disease: the cellular or phagocytic theory championed by Metchnikoff and his followers, and the humoral theory developed by Ehrlich.

Both of these theories, the cellular and the humoral, have in com-

[1] Hektoen, Jour. Am. Med. Assn., 1906, xlvi, 1407.

mon, tacitly at least, two important features: the specificity of the protective substances (antibodies) formed as the result of infection, and the principle that no new mechanism is evolved *de novo* to meet the conditions existing during an infection; rather, there is an increase in activity along definite lines in the preëxistent, latent or reserve mechanism of defense.

Neither theory affords a satisfactory explanation of all the features of immunity following infection and it is very probable that cellular

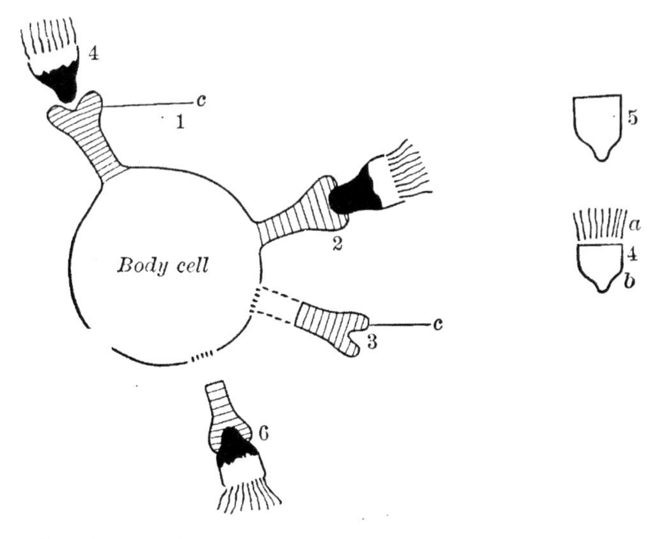

Fig. 5.—Side-chains, first order (antitoxins and antiferments). 1, side-chain attached to cell; *c*, haptophore group; 2, side-chain to which is attached a toxin molecule; 3, a cast off side-chain of the first order: antitoxin or antiferment; 4, a toxin or enzyme molecule; *a*, toxophore group; *b*, haptophore group; 5, a toxoid: the toxophore group is destroyed, leaving the haptophore group (*b*) intact; 6, a toxin molecule attached to a cast-off side-chain (antitoxin), illustrating the neutralization of toxin by antitoxin in the blood stream.

activity and the production of specific antibodies is more important in certain types of infection, while phagocytic activities are more intimately concerned in other types.

A. **The Humoral, Side-chain or Ehrlich Theory of Immunity.**—According to Ehrlich's conception, every cell of the body has two functions: a physiological function, which constitutes a special type of activity of the cell—secretory for a glandular cell, contractile for a muscle cell, or conductive for a nerve cell—and a nutritional function, which is concerned with the removal of the necessary food substances from the general supply circulating in the blood or lymph channels, and the appropriation and eventual utilization of these specific food materials by the cell. These nutritional substances undoubtedly serve two purposes: Structural, to replace cellular waste, and Fuel, to supply cellular energy.

The nutritional requirements of the individual cell are varied as their activities are varied, and Ehrlich conceives that each cell possesses a number of chemical affinities or receptors, for convenience of discussion designated as "side-chains" or "haptines," which are

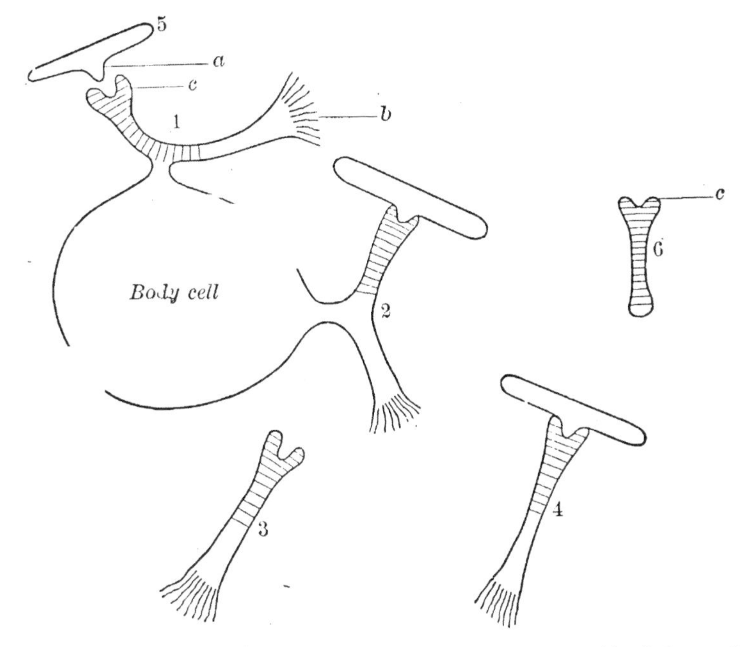

FIG. 6.—Side-chains, second order (agglutinins and precipitins). 1, side-chain attached to cell; *c*, haptophore group; *b*, zymophore group (agglutinophore or precipitinophore group); 2, side-chain to which is attached a bacterial cell; *a*, haptophore group of bacterial cell; 3, a cast-off side-chain of the second order, agglutinin or precipitin; 4, a side-chain attached to a bacterial cell (agglutination); 5, a bacterial cell; *a*, haptophore group; 6, an agglutinoid; the zymophore group is destroyed, leaving the haptophore group intact.

the means of attaching to the cell by chemical union, the essential nutritive substances preparatory to their assimilation. When the particular food attached to the cell by chemical affinity—anchored by the side-chain, to use Ehrlich's terminology—has been assimilated, more of the same kind of food is removed from the blood stream and attached to the cell, in accordance with its normal physiological requirement. The cell, acting through its side-chain, does not exhibit discrimination between nutritive substances and irritating or harmful substances which may accidentally possess the same combining affinity for the cell. Consequently, when poisonous substances possessing chemical affinities similar to those of the normal food substances circulate in the blood stream, they may become attached to the cell in place of the normal physiological nutrients. The anchoring of these poisonous substances, unlike the attachment of normal nutrient substances, is followed by damage to the cell, or, in extreme cases, by the death of the cell.[1] If the cell is not actually killed by the presence of the toxic substance acting upon it through the side-chains, it is irritated, as it were, and the toxic substance imposes a twofold burden upon the cell—loss of the side-chains to which it is attached and which are essential to maintain the nutrition of the cell, and greater or lesser damage to the function of the cell, due to toxic inhibition of its normal activities. A cell cannot disembarrass itself of the poison, nor can it assimilate it. It can, however, throw off the side-chain with the poison still firmly united to it chemically; the extruded poison cannot enter into chemical combination with other cells possessing the same chemical affinity, for it is already attached to a side-chain. Its combining power is saturated.

Side-chains are a necessity to the cell, however; without them the cell would starve. Consequently the cell regenerates new side-chains of precisely the same kind to replace those thrown off after being bound to non-assimilable substances. If enough of the soluble poison or toxin circulates in the blood stream, this process of union of toxin to the cell by its side-chains and its expulsion from the cell with the side-chains attached to it is so frequently repeated that the cell regenerates side-chains in excess of the normal requirements, in accordance with the Weigert theory of overproduction. This casting off of supernumerary side-chains is important. Were they not cast off the cell would be vulnerable to toxin in direct proportion to the extra number of side-chains, which would furnish extra bonds for its attachment. As the cast-off side-chains circulate in the blood stream, however, they are an element of protection to the cell, for they retain their original combining power for the toxin and unite with it and neutralize it as it circulates in the blood stream; that is, before it can reach the cell itself. It will be seen, therefore, that the same

[1] If the toxic material circulates in the blood stream but does not become attached to the body cells, it is harmless to the host, according to this theory, and the host is naturally immune.

mechanism of the living body which is susceptible of being poisoned becomes the protective agent if it circulates in the blood stream. It is obvious that the cast-off side-chains constitute antitoxin. The body as a whole is qualitatively the same after as before these side-chains are formed in excess of the normal cellular needs; the difference is a quantitative one. An animal is naturally immune, according to

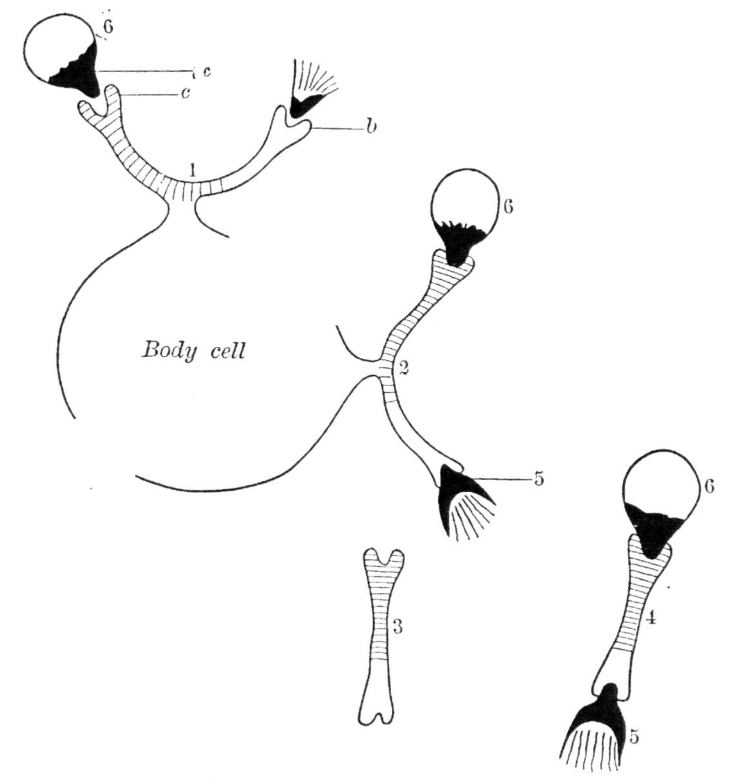

FIG. 7.—Side-chains, third order (bacteriolysins, hemolysins and cytolysins). 1, side-chain attached to cell; *c*, haptophore group; *b*, complementophile group; 2, side-chain to which is attached a bacterial-cell (6) and complement (5); 3, a cast-off side-chain of the third order; amboceptor; 4, a cast-off side-chain to which are attached a bacterial cell (6) and complement (5) illustrating lysis; 5, complement.

this theory, if the cells of the body do not unite with toxin, that is, if they do not contain side-chains which fit the toxin "as a key fits a lock," to use Emil Fischer's analogy. Toxin may circulate in the blood stream of such animals, but it does not unite with the cells.

Side-chains of the First Order.—From the standpoint of the side-chain theory, the toxin molecules consist of two groups—a combining or haptophore group, and a poisoning or toxophore group. The former is relatively thermostabile, the latter thermolabile. If toxin is heated to 70° C. for a few minutes, or allowed to stand for several weeks, it will be found that the poisonous property of the toxin has disappeared, or has been materially reduced. It still retains its original power of uniting with and neutralizing antitoxin, however. The thermolabile toxophore group has been destroyed or weakened by the heating process, or on standing. The thermostabile group—the haptophore group—has not been impaired. Toxin which has lost part or all of its original poisoning properties, but which still unites with antitoxin is called *toxoid.*

The soluble toxins of the diphtheria and tetanus bacilli are not simple substances; they contain at least two physiologically separate poisons. Thus, the toxin of the diphtheria bacillus contains in addition to the poison which produces acute symptoms, a second poison which acts slowly and appears to be responsible for postdiphtheritic paralyses and emaciation. This second poison has less affinity for antitoxin than the acute poison, and it is called a *toxone.* Similarly, the tetanus toxin appears to consist of at least two distinct poisons —tetanospasmin, which has an especial affinity for nerve cells and which elicits the acute symptoms of tetanus, and tetanolysin, which causes hemolysis of erythrocytes. The injection of soluble or exotoxins produced by bacteria leads to the formation of soluble specific antibodies which are called antitoxins. Antitoxins are supernumerary side-chains which have been produced in excess of the physiological needs of the cell, in response to the stimulus of a specific toxin, and cast off into the blood stream.

It has been shown that repeated injections of solutions containing active enzymes—as, for example, rennin—into animals, is followed by the appearance in the blood stream of specific antibodies which will prevent the activity of the homologous enzyme. These antibodies, or anti-enzymes, as they are called, exhibit the specificity and other characteristics which distinguish antitoxins.

Antitoxins and anti-enzymes are called side-chains of the first order by Ehrlich. They possess the property of combining with and neutralizing their respective toxins or enzymes.

Side-chains of the Second Order.—If substances of greater complexity than those just described are needed for the nutrition of the cell, some preliminary treatment, probably in the nature of digestion, may be required to prepare these substances for assimilation after they are bound to the cell. A side-chain of the first order, which possesses simply a combining group, does not provide the requisite power of digestion, according to Ehrlich, and to effect this digestion side-chains of somewhat more complex structure are required. Side-chains of this more complex type, side-chains of the second order, possess not only a combining group for the foodstuff, but a digestive group as well. This digestive or zymophore group, as it is called, acts upon foodstuffs after they are anchored to the cell by the combining or haptophore group. The complete side-chain of the second order, therefore, is composed essentially of a combining or haptophore group, and a zymophore group as well. The haptophore group of the second order side-chain is relatively stabile, but the zymophore group is labile and readily becomes inactive without, however, impairing the original combining ability of the side-chain. Side-chains of the second order are as vulnerable to pathological substances possessing the requisite chemical affinity as side-chains of the first order, and repeated irritation of a cell by such pathological substances leads eventually to an overproduction of side-chains of the second order and an elimination of the supernumerary side-chains in excess of the physiological need of the cell into the blood stream. Side-chains of the second order which are thus cast off from the cell in response to the stimulation of bacterial or other alien protein are of importance immunologically. If the serum of an animal containing such side-chains is brought into contact with a suspension of the homologous bacterium, the organisms are sooner or later clumped together or agglutinated. If, on the contrary, the serum is brought into contact with a clear solution of the homologous protein, a precipitate forms. These reactions are highly specific and those side-chains which cause agglutination of the specific bacterium or precipitation with the homologous protein solution are called respectively, agglutinins and precipitins.

The relative instability of a zymophore group of a side-chain of the second order may be inferred from the following experiment:

A serum obtained by injecting a horse with repeated graduated doses of typhoid bacilli will clump or agglutinate the specific organism in high dilution. If the serum is heated to 60° or 70° C. for a few minutes, or if it has been kept for a long time, it will no longer clump the bacilli, or, at least, it will clump them imperfectly. If such a serum is allowed to stand in contact with typhoid bacilli for an hour or two then removed by centrifugalization, it will be found that the bacilli will no longer agglutinate with a fresh, highly potent agglutinating serum. The bacteria are saturated with the combining group of the serum whose agglutinophore group had been inactivated by heating. This experiment shows that the combining group is relatively stabile, and that it is active even though the zymophore group is inactive. A side-chain of the second order which has lost its ability to cause agglutination with a specific organism, but which still retains its combining power, is called an *agglutinoid.* It bears a striking resemblance to a toxoid in that the active or ergophore group is destroyed, but the combining group remains intact.

Sera containing specific precipitins readily lose their ability to form precipitates with the homologous protein. The precipitins have changed to *precipitinoids,* due to a functional loss of their precipitinophore group.

The part played by side chains of the second order, agglutinins and precipitins, in immunity is not well understood. Their relation to immunity is less clear than the relation of antitoxin to immunity.

Side-chains of the Third Order.—Nutritive substances of large molecular aggregation may require considerable modification to fit them for cellular assimilation. Such substances are removed from the blood stream and bound to the cell by side-chains of the third order. They are then acted upon by an enzyme (complement) which is also present in the blood stream. It will be seen that both the nutritive element and a digestive enzyme circulate in the blood, but that no reaction occurs between them until they are both united by a side-chain of the third order, which must therefore consist essentially of two combining groups. One of these, the cytophilic group or haptophore, unites specifically with the nutritive element. The other combining or haptophore group, the complementophilic group, unites with the enzyme or complement which is present in the blood stream.

Side-chains of the third order are called amboceptors because they possess two combining groups. An excessive irritation of a cell by a substance capable of uniting with the cytophilic group of a side-chain of the third order will lead to overproduction and elimination of these side-chains precisely as toxins lead to an overproduction of side-chains of the first order (antitoxin formation). The side-chains of the third order, furthermore, exhibit specificity for the substance which led to their overproduction, just as antitoxins exhibit specificity for their homologous toxins.

It has been shown that the zymophoric group of a side-chain of the second order is permanently a part of the structure. The complement, which is analogous to the zymophore group of the second order, is not attached to a side-chain of the third order until the cytophilic group of the latter has combined with its antigen. The zymophore group of the second order side-chain is readily destroyed and it cannot be replaced. The zymophoric group of the third order side-chain is not an integral part of the structure, and it can be introduced under appropriate conditions.

Third order side-chains or amboceptors are cytolysins. Those specific for bacteria are called bacteriolysins; those specific for blood are called hemolysins; and those specific for the cells of various tissues or organs are called cytolysins.

The activity of the lysins, according to the Ehrlich theory, depends on the union of non-specific complement and a specific antigen by the specific amboceptor. A union of antigen and amboceptor may take place in the absence of complement, but a union of antigen and complement cannot take place in the absence of amboceptor. The amboceptor, like other haptophore groups, is relatively thermostabile. The non-specific complement (found in fresh blood serum from any animal) is thermolabile and readily destroyed.

Thus far it has been assumed that the cells of the body defend themselves against toxins, alien protein or alien cells by the formation of specific antibodies or side-chains. Welch[1] has made the important suggestion, which has experimental evidence in its favor, that bacteria may also produce side-chains which are specific for certain cells of the host. A struggle between host and microbe, therefore, would not be one-sided; a dual attempt at immunization is going on during a bacterial invasion, in which the microbe attempts to protect itself against the specific weapons of the host as the host attempts to protect itself against the weapons of the invading microorganism. Thus, bacteria grown in media containing agglutinating sera gradually lose their agglutinability, but this acquired loss of agglutinating power is not exhibited by descendants of the inagglutinable strain grown for some time in media not containing agglutinins.

The side-chain theory, originally formulated to explain antitoxin immunity, but enlarged in its scope to include the phenomena of agglutination, precipitation and cytolysis, has been subjected to much adverse criticism. It was assumed that toxin and antitoxin, for example, united in simple proportions as a strong acid and a strong

[1] Huxley Lecture, 1902.

base unite; the chemical analogy of toxin-antitoxin union to form an inert mixture comparable to a salt was further accentuated by the effect of moderate degrees of heat in hastening the reaction between the two. A very thorough investigation of the quantitative neutralization of toxin by antitoxin revealed the error of this supposition and Ehrlich was led to assume a very complex structure for the toxin molecule, in which there existed several fractions possessing individually, different affinity for antitoxin.

Madsen and Arrhenius[1] studied the toxin-antitoxin union from the standpoint of physical chemistry and found that the slightly dissociated reactive substances united in conformity with the law of mass action of Guldberg and Waage. Their conclusion was that toxin and antitoxin react like a weak acid and weak base, and that it is a reversible reaction, so that a mixture of toxin and antitoxin always contains free toxin, free antitoxin and toxin-antitoxin, the relative amounts being calculable according to the law of mass action. The observations of Theobald Smith[2] and of many other observers that neutral mixtures of toxin and antitoxin would induce active immunity in experimental animals are in harmony with this view. Biltz[3] has advanced an hypothesis, based upon the assumption that toxin and antitoxin are colloids, which in essence assumes that the toxin-antitoxin reaction is a phenomenon of adsorption, quite unlike the reaction of a weak acid and a weak base.

The humoral theory of immunity fails to attribute to phagocytic cells any prominent part in immunity. No theory has been advanced, up to the present time, which explains all the phenomena of humoral immunity; whatever the final solution may be, the side-chain theory as developed and defended by Ehrlich must, and always will be, a worthy monument to a great man.

B. **The Cellular or Phagocytic Theory of Immunity.**—The cellular theory of immunity, formulated and championed by Metchnikoff, had its inception in observations of the nutritive activities of amebæ, which could be watched under the microscope. It was observed that these simple, transparent protozoa, when about to feed, approached and flowed around a minute organism, as a bacterial cell. Shortly after engulfment the contour of the ingested bacterium lying within the substance of the ameba became less and less distinct and

[1] See Arrhenius, Immunochemie, Leipzig, 1907, for full details.

[2] Jour. Exp. Med., 1909, xl, 241, Active Immunity Produced by So-called Balanced or Neutral Mixture of Diphtheria Toxin and Antitoxin.

[3] Ztschr. f. physiol. Chem., 1904, 615.

finally disappeared entirely. His attention was soon directed to a small, transparent crustacean, *daphnia,* within whose body cavity could be distinguished minute wandering cells which exhibited ameboid movements. The physiological significance of these ameboid cells—which are potentially leukocytes—was not clear until it was found that they engulfed and digested certain yeast spores that occasionally gained entrance to the body cavity of the crustacean. If the yeast spores were not too numerous the wandering cells flowed around and eventually destroyed them; if, on the contrary, the number of yeast spores was too great, the wandering cells could not remove the entire

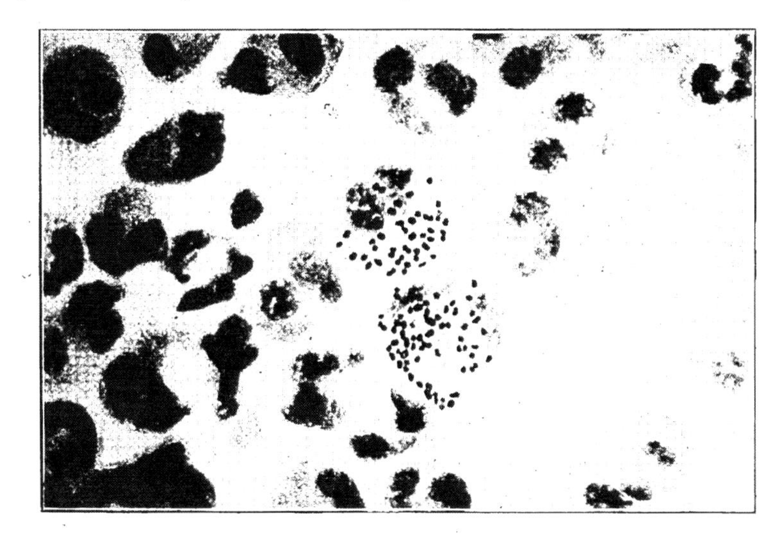

Fig. 8.—Phagocytosis of gonococcus.

number and the residual spores germinated and killed the host. It was evident that the phagocytic activity of the ameboid cells played a prominent part in protecting daphnia from an infection with the yeast.

Next Metchnikoff injected anthrax bacilli into the lymphatic sac of frogs and found again that wandering cells—leukocytes—engulfed and destroyed the bacteria, thus preventing infection and death of the frog. This line of observation was followed through an extensive series of lower animals, mammals, and finally in man, where the engulfment of the meningococci, gonococci, pneumococci, and staphylococci by polymorphonuclear leukocytes during the course of acute infections with these organisms afforded a striking demonstration of the phagocytic activity of leukocytes which circulate normally in the blood and lymph streams. These and many other observations and experiments led to the formulation of the phagocytic theory of immunity. Natural immunity, according to this theory, is leukocytic immunity—that is, the natural barriers of the body, reënforced by the activity of leukocytes in the blood and lymph streams which bathe the intact skin, mucous membranes, etc., suffice to protect the body against invasion by moderate numbers of bacteria or other microörganisms. Infection of the body, according to this view, is attributable to a failure of the leukocytic defense, or to too large numbers of invading organisms, or both factors combined.

Metchnikoff classified phagocytic cells of the body into two groups:

1. *Macrocytes or Macrophages.*—Large mononuclear cells and certain fixed tissue cells, particularly of the spleen, liver, lungs, and lymph nodes. Macrophages are active in the removal of necrotic tissue, injured blood cells, and similar abnormal cellular elements of the body, and in chronic bacterial infections, notably in tuberculosis, leprosy, and actinomycosis. They contain a digestive enzyme—macrocytase—which dissolves or digests these abnormal cells.

2. *Microcytes or Microphages.*—Chiefly polymorphonuclear leukocytes which occur in the blood stream. They engulf bacteria and similar cells. Microcytes contain a digestive enzyme—microcytase—which dissolves or digests bacteria.

The substance which Ehrlich regards as complement is normally present in the leukocytes as macro- and microcytase, according to Metchnikoff. These cytases are liberated into the blood stream when the leukocytes are destroyed (phagolysis).

The phenomenon of phagocytosis may be divided into three separate and distinct phases: the method of approach of the phagocytic cell to its prey (chemotaxis), the engulfment, and finally the digestion or destruction of the latter.

The Method of Approach.—It was a matter of observation by Metchnikoff and his followers that phagocytosis was more marked in mild bacterial infections and during recovery than in severe infections and the early acute stages of the disease. The importance of chemotaxis as the attractive force of leukocytes to bacteria, however, was not clearly realized until Massart and Bordet[1] showed by ingenious experiments that non-virulent bacteria apparently secrete substances

[1] Ann. Inst. Past., 1891, v, 417.

which draw phagocytic cells to them.[1] Virulent organisms of the same strain not only do not appear to attract leukocytes, but they appear to repel them. Bordet explained the increase of virulence of bacteria through passage in experimental animals on the ground that the less virulent individuals were engulfed and killed; the more virulent members survived and produced a thoroughly virulent strain. Vaillard and Vincent[2] and Vaillard and Rouget[3] showed that bacterial toxins may repel or paralyze leukocytic activity; if tetanus spores are bathed with tetanus toxin before injection into the animal body, the leukocytes do not collect at the point of injection, the spores germinate and the animal dies of tetanus. If, however, the spores are washed free from tetanus toxin and then injected, leukocytes appear at the site of inoculation, engulf the spores, and either destroy them or prevent their germination.

The mechanism of chemotaxis has been a subject of much discussion. Evidence is accumulating which would suggest that chemotactic stimuli of bacterial origin which reach leukocytes enter the phagocytic cell in greater concentration on that side which is nearer the source of the chemotactic substance, lowering the surface tension at that point. A flow of protoplasm in this direction, in obedience to the lowered resistance, will result in the protrusion of a pseudopodium, which will continue to advance until the surface tension is equalized.[4] This generally occurs when the leukocyte has flowed around or engulfed the organism.

Engulfment.—The earlier view associated the protrusion of pseudopodia and the subsequent engulfment of bacteria or other cell as an autovoluntary act of the leukocyte. The inclusion of inert particles, as dust or other minutely comminuted granules, would appear to discredit this hypothesis. The engulfment of living or inert bacteria or other minute bodies is, as Wells aptly expresses it,[5] "but an extension of the phenomena of chemotaxis. When the substance toward which the leukocyte is drawn is small enough, the leukocyte simply continues its motion until it has flowed entirely about the particle."

Digestion.—The ultimate solution of engulfed substances other than purely inert particles is by intracellular enzymes contained within

[1] Inert particles, as coal dust, are engulfed by phagocytic cells; it is difficult to explain this phenomenon on the basis of chemotaxis.

[2] La sémaine médicale, 1891, xi, No. 5.

[3] Ann. Inst. Past., 1892, vi, No. 6.

[4] See Well's Chemical Pathology, 1914, 2d ed., pp. 230-251 (Saunders & Co.), for an excellent résumé of the literature.

[5] Well's Chemical Pathology, 1914, 2d ed., p. 238 (Saunders & Co.).

the phagocytic cells. These enzymes are of two kinds: macrocytase, present in the macrophages, and microcytase, found in the microphages.[1] Van de Velde,[2] Buchner,[3] Hahn,[4] and Bordet[5] have demonstrated such endo-enzymes. The solution of bacteria engulfed in leukocytes can be shown by appropriate staining methods; the organisms gradually lose their ability to take up stain and eventually disappear.

At this stage of the development of the phagocytic theory of immunity, the important part played by the blood serum in preparing bacteria for phagocytosis was prominently set forth in the investigations of Wright and Douglas,[6] although foreshadowed by the excellent and comprehensive observations of Denys and LeClef[7] and Neufeld and Rimpau.[8] Wright and Douglas showed that leukocytes, freed carefully from adherent serum by washing with salt solution, would not engulf bacteria, or, at least, but slowly. The addition of serum from a normal or immunized animal caused active phagocytosis to take place. The substances in the blood serum which prepare bacteria for engulfment by leukocytes were called "opsonins" by Wright and Douglas: the immune opsonins—which are specifically increased in immunized animals—are almost certainly identical with the substances called bacterial tropins by Neufeld and Rimpau. That the opsonic substances of the serum act primarily upon the bacteria rather than upon the leukocytes was clearly shown by the observations of Hektoen and Reudiger.[9] Streptococci suspended in plasma, blood serum or defibrinated blood were engulfed by leukocytes. Leukocytes, washed free from serum or plasma, were without phagocytic action upon the same bacteria. If the streptococci, however, were allowed to stand in contact with serum, plasma, or defibrinated blood for a short time at 37° (a much longer exposure at 0° to 4° C. was necessary), then washed free from adherent serum or plasma, and exposed to washed leukocytes, active phagocytosis took place.

The present tendency is to ascribe to phagocytosis an important part both in the destruction of many kinds of invading bacteria and in the removal of alien or abnormal cells as well. The importance

[1] For a detailed discussion of leukocytic enzymes, see Opie, Jour. Exp. Med., 1905, viii, 410.
[2] La Cellule, x, 2; Cent. f. Bakt., 1898, xxxiii, 692.
[3] München. med. Wchnschr., 1894, 718.
[4] Arch. f. Hyg., 1895, xxviii, 312.
[5] Ann. Inst. Past., 1895, ix, 398.
[6] See Wright, Studies in Immunization, 1909, Constable.
[7] La Cellule, 1895, xi.
[8] Deutsch. med. Wchnschr., 1904, 1458.
[9] Jour. Inf. Dis., January, 1905, ii, No. 1.

of substances contained within the plasma or blood serum, which prepare bacteria for phagocytosis—to use Wright's terminology—has modified somewhat the original conception of phagocytosis as proposed by Metchnikoff.

The phagocytic theory and the humoral theory of immunity would appear to be in direct opposition. Metchnikoff maintained that the fundamental basis of immunity resides in the phagocytic activity of macro- and microphages. He believed that the humoral immune bodies are derived either from leukocytes or the organs in which they are formed—the bone marrow and lymphatic system. The champions of the humoral theory, on the other hand, would attribute the healing principle to soluble substances contained in the body fluids. The leukocytes and other phagocytic cells, according to the extremists who advocate this theory, would be regarded as scavengers merely, whose function it is to remove the débris—dead bacteria or disabled bacteria—after they are overwhelmed by the activity of the soluble natural and immune antibodies.

A final decision of the importance of cellular and humoral factors in immunity cannot be made at the present time. It is not unlikely that both theories will be modified somewhat as additional evidence accumulates.

CHAPTER VII.

ANAPHYLAXIS, ALLERGY OR HYPERSENSITIVENESS.[1]

PROTEIN fed to man or animals is reduced to simple compounds, chiefly amino-acids, by the action of gastro-intestinal enzymes before it is absorbed from the alimentary canal. These gastro-intestinal enzymes act rapidly under normal conditions, and without an appreciable latent period. One noteworthy result of digestion is a complete denaturization of all ingested protein before it enters the tissues of the host; absorption of unaltered or partially-digested protein is prevented or reduced to a minimum.

The importance of a denaturization of protein before it enters the tissues becomes apparent when a comparison is made between the effects of parenteral injections of the end-products of protein digestion on the one hand, and of the unaltered protein itself on the other hand. Repeated parenteral injections of amino-acids in moderate amounts appear to be without serious or noteworthy effects upon experimental animals. A single parenteral injection of an unaltered protein is also without visible effect, as a rule. A second parenteral injection of the same protein, after an interval of ten to fourteen days, frequently is followed by a rather definite train of symptoms, severe in character and wholly unlike the negative response to a corresponding treatment with amino-acids or normal end-products of gastro-intestinal digestion.

Sensitization.—The first parenteral injection of a protein[2] which is foreign to the body, or in some instances, natural for the body but alien for the blood, is without visible effect upon the animal, but leads to its sensitization to the specific protein. The sensitizing agent is variously referred to as a sensitizer, sensibilisinogen, or anaphylactogen, and may be effective in very small doses. Rosenau and Anderson[3] were able to sensitize guinea-pigs with one-millionth of a cubic centimeter of horse serum; Wells[4] has sensitized the same animal

[1] For an excellent résumé of the literature of anaphylaxis complete to 1912, see Hektoen, Jour. Am. Med. Assn., 1912, lviii, 1081.

[2] Proteins deficient in tryptophane or tyrosin are said not to sensitize.

[3] Bull. 29 and 36, Hygienic Laboratory, Washington, D. C., 1906, 1907.

[4] Wells' *Clinical Pathology*, 1914, 2d ed., 180.

with one twenty-millionth of a gram of crystallized egg albumen. Usually 0.001 to 0.1 c.c. of serum is an effective sensitizing dose.

A latent period intervenes between the initial injection of the animal with sensitizing protein and sensitization—on the average this is about ten to fourteen days. Gay and Southard[1] showed, however, that the time necessary to effect sensitization depends somewhat upon the size of the sensitizing dose, larger amounts requiring longer periods than smaller amounts. White and Avery[2] have found that a relation exists between the minimum sensitizing and the maximum intoxicating dose, larger amounts of protein being required on reinjection to elicit a reaction when the sensitizing dose is very small, and *vice versa*.

Reinjection of the Homologous Protein.—Repeated injections of the homologous protein spaced at intervals less than ten days do not, as a rule, cause symptoms of acute anaphylaxis—after a third or a fourth injection, however, there appears at the site of the first injection a swelling, usually indurated and more or less edematous, which may lead to extensive necrosis and sloughing. These local reactions, the so-called Arthus[3] phenomenon, are closely related phylogenetically to the anaphylactic symptoms described below.

If the second parenteral injection is made after sensitization is established—usually after ten to fourteen days—symptoms follow almost immediately, which vary somewhat according to dosage and the site of inoculation. A very large dose frequently results in rapid death, the Theobald Smith phenomenon.[4] Very broadly speaking, it requires 200 to 2000 times as much protein to cause acute anaphylaxis as to effect sensitization.

Intravenous or intracerebral injections of moderate doses are followed very soon by a period of excitement (in dogs, followed by a period of depression),[5] the animal is restless and moves about in a bewildered manner and shows signs of respiratory embarrassment. It coughs (a normal guinea-pig rarely or never coughs) and scratches the corners of its mouth. This state is followed by dyspnea, with involvement of the diaphragm and bronchial musculature leading to

[1] Jour. Med. Res., 1908, xviii, 407.
[2] Jour. Inf. Dis., 1913, xiii, 103.
[3] Compt. rend. Soc. Biol., 1903, lv, 20; 1906, lx, 1143.
[4] Theobald Smith, Jour. Med. Res., 1905, xiii, 341; Otto, Leuthold-Gedenkschrift, 1096, i, 153.
[5] Guinea-pigs in general react most strikingly to anaphylactic stimuli; man is less sensitive. Rabbits, sheep, goats, horses, and birds, in the order mentioned, are less susceptible than man. Cold-blooded animals appear to be refractory.

bronchial spasm and later to paralysis of respiration,[1] lowered blood-pressure, frequently cyanosis, and death. Smaller intravenous injections are followed by the same symptoms of excitement and respiratory involvement, but to a lesser degree. Frequent micturition and fluid, often bloody stools together with great prostration and dyspnea are usually observed. The animal cannot stand and may die after several hours, or eventually recover.

Intraperitoneal injections elicit similar symptoms. Subcutaneous injections rarely cause acute death; as a rule the animal has a febrile reaction and repeated injections may be followed by the Arthus phenomenon. If the animal survives an anaphylactic reaction it is frequently observed to be more refractory or even temporarily immune to subsequent injections of the same protein. This refractory state is called anti-anaphylaxis by Besredka and Steinhardt.[2] This period of refractoriness is of variable duration.

The postmortem appearance of guinea-pigs which have died from the effects of acute anaphylaxis is usually striking and characteristic. The lungs remain fully distended when the thorax is opened, the cut surface is rather dry, and death appears to have resulted from asphyxiation due to a tonic spasm of the bronchial musculature.[3]

Severe but non-fatal anaphylactic reactions are accompanied by a lowering of the body temperature, lowered arterial pressure, leucopenia, frequently with a temporary partial or complete loss of coagulability of the blood,[4] followed by a secondary febrile rise of temperature and a leukocytosis in which polymorphonuclear leukocytes and frequently eosinophiles[5] are increased. Animals killed during the early acute symptoms show but little distention of the lungs—the lesions may resemble those of an acute toxic gastro-enteritis. Ecchymoses and ulcers may be found occasionally in the stomach and intestines, together with parenchymatous degeneration of the liver and particularly the kidneys, which may lead eventually to fatty degeneration of these organs.

The symptoms of anaphylaxis may be masked or even prevented by the administration of certain drugs immediately before the reinjection—of these atropin, chloral hydrate and similar narcotics are considered particularly efficient.

[1] Auer and Lewis, Jour. Am. Med. Assn., 1909, liii, 6; Biedl and Kraus, Wien. klin. Wchnschr., 1910, 844.

[2] Ann. Inst. Past., 1907, xxi, 117, 384.

[3] Auer and Lewis, loc. cit.

[4] Biedl and Kraus, Wien. klin. Wchnschr., 1909, 363; Friedberger and Grober, Zeit. f. Immunitätsforsch., 1911, ix, 216.

[5] Moschowitz, New York Med. Jour., 1911, lxxxxiii, 15.

THE NATURE OF THE POISON, ANAPHYLATOXIN.

The anaphylactic reaction, like other serological reactions, appears to depend upon the elaboration of a specific antibody in the sensitized animal. The specificity of the reaction is very striking in the physiological sense—the serum of one animal fails to sensitize for the serum of an unrelated animal. Egg protein of one species also fails to sensitize an animal against the egg protein of another species. Osborne and Wells,[1] using vegetable proteins which can be obtained in a state of relative purity, have shown that sensitization, in the last analysis, depends chiefly upon the chemical composition of the sensitizer. Thus, one vegetable protein fails to sensitize against a second, unlike protein, even though they be derived from the same seed.

The specificity of the reaction is striking—it takes place only in response to a second injection of the homologous protein, but the symptomatology is essentially the same, irrespective of the sensitizer. The promptness with which the reaction appears after the reinjection suggests at once that the poison is radically different from a true bacterial toxin, which invariably requires a definite latent period before symptoms can be detected. In this respect the anaphylatoxin resembles somewhat an alkaloidal poison. Up to the present time no antitoxins have been prepared. The action of the poison is peripheral rather than central, according to Auer and Lewis.[2] Schultz[3] and others have shown that it acts powerfully upon smooth muscle fibers; Biedl and Kraus[4] and others have shown that an injection of peptone into dogs elicits symptoms and pathological changes indistinguishable from those of anaphylaxis. They were inclined to regard the anaphylatoxin as similar to, or possibly identical with peptone. Animals immune to anaphylactic reactions react slightly or not at all to peptone injections.

Passive anaphylaxis may be induced in a non-sensitized animal by an injection of the serum of a sensitized animal. Usually a few hours elapse before the recipient of the specific antibody is reactive, however. The experiments of Pearce and Eisenbrey,[5] of Weil,[6] Dale,[7]

[1] Jour. Inf. Dis., 1913, xii, 341.
[2] Loc. cit.
[3] Hygienic Laboratory Bulletin, 1912, No. 80.
[4] Wien. klin. Wchnschr., 1901, No. 11.
[5] Journ. Inf. Dis., 1910, vii, 565.
[6] Jour. Med. Res., 1913, xxvii, 497; 1914, xxx, 87, 299.
[7] Jour. Pharm. and Exp. Therap., 1913, iv, 167.

Schultz[1] and others indicate that the reaction occurs within the cells of the body rather than in the blood stream. The urine of anaphylactic animals is toxic and 2 c.c. is frequently sufficient to kill guinea pigs with anaphylactic symptoms, according to Pfeiffer.[2]

Anaphylaxis may be defined as a congenital or acquired condition of hypersensitiveness of man or animals to the parenteral introduction of proteins, which is incited by one or more injections of bacterial, plant, animal or human protein. Active acquired hypersensitiveness can be transmitted to non-sensitized individuals by the injection of the serum of an anaphylacticized individual, inducing in the recipient of the serum a condition of passive anaphylaxis. Anaphylaxis, therefore, belongs to the group of immunological reactions.

Theories.—Vaughan[3] has shown that all proteins may be split into two fractions if they are heated with alcoholic potassium hydroxide; one portion, insoluble in alcohol, when injected into animals gives symptoms indistinguishable from those of anaphylaxis, irrespective of the protein. The alcohol-soluble fraction is not toxic. The alcohol-insoluble fraction obtained from various animal, vegetable, and bacterial proteins always reacts the same, not only symptomatically, but quantitatively as well. His theory is that the protein molecule consists of two parts: an archon or nucleus, which is poisonous and elicits the symptoms of anaphylaxis when it is injected parenterally into animals, and common to all proteins; and additional groups which are non-poisonous, but confer upon a protein by their number and arrangement, its specificity. When a protein is injected parenterally into an animal, the cells of the animal elaborate an enzyme which will specifically disintegrate it. Among the products of disintegration is the poisonous nucleus or archon in a more or less free state. The liberation of this substance causes acute poisoning of the host. This substance, for which no antibody or antitoxin has been prepared so far, is the "endotoxin" of bacteria.

Many of the phenomena of anaphylaxis are readily explained in the light of Vaughan's work. The latent period or pre-anaphylactic state which intervenes between the injection of a protein and the appearance of sensitization is the time required to mature the specific enzyme. The specificity of the enzyme (called forth by the stimulus of alien protein in the tissues) is determined by the arrangement and

[1] Loc. cit.
[2] Zeit. f. Immunitätsforsch., 1911, x, 550.
[3] Protein Split Products, 1913, for full discussion.

number of groups arrayed around the poison group of the protein; the similarity or identity of the symptoms of anaphylaxis irrespective of the protein depends upon the liberation of the poison nucleus (common to all proteins) in a relatively free state. The induction of passive sensitization depends upon the injection of this specific enzyme, which is present in the serum of a sensitized animal, into a non-sensitized animal.[1]

Vaughan regards the formation of a specific proteolytic enzyme in response to the injection of alien protein into the tissues as a protective mechanism to rid the body of foreign substance; the theoretical importance of this conception as a purposeful reaction is clearly shown in bacterial infections. The incubation period of many bacterial infections is about two weeks, during which clinical symptoms are not pronounced. This is interpreted as the time required by the cells of a host to mature a specific enzyme capable of disintegrating the alien protein (bacterial cells). The symptomatology of bacterial disease is caused largely by the liberation of the poisonous nucleus of the bacterial protein in special tissues or organs. Natural immunity to bacterial disease, according to this theory, is due to the inability of the organism to grow in the tissues of the host; active immunity is conferred on the host by the presence of a persistent enzyme which will disintegrate the specific organism whenever it is reintroduced into the body.

Chemically, the poison nucleus or endotoxin is stated by Vaughan to resemble beta-imidazoleethylamine, described previously.[2] The specificity of the anaphylactic reaction depends upon the cleavage of the protein molecule by a specific proteolytic enzyme with the liberation of a non-specific poisonous product of protein degradation. Abderhalden[3] and his associates have demonstrated proteolytic enzymes in the blood stream.

Friedberger[4] has shown that a poison may be obtained by incubating the inactivated serum of a sensitized animal with an excess of complement and homologous (sensitizing) protein, which, when injected into guinea-pigs, elicits the symptoms of anaphylaxis. It is not true toxin, for no antibody is produced in response to repeated, sublethal injections; it appears to differ from Vaughan's poison in that

[1] The importance of the degradation of protein in the alimentary tract can be appreciated in the light of what has been stated about anaphylaxis.

[2] See page 76.

[3] Zeit. f. physiol. Chem., 1912, lxxxii, 109; Abwehrfermente des tierischen Organismus, Berlin, 1913.

[4] Zeit. f. Immunitätsforsch., iv, 636; vii, 94; Ueber Anaphylaxie, Ibid., 1911, ix, 394 (in collaboration with Goldschmidt, Schmanowsky, Schültze, and Nathan).

it is destroyed or inactivated at a temperature above 65° C. The poison does not form if complement is not present in solution with the inactivated serum and antigen, which would suggest a resemblance to other cytolytic reactions in which the specific amboceptor is activated by complement.

The essential distinction between the theory of Vaughan and that of Friedberger would appear to rest upon the nature of the poisonous substance liberated; Vaughan would maintain the specificity of the enzyme and the identity of the poisonous substances formed from various proteins. Friedberger's theory, which was developed several years after Vaughan's first work, would emphasize the distinction between an enzyme and the specific amboceptor, which requires complement for its activation. Keysser and Wassermann[1] and more recently, Jobling and Petersen[2] have found that serum shaken with kaolin, chloroform, and other agents will absorb substances from serum, leaving the remainder toxic for guinea-pigs; the reaction induced by the injection of small amounts of altered serum resembles closely that of anaphylaxis. Jobling and Petersen believe that the toxic substance originates not from bacteria necessarily, but from serum itself. Under normal conditions, anti-enzymes prevent the normal serums from causing auto-autolysis; kaolin, bacteria, etc., added to the serum, absorb and thus remove the anti-enzymes, thus permitting the serum to digest itself. In other words, the poisonous substance may originate in the serum rather than in the bacteria or other alien protein. These facts do not necessarily detract from Vaughan's theory, but until more is known of the entire subject, a final discussion of the mechanism of anaphylaxis must be postponed.

ANAPHYLAXIS IN MAN.

Natural Hypersensitiveness.—It has long been known that the inhalation of organic substances—as the pollen of various plants, or emanations from horses or guinea-pigs, of peptone or other similar material—may excite acute coryza and that train of symptoms popularly recognized as "hay-fever" or "pollen fever" in some, but by no means all, individuals. If the specific pollen or dust is rubbed on the nasal mucosa of these sensitized individuals a violent reaction will take place. Other individuals develop a severe urticaria if they eat certain proteins: the flesh of arthropods, particularly crabs and lobsters, vegetables, eggs, milk are known to excite symptoms in indi-

[1] Ztschr. f. Hyg., 1911, lxviii, p. 535.
[2] Jour. Exp. Med., June, 1914, xix, p. 480.

viduals who exhibit an "idiosyncrasy" to one or another of these substances. This idiosyncrasy to foreign protein may be either congenital or post-natal; the protein is supposed to have passed unchanged through the intestinal tract in the latter case. The phenomena in these instances are explained on the basis of sensitization with specific protein; a mild anaphylactic reaction occurs when the specific dust reaches the nasal mucous membrane or the specific protein enters the digestive tract.

The tendency at the present time is to regard certain clinical and pathological symptoms of bacterial infections—particularly fever—and the production of specific pathological lesions as manifestations of anaphylaxis as outlined by Vaughan.[1] The body is sensitized to the alien protein, be it organic dust, protein of the food, or invasive bacteria; the anaphylactic reaction takes place when the homologous protein is brought into contact with the sensitized individuals through the proper channels. It will be remembered that the incubation period in many bacterial infections was explained as the time elapsing between the arrival of the alien protein (bacterial cells) in the tissues of the host and the maturing of a specific proteolytic enzyme that would effect their disintegration. The symptomatology of bacterial infections, according to Vaughan, is largely due to the liberation of the anaphylatoxin incidental to the lysis of the residual organisms.

Artificial or Acquired Hypersensitiveness. — The phenomena grouped for convenience as acquired hypersensitiveness are met with chiefly in connection with the administration of the sera of animals immunized for therapeutic purposes. Three types of anaphylactic reaction may be recognized:

1. **Sudden Death.**—A very few cases are on record in which the administration of antitoxin for therapeutic purposes, either for immunization or curatively, has been followed within a few minutes or hours by death. Already, in 1896, Gottstein[2] had collected 12 which followed the injection of diphtheria antitoxin, 8 of whom were diphtheritic, 4 healthy individuals. About 1 in every 50,000 appears to be the proportion of deaths due to an injection of therapeutic sera. The symptoms are essentially those observed in sensitized experimental animals which die shortly after the injection of the homologous protein. Behring, Kitasato and other observers had noticed many years ago, when antitoxin was first prepared on a large scale, that animals immunized with large amounts of tetanus or diphtheria toxin

[1] Loc. cit.
[2] Therap. Monatschr., 1896, Heft 5.

occasionally succumbed to a subsequent small dose of the homologous toxin, although the blood serum of these animals contained much specific antitoxin.

2. **Serum Sickness or Serum Disease.**—Attention was first directed to serum sickness by von Pirquet and Schick,[1] who noticed that there occasionally developed in individuals who had received an injection of antitoxic sera, usually after seven to fourteen days, fever and a rash which might be urticarial, scarlatinal, or, in the more severe cases, morbilliform; enlargement of lymph glands, particularly those near the site of inoculation; and joint pains, more frequently of the metacarpal joints. A slight edema, frequently of the angioneurotic type, was also occasionally observed. The fever is usually slight and there are signs of respiratory embarrassment, not as a rule marked, but occasionally severe. These reactions, sudden death and serum sickness, are more common in asthmatics, and in those individuals presenting the pathological syndrome known as status lymphaticus.

According to Moschowitz,[2] these individuals, particularly the asthmatics, present an eosinophilia. The exact cause of sudden death following the administration of diphtheria antitoxin is not definitely known, but it has been assumed that respiratory involvement is a potent factor. The appearance of serum disease seven to fourteen days after the administration of antitoxin is supposed to depend upon the fact that some of the alien protein (serum) remains in the body during the period of pre-anaphylaxis (period of sensitization), and that this residual protein is broken down by the mature specific enzyme or enzymes with the liberation of a poisonous substance which causes the anaphylactic shock.

3. **Arthus Phenomenon.**—During the course of immunization against rabies by the Pasteur method it is frequently noticed that after three or four injections a subsequent injection causes symptoms of inflammation at the site of the first injection, and that this phenomenon is repeated, usually, but not always, with diminishing intensity at the site of earlier injections as the treatment progresses. This inflammatory reaction at the site of injection is not due to bacterial infection ordinarily, but is rather an expression of anaphylaxis. It is comparable to the Arthus phenomenon produced in rabbits by successive injections of serum referred to above. Also in revaccination (vaccinia) a so-called accelerated reaction may occur the second time the indivi-

[1] Die Serumkrankheit, Leipzig, 1905.
[2] New York Med. Jour., 1911, lxxxxiii, 15.

dual is vaccinated. This accelerated reaction again is a mild edition of the Arthus phenomenon.

4. **Prophylaxis.**—At first sight it might appear that the administration of diphtheria and tetanus antitoxin for therapeutic purposes would be a dangerous procedure. If there is reason to suspect that the patient would react to the injection of antitoxin it is advisable to inject 0.1 or 0.2 c.c. subcutaneously and wait half an hour. If no symptoms develop, the full dose may be given without danger; it is generally believed that even if mild symptoms do follow the initial injection, the full dose may be given with safety after half an hour; the first injection appears to abort what otherwise might be a reaction dangerous to the patient

The present method of concentrating diphtheria antitoxin by fractional precipitation of the globulin[1] appears to reduce very materially the incidence of serum sickness. According to German investigators, antitoxin which has stood for one or two months has lost to a very considerable extent the substance or substances which cause the symptoms of serum sickness.

Practical and Theoretical Considerations.—A. Advantage is taken of the sensitization of individuals by bacterial protein during certain bacterial infections, particularly those with the tubercle bacillus, B. mallei, and in syphilis, for diagnostic purposes. It has been shown almost beyond doubt that individuals suffering from these diseases are sensitized to the bacterial protein, and it is possible to make a fairly definite clinical diagnosis by introducing extracts of the specific organisms into the skin and inducing there an anaphylactic reaction which, if the dose is small, is local in character, but which may be general and severe if the dose is increased in amount. The von Pirquet, Calmet, Moro, and Koch methods of utilizing tuberculin for diagnostic purposes are directly dependent upon this reaction of hypersensitiveness. The diagnostic use of mallein and luetin depend upon the same phenomenon.

B. Advantage is also taken of the specificity of the anaphylactic reaction for the recognition of proteins. Wells and Osborn[2] and many others have sensitized guinea-pigs with proteins and then injected into these sensitized animals proteins which are to be identified either specifically or phylogenetically. The nature and extent of the anaphylactic reaction in these animals furnishes the most delicate test (except possibly the precipitin test) which is available for such investigations.

[1] Banzhof, Johns Hopkins Hosp. Bull., 1911, xxii, 241.
[2] Loc. cit.

CHAPTER VIII.

ANTIGENS AND THE TECHNIC OF SERUM REACTIONS.

NATURE OF ANTIGENS AND ANTIBODIES.

THOSE substances which cause specific antibody formation when they are introduced into the tissues or the body fluids of the host are called antigens. Their chemistry is as yet unknown, but available evidence would indicate that they are protein in nature and highly organized chemically. Degradation products of proteins, as albumoses and peptones and carbohydrates and fats, are not ordinarily antigenic, that is, they do not lead to antibody formation when they are introduced into the animal body.[1] The antigenic properties of lipoids are still a subject of controversy: lipoids appear to play a prominent part in certain types of immunological reactions, but their ability to stimulate specific antibody formation cannot be regarded as proven at the present time.[2]

The function of antibodies as specific offensive weapons of the host against alien organisms or their products has long been recognized in bacteriology, and most important laboratory diagnostic methods have been elaborated through a study of the reactions between specific antigens and their respective antibodies. Antibodies are soluble and are found in various concentrations in blood serum derived from immunized animals. Many attempts have been made to determine changes in the chemical composition or physical properties of immune sera from those of normal serum. Atkinson,[3] Gibson,

[1] The injection of carbohydrates and fats may, however, lead to specific enzyme formation. See Röhmann, Antigene Wirkung der Kohlenhydrate, Deutsch. med. Wchnschr., 1914, xl, 204.

[2] See Pick, Kolle, and Wassermann, Handbuch der pathogenen Mikroörganismen, 2d ed., Bd. I, for discussion of the chemistry of antigens.

[3] Jour. Exp. Med., 1899, iii, 649.

and Banzhaf[1] and others have found that the sera of horses immunized to diphtheria toxin show a marked increase in globulin content, with a decrease in albumin content. Beljaeff[2] could find no appreciable change in the refractive index, specific gravity, freezing point or reaction of the serum of an immune animal above that of a normal animal.

The chemical nature of antibodies, aside from their apparently close relation to globulin, has not been determined. There is evidence that antitoxin molecules may be larger than toxin molecules, however. Martin and Cherry[3] found that toxins could be forced through dense porcelain filters impregnated with gelatin, which would restrain antitoxin, and Arrhenius and Madsen[4] determined that the toxin molecule diffused several times as rapidly as the antitoxin molecule, from which observation they assumed that the antitoxin molecule was larger than the toxin molecule.

AGGLUTININS. AGGLUTINOIDS AND PROAGGLUTINOIDS.

Gruber and Durham[5] appear to have been the first to clearly demonstrate specific clumping in broth cultures of typhoid and cholera organisms when their respective sera were added to them. Somewhat later Widal,[6] and independently Grünbaum,[7] utilized the principle of the specific agglomeration of bacteria by their immune sera for the diagnosis of typhoid fever. They found that relatively early in the disease, sera of typhoid patients clumped typhoid bacilli from broth cultures. Pfaundler[8] observed that typhoid bacilli grown in broth containing low concentrations of specific sera grew out into long, tangled filaments, the "thread" reaction. Originally this phenomenon was regarded as highly specific, but it has largely given way to the macroscopic or microscopic agglutination test.

Agglutination in the bacterial sense may be defined as a clumping or agglomeration of bacteria from a uniform suspension in a fluid medium, brought about by the addition of specific antibodies—agglutinins. It takes place in two stages if motile bacteria are concerned. First there is loss of motility—"immobilization"—and

[1] Jour. Exp. Med., 1910, xii, 411.
[2] *Cent.* f. Bakt., Orig., 1903, xxxiii, 293, 396.
[3] *Proc.* Royal Soc., 1898, lxiii.
[4] Festskrift Statens Serum Institute, 1902.
[5] München. med. Wchnschr., 1896, No. 13.
[6] La Sémaine Médicale, 1896, No. 13.
[7] Brit. Med. Jour., 1897, May 1, and München. med. Wchnschr., 1897, 330.
[8] *Cent.* f. Bakt., 1898, xxiii, 9, 71, 131.

eventually clumping. Smith and Reagh[1] working with a non-motile hog cholera bacillus have demonstrated both flagella and somatic agglutinins, the former paralyzing the activity of the flagella, the latter agglomerating the organisms themselves. Non-motile bacteria usually agglutinate somewhat more slowly than motile organisms. Small amounts of neutral salts are necessary for the clumping of bacteria,[2] although a union of the specific organism and its agglutinin will take place even if salts are absent. The specific substance (or substances) of the bacterial cell which reacts with the specific antibody of the serum (agglutinin) is known as agglutinogen. Closely related bacteria, as typhoid and paratyphoid bacilli, may possess a certain amount of agglutinogen in common, but, as a rule, the specific organisms are clumped in immune sera at much greater dilution than related organisms are clumped. Also, the specific organisms will remove the agglutinin completely from immune sera, while closely related bacteria only remove that portion of the agglutinating substance which is common to both organisms, leaving behind the specific agglutinin which will then agglutinate the specific organism, but not its closely related fellow; that is to say, closely related bacteria will react with the common or group agglutinin, but fail to absorb the specific agglutinin.

Experience has shown that the sera of normal adults frequently contain agglutinin which will clump various bacteria and the potency of these "normal" or natural agglutinins may even be sufficient to clump moderate numbers of typhoid bacilli in dilutions as great as 1 to 30. The sera of normal nurslings contain only minimal amounts of normal agglutinins as a rule, and the conclusion has been drawn that normal agglutinin may be either

(*a*) Group agglutinin, derived from mild infection with closely related organisms, or

(*b*) True immune agglutinins resulting from mild or unrecognized infection with the specific organism.

No definite distinction has been noted between natural and immune agglutinins; the latter are usually present in sera, however, in much greater concentration than the former.

The site of formation of agglutinins in the body is not definitely known, although lymphoid tissues appear to be intimately concerned, especially bone-marrow and the spleen. Pryzgode[3] states that

[1] Jour. Med. Res., August, 1903, x, No. 1.
[2] Bordet, Collected Studies in Immunity, 1909 (translation by Gay).
[3] Wien. klin. Wchnschr., 1913, xxvi, 84.

cultures of spleen tissue *in vitro* will form specific agglutinins for typhoid bacilli if the virus is brought into contact with the tissue cells. As a general rule the concentration of a specific agglutinin is greater in the blood stream than in the tissues of the body.

Preparation of Specific Agglutinating Sera.—Specific agglutinating sera for experimental purposes are best obtained from rabbits, whose serum normally contains no agglutinin. Several, usually three to five intravenous injections of 1, 2, 3 and 5 loopfuls respectively of killed cultures of typhoid bacilli at eight-day intervals, produce powerful agglutinating sera. The animal is bled about two weeks after the last injection. For large amounts of agglutinating sera horses or asses must be used.

Properties of Agglutinins.—Agglutinins are of unknown chemical composition, but they may be separated from solution by those precipitants which throw down globulins, and they may be removed from solution by absorption in animal charcoal. Toward heat they are moderately resistant, usually remaining active after an exposure of twenty minutes to 55° C., a degree of heat sufficient to inactivate complement. Agglutinins, therefore, appear to be quite distinct from bacteriolysins. The temperature at which agglutinins are destroyed depends upon their specificity, agglutinins for plague bacilli being more sensitive than typhoid agglutinins. The reaction of the medium also affects their stability. Alkalis, even in dilute solution, rapidly destroy agglutinins; acids are somewhat less harmful. Naturally the duration of exposure to these various agents exercises an important influence upon their resistance. Agglutinins do not appear to pass through parchment membranes, but it is stated that agglutinogen will slowly diffuse under similar conditions. This would suggest that the agglutinin molecule is larger than the agglutinogen molecule. Preserved in a dry state, in a cool place away from light, agglutinins preserve their properties unimpaired for days. In solution and upon standing agglutinins rapidly lose their property of clumping bacteria, but they still retain their original ability to unite firmly with bacteria. Ehrlich designates agglutinins which have lost their ability to cause clumping but still retain their combining power for agglutinogen, agglutinoids. In his terminology they are side-chains of the second order which have lost their agglutinophore (ergophore) group. Agglutinins acting in neutral salt-free media also fail to cause clumping of bacteria, but in this case the addition of a small amount of NaCl or even some weak acid very soon brings about a typical reaction.[1] This and similar observations have attracted attention to the similarity between the precipitation of bacteria to which agglutinin is anchored by neutral salts, and the precipitation of finely suspended clay by the addition of neutral salts; the inference has been drawn that the phenomenon of agglutination is one of physico-chemistry.

Specificity of Agglutination Reactions: Group Agglutinins.—The composition of the agglutinogen—that constituent of the bacterium which stimulates agglutinin formation—is unknown, but it appears to be complex and probably not a single chemical compound. Closely related bacteria may possess in common a small amount of agglutinogen—a least common multiple, as it were—which stimulates the production of "group agglutinin" that reacts with related bacteria more or less in proportion to their content of the common antigen or agglutinogen. The specific agglutinin produced by the entire agglutinogen content of an organism is more potent and fails to react with related bacteria. Thus, the serum of an animal immunized against B. typhosus may agglutinate that organism in a dilution of 1 to 3000; B. paratyphosus will be agglutinated in a dilution of 1 to 300 by the same serum, and B. coli would agglutinate only in a dilution of 1 to 50. The group agglutinin in this example would be effective for B. paratyphosus in a dilution of 1 to 300, but in greater dilutions it would be ineffective. For B. coli in the instance cited, the group agglutinin is ineffective in dilutions above 1 to 50.

The common or group agglutinin for B. paratyphosus in this typhoid serum could be quantitatively removed by leaving it in contact with a large number of paratyphoid bacilli for a few hours, then centrifugalizing to remove the organisms. The residual serum would contain only agglutinin specific for B. typhosus. If B. typhosus were added to the serum, all the agglutinin—both "group" and specific—would be removed.

As a general rule, group agglutinins constitute a minor fraction of the total agglutinin and in practice the degree of dilution of the serum used in specific cases is ample to exclude error. It occasionally happens that sera of low dilution, especially those rich in agglutinoids, fail to clump the specific organism; as the serum is diluted more and more the phenomena of clumping become more and more marked; finally a degree of dilution is reached beyond which the serum again becomes ineffective. The initial negative agglutination in concentrated serum

[1] Bordet, Ann. Inst. Past., 1899, xiii, 225.

is known as a "proagglutinoid" reaction; it is attributed by Ehrlich to the presence of "agglutinoids" in the serum—side-chains of the second order which have lost their agglutinophore group, but still retain their combining group (haptophore group). These "agglutinoids," which are deteriorated agglutinins, have a greater affinity for the agglutinogen of the bacteria than have the unchanged agglutinins, and consequently prevent the latter from becoming attached to the organisms. If the serum is diluted a point is reached where the agglutinoids are numerically too few to interfere with the action of the agglutinins, which usually far outnumber the agglutinoids. As the serum is more and more diluted a point is eventually reached where the content of agglutinin is insufficient to react with the bacteria. If, however, bacteria are cultured in this dilute serum, they frequently develop into long, thread-like, interwoven filaments, the so-called "thread-reaction" of Pfaundler. It is obvious that the maximum dilution at which a serum will agglutinate bacteria depends somewhat upon the number of organisms; there is, in other words, a relation between the amount of agglutinin in the serum and agglutinogen in the bacteria.

Non-agglutinable Bacteria.—Occasionally strains of bacteria, as B. typhosus, freshly isolated from the body, may not agglutinate with the specific serum. This resistance to agglutination is supposed to result from some unknown change in the agglutinogen of the bacterium during its development in the body. A similar loss of agglutinability may be experimentally brought about by growing the bacteria in gradually increasing concentrations of specific agglutinating serum outside the body. This inagglutinability is usually lost after a few days' development on artificial media; the organisms will then clump in a characteristic manner in a serum that originally was ineffective.

The Reaction of Agglutination.—The practical value of the reaction of agglutination depends upon the visible clumping or agglomeration of a suspension of bacteria in a fluid medium containing some neutral salt, when a relatively small amount of immune serum specific for the organism is added to it. The reaction may be expressed thus:

Organism (Agglutinogen) + Specific Serum (Specific Agglutinin) = Agglutination.

If a specific organism is added to an appropriate dilution of unknown serum with proper precautions, and characteristic clumping takes place, or if a known specific serum is added with suitable precautions to a suspension of an unknown organism, and characteristic clumping takes place, a specific diagnosis of the serum or of the organism can be arrived at. In the first instance, a diagnosis of disease may be made; in the second instance the identity of an organism may be established. The laboratory diagnosis of typhoid and paratyphoid fever, of the various types of bacillary dysentery and of other bacterial infections is frequently made by testing the serum of the patient for agglutination with a known culture of the organism.[1] The laboratory identification of specific bacteria, conversely, is frequently established or corroborated through their agglutination with known specific agglutinating sera.

The reaction of agglutination may be made either microscopically or macroscopically.

1. **Microscopic Method.**—A drop of serum from a patient, diluted to the proper degree, is mixed with an equal amount of a broth culture of the desired organism on a clear cover-glass,[2] and then suspended over the cavity of a hollow ground slide, ringed with vaseline to prevent evaporation, and examined under the microscope. Motile bacteria, as for instance B. typhosus, soon lose their motility (immobilization) and gradually collect in small groups which tend to coalesce into larger and larger clumps, leaving the field between them practically free from organisms. The bacteria are not necessarily killed by agglutination. The reaction ordinarily is complete within two hours. Killed cultures of bacteria may be used in place of living cultures but the reaction is usually less clear-cut.

The advantage of the microscopic method lies chiefly in the small amount of serum required to perform the test. One of its chief disadvantages lies in the relative inaccuracy of the dilution of the serum. (See chapter on B. typhosus for full discussion of technic.)

2. **Macroscopic Method.**—Various dilutions of serum, accurately measured by volumetric pipettes, are brought into small, sterile test-tubes, together with suspensions or broth cultures of the bacteria. Agglutination is manifested by the gradual accumulation of a flocculent sediment of bacteria, leaving the supernatant liquid perfectly clear. Control tubes without serum remain uniformly clouded.

The part played by agglutinins in immunity is unknown; the

[1] The technic and precautions to be observed are discussed individually in the chapters upon specific pathogenic bacteria.

[2] For a majority of bacteria, eighteen-hour cultures in 0.1 per cent. dextrose broth are particularly advantageous. Cultures grown in plain broth are usually much less actively motile and agglutinate less readily.

concentrations of agglutinins in immune sera, as measured by present-day methods, throws no light upon the degree of immunity or the prognosis. Very severe typhoid infections, for example, may show little agglutinin in their sera, and mild cases may exhibit sera comparatively rich in agglutinin content. Their chief value at the present time lies in their relation to the diagnosis of disease.

PRECIPITINS. PRECIPITINOIDS.

In the preceding section it was shown that the sera of animals immunized with various bacteria contained substances—agglutinins—which agglutinated the specific organisms. Kraus[1] showed that these immune sera would cause a precipitate when they were added to clear filtrates of the specific organisms. During the process of immunization, therefore, specific antibodies, termed precipitins, are formed, which react with the specific soluble antigen, precipitinogen, in germ-free filtrate of broth cultures of the specific organisms, to form a precipitate. Later investigations have shown that any soluble protein, as egg-albumen, injected into experimental animals may stimulate the production of specific precipitins which will cause a precipitation in clear solutions of the homologous protein. These reactions have a marked specificity: The sera of animals immunized against casein of cows' milk, for example, will cause precipitation in clear solutions of this protein, but will fail to cause precipitation in solutions of casein from human milk. The sera of closely-related animals may contain small amounts of "group" precipitins, and biological relationships have been established, based upon the community of these antibodies. Thus, the sera of certain anthropoid apes[2] are said to be precipitated by the sera of animals immunized to the serum of man; sera from the lower monkeys fail to react with the human serum. From these observations the inference has been drawn that these anthropoid apes are more closely related to man than are the lower monkeys.[3]

Precipitins closely resemble agglutinins in their method of formation, their resistance to physical agents and their reactions. Like the agglutinins, they possess both a thermostabile haptophore or combining group and a thermolabile ergophore group. The precipitinophore

[1] Wien. klin. Wchnschr., 1897, 736.

[2] Grünbaum, Lancet, January, 1902.

[3] See Nuttall, Jour. Hyg., 1901, i, No. 3; Proc. Royal Acad., November, 1901, lxix; Proceedings Cambridge Philosophical Society, January, 1902; Brit. Med. Jour., April, 1902, i, for full details.

group is very labile and readily becomes non-functionating, but the combining group is relatively stabile. A precipitin which has lost its ergophore group is called a precipitinoid.

The precipitate formed by a specific serum acting upon a clear solution of the antigen (precipitinogen) probably is derived from the serum, because very dilute solutions of the immunizing protein will throw down a relatively bulky precipitate, far too great in amount to come from the antigen in the dilution used.[1]

Precipitins have been extensively studied in their relation to certain aspects of Forensic Medicine, but they have little practical value in the laboratory diagnosis of bacterial disease. They are found in sera under the same conditions as agglutinins, but the technic for their demonstration is more involved than that for agglutinins. Their relation to immunity is unknown, but probably similar to that of agglutinins.

LYSINS.

Mention has been made (see preceding section) of the bactericidal power of fresh blood serum of a normal animal and man. This important discovery, that normal sera contain substances that will destroy moderate numbers of bacteria, was made by Nuttall,[2] who also observed that there was a limit to this destructive activity and that this property was lost upon standing, or rapidly destroyed by an exposure of the serum to 55° C. for half an hour. Buchner[3] corroborated and extended these observations and designated the unknown stabile component "alexin." Pfeiffer[4] then showed that the destructive action of normal sera could be increased many fold above its original level *for a specific organism* if that organism were repeatedly injected into an animal in sublethal, but gradually increased doses. The serum of such an animal would still destroy only moderate numbers of heterologous bacteria, but relatively great numbers of the homologous bacteria. This observation opened the way for the highly important study of active acquired immunity against bacteria. Pfeiffer observed that heating immune sera to 50° to 56° C. for half an hour destroyed their bactericidal properties, precisely as Nuttall had found that natural, non-specific bactericidal properties were destroyed under similar conditions. Bordet[5] then discovered that the

[1] Welsh and Chapman, Ztschr. f. Immunitäsforsch., 1911, ix, 517.
[2] Ztschr. f. Hyg., 1888, iv, 353.
[3] Cent. f. Bakt., 1889, v, 817; vi, 1, 561.
[4] Ztschr. f. Hyg., 1894, xviii, 1; 1895, xix, 75–100.
[5] Ann. Inst. Past., 1895.

addition of a small amount of unheated blood serum from a non-immune animal would "reactivate" the heated inactive immune serum and restore its bactericidal power to its original level. These experiments collectively demonstrated clearly that:

1. Normal sera had an inherent but limited destructive action upon a variety of bacteria.

2. That this destructive or bactericidal action could be greatly increased for specific organisms through repeated injections of sub-lethal doses of them.[1]

3. That both normal and immune sera lost their bactericidal properties by heating them to 55° C. for half an hour.

4. That immune sera would regain their specific bactericidal power if a small amount of fresh normal blood serum of a non-immune animal were added to them.[2]

Bordet[3] showed similarly that the red blood cells of an alien animal were also destroyed to a limited degree by the serum of a normal animal, but that the destruction could be greatly increased for specific erythrocytes if they were repeatedly injected into an experimental animal. The blood serum becomes specifically hemolytic. Here again Bordet[4] found that heating an immune serum to 55° C. for thirty minutes destroyed its activity, but that a small amount of fresh serum from a non-immune animal (whose serum *per se* would not dissolve the homologous cells) would reactivate the serum. Thus, both specific bacteriolytic sera and specific hemolytic sera must contain two distinct components—a thermostabile component resisting an exposure to 55° C. for half an hour and contained *only* in the immune serum, and a thermolabile component destroyed or inactivated at 55° C., which is present both in active immune bacteriolytic and hemolytic sera, and also in normal sera. To the thermolabile substance present in unheated normal and immune sera, Bordet gave the name "alexin;" to the thermostabile specific substance in immune sera he gave the name "substance sensibilitrice." He regarded the "substance sensibilitrice" as a sensitizer or mordant which made bacteria or blood cells vulnerable to the ferment-like or digestive action of the "alexin."

Ehrlich and Morgenroth[5] studied the phenomena of hemolysis in

[1] Presumably leaving the original non-specific bactericidal power at its initial level for all except the specific organism, and possibly for closely related forms.
[2] Moxter, Cent. f. Bakt., 1899, xxvi, 344.
[3] Loc. cit.
[4] Ann. Inst. Past., 1898, xii, No. 10.
[5] Berl. klin. Wchnschr., 1899, No. 1 and 22. See also Collected Studies on Immunity, Ehrlich, translated by Bolduan, 1910.

great detail and demonstrated by very careful and ingenious experiments that the phenomena observed by Bordet were fundamentally correct. They showed:

1. That inactivated specific hemolytic serum (heated to 55° C.) was absorbed by the homologous red blood cells, and that these "sensitized" cells, separated from the serum after a few hours and washed carefully, were readily hemolyzed when resuspended in salt solution to which was added a small amount of fresh, unheated, normal guinea-pig serum.

2. The supernatant residual fluid from which the red blood cells had removed all the immune body was incapable of causing hemolysis of the homologous red blood cells when fresh normal serum was added to it. The erythrocytes, in other words, quantitatively removed the thermostabile "substance sensibilitrice" from solution.

3. If normal sera were allowed to remain in contact with the same red cells for an equal length of time, and these red cells were then removed by centrifugalization and resuspended in salt solution containing normal fresh serum, no hemolysis took place, leading to the conclusion that the thermolabile substance (alexin of Bordet) is not removed from solution by erythrocytes. Apparently alexin is not bound or anchored directly to the red cells.

4. Finally, it was shown that inactivated immune serum, red blood cells and fresh normal serum could be maintained at 0° C. without apparent hemolysis. At 37° C. the same solution soon exhibited complete hemolysis. Thus, at the lower temperature, the normal serum failed to cause hemolysis. If the mixture maintained at 0° C. were centrifugalized, however, after some hours, and the red blood cells washed thoroughly and resuspended in salt solution, hemolysis promptly occurred when a small amount of normal serum was added to the suspension, thus showing clearly that the inactivated immune serum was bound or anchored by the red blood cells at 0° C., even though activation did not take place.

Ehrlich substituted the term "amboceptor" for Bordet's term "substance sensibilitrice" and complement for the term "alexin," and conceived that the immune body—amboceptor—consisted essentially of two combining or haptophore groups—one the cytophilic group, possessing a specific combining power for the specific cell (bacterium or erythrocyte), the other, complementophilic group, combining with the non-specific complement. According to this theory the union of complement to specific cell takes place through the amboceptor; Bordet maintains that neither the specific cell (antigen) of itself nor the substance sensibilitrice (amboceptor) of itself unites with alexin (complement). When both are simultaneously present, however, alexin is absorbed. In other words, amboceptors as such do not exist, according to this view, and consequently complement cannot be bound to the specific cell by a complementophile (haptophore) group.

Multiplicity of Amboceptors and Complement.—The researches of Nuttall and Buchner and of Moxter[1] have shown that fresh normal serum possesses definite but limited bactericidal powers, apparently not specific (for a variety of bacteria may be destroyed) which are destroyed by an exposure of thirty minutes to 55° C. Furthermore, the "inactivated" serum appears to regain its original bactericidal value for various organisms when it is mixed with a relatively small amount of normal serum. In other words, normal serum and specific immune serum (unheated) alike appear to depend upon thermostabile amboceptor and thermolabile complement for their bacteriolytic and hemolytic activities. They differ in the highly specific potency of the immune serum for its homologous cell. Ehrlich and Morgenroth[2] believe that the normal or natural cytolytic activities of sera depend upon a multiplicity of specific amboceptors, each for its specific red blood cell or other cell, and Pfeiffer[3] has made similar observations for the normal bactericidal powers of blood. Ehrlich and Morgenroth have attempted to demonstrate a multiplicity of complements in normal sera also; heated normal sera injected into normal animals are claimed by the Ehrlich school to give rise to anticomplementoids, the supposition being that the heat has destroyed the ergophore group of complement but not its combining group, giving rise to a "complementoid," precisely as a toxin which has lost its toxophore group becomes a toxoid. There appears to be no theoretical limit to the anti- and anti-antibodies which may thus be produced by various increasingly complicated investigations. Bordet and Gay[4] deny the multiplicity of complement.

Fixation of Complement.—Bordet and Gengou,[5] in a series of experiments, brought forth experimental evidence of the unity of complement and, incidentally, developed a method of investigation now

[1] Loc. cit. [2] Loc. cit.
[3] Harben Lecture, Jour. Royal Inst. Public Health, 1909, xvii, 385.
[4] Collected Studies in Immunity by Bordet and his associates (translated by Gay, 1909).
[5] Ann. Inst. Past., 1901, xv.

extensively utilized to demonstrate the presence of various specific immune antibodies. If a specific immune body (as for example, the serum of an animal immunized to typhoid bacilli) is heated to 55° C. for half an hour, then added to a suspension of typhoid bacilli together with normal unheated serum, a union between the bacilli, the specific antibody of the serum (amboceptor, substance sensibilitrice) and the complement (alexin) will take place. If the proportions of the three reactive bodies are correct, all the complement or alexin will be bound, provided the mixture is incubated a few hours at 37° C. If, now, red blood cells and inactivated immune serum specific for the red blood cells are added to the mixture of bacteria, immune body and complement, no hemolysis should be noticed, because the complement is quantitatively anchored to the bacteria-immune serum complex. If, on the other hand, the inactivated immune serum added to the suspension of typhoid bacilli be not typhoid immune serum, complement will not be bound to the bacteria, for the specific amboceptor or substance sensibilitrice will not be present. The complement or alexin, therefore, is not anchored to the bacteria, and it is free to act when the red blood cells and their specific inactivated serum are added to the mixture of bacteria and serum. Under this condition hemolysis occurs, because the red blood cells, inactive immune body and complement unite. The production of hemolysis being visible, it acts as an indicator in such instances. Wassermann and his associates have utilized this method of "fixation of complement" for the serologic diagnosis of syphilis, and gradually a relatively large number of diagnoses of clinical importance have been developed along the same lines.

The Determination of Specific Antibodies by the Method of Complement Fixation.—*Principle Involved.*—When an antigen (bacteria, erythrocytes, tissue cells, protein, or other substance which stimulates specific antibody formation) is mixed intimately with its specific inactivated immune serum and fresh normal complement a firm union of the three components takes place.[1] The result of this union is an injury or destruction of the antigen. If the antigen be bacterial cells or tissue cells there is usually no visible change in the gross appearance of the mixture, and cultural or chemical investigation must be relied upon to demonstrate the lytic process. Erythrocytes, on the other hand, undergo changes in the presence of inactivated specific immune serum and complement which result in the liberation of hemoglobin,

[1] Bordet and Gengou, Ann. Inst. Past., 1901, xv, 290.

which colors the solution deep red. This change is clearly visible and requires no additional procedure for its demonstration; the liberation of hemoglobin is in itself an indicator of the reaction which has taken place.

The relation between antigen, immune serum, and complement is quantitative; consequently, if the respective amounts of the three components are correctly proportioned, no free unattached complement will be present in a mixture of them after an appropriate incubation at body temperature is practiced to allow of their union. Usually an hour at 37° C. suffices for this union to take place quantitatively.

These very important observations of Bordet and Gengou have led to the development of a technic for the diagnosis of infection, and the identification of antigens by the method of complement fixation.

The underlying principles of the reaction of complement fixation are three:

(*a*) The union of specific inactivated immune serum and homologous antigen.

(*b*) The quantitative activation of the antigen—inactivated specific immune serum complex by non-specific complement; and

(*c*) The visible hemolysis that results from the activation of an erythrocyte—inactivated specific immune serum complex by non-specific complement.

The general plan of procedure is to incubate an antigen (as bacterial cells) and inactivated serum and complement in proper proportions for an hour, to permit the three components to unite. A mixture of erythrocytes and specific inactivated hemolytic serum is now added. If the reactive substances are properly proportioned and the inactivated serum first added is specific for the antigen (bacteria), no hemolysis will occur when the hemolytic system is added, because all the complement present is bound by the bacteria-immune serum complex. On the contrary, if the inactivated serum is not specific for the bacterial antigen, no union between the two will take place, complement will not be bound, and it is free in the mixture. It will activate the erythrocyte-inactivated immune serum complex, and hemolysis will occur.

It will be seen that the hemolytic system is added as an indicator; an absence of hemolysis shows a union of bacterial antigen, inactive specific bacterial immune serum and complement. Hemolysis shows that the union has not been formed, the complement was free in the mixture and it united with the hemolytic system, causing hemolysis in the erythrocyte antigen through the specific amboceptor or hemolysin.

The method of complement fixation may be employed to examine sera for specific antibodies, using a known antigen, or to test suspected antigens with sera containing specific antibodies. The most practical application of the method in medicine is the serum diagnosis of syphilis, glanders, and other bacterial infections.

The Technic of Complement Fixation.—The technic of complement fixation is simple in principle, but it requires the most scrupulous attention to details. All glassware must be neutral in reaction, chemically clean, and bacteriologically sterile. Physiological salt solution (0.85 to 0.90 per cent. C.P. NaCl in neutral distilled water) used for washing red blood cells and for dilutions should be sterile and stored in clean containers.

The Wassermann Serum Diagnosis of Syphilis.—Five elements enter into the Wassermann test for syphilis: the antigen, suspected syphilitic serum, complement, and a hemolytic system consisting of red blood cells and specific immune hemolytic serum (hemolysin).

Preparation and Standardization of Antigen.—The antigen originally employed by Wassermann and his collaborators was an aqueous extract of syphilitic tissue which was prepared by suspending one part by weight of finely comminuted liver of a syphilitic fetus[1] in five parts of physiological salt solution containing 0.5 per cent. phenol as a preservative. After several days' violent agitation in the dark it is strained through several layers of cheesecloth to remove coarser particles and stored in amber bottles in the refrigerator. Sedimentation takes place until a brownish, slightly opalescent fluid remains, which is the luetic antigen.

Later work[2] showed that alcoholic extracts of luetic liver were more stable than watery extracts. The specific reacting component, according to Porges and Meier, is lipoidal in nature, and in this sense it is not biologically specific. The fixation of complement appears to depend upon a substance in the antigen, lipoidal in nature, which effects a union of antigen, immune body and complement. Citron has proposed the term "lues reagine" for this substance. Alcoholic extracts of syphilitic liver are prepared by shaking finely comminuted liver with ten times the weight of absolute alcohol for a few days,

[1] The tissue is examined for the specific organism; if Treponemata are abundant it is converted into antigen, otherwise it is discarded.

[2] Especially by Porges and Meier, Berl. klin. Wchnschr., 1908, No. 15.

then digesting the mixture at 37° C. for a week. The extract is filtered through filter paper and placed in the refrigerator.

Alcoholic extracts of normal organs, prepared in the same manner as luetic livers, have been found to be quite as good as alcoholic extracts of syphilitic livers for the diagnosis of syphilis. In practice heart-muscle of normal guinea-pigs, freed from all fat, is used.

Noguchi's Acetone-Insoluble Lipoidal Antigen.[1]—Noguchi and others have shown that alcoholic extracts of organs may, and frequently do, contain sufficient amounts of neutral fats, or their hydrolytic cleavage products, to make the antigen hemolytic or anticomplementary. These substances are for the most part soluble in acetone, while the antigenic fraction is insoluble in acetone. One part of fat-free heart muscle or liver from a guinea-pig is cut into very fine pieces, mixed with ten parts of absolute alcohol, and extracted in the incubator at 37° C. for a week or ten days, being thoroughly shaken every day. The soluble substances are freed from the fragments of tissue by filtration through fat-free filter paper, and rapidly evaporated to dryness.[2] Sufficient ether is then added to take up the brownish residue, and it is then allowed to stand until a clear, slightly colored, ethereal solution is obtained, free from suspended material. The ethereal solution is concentrated by evaporation to a point where separation of a sediment begins, then it is poured into several volumes (usually ten) of pure acetone. A voluminous precipitate forms at once, and settles out as a tenacious gummy mass. This is retained, under acetone, as the antigen. The acetone-soluble solution is discarded. The antigen thus prepared consists largely of lecithins and related substances. It keeps well and appears to be very sensitive and reliable. From 0.2 to 0.3 gram are dissolved in a mixture of 1 c.c. of ether (free from alcohol and having a neutral reaction) and 10 c.c. of neutral absolute methyl alcohol. This solution is kept in an amber bottle in the refrigerator as a stock antigen. One cubic centimeter of this stock antigen is added to 19 c.c. of physiological salt solution; this is the antigen used for the test.

Before making a test it is necessary to standardize the antigen. It is essential to know the anticomplementary titer, that is, that maximum amount of antigen which will inhibit hemolysis in the presence of syphilitic serum, but which will not inhibit hemolysis

[1] Noguchi, Serum Diagnosis of Syphilis.

[2] Best by exposing the filtrate in a broad shallow dish to an air current from an electric fan.

when non-syphilitic serum is used. In addition, the following determinations are sometimes desirable.

The hemolytic titer, that amount of antigen which will of itself cause lysis of red blood cells, and the antigenic titer, the amount of complement it will absorb or "fix" in the presence of a definite amount of specific syphilitic serum.

The anticomplementary titration is made by mixing graded amounts of antigen and a constant amount of complement (0.1 c.c. of a 10 per cent. solution[1]) with constant amounts (0.1 c.c.) of known syphilitic serum and normal serum, both inactivated.

The various mixtures are incubated in a water-bath at 37° C. for an hour, then 0.2 c.c. of red blood cell suspension and inactivated hemolytic serum are added and again incubated in the water-bath at 37° C. The maximum amount of antigen which will give complete inhibition of hemolysis with syphilitic serum and no inhibition of hemolysis in the non-syphilitic serum is regarded as the unit.

EXAMPLE OF AN ANTICOMPLEMENTARY TITRATION OF ANTIGEN.

Tube.	Antigen.	Normal serum Inactive, c.c.	Complement, 10 per cent. c c.		Red blood cells, c.c.	Hemolytic serum inactivated units.		Result.
1	0.2	0.1	0.10	Water-bath 37° C.: one hour, then add	1.0	1.5	Water-bath 37° C.—one hour.	Complete hemolysis.
2	0.4	0.1	0.10		1.0	1.5		" "
3	0.6	0.1	0.10		1.0	1.5		" "
4	0.8	0.1	0.10		1.0	1.5		" "
5	1.0	0.1	0.10		1.0	1.5		Partial inhibition.
6	1.5	0.1	0.10		1.0	1.5		Marked inhibition.
7	2.0	0.1	0.10		1.0	1.5		Complete inhibition.
8[2]	—	0.1	0.10		1.0	1.5		Complete hemolysis.
9[3]	—	—	0.10		1.0	1.5		Complete hemolysis.

Tube 5, containing 1.0 c.c. antigen, shows beginning inhibition of hemolysis. This is regarded as the anticomplementary titer of the antigen.

As a general rule, the hemolytic titer is higher than the anticomplementary titer. The test is readily made, if desired, by using the same amounts of antigen mixed with 1 c.c. of red blood cell suspension and sufficient salt solution to bring the volume to 4 c.c.

It is customary to use one-fourth the anticomplementary titer as the standard amount of antigen to be used in the actual test. In the

[1] Prepared by adding fresh normal guinea-pig serum to physiological salt solution in the proportion of one part serum to nine parts salt.

[2] Serum control.

[3] Hemolytic control.

example cited, 1.0 c.c. of the antigen was found to be anticomplementary, consequently 0.25 to 0.3 c.c. would be the proper amount of antigen to employ in the test.

Complement.—Fresh guinea-pig serum is the usual source of complement for fixation reactions. The animal should be healthy and not previously injected with protein of any nature. The serum of pregnant pigs is not trustworthy. Blood may be obtained directly from the heart of the living animal by aspiration through a hypodermic needle, from a severed carotid artery, or, more expeditiously by cutting the throat of the animal, avoiding the esophagus, and collecting the blood in sterile Petri dishes. The freshly drawn blood is allowed to stand for a few hours at a low temperature and the serum is pipetted off. Complement must be kept cold (below 16° C.) and in the dark. It must be used fresh, for it deteriorates rapidly. In a frozen condition, however, it will remain active for two or three weeks. Both the "activating" and combining properties of normal fresh guinea-pig serum are sufficiently constant for the reaction of complement fixation.

Hemolytic System.—(*a*) *Hemolytic Serum* (*Hemolysin*).—Hemolytic serum is obtained from rabbits which have been injected with 2 c.c., 4 c.c., and finally 6 c.c. of a 50 per cent. solution of washed sheep red blood cells[1] at intervals of two or three days. The injections may be made intraperitoneally or intravenously, the latter being preferable. Not less than nine days after the injection the animal is bled to death from the carotid artery under anesthesia, the blood being received in sterile test-tubes, which are placed in an inclined position in the ice-box. The serum is removed, centrifugalized if not wholly free from blood corpuscles, and placed in small amber bottles with aseptic precautions. These are heated to 56° C. for half an hour to effect inactivation (to destroy complement).

(*b*) *Red Blood Cells.*—Erythrocytes of the sheep are used. The blood of a sheep is collected either in small sterile flasks containing one volume of 0.85 per cent. salt solution and 0.5 per cent. sodium citrate, or in sterile centrifuge tubes. If the former is used, nine volumes of blood are allowed to flow into the flask and immediately mixed intimately with the citrate solution, which prevents clotting. This method is applicable if the blood cannot be centrifuged imme-

[1] Fresh red blood cells of the sheep are freed from serum by repeated washings with physiological salt solution—usually five washings suffice. The corpuscles are then suspended in a volume of salt solution twice that of the corpuscles themselves.

diately. If centrifuge tubes are used, an amount of blood not more than one-third the capacity of the tube (about 5 c.c.) is collected and twice the volume of sterile salt solution is added to it. The corpuscles are sedimented, the supernatant solution is pipetted off, fresh salt solution is poured in, and the corpuscles resuspended by careful stirring with a clean glass rod. This process is repeated five times, each time discarding the washings. The last time the volume occupied by the erythrocytes is read off on the graduations of the tube and they are suspended in a volume of salt solution twenty times that occupied by the erythrocytes. This makes a 5 per cent. suspension. Erythrocytes are obtained by centrifugalization from the citrated blood in precisely the same manner. This suspension of red blood cells, kept in a cool, dark place, may be used for two days, but not longer. Beyond that time the cells deteriorate and hemolyze with abnormal readiness, thus vitiating the value of the test.

(*c*) *Standardization of Hemolytic System.*—It is very important to know with exactness the amount of hemolytic serum (inactivated, of course) which will effect complete hemolysis of 1 c.c. of a 5 per cent. suspension of sheep erythrocytes in the presence of a constant amount of complement. The determination of this factor gives the hemolytic titer of the hemolytic serum. It is readily determined by adding to a series of tubes, 0.1 c.c. of fresh guinea-pig serum (complement), 1 c.c. of erythrocyte suspension, and varying amounts of the inactivated hemolytic serum. The smallest amount of hemolysin which will effect hemolysis under the conditions stated is the hemolytic titer or unit. Thus, the following tubes incubated at 37° C. for one hour showed:

Tube.	Complement.	5 per cent. suspension sheep erythrocytes.	Inactivated hemolytic serum.	Result.
1	0.1 c.c.	1 c.c.	0.10 c.c.	Complete hemolysis.
2	0.1 c.c.	1 c.c.	0.075 c.c.	" "
3	0.1 c.c.	1 c.c.	0.050 c.c.	" "
4	0.1 c.c.	1 c.c.	0.025 c.c.	
5	0.1 c.c.	1 c.c.	0.010 c.c.	
6	0.1 c.c.	1 c.c.	0.0075 c.c.	"
7	0.1 c.c.	1 c.c.	0.0050 c.c.	" "
8	0.1 c.c.	1 c.c.	0.0025 c.c.	Partial hemolysis.
9	0.1 c.c.	1 c.c.	0.0010 c.c.	No hemolysis.
10	0.1 c.c.	1 c.c.	0.00075 c.c.	" "
11[1]	0.1 c.c.	1 c.c.	0. c.c.	" "
12[2]	0.0 c.c.	1 c.c.	0. c.c.	" "

0.0025 of this serum is one unit; the hemolytic titer is 0.0025 c.c., in other words. It is customary to use two units in the actual test, consequently 0.005 c.c. would be the amount used.

[1] Complement control. [2] Erythrocyte control.

It must be emphasized that precision of measurement is an absolute requirement for success; the activating power of complement for hemolysin does not follow the law of multiple proportions—it is rather an inverse ratio, as Noguchi[1] has pointed out. Relatively less complement is required to induce complete hemolysis in a system containing four units than is required for a system containing but a single hemolytic unit.

The serum to be examined for specific antibodies by the method of complement-fixation must be sterile and free from hemoglobin. The products of bacterial growths in serum may be anticomplementary and the presence of hemoglobin in serum also tends to inhibit hemolysis. Blood, therefore, should be withdrawn with aseptic precautions from the median basilic vein of the patient into sterile test-tubes, and either centrifugalized at once and the serum removed from the clot, or placed in an inclined position in a cool place until the serum has separated. The serum must be clear[2] and free from erythrocytes or hemoglobin.[3] It is inactivated at 54° to 55° C. for half an hour in a water-bath.[4]

The Technic of the Test.—It is essential that the hemolytic system —erythrocytes, hemolysin, complement—be standardized daily. Varying amounts of hemolysin are added to constant amounts of erythrocyte suspension and complement, as outlined above. A known positive syphilitic serum and a known negative syphilitic serum, together with suitable controls, must be examined along with the unknown serum to be tested.

The following reagents are required:

1. Sterile physiological salt solution.
2. Fresh guinea-pig sérum (complement)—add 0.1 c.c. to each tube.
3. Five per cent. suspension of washed sheep erythrocytes in normal salt solution—use 1 c.c. to each tube.
4. Hemolysin (amboceptor)—use twice the hemolytic unit (the unit must be determined daily).
5. Known syphilitic serum—inactivate and use 0.2 c.c.
6. Known normal (non-syphilitic) serum, inactivated—use 0.2 c.c.
7. The serum to be tested—inactivate, use 0.2 c.c.

[1] Serum Diagnosis of Syphilis.

[2] Blood is best obtained early in the morning, before the patient has eaten; blood obtained at the height of digestion frequently contains fats which make the serum turbid.

[3] Small amounts of blood, yielding a few drops of serum, may be obtained from the finger-tip or the lobe of the ear. Massage must not be practised, for there is danger of damaging erythrocytes with the liberation of hemoglobin.

[4] Noguchi, Serum Diagnosis of Syphilis, states that inactivation at 54° *C.* should be practised—the higher temperature weakens the reactive substance somewhat.

Unknown serum.	Known positive syphilitic serum.	Known normal non-syphilitic serum.	Controls.
Tube 1. Serum, 0.2 c.c. Complement, 0.1 c.c. Salt solution, 2.7 c.c.	Tube 3. Serum, 0.2 c.c. Complement, 0.1 c.c. Salt solution, 2.7 c.c.	Tube 5. Serum, 0.2 c.c. Complement, 0.1 c.c. Salt solution, 2.7 c.c.	Tube 7. Complement, 0 1 c.c. Salt solution, 2.9 c.c.
Tube 2. Serum, 0.2 c.c. Complement, 0.1 c.c. Antigen,[1] 1 c.c. Salt solution, 1.7 c.c.	Tube 4. Serum, 0.2 c.c. Complement, 0.1 c.c. Antigen, 1 c.c. Salt solution, 1.7 c.c.	Tube 6. Serum, 0.2 c.c. Complement, 0.1 c.c. Antigen, 1 c.c. Salt solution, 1.7 c.c.	Tube 8. Complement, 0.1 c.c. Antigen, 1 c.c. Salt Solution, 1.9 c.c.

After mixing the tubes are placed in a water-bath maintained at 37° C. for one hour, to permit of the fixation of complement; 1 c.c. of a 5 per cent. suspension of erythrocytes and two units of hemolysin are then added to each tube, mixed and reincubated for one hour, then read. Tubes 1, 3, 5, 7, 6 and 8 should show complete hemolysis. Tube 4 should show complete inhibition of hemolysis (positive reaction). If such be the case all the reagents are properly adjusted, and Tube 2, containing the unknown serum, is read. If hemolysis is absent the reaction is positive; if hemolysis is complete the reaction is negative.[2]

The Method of Noguchi.[3]—A rigorous standardization of reagents is a prerequisite for accuracy in the serum diagnosis of syphilis, and Noguchi has pointed out that a variable inherent inaccuracy exists in the Wassermann method. He has shown that human sera may contain variable amounts of hemolysin specific for sheep erythrocytes. Human sera, however, contain no hemolysin for human erythrocytes. The Noguchi modification, therefore, substitutes human red blood cells (obtained from placenta or at autopsies) for sheep red blood cells. Rabbits are immunized to carefully washed human erythrocytes and the hemolytic unit of the rabbit serum is determined in the usual manner. The following reagents are required to perform the Noguchi test:

1. Complement—Fresh guinea-pig serum in 40 per cent. dilution (one part clear fresh serum to 2.5 parts sterile salt solution).

2. Hemolytic Serum—Rabbit serum, immunized against human erythrocytes, is titrated against human erythrocytes to determine the hemolytic unit. Two units are used in the test.

[1] Twice the antigen titer, determined by titration, diluted with salt solution; thus, if the antigenic titer of the acetone insoluble extract is 0.2 c.c., and the anticomplementary titer is found to be 1.75 c.c., 0.4 c.c. of the extract are diluted with 0.6 c.c. salt solution and used in the diluted state. In practice, enough extract should be diluted to last one day.

[2] For a discussion of results, see section on Treponema pallidum.

[3] Noguchi, Serum Diagnosis of Syphilis.

PLATE I

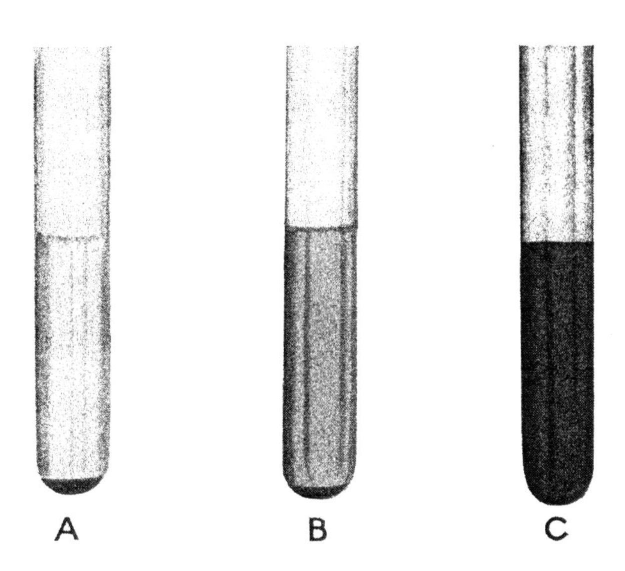

Wassermann Reaction. (Simon.)

A, positive; *B*, partial; *C*, negative reaction.

Note undissolved blood corpuscles in *A*, partial hemolysis in *B*, and complete hemolysis in *C*.

3. Human Erythrocytes—Red blood cells are obtained from a normal individual, washed thoroughly with salt solution, and made up as a 1 per cent. suspension in salt solution. 1 c.c. of the suspension is used in the test.

4. Antigen—The acetone-insoluble antigen is used.

5. Patient's Serum—Obtained *fresh*, from 2 to 5 c.c. of blood. It is used unheated.

6. Known syphilitic serum.

7. Known normal (non-syphilitic) serum.

The test is performed as follows:

Unknown serum.	Known positive serum.	Known negative serum.	Controls.
Tube 1. Serum, 1 drop. Complement,[1] 0.1 c c. Erythrocytes, 1.0 c.c	Tube 3. Serum, 1 drop Complement, 0.1 c.c. Erythrocytes, 1.0 c.c.	Tube 5. Serum, 1 drop. Complement, 0.1 c.c. Erythrocytes, 1.0 c.c.	Tube 7. Complement, 0.1 c.c. Erythrocytes, 1.0 c.c.
Tube 2. Serum, 1 drop. Complement, 0 1 c c. Antigen, 2 units. Erythrocytes, 1.0 c.c.	Tube 4. Serum, 1 drop. Complement, 0.1 c.c. Antigen, 2 units. Erythrocytes, 1.0 c.c.	Tube 6. Serum, 1 drop. Complement, 0.1 c.c. Antigen, 2 units. Erythrocytes, 1.0 c.c.	

Mix and incubate one hour in water-bath at 37° C. Remove and add 2 units hemolysin to each tube and incubate in water-bath for one hour. Tubes 1, 3, 5, 6 and 7 should show complete hemolysis. Tube 4 should show no hemolysis (positive control). If such be the case the reagents are correctly adjusted and a reading of Tube 2 will be positive (no hemolysis) or negative (hemolysis).

A further simplification of the method has been made by Noguchi. The hemolysin and antigen respectively may be absorbed on squares of filter paper, dried, and standardized. In this state they retain their potency for several weeks. In practice the squares of paper are added directly to the tubes, thus saving much time.

Complement-fixation in Bacterial Infections.—*Preparation of Antigen from Bacteria.*—Experience has clearly shown that bacterial antigens should be polyvalent—prepared by mixing in equal amounts, several strains of the same organism. The antigen may be prepared in one of several ways.

The simplest method is to wash off bacteria from agar slants, at the period of maximum growth, with salt solution and shake thoroughly to make a uniform suspension. A small amount of phenol (0.5 per cent.) and 3 per cent. glycerin are then added and the whole sterilized at 56° to 60° C. for one hour. Relatively more of the proteins of the bacterial cell may be obtained in solution if the bacterial emulsion is shaken in a shaking machine with sterile, sharp quartz-sand for twenty-four hours: filtration through coarse Berkefeld filters removes

[1] Forty per cent. solution of fresh guinea-pig serum in salt solution.

the sand and broken bacterial cells, and the filtrate is preserved with 0.5 per cent. phenol. Besredka prepares a bacterial antigen from dried bacterial cells, which are obtained by drying bacteria scraped from agar slants or other solid media over sulphuric acid or calcium chloride. The dried organisms are ground in agate mortars with crystals of NaCl to an impalpable powder, which is then gradually rubbed up in successive portions of water until a physiological salt solution is obtained (corresponding to 8.5 grams NaCl in a liter of distilled water).

It has been found that much of the antigenic substance of bacteria is precipitated by an excess of alcohol; a considerable excess of alcohol is added to a suspension of bacteria, or to an emulsion of the cell substance prepared according to Besredka's process, outlined above. The precipitate from the alcoholic solution is separated by filtration, dried, and ground to an impalpable powder with NaCl crystals. The powder is gradually brought into solution by the addition of water in successive amounts until isotonicity is reached. An attempt is made to create a definite concentration of antigen by starting with a known quantity of dried bacteria and a corresponding amount of NaCl crystals. Thus, 1 gram of dried bacterial substance, ground in a mortar with 0.85 gram NaCl crystals and gradually brought to a volume of 100 c.c. with distilled water, would yield, theoretically, an antigen of 1 per cent. strength. Bacterial antigens must be kept cold and in a dark place, preferably in sealed amber bottles. Deterioration gradually occurs and all bacterial antigens suspended or dissolved in liquids are relatively unstable.

Standardization of Bacterial Antigens.—The standardization of bacterial antigen differs in no respect from that of a syphilitic antigen. The anticomplementary titer and the antigenic titer are determined, the latter by titration with a specific immune serum.

The Diagnosis of Glanders by the Method of Complement-fixation.—The antigen is prepared from glycerin-agar cultures[1] of several strains of B. mallei incubated at 37° C. for forty-eight hours. The organisms are autolyzed in distilled water for several hours at a relatively high temperature (70° to 80° C.), then freed from suspended particles by filtration through coarse Berkefeld filters. The filtrate is stored in amber bottles in the ice-box after the addition of 0.5 per cent. phenol.

The anticomplementary titer is determined from a series of tubes containing constant amounts of complement and graduated amounts

[1] Reaction 1.5 per cent. acid to phenolphthalein.

of antigen (1 to 20 dilution in salt solution).[1] The total volume of complement and antigen is brought to 3 c.c. by the addition of salt solution. After one hour's incubation in the water bath at 37° C., 1 c.c. of sheep erythrocyte suspension and 1.5 units sheep erythrocyte hemolysin are added and reincubated. That dilution of antigen which shows the slightest inhibition of hemolysis is taken as the anticomplementary titer of the antigen. Not more than one-half this amount, and preferably one-fourth of the anticomplementary titer, is used in the test.

The actual determination is made in the same manner as for the Wassermann test.[2] It is well to include a known positive and known negative glanders serum of the same animal species as the unknown, together with suitable controls of the hemolytic system. The length of incubation is determined by the time it takes to effect complete hemolysis in the known negative and the hemolytic controls. Frequently ten or more hours will elapse before this occurs.

AGGRESSINS.

Progressively pathogenic bacteria appear to differ from parasitic bacteria or "opportunists" in that they are able to force an entrance to the underlying tissues of the host through natural, non-specific barriers which ordinarily suffice to restrain the more parasitic types of microbes. Bail[3] has advanced an hypothesis, based upon experimental evidence, which attributes the invasiveness of pathogenic bacteria and their ability to develop in the tissues of the host to "aggressins." These aggressins, according to Bail, are present and may be demonstrated in exudates resulting from bacterial infection, but they are not, as a rule, found in artificial cultures of the same organism. To demonstrate the action of aggressins, Bail removed bacteria from exudates by centrifugalization and injected the clear supernatant fluids, together with a sublethal dose of the homologous bacterium, into experimental animals. Rapidly fatal infections developed. The aggressin-containing exudates were not inactivated by prolonged exposure to 50° C., and it was shown, furthermore, that

[1] Usually a range of antigens from 2 c.c. to 0.05 c.c. will be found sufficient.

[2] For full discussion of results, see Mohler and Eichhorn, Bureau of Animal Industry Bulletin 136, April 7, 1911.

[3] See Der Problem der bakteriellen Infektion, Bail, in Bibliothek medizinischer Monographien, xi; see also Müller in Oppenheimer's Handbuch der Biochemie, 1909, ii, 1, 681.

their injection into susceptible animals stimulated the formation of "antiaggressin," which greatly increased the resistance of the animal to subsequent infection. The sera of animals immunized with aggressin-containing fluids conferred a limited degree of immunity to specific infections in non-immune animals (passive immunity). It has been claimed by Doerr[1] and others that the aggressins are of the nature of bacterial endotoxins and that the immunizing properties of aggressin fluids are due to their content of specific substances derived from the autolysis of bacterial cells.

The aggressin theory must, for the present, be regarded as not definitely proved.

OPSONINS. TROPINS. BACTERIAL VACCINES.

A most important contribution to the literature of immunity is the work of Denys and his associates,[2] who showed that the sera of rabbits immunized to Streptococcus pyogenes possessed two properties not exhibited by the serum of a normal animal, namely, the property of restricting the development of the organism, and the property of stimulating phagocytosis. Their very comprehensive studies demonstrated that the leukocytes of normal animals, suspended in the serum of immunized animals, phagocytized streptococci energetically, but the leukocytes of immunized animals suspended in normal serum failed to exhibit phagocytic activity. Their conclusion was that the immunity of rabbits to the streptococcus resides in the serum. These observations not only added materially to the restricted field in which they were cast—they brought sharply into focus the interrelation of the humoral and cellular aspects of immunity.

Wright and Douglas,[3] using a modification of the technic of Leishman,[4] were able to study phagocytosis *in vitro:* by an ingenious series of experiments they showed that normal serum contains substances —opsonins—which prepare bacteria for phagocytosis, as described in a preceding section (Cellular Immunity). The technic of measuring the potency of opsonins in the sera of normal and infected individuals, as practised by Wright and his associates, consisted

[1] Wien. klin. Woch., 1906, No. 25.

[2] Denys and Le Clef, La Cellule, 1895, xii; Bull. de l'Acad. roy. de Belgique, 1895; Denys and Marchand, Ibid., 1896; Van de Velde, Ann. Inst. Past., 1886, x; Marchand Arch. de Méd. exp., 1898; Denys, Cent. f. Bakt., 1898, xxiv, 685.

[3] See Studies in Immunization, Constable, 1909, for complete biography.

[4] Brit. Med. Jour., 1902, i, 73.

essentially in mixing intimately equal volumes of bacterial emulsion, serum, and leukocytes; after incubation at body temperature the mixture was spread evenly upon microscopic slides, stained, and examined with the microscope. The average number of bacteria per polymorphonuclear leukocyte was determined by direct count. A comparison, under parallel conditions, of the phagocytic activity of leukocytes for a specific organism in the serum of a normal individual and that of an individual infected with the specific organism, according to the technic outlined below, was called by Wright the opsonic index.

Procedure.—1. **Leukocyte Suspension.**—About 0.5 c.c. of blood, drawn from the lobe of the ear or the tip of the finger, is collected in a centrifuge tube containing 10 c.c. of sterile physiological salt solu-

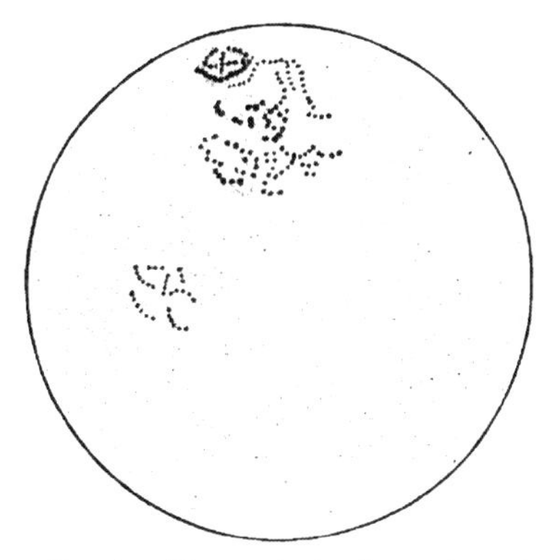

Fig. 9.—*Phagocytosis of streptococci.*

tion in which has been dissolved 1 per cent. of sodium citrate; this mixture is centrifuged at moderate speed until a sharp separation of blood cells and clear supernatant fluid is obtained. The supernatant fluid is carefully poured off and the top layer of blood cells, which contains practically all the leukocytes, is removed to a fresh centrifuge tube containing 10 c.c. of physiological salt solution.

A second centrifugalization is made, and again the supernatant fluid, containing the last traces of blood serum, is discarded. The sediment, rich in leukocytes, is used as the leukocyte suspension in the test.

2. **Suspension of Bacteria.**—Bacteria from a culture on solid media are suspended in sterile salt solution and agitated until a fine opalescent emulsion is obtained. This is most conveniently accomplished

in a shaking machine, but repeated shaking in a stoppered test-tube containing glass beads will usually suffice. The coarser clumps of bacteria are removed by filtration through a coarse filter paper. The density of the bacterial suspension should be such that not more than ten bacteria per leukocyte will be taken up as the average.

3. **Serum.**—(*a*) Blood from three or four normal individuals is collected in capillary tubes; after the serum has separated a "pool" or mixture is made, composed of equal volumes of each serum. Experience has shown that "pooled" serum furnishes a more reliable normal opsonic index than that obtained from a single individual.

(*b*) *Serum from the Patient.*—This is prepared in the manner described above.

The Test.—A capillary pipette of 1 to 1.5 mm. bore is made by drawing out a piece of glass tubing previously softened in the flame. If the tubing is heated in the center until it softens, then, after a few seconds, drawn slowly and steadily out, the desired size and shape is readily obtained. A close-fitting rubber bulb attached to the larger end is a convenience.

A mark about 1 to 1.5 cm. from the capillary end is made with a wax pencil, and a volume each of the leukocytes, pooled serum, and bacterial suspension are drawn into the pipette. It is convenient to separate each ingredient by a small air bubble, to insure uniformity of volume. Mixing is accomplished by carefully expelling and drawing back the respective elements into the pipette. Finally, the mixture is drawn well up into the pipette, the end is sealed in the flame of a Bunsen burner, and the charged pipette is placed in the incubator at 37° C. This is the normal or control.

A precisely similar preparation is made, using the serum of the patient in place of the pooled serum.

Incubation is maintained for fifteen minutes.

The ends of the pipettes are now broken off, and the contents of each pipette mixed as before. A large drop of each respective mixture is spread upon clean glass slides, using the same technic as that for preparing a blood smear, and air-dried. The preparations are stained with Löffler's methylene blue, Wright's stain, or other stain suitable for the organism used.

The number of bacteria in fifty, one hundred, or two hundred leukocytes are determined by direct count, and the average number of bacteria per leukocyte of the normal serum compared with the average number of bacteria per leukocyte in the pathological serum:

EXAMPLE.

	Bacteria in 100 leukocytes.	Bacteria per leukocyte.
Staphylococcus suspension + pooled serum and leukocytes	750	7.5
Staphylococcus suspension + patient's serum and leukocytes	250	2.5

Opsonic index, patient's serum $= \frac{2.5}{7.5}$ or 0.33 per cent.

Numerous observers have been unable to obtain uniform results with the technic of Wright for opsonic index determination, and this is not surprising when the many variable factors entering into the method are reviewed. Attempts have been made to eliminate or limit the variable factors: Simon proposed a dilution method in which the pooled and patient's serum are diluted 1 to 10, 1 to 100, etc., before incubation with the bacteria and leukocytes. That dilution of serum at which phagocytosis practically ceases in the normal and patient's serum respectively is taken as a basis for comparison. Inasmuch as the opsonic index is rarely determined as a guide for treatment of bacterial disease with bacterial vaccines at the present time, however, a discussion of these modifications, which are too involved for practical use, is left for more pretentious volumes.

The Nature of Opsonins.—There appears no doubt that the hypothetical substance or substances called opsonin by Wright exist in normal sera, and it is equally certain that they may be diminished during infection. Furthermore, opsonin may be increased either in amount or in potency by careful immunization. The relation of opsonins to other antibodies, normal or specific, is a subject of controversy at present. The researches of Neufeld and Rimpau,[1] Hektoen[2] and others indicate that the normal opsonins—those of normal sera—are thermolabile, but those developed during immunization to a specific organism—bacteriotropins—are relatively thermostabile.

It has been suggested that opsonins or bacteriotropins are not to be distinguished from other immune bodies—as normal and specific amboceptors or agglutinins. The rapidity with which the opsonic index may be increased or diminished within a few hours following injections of bacteria, however, would suggest a possible distinction between these antibodies and the slowly developing specific bactericidal and agglutinating antibodies.

Vaccine Therapy.—The value of vaccines and of autogenous vaccination in bacterial prophylaxis and bacterial immunization as set

[1] Deutsch. med. Wchnschr., 1904, 1458.
[2] Jour. Inf. Dis., 1906, iii, 434; 1909, vi, 78; 1913, xii, 1.

forth by Wright marks a distinct epoch in bacterial therapeutics in spite of the practical failure of his opsonic index determination as a theoretical guide to immunization and treatment. He has used bacterial vaccines both for prophylaxis—to prevent infection with specific bacteria—and therapeutically—to arrest infection.

Prophylactic Vaccination.—The object of prophylactic vaccination is to increase the resistance of the recipient to specific infection. This is accomplished by reinforcing the natural initial defenses of the body with specific antibodies, generated in the host in response to the injection of the specific microörganism as a vaccine. In prophylactic vaccination the host has ample time to work over the vaccine, and by prolonging the treatment through repeated graduated doses the maximum degree of immunity may be expected. To attain the maximum immunizing effect the bacteria of the vaccine should be as near their normal state as possible, that is, they should be endowed with all the antigenic properties they possess in the natural disease produced by them in the host.

Following the brilliant work of Jenner with cowpox vaccine and the epoch-making observations of Pasteur, observers are fairly agreed that the best results from prophylactic vaccination are obtainable only by the use of an attenuated living virus. The action of such a living virus is, as Theobald Smith[1] has aptly expressed it, "a multitude of feeble blows, each of which produces an immunological response." The dangers attending the use of attenuated viruses, however, ordinarily preclude their employment, due to inability to control the virulence of attenuated cultures. The possibility of creating carriers cannot be overlooked. For this reason killed cultures are almost invariably selected.

It is, of course, impossible to utilize an autogenous vaccine, but for purposes of immunization a polyvalent vaccine is indicated. The action of a dead virus is limited practically to a single immunological response, hence the need of repeated inoculations.

Therapeutic Vaccination.—In chronic, long-drawn out focal or local infections, the invading microbes are either holding their own or gaining the ascendency and the object of bacterial vaccination is to turn the tables on the invaders. The products of immunization must be used at once, and the organisms comprising the vaccines for this purpose cannot ordinarily be as resistant as their originals in the host. The underlying principle of therapeutic vaccination, according to

[1] Jour. Am. Med. Assn., 1913, lx, 1591.

Wright,[1] is to exploit the normal tissues of the body in the interest of the infected tissue. For this purpose, microbes similar to those causing the infection (autogenous organisms) are inoculated into some other part of the body. This inoculation is not, to use Wright's phraseology, a mere replica of the original infection; there are two important points of difference: (1) the microbes of the vaccine are killed, so that their multiplication within the host is impossible; (2) the dose of vaccine must be so regulated that the tissues of the host at the site of inoculation and elsewhere must inevitably win. Victory of the host is brought about through the elaboration of specific antibodies generated in the healthy tissues on a scale more than adequate to bring about a destruction of the organisms introduced into the healthy tissue. The surplus of the specific antibodies will find its way, through blood and lymph channels, to the focus of infection, and will reinforce the partially depleted defensive forces which have ineffectually opposed the invading organisms.

It should be borne in mind that vaccine therapy cannot be reasonably applied unless an exact bacteriological diagnosis has been made. The immunizing effects of vaccines are definitely limited by the ability of the normal tissues of the patient to produce antibodies; to inject too frequently or in too large doses may not only be barren of results—it may result in a decrease rather than an increase of resistance to infection.

It is essential for the best results of vaccination that the focus of infection be so situated anatomically that the newly formed antibodies be drawn to the infected area by the production of local hyperemia. Infections of long standing naturally respond to treatment more slowly than newly acquired infections.

Preparation of Vaccines.—Much discussion has arisen concerning the use of autogenous vaccines as compared with stock or polyvalent vaccines. So little is actually known of what vaccines may accomplish in the body that it is impossible to answer this question definitely. It is desirable, however, to retain in the vaccine all possible antigenic properties which were possessed by the organism in the body. It is a well-known fact that certain kinds of organisms rapidly lose their ability to produce disease when they are grown for any length of time outside the body. Others retain their virulence for some time. This would appear to indicate that stock vaccines of the former would be unsatisfactory, while stock vaccines of the latter might be

[1] *Proc. Roy. Soc. of Med.*, London, 1910, iii.

more successful. It is a safe general rule to state that an autogenous vaccine is desirable.

The preparation of vaccine is carried out as follows:

1. Obtain pure cultures of the organisms from the lesion or whatever material is available. The details of culture vary with the type of organism that is expected.

2. Inoculation of the pure culture, or cultures in the event of multiple infection, in suitable media to furnish the desired amount of growth.

3. Removal of the growth, with sterile precautions, to a sterile container, such as a test-tube containing sterile glass beads. This is accomplished by washing the growth from the medium into sterile saline solution: 5 to 10 c.c. of salt solution are required for an ordinary agar slant culture. When enough growth is accumulated it is transferred to the sterile test-tube, being careful that no organisms contaminate the upper part, else they may escape sterilization.

4. Sterilization: Heat the bacterial suspension in a water-bath. Usually one hour at 60° to 65° C. suffices. Care must be taken that the level of water in the water-bath is well above that of the level of the suspension in the test-tube.

5. Test sterility of the suspension. Inoculate suitable media and observe the absence of growth. In skin infections it is sometimes desirable to exclude the presence of the tetanus bacillus.

6. Shake the suspension vigorously to distribute the organisms uniformly in it.

7. Standardize: Determine the number of bacteria in a cubic centimeter. This is very simply accomplished by thoroughly mixing equal volumes of freshly drawn blood and bacterial emulsion in a pipette, spreading the mixture on a microscope slide, drying and staining it with Wright's or Jenner's stain. Determine by actual counting in a number of fields the proportion of bacteria to red cells. Knowing the number of red blood cells in a cubic centimeter of blood (5,000,000,000) and the proportion of bacteria to red blood cells, it is a simple matter to determine the number of bacteria in the suspension.

A more accurate procedure is to draw up one volume of vaccine in the erythrocyte pipette of a hemocytometer, dilute to the 101 mark with a dilute solution of fuchsin or other suitable stain, mix and transfer to the counting chamber. An enumeration of the bacteria is made in precisely the same manner that a blood count is made.

8. Dilute the suspension to the required degree with phenol, so that the finished vaccine shall contain 0.25 to 0.5 per cent. of it. This is the finished vaccine.

9. Redetermine sterility if necessary.

Sensitized Vaccines.—Killed bacteria which have been immersed in a specific serum—sensitized vaccines—are said to be less liable to produce general and local reactions. The immunity developed in response to the injection of these sensitized vaccines is said to appear more rapidly, and doses thirtyfold those of unsensitized vaccines may be injected without serious effect.

The Injection.—The skin at the site of injection is cleaned with soap and water and then with alcohol; or better, after carefully drying it is painted with tincture of iodin. The required amount of vaccine is injected subcutaneously through this area, from a sterile syringe.

The Dosage and Frequency of Injection.—It is advisable to begin with small doses of vaccine, quantities which past experience has shown to do no harm so far as can be determined by clinical evidence, and to increase the size of the dose gradually, the injections usually being given at intervals of about a week. If no change results from the treatment, larger doses may be tried. If the symptoms become aggravated the doses should be diminished and given at less frequent intervals. Generally speaking, in the more acute cases smaller doses should be selected to begin with, larger doses being reserved for the more chronic cases. The amounts of vaccine to be injected vary widely according to different investigators. Generally speaking, the following figures are fairly representative:

	Minimum.[1]	Maximum.[1]	Average.[1]
Staphylococcus	5.0	1000	25
Streptococcus	2.5	100	25
Pneumococcus	2.5	100	25
Gonococcus	2.5	300	30
Coli	5.0	1000	100
Pyocyaneus	5.0	1000	100

Indications for the Use of Bacterial Vaccine.—Generally speaking, bacterial vaccines are contraindicated in acute disease, but may be employed in practically any localized infection, or an infection which has become chronic.[2]

[1] Figures represent millions of organisms.

[2] An excellent discussion of the present status of vaccine therapy is that of Theobald Smith, An Attempt to Interpret the Present-day Use of Vaccines, Jour. Am. Med. Assn., 1913, lx, 1591.

Results.—Opinions differ widely as to the value of vaccines. According to the theory of bacterial vaccination, subacute and chronic infections which are localized should give the best results, and such indeed appears to be the case. For example, a streptococcus septicemia abates and leaves a joint involvement or a heart valve vegetation. Vaccine therapy has a better chance of producing results during this secondary stage than during the earlier acute septicemic stage. Gonorrheal arthritis, pneumonias which resolve by lysis, pus sinuses, and localized colon infections are suitable for treatment. In acute inflammations of the mucous membranes of the intestines, bladder, throat, etc., the results have been either negative or unsatisfactory.

So far as specific organisms are concerned, staphylococcus vaccines give the most constant and satisfactory results. Furuncles, severe carbuncles, some cases of acne, and even low-grade staphylococcus septicemias yield rather readily to vaccine therapy with this organism. Streptococcic and pneumococcic infections are much more resistant, generally speaking, to vaccine treatment than are staphylococcus infections.

CHAPTER IX.

THE MICROSCOPIC AND CULTURAL STUDY OF BACTERIA.

METHODS FOR THE MICROSCOPIC STUDY OF BACTERIA.

BACTERIA may be examined directly under the higher powers of the microscope for their morphology, motility, arrangement, method of reproduction, and their behavior in specific sera, or they may be stained with various anilin dyes and chemicals to bring out details of structure or composition, and their relation to various tissues in pathological processes.

Glass slides and cover-glasses are conveniently used for this purpose. Microscopic slides should be made from clear, colorless glass. Cover-glasses should be made of thin glass. The available working distance of oil-immersion lenses is somewhat less than 1.5 mm., consequently cover-glasses should not measure more than 1 mm. in thickness as a maximum limit. Number 1 cover-glasses are suitable for bacteriological work.

Glass slides and cover-glasses are best cleaned in a mixture of potassium bichromate, 1 part; water, 4 parts; sulphuric acid, 6 parts. The bichromate is dissolved in the water with the aid of heat and cooled; the acid is added slowly with constant stirring. Immersion in this mixture for twenty-four hours removes dirt and grease from both slides and cover-glasses. The cleaned glassware is removed from the cleansing bath and washed with running water until neutral to litmus paper. It is stored either in slightly ammoniacal alcohol, or dried with a soft cloth, previously freed from grease by boiling in a 5 per cent. sodium carbonate solution.

I. **Examination of Living Bacteria.**—A. **Hanging Drop.**—The motility, shape, and size of bacteria may be studied in a "hanging-drop" preparation. A drop of fluid from a bacterial culture in liquid media

FIG. 10.—Hollow-ground slide for hanging drop.

is transferred to the center of a thin cover-glass. If the growth is upon solid media a drop of physiological salt solution[1] is placed upon the center of the cover-glass as before, and a very small amount of the culture is removed with a platinum needle and emulsified in it. Next, the rim of the concavity in a "hollow-ground slide" is ringed with vaselin and the cover-glass is inverted over it in such a manner that the drop is suspended in the hollow, but touches neither the sides not the bottom. The vaselin seals the preparation, causing it to adhere to the slide, and also prevents evaporation. The preparation is now ready for microscopic examination. The one-sixth or one-eighth-inch objective should be used, with the diaphragm partly closed to reduce the intensity of illumination. It is desirable to focus first upon the edge of the drop; the edge is sharply defined and readily located. Bacteria are usually more numerous at the edge than in the center of the drop.

B. **Hanging Block.**—It is desirable occasionally to follow the development of bacteria through several generations, to study the germination of spores, or to examine special structures within the bodies of individual organisms. The hanging-drop method is unsuited for this purpose, which presupposes immobilization of the organism. Hill[2] has invented an ingenious modification of the hanging-drop method, the hanging block, which fulfils this requirement. His directions for preparing it are:

"Pour melted nutrient agar into a Petri dish to the depth of about one-eighth or one-quarter inch. Cool this agar and cut from it a block

[1] Physiological salt solution is prepared by dissolving 8.5 grams NaCl in distilled water 1000 c.c.

[2] Jour. Med. Research, March, 1902, vii, 202.

about one-quarter inch to one-third inch square and of the thickness of the agar layer in the dish. This block has a smooth upper and under surface. Place it, under side down, on a slide and protect it from dust. Prepare an emulsion, in sterile water, of the organism to be examined if it has been grown on a solid medium, or use a broth culture; spread the emulsion or broth upon the upper surface of the block as if making an ordinary cover-slip preparation. Place the slide and block in a 37° C. incubator for five to ten minutes to dry slightly. Then lay a clean sterile cover-slip on the inoculated surface of the block in close contact with it, carefully avoiding air-bubbles. Remove the slide from the lower surface of the block and invert

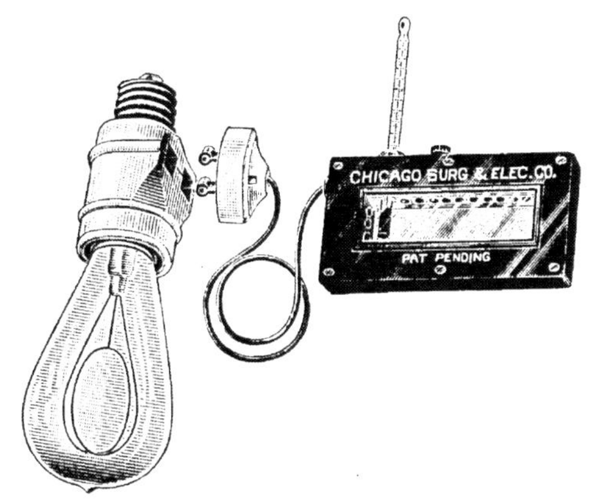

FIG. 11.—Warm stage, electrically heated, for the cultivation of bacteria.

the cover-slip so that the agar block is uppermost. With a platinum loop, run a drop or two of melted agar along each side of the agar block, to fill the angles between the sides of the block and the cover-slip. This seal hardens at once, preventing slipping of the block. Place the preparation in the incubator again for five or ten minutes, to dry the agar-agar seal. Invert this preparation over a moist chamber and seal the cover-slip in place with white wax or paraffin. Vaselin softens too readily at 37° C., allowing shifting of the cover-slip. The preparation may then be examined at leisure.[1]

[1] A light, detachable, electrically heated warm-stage incubator, manufactured by the Chicago Surgical and Electrical Company according to specifications furnished by the writer is very satisfactory for this purpose. Bacteria may be maintained constantly at any desired temperature between that of the room and 45° C. for several days, and observed continuously without difficulty. If the warm-stage incubator is attached to a graduated mechanical stage, many individual bacteria may be observed in the same preparation by recording their respective positions as indicated on the graduated rectilinear stage verniers.

C. **Dark Field Illumination and Ultramicroscopic Examination.**—For the study of very minute particles in suspension, the ultramicroscope of Siedentoff and Zsigmondy[1] has been used, but the dark-ground illumination apparatus of Reichert,[2] a much simpler device, readily adjusted to any microscope, has largely supplanted it for bacterial examinations. With the Reichert apparatus the flagella of bacteria and other structures of low-refractive index may be observed. Treponema pallidum in fresh smears from lesions is readily seen with the dark ground illuminating apparatus.

D. **Intra Vitam Staining.**—Nakanishi[3] has applied the method of *intra vitam* staining to the study of spores and granules in living bacterial cells. The method consists essentially in emulsifying a small amount of bacterial growth in normal salt solution containing sufficient aqueous methylene blue to impart a distinct blue color to the solution. The preparation is viewed as in the hanging-drop slide. The organisms absorb sufficient dye to impart to them a faint color, and granules within their bodies frequently stain with moderate intensity. The development of spores from pre-sporogenic granules may be studied by this method.

II. **Staining of Bacteria.**—A. **Chemistry of Stains.**—The stains of value for coloring bacteria are almost exclusively anilin dyes which contain one or, more commonly, several benzene rings. Their coloring properties have been shown to depend upon two distinct radicals; double-bonded atoms as $C = C$, $C = O$, $C = N$, $N = N$, known as chromophoric groups, and auxochromic groups, which impart to or intensify the color. Of the chromophoric groups, NH_2 and $-OH$ are the more important. The latter form salts which may be either basic or acid in character. Bacteria usually stain best with basic dyes, as do nuclei of higher plant and animal cells.

The chemistry of the staining process itself is a matter of discussion. It was formerly held that the cell protoplasm united chemically with the stain as an acid unites with a basic salt, but later investigations, particularly those of Michaelis,[4] are not in harmony with this view. It is probable that the physical state of the cell membrane as well as the composition of the cytoplasm play a part in the staining process.

B. **Preparation of Stains.**—Stains prepared by Grübler or Merck are commonly used for the staining of bacteria. They are conveniently

[1] Zeit. f. wissenschaftl. Mikroskopie, 1909, xxvi, 391.
[2] München. med. Wchnschr., 1906, 2351; Hyg. Rund., 1907, No. 18; *Cent.* f. Bakt. Orig., 1909, li, 14.
[3] Munchen. med. Wchnschr., 1900, No. 6; *Cent.* f. Bakt., 1901, xxx, 97, 145, 193, 225.
[4] Einfuhrung in die Farbstoffchemie, 1902, Berlin.

kept in stock as saturated aqueous or alcoholic (96 per cent.) solutions. The solubility of stains in water and in alcohol respectively varies, but, as a rule, the solubility in alcohol is greater than that in water. Saturated solutions of anilin dyes are unsuited for the staining of microörganisms, but they are more stable than diluted solutions provided they are kept in tightly-stoppered bottles away from the light. Dilutions of saturated solutions are prepared as they are needed for current use.

C. **Technic of Staining Bacteria.**—1. Preparation of a film of bacteria for staining: A drop of a culture of bacteria from a fluid medium —as broth—is removed with a platinum loop and spread upon a clean cover-glass or glass slide. Bacteria from a solid medium are emulsified in a small drop of water on the slide.[1]

2. The film of bacteria is allowed to dry in the air; evaporation may be hastened in the incubator at 37° C.

3. The air-dried film is next fixed by passing it *once* slowly through the flame of a Bunsen burner, film side upward; about one-half second's exposure to the flame suffices; a longer exposure destroys or changes the staining properties of the organisms.

4. Staining: A 5 per cent. aqueous solution of methylene blue, fuchsin, or gentian violet, prepared by adding 5 c.c. of filtered saturated stock solution to 95 c.c. of distilled water, is used. The slide or cover-glass is flooded with the desired stain, and after one to five minutes, depending upon the intensity of the stain used, the excess is poured off and the preparation is washed thoroughly with water. The residual moisture is removed with filter paper or by air-drying, and a small drop of Canada balsam (dissolved in xylol) is placed in the center of the stained area. The film is finally enclosed between a slide and a cover-glass.

D. **Intensive Stains for Bacteria.**—Simple aqueous or alcoholic solutions of anilin dyes are frequently inefficient for staining bacteria and resort is made to intensified stains. One of the most useful of the intensified stains is Löffler's alkaline methylene blue, prepared in the following manner:

1 to 10,000 aqueous solution of potassium hydroxide[2]	70 c.c.
Saturated alcoholic solution methylene blue	30 c.c.

[1] It is essential that the emulsion shall be but faintly opalescent when viewed by reflected light; a distinct clouding indicates that too many organisms have been added, in which event the preparation will be found to be unsatisfactory.

[2] Conveniently prepared by dissolving 1 gram of KOH in 100 c.c. distilled water and adding 1 c.c. of this solution to 99 c.c. of distilled water.

Fixed films of bacteria are stained from one to five minutes with this stain, or the films are flooded with the stain and heated until steam rises (not boiled) for one to three minutes. It is difficult to overstain with Löffler's methylene blue unless evaporation takes place to such a degree that the stain dries on the slide. The stain is washed off with water, dried, and mounted.

E. **Stains for Special Structures of the Bacterial Cell.**—1. *Spores.*—(*a*) Flood fixed film of bacteria with carbol-fuchsin[1] and steam (not boil) for five minutes.

(*b*) Wash thoroughly in running water.

(*c*) Decolorize with 1 per cent. sulphuric acid until excess stain is removed.

(*d*) Wash thoroughly in running water.

(*e*) Flood with saturated aqueous solution methylene blue (or Löffler's alkaline methylene blue) and allow to stain one minute.

(*f*) Wash in water, dry, and mount.

Spores stain red, vegetative cells blue.

Möller's Spore Stain:[2]—(*a*) Suspend the fixed film of bacteria in chloroform for two minutes.

(*b*) Wash with water.

(*c*) Flood with 5 per cent. chromic acid solution for two minutes.

(*d*) Wash thoroughly in running water.

(*e*) Flood with carbol-fuchsin and steam for five minutes.

(*f*) Wash thoroughly in water.

(*g*) Decolorize with 1 per cent. sulphuric acid until excess stain is removed.

(*h*) Wash thoroughly in water.

(*i*) Flood with Löffler's alkaline methylene blue and allow to stain one minute.

(*j*) Wash in water, dry, and mount.

Spores stain red, vegetative cells blue.

2. *Capsule Stains.*—*Welch Method.*[3]—(*a*) Fixed films of bacteria are flooded with glacial acetic acid for a few seconds.

(*b*) The acid is poured off and the preparation is washed two or three times with anilin oil gentian violet, then flooded with the stain, which is allowed to act for three to five minutes.

(*c*) Wash with 2 or 3 per cent. aqueous solution of sodium chloride.

(*d*) Mount in salt solution and examine.

Capsules faint purple, bacterial body deep purple.

[1] Saturated alcoholic solution of basic fuchsin, 10 c.c.; 5 per cent. aqueous phenol solution, 90 c.c.

[2]

[3]

Hiss's Method.[1]—(*a*) Place a drop of sterile blood serum upon a slide and emulsify bacteria in it.

(*b*) Dry in the air and fix by heat.

(*c*) Flood smear with 5 per cent. solution[2] of gentian violet or fuchsin; steam for thirty seconds.

(*d*) Remove excess of stain by washing in a 20 per cent. solution of copper sulphate.

(*e*) Dry with filter paper. Mount and examine.

Capsule faint pink or purple; body of organism deep red or purple.

Rosenow Method.[3]—(*a*) Prepare the smear on perfectly clean cover-glass.

(*b*) When smear is nearly dry, cover with 10 per cent. aqueous solution of tannic acid for twenty seconds.

(*c*) Wash with water; remove moisture with filter paper.

(*d*) Flood with anilin-oil gentian violet and steam gently for thirty to sixty seconds.

(*e*) Wash thoroughly in water.

(*f*) Cover with Gram-iodin solution, one minute.

(*g*) Decolorize with 96 per cent. alcohol.

(*h*) Stain one or two minutes with a saturated (60 per cent.) alcoholic solution of Grübler's eosin.

(*i*) Wash in water; dry and mount in balsam.

Capsules pink; bacteria blue.

3. *Polar Bodies.—Neisser Stain.*[4]—

Preparation of Stain.—

Solution A—	Methylene blue	1 gram
	Ninety-six per cent. alcohol	20 c.c.
	Glacial acetic acid	50 c.c.
	Distilled water	950 c.c.
Solution B—	Bismarck brown	1 gram
	Distilled water	500 c.c.

(*a*) The air-dried film, fixed by heat, is flooded with solution A for three to five seconds.

(*b*) Wash with water.

(*c*) Flood with solution B for five seconds.

(*d*) Wash with water, dry, and mount.

Polar bodies stain blue; bacterial cells brown.

[1] Jour. Exp. Med., 1905, vi, 338.
[2] Saturated alcoholic solution of the dye, 5 c.c.; distilled water, 95 c.c.
[3] Jour. Infect. Dis., 1911, ix, 1.
[4] Ztschr. f. Hyg., 1897, xxiv, 443.

4. *Flagella.—Preparation of Film.*[1]—(*a*) Add enough of an eighteen to twenty-four-hour agar culture to a test-tube containing 5 c.c. of sterile salt solution to produce a faint turbidity in the upper half of the solution.

(*b*) Incubate at 37° C. for thirty to sixty minutes.

(*c*) Place two or three loopfuls of the suspension upon a perfectly clean cover-glass and allow to dry spontaneously in the air or in the incubator.

Do not attempt to spread the films with the platinum loop; agitation breaks off flagella.

Staining Flagella.—Pittsfield's Flagella Stain.—[2]

Preparation of Stain.—

(*a*) Mordant:

Tannic acid, 10 per cent. aqueous solution	10 c.c.
Mercuric chloride, saturated aqueous solution	5 c.c.
Alum, saturated aqueous solution	5 c.c.
Carbol fuchsin	5 c.c.

(*b*) The Stain:

Alum, saturated aqueous solution	10 c.c.
Carbol fuchsin	5 c.c. or,
Gentian violet	2 c.c.

Flood the dried and fixed film with the mordant and steam gently for one minute. Wash in running water, air-dry and flood with the stain. Heat gently two minutes, wash thoroughly in water, air-dry and mount.

F. **Differential Stains for Bacteria.**—1. *Gram Stain.*[3]—A most important differential method of staining bacteria is that of Gram. Bacteria may be divided into two groups: those which retain the initial stain —Gram-positive organisms—and those which fail to retain the initial stain but color with the counter stain—Gram-negative bacteria.

It was believed formerly that the organisms which retained the initial stain—the Gram-positive bacteria—contained within their protoplasm, a substance of unknown composition which united chemically with gentian violet (or other pararoseanilin dye) and iodin to form a compound relatively insoluble in alcohol. Gram-negative bacteria did not contain the hypothetical substance, which, in association with the dye and iodin, was insoluble in alcohol. Treatment of the latter group with alcohol, therefore, would remove the

[1] Kendall, Jour. Applied Microscopy, 1901, v, 1836.
[2] Medical News, September 7, 1895.
[3] Gram, Fortschr. d. Med., 1894, ii.

gentian violet, leaving them unstained. In the unstained condition the organisms were colored with the second or counter stain. Subsequent investigation has largely discredited this view. It has been shown by Kruse[1] that the cytoplasm of Gram-positive bacteria is more resistant to autolysis, to the action of trypsin, and to solution in dilute KOH than that of Gram-negative organisms, probably because the cytoplasm of the former is less permeable to these various reagents than is that of the latter. Eisenberg,[2] and Guerbet, Mayer and Schaeffer[3] have advanced the hypothesis that Gram-positiveness is due to the lipoidal content of the cell membrane (ectoplasm) and specifically to unsaturated fatty acids and phosphatids. The addition of iodin, according to this theory, through the formation of alcohol-insoluble combinations with the lipoids in the ectoplasm, renders the cell wall impermeable to alcohol and thus prevents removal of the dye which has already penetrated into the cell contents.

Preparation of Stain:

Solution A—	Saturated aqueous solution of anilin[4]	90 c.c.
	Saturated alcoholic solution of gentian violet	10 c.c.
	Five per cent. aqueous solution of carbolic acid	90 c.c.
	Saturated alcoholic solution of gentian violet	10 c.c.

The above solutions are unstable, but retain their tinctorial value for two or three days if they are kept stoppered.

Solution B[5]—	Distilled water	300 c.c.
	Potassium iodide	2 grams
	Iodin crystals	1 gram
Solution C—	Bismarck brown, saturated aqueous solution	10 c.c.
	Distilled water	90 c.c.

Procedure.—(*a*) Prepare and fix film of bacteria in the usual manner.

(*b*) Flood with anilin-oil gentian violet (or carbolic gentian violet) and stain five minutes.

(*c*) Pour off excess of stain and flood with iodin solution.

(*d*) Decolorize with 96 per cent. alcohol until no more stain can be removed.

(*e*) Wash thoroughly in water.

(*f*) Counterstain with Bismarck brown[6] for two minutes.

(*g*) Wash in water, dry, and mount.

[1] München. med. Wchnschr., 1910, p. 685.

[2] Cent. f. Bakt., 1909, xlix, 465; 1910, li, 115; liii, 481, 551; lvi, 183.

[3] Compt. rend., Soc. biol., lxviii, 353.

[4] Three c.c. of anilin oil are shaken for several minutes in 100 c.c. of distilled water. The solution is filtered through filter paper to remove the undissolved anilin.

[5] This iodin solution is variously known as Gram's iodin solution or Lugol's solution.

[6] Dilute aqueous fuchsin, 1 to 10, may be used in place of Bismarck brown.

Bacteria which retain the initial stain—Gram-positive bacteria—are colored dark purple or blue; those which fail to retain the initial stain—Gram-negative bacteria—are brown, or bright pink if fuchsin is used as a counterstain.

2. *Stains for Acid-fast Bacteria.—Ziehl-Neelsen Method.*[1]—(*a*) Stain dried and fixed smear with carbol fuchsin, as described on page 180.

(*b*) Wash thoroughly with water.

(*c*) Decolorize with acid alcohol[2] until no more color can be washed out.

(*d*) Wash with water.

(*e*) Counterstain lightly with Löffler's alkaline methylene blue.

(*f*) Wash, dry, and mount.

Acid-fast bacilli and spores red; all others blue.

3. *Fränkel-Gabbet Method.*[3]—(*a*) Stain with carbol fuchsin as in the Ziehl-Neelsen method and wash in water.

(*b*) Decolorize and counterstain simultaneously with the following solution:

Methylene blue	2 grams
Water	75 c.c.
Sulphuric acid	25 c.c.

The counterstain is allowed to act for one minute.

(*c*) Wash, dry, and mount.

4. *Polychrome Stains.*—Polychrome stains are of special value for the examination of exudates, body fluids or tissues in which the histological relations of bacteria are to be investigated. These stains, or modifications of them, are also useful in the study of treponemata, spirochetes, and protozoa.

Wright's Stain.[4]—*Preparation.*—"To a 0.5 per cent. aqueous solution of sodium bicarbonate add methylene blue (B.X., or 'medicinally pure') in the proportion of 1 gm. of the dye to each 100 c.c. of the solution. Heat the mixture in a steam sterilizer at 100° C. for one full hour, counting the time after the sterilizer has become thoroughly heated. The mixture is to be contained in a flask, or flasks, of such size and shape that it forms a layer not more than 6 cm. deep. After heating the mixture is allowed to cool, placing the flask in cold water if desired, and is then filtered to remove the precipitate which

[1] Ziehl, Deutsch. med. Wchnschr., 1882, 451; Neelsen, Fort. d. med., 1885, 200.
[2] Ninety per cent. alcohol containing 3 per cent. by volume of hydrochloric acid.
[3] Fränkel, Berl. klin. Wchnschr., 1884; Gabbet, Lancet, 1887.
[4] Mallory and Wright, Pathological Technic, 1915, 6th ed., p. 382.

has formed in it. It should when cold have a deep purple red color when viewed in a thin layer by transmitted yellowish artificial light. It does not show this color while it is warm.

"To each 100 c.c. of the filtered mixture add 500 c.c. of a 0.1 per cent. aqueous solution of 'yellowish, water-soluble' eosin and mix thoroughly. Collect on a filter the abundant precipitate which immediately appears. When the precipitate is dry, dissolve it in methylic alcohol (Merck's 'reagent') in the proportion of 0.1 gm. to 60 c.c. of the alcohol. In order to facilitate solution the precipitate is to be rubbed up with alcohol in a porcelain dish or mortar with a spatula or pestle.

"This alcoholic solution of the precipitate is the staining fluid. It should be kept in a well-stoppered bottle because of the volatility of the alcohol. If it becomes too concentrated by evaporation and thus stains too deeply, or forms a precipitate on the blood smear, the addition of a suitable quantity of methylic alcohol will quickly correct such faults. It does not undergo any other spontaneous change than that of concentration by evaporation.

"A most important fault met with in the working of some samples of this fluid is that it fails to stain the red blood corpuscles a yellow or orange color, but stains them a blue color which cannot readily be removed by washing with water. This fault is due to a defect in the specimen of eosin employed. It can be eliminated by using a proper 'yellowish, water-soluble' eosin."

Method of Staining.—(*a*) Unheated air-dried films[1] are covered with the stain, which is allowed to act for one minute.

(*b*) Add an *equal volume* of distilled water to the stain and allow to stand for three minutes.

(*c*) Wash in water for thirty seconds, or until a pink color develops.

(*d*) Dry rapidly with filter paper and mount in balsam.[2]

Giemsa Method.[3]—*Preparation of Stain:*

Azur II (eosin)	3.0 grams
Azur II	0.8 grams
Glycerin, *C. P.*	250 c.c.
Neutral absolute methyl alcohol	250 c.c.

The dyes are dissolved in the glycerin at 60° C.; the alcohol, warmed to 40° C., is then added, thoroughly mixed by shaking, and allowed to cool slowly to room temperature, then filtered. Immediately before

[1] Films more than a few hours old do not stain as well as fresh ones.
[2] The balsam must be neutral in reaction.
[3] Giemsa, *Cent. f. Bakt.*, 1904, xxxvii, 308.

use, 10 c.c. of distilled water are slightly alkalinized by the addition of two drops of a 10 per cent. solution of sodium carbonate, and exactly ten drops of the stain are then added.

Staining with Giemsa Solutions.—(*a*) Films are fixed by immersion in neutral absolute methyl alcohol for one minute, air-dried, and covered with the diluted stain, which is allowed to act for fifteen to twenty minutes when ordinary exudates and bacteria are used; for one to three hours if Treponemata or Negri bodies are sought for.

(*b*) Wash in water, dry and mount.

5. *W. H. Smith's Solution Stain.*—(*a*) Stain the fixed smear with anilin oil gentian violet for one minute.

(*b*) Wash with water.

(*c*) Flood with Gram-iodin solution for thirty seconds.

(*d*) Decolorize with 95 per cent. alcohol.

(*e*) Wash with ether for a few seconds.

(*f*) Flood with absolute alcohol for five seconds.

(*g*) Stain with saturated aqueous solution eosin for one to two minutes.

(*h*) Wash with absolute alcohol for a few seconds.

(*i*) Clear with xylol.

(*j*) Mount in balsam.

III. **Staining Bacteria in Tissues.**—Paraffin sections are preferable, partly because very thin sections may be cut; chiefly because celloidin stains somewhat with the stains ordinarily used.

The Gram-Weigert Stain for Bacteria in Tissues.[1]—(*a*) Stain paraffin sections with anilin oil methyl violet for five to twenty minutes.

(*b*) Wash in water to remove excess of stain.

(*c*) Gram-iodin solution for one minute.

(*d*) Wash in water to remove excess of iodin.

(*e*) Decolorize with several changes of absolute alcohol until no more color comes out.

(*f*) Clear section in xylol.

(*g*) Mount in neutral xylol balsam.

Mallory and Wright Modification for Celloidin Sections.[2]—(*a*) Stain sections with lithium carmine for two to five minutes.

(*b*) Remove excess of stain with acid alcohol.

(*c*) Wash in water.

(*d*) Dehydrate in 95 per cent. alcohol.

[1] Mallory and Wright, Pathological Technic, 6th ed., 1915, p. 432.
[2] Ibid.

(*e*) Expose to ether vapor to fix section to slide.

(*f*) Stain with anilin oil methyl violet for five to twenty minutes.

(*g*) Remove excess stain with normal salt solution.

(*h*) Gram-iodin solution for one minute.

(*i*) Remove excess iodin with water.

(*j*) Remove moisture as thoroughly as possible with filter paper.

(*k*) Dehydrate in several changes of anilin oil.

(*l*) Clear with several changes of xylol.

(*m*) Mount in neutral xylol balsam.

Staining Tubercle Bacilli in Tissues.—(*a*) Paraffin sections are covered with carbol-fuchsin and steamed gently for five minutes.

(*b*) The excess stain is removed with water.

(*c*) Decolorize and counterstain with Gabbet methylene-blue sulphuric acid stain about one minute.

(*d*) Remove excess of stain and acid with water.

(*e*) Dehydrate with absolute alcohol.

(*f*) Clear section in xylol.

(*g*) Mount in xylol balsam.

Staining Actinomyces in Tissues—Mallory Method.[1]—(*a*) Stain paraffin sections with saturated aqueous eosin for ten minutes.

(*b*) Remove excess stain with water.

(*c*) Stain with anilin oil methyl violet for two to five minutes.

(*d*) Remove excess stain with normal salt solution.

(*e*) Remove excess water with filter paper.

(*f*) Clear in anilin oil.

(*g*) Remove anilin oil with several changes of xylol.

(*h*) Mount in neutral xylol balsam.

The clubs stain pink, the filaments blue.

IV. **Methods and Media for the Cultivation of Bacteria.**—One of the most important procedures in bacteriology is the preparation of nutritive media in which the morphology, chemistry, and cultural characteristics of the organism may be studied; furthermore, it is possible by cultural methods to separate one type of bacterium in pure culture from associated organisms, and to study its reactions apart from all contaminating microörganisms. The technic of isolating and cultivating bacteria is exacting at every step of the process, from the preparation of glassware to the selection of suitable nutritive media, and their preparation requires not only scrupulous cleanliness; it necessitates a most rigorous maintenance of sterility.

[1] Loc. cit., p. 433.

Bacterial cultivation is usually carried out in glass vessels—test-tubes, flasks, fermentation tubes, and Petri dishes—because glass is transparent and permits an unobstructed view of the reactions taking place within. It is obvious that glassware employed in bacterial laboratories must be chemically and bacteriologically clean.

Preparation of Glassware.—The method to be employed in the cleaning of glassware depends somewhat on the purpose for which it is

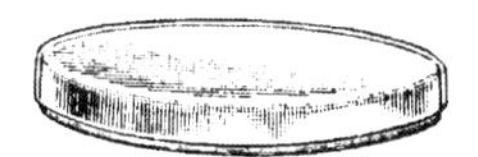

FIG. 12.—Petri dish.

used. New glassware frequently contains alkali, which is readily neutralized by diluted acid, hydrochloric or sulphuric. Glassware that has contained cultures of bacteria is first sterilized in the autoclave to remove all danger of infection, then immersed in a strong solution of soap-powder and soap-suds maintained at a boiling temperature for half an hour. The adherent media is removed with a brush or swab; a final thorough rinsing in clear water removes all

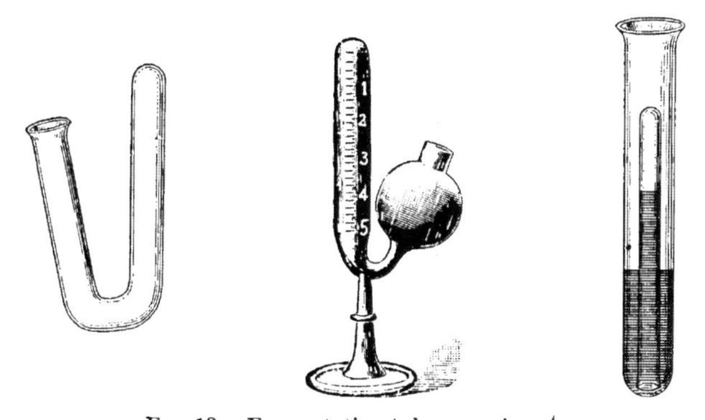

FIG. 13.—Fermentation tubes—various types.

traces of soap. Very dirty glassware or glassware in which chemical determinations are to be made should be cleaned in chromic acid solution, which is prepared by adding a saturated aqueous solution of potassium bichromate to a 1 to 3 dilution of sulphuric acid. Twenty-four hours' exposure to chromic acid removes all traces of organic matter, as a rule. Following the acid bath the glassware is thoroughly rinsed in clear water and dried.

The cleaned glassware—test-tubes, flasks, or fermentation tubes—is then stoppered with non-absorbent cotton—cotton batting—which has a long staple or fiber. The cotton plugs must be carefully fitted—neither too loose, which would permit of the passage of adventitious microörganisms, nor too tight, for obvious reasons. The cotton plugs are conveniently prepared from a layer of cotton batting about two inches square (for a test-tube of ordinary diameter, about 15 mm.), which is laid squarely over the orifice. The center of the square is gently pushed down into the neck of the tube for a distance of about three-fourths to one inch; sufficient cotton protrudes from the tube to be conveniently grasped by the fingers and removed. It is frequently advisable to cover the cotton plugs with two or three layers of filter paper, which prevents an accumulation of dust on the cotton. Wide-mouthed containers are sealed with several layers of unglazed paper fastened in place with a piece of twine. Flasks are frequently not plugged with cotton; the neck is simply covered by an inverted beaker of appropriate size.

Glassware should always be sterilized before media is placed in it; this is readily accomplished by dry heat. A hot-air sterilizer is used, in which a temperature of 180° C. is maintained for one hour. A higher temperature must be avoided, to prevent charring of cotton plugs. The heat must be increased gradually and diminished gradually, to prevent cracking of the glass. By this process not only is the utensil rendered sterile, the plugs of cotton retain their shape when withdrawn, as well.

A majority of the bacteria pathogenic for man and many parasitic and saprophytic forms as well require relatively complex organic compounds containing carbon, hydrogen, nitrogen, and oxygen, together with other elements for their nutrition. These foodstuffs provide both the structural and fuel requirements of the organism, as explained in the chapter on Bacterial Metabolism. Experience has shown that a medium containing meat infusion, peptone, and salt is a satisfactory one for many bacteria. This medium may be enriched by the addition of various ingredients to meet the requirements of the more fastidious organisms.

Meat infusion is prepared from finely comminuted lean meat[1] freed from fat. 500 grams of meat are intimately mixed with 1000 c.c. water and allowed to infuse over night in the refrigerator. It is then strained

[1] Beef hearts make a very satisfactory meat infusion and their cost is much less than the better cuts of meat.

through several layers of cheese-cloth, the volume recorded, then heated to boiling. The coagulum which forms is removed by filtration through filter paper and the clear, amber-colored fluid, after restoring the loss due to evaporation, is run into flasks and sterilized in an autoclave at 15 pounds' pressure for fifteen minutes. This plain meat infusion contains but little protein; it is relatively rich, however, in soluble meat extractives, soluble salts and muscle-sugar—dextrose. It is not suitable in itself as a complete nutritive medium for most bacteria, but it forms the basis of many of the commonly used nutritive media. Meat extract (Liebig's or other kinds) is frequently substituted for meat infusion. Three grams of meat extract are dissolved in a small volume of water, filtered through a cold wet filter paper to remove fat, and made up to a volume of a liter. The solution contains some meat extractives, including a relatively large proportion of xanthin bases and a very considerable amount of salts, particularly sodium chloride. Little or no muscle-sugar is present. It is distinctly inferior to meat infusion, however, as a basis for cultural media, especially for the more delicate pathogenic organisms.

The Reaction of Media.—Bacteria are relatively sensitive to comparatively slight changes in the reaction of their nutritional environment, and it is essential to create a suitable initial degree of acidity or alkalinity in media to favor their growth. A reaction neutral to phenolphthalein—slightly alkaline to litmus—is suitable for most of the bacteria pathogenic for man—human tissues and blood are slightly alkaline to litmus. A reaction of 1 per cent. acid (+1.0), using phenolphthalein as an indicator, has been recommended by the Laboratory Section of the American Public Health Association for the routine bacterial examination of water, ice, sewage, milk, cream, and ice-cream. A reaction of 1 per cent. signifies that 1 c.c. of normal NaOH would be required to neutralize the acid in 100 c.c. of the medium. Ten c.c. of $\frac{N}{1}$ NaOH would be required to exactly neutralize one liter of medium having an acidity of 1 per cent.

The reaction may be determined accurately in the following manner: to 45 c.c. of distilled water, contained in a porcelain evaporating dish of 100 c.c. capacity, are delivered exactly 5 c.c. of the medium from a graduated pipette. The solution is brought to the boiling-point over the free flame to expel CO_2 and 1 c.c. of a solution of phenolphthalein[1]

[1] Made by dissolving 0.5 gram phenolphthalein in 100 c.c. 50 per cent. alcohol. This indicator is colorless in acid solution—pink in an alkaline solution. CO_2 interferes with its accuracy as an indicator. It is especially sensitive to organic acids which occur in ordinary media, hence its value in media titrations.

is added as an indicator. The solution usually remains colorless, because ordinary media are acid in reaction; $\frac{N}{20}$ NaOH is added slowly from a burette until a faint pink color appears and persists after one minute's boiling. From the amount of $\frac{N}{20}$ alkali required to neutralize 5 c.c. of medium, the reaction of the entire amount is readily computed. Thus:

5 c.c. media are neutralized by 3 c.c. $\frac{N}{20}$ NaOH.
100 c.c. media would be neutralized by 3 c.c. normal $(\frac{N}{1})$NaOH.
1000 c.c. media would be neutralized by 30 c.c. normal $(\frac{N}{1})$NaOH.

To reduce the reaction of a liter of medium whose initial reaction is + 3.0 to + 1.0, 20 c.c. of normal NaOH would be required. It is necessary to heat the medium after adding the alkali, in order to promote the reaction between the acids of the medium and the neutralizing solution and a redetermination of the reaction should be made to make certain that the desired change in acidity has taken place. Frequently a second addition of alkali is necessary to create the desired final reaction.

A satisfactory reaction for cultural media designed for most pathogenic bacteria may be created by adding $\frac{N}{10}$ NaOH solution—a few drops at a time, to the entire volume, using filter paper dipped in phenolphthalein solution, and dried, as an indicator. When the paper shows a faint pink color the addition of alkali is discontinued. The reaction is practically neutral under these conditions.

The Clarification of Media.—It is desirable, in the preparation of culture media, to remove all insoluble substances. This is accomplished by filtration methods, with or without preliminary treatment, to flocculate the substances in suspension. The addition of non-heat-coagulable proteins, as gelatin, frequently requires clarification with a coagulable protein, as egg-albumen, to remove the finely divided suspended matter.

Clarifying with Eggs.—For each liter of medium to be clarified, two eggs thoroughly whipped in a small amount of water are added. The temperature of the medium should not exceed 50° C. The eggs are thoroughly stirred in and the entire mixture is slowly heated to 100° C., either in a double boiler or in the Arnold sterilizer. A firm coagulum forms during the heating process, which enmeshes the suspended particles it is desired to remove. The medium should never be disturbed during the coagulating process. The clear underlying medium is drawn off and filtered through cotton.

Filtration through Cotton.—A large glass funnel is lined with a double layer of absorbent cotton; the layers are placed at right angles, thus laying the fibers of cotton at right angles. The cotton is moistened with a small amount of hot water if agar or gelatin is to be filtered. The medium is then carefully poured into the funnel, care being taken that the cotton is not displaced by the force of the inflowing fluid. The first portion of filtrate may not be clear and it is somewhat diluted

FIG. 14.—Hot-air sterilizer. Lautenschläger form. (*Park.*)

with the water originally used to wet the cotton—hence it should be returned and refiltered. Agar and gelatin filter slowly, which may lead to congelation, therefore the top of the funnel should be covered to prevent undue loss of heat. Funnels surrounded by a hot water jacket are sometimes used in the filtration of these media. Media that are fluid when cold may be often advantageously clarified by filtration through a good grade of heavy filter paper, with or without a preliminary clarification with eggs, as occasion demands.

The Distribution of Media.—The clarified media are either stored in flasks or transferred to smaller containers for immediate use. Then they are sterilized. Most media are used in test-tubes. Test-tubes are filled from a reservoir, usually a large funnel, the smaller end of which is provided with a short length of rubber tubing, into which a short glass tube, constricted somewhat at the outer end, is introduced. The flow is controlled by a pinch cock, which constricts the rubber tubing midway between the funnel and the delivery tube. The cotton plug is removed from a test-tube and the delivery tube is introduced into the open end of it to a depth of about two inches. The pinch

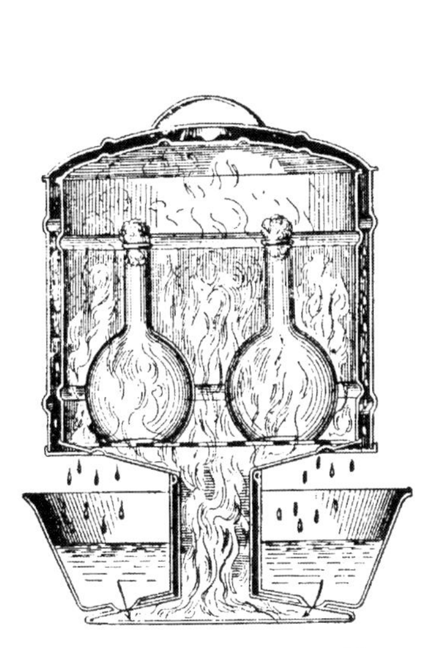

Fig. 15.—Arnold steam sterilizer. (Abbott.)

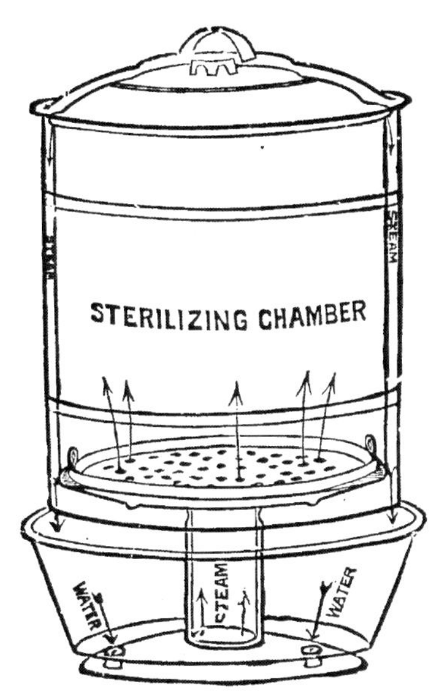

Fig. 16.—Arnold steam sterilizer Ordinary type. (*Park.*)

cock is opened somewhat and the desired volume is allowed to flow in. The pinch cock is then released to stop the flow, the delivery tube removed, care being taken that no media touches that part of the test-tube where the cotton fits, so that it will not adhere to the sides of the tube, and the cotton plug is replaced. Usually about 8 to 10 c.c. of media are added to a tube.

Sterilization of Media.—Media which do not contain coagulable proteins, gelatin or carbohydrates are sterilized for fifteen minutes in an autoclave at a live steam pressure of fifteen pounds (121.3° C.). Media containing gelatin or carbohydrates are sterilized at a lower temperature by discontinuous sterilization—half an hour on three successive days, in flowing steam in an Arnold sterilizer. After each sterilization the medium is kept at room temperature to permit of the germination of spores. Lower temperatures are occasionally employed, particularly for the sterilization of blood serum or other native proteins—an exposure of to 70° C. for an hour on each of six successive days usually suffices. Bacteria may be removed from fluid media and from various sera and solutions containing thermolabile toxins or similar products by filtration through sterile porous filters made of unglazed porcelain or diatomaceous earth—Pasteur or Berkefeld filters. These filters are made with varying degrees of porosity,

FIG. 17.—Arnold steam sterilizer. Boston Board of Health type. (*Park.*)

regulated largely by the thickness of their walls to accommodate varying needs. Usually the fluid is forced through the walls of the filter into the center, which is hollow, by suction. The clear, bacteria-free filtrate passes into a sterile container attached to the filter. The filters and their necessary accessory parts are sterilized in the autoclave for fifteen minutes at fifteen pounds live-steam pressure. Turbid fluids should be passed through several layers of filter paper prior to filtration, to remove the grosser particles which otherwise would soon clog the filter. A time limit, usually not exceeding two hours as a maximum, should be set, beyond which filtration should be stopped

—bacteria may be forced through filters and contaminate the filtrate if the process is carried much beyond this interval.

New, unused filters should be cleaned by running several liters of clean water through them and they should invariably be tested before use to guard against "pin-holes."

After filtration the filter is sterilized to kill whatever bacteria have contaminated it. Then the surface is thoroughly scrubbed with a brush and 1 per cent. alkaline permanganate solution (potassium permanganate 10 grams, water 1000 c.c.) is run through to remove organic matter. Five per cent. oxalic acid is then passed through to remove

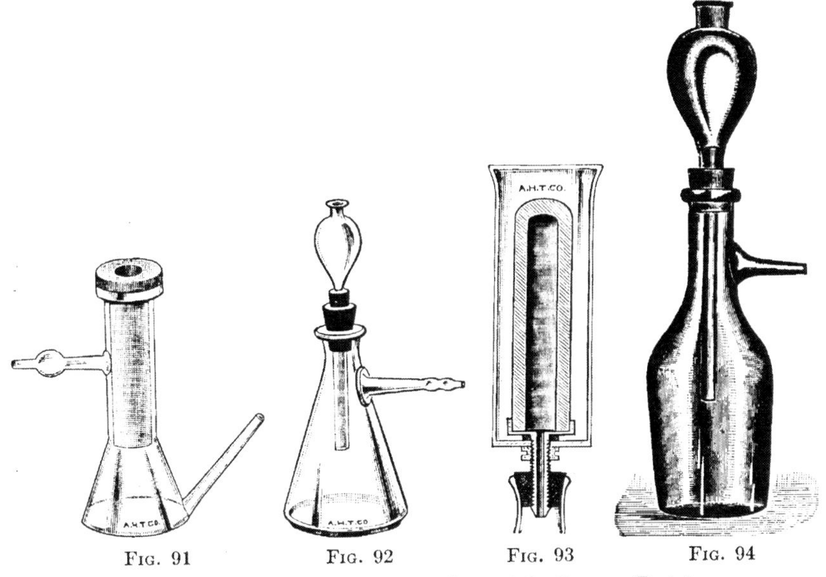

Fig. 91 Fig. 92 Fig. 93 Fig. 94

Figs. 91 to 94.—Types of unglazed porcelain filters. (*Park.*)

the permanganate solution and the acid finally removed by repeated washings with water. If the filter becomes so clogged with organic matter that it can no longer deliver a reasonable amount of filtrate, the filter is placed in a muffle-furnace, gradually heated to about 250° C., and as gradually cooled. It is then cleaned as before with permanganate solution, to remove the last traces of organic matter.

Storage of Media.—If media are not to be used at once it is necessary to protect them from evaporation and contamination. Flasks of media are preserved best by tying paper caps over the cotton plugs if the period of storage does not exceed a few days, or by pouring melted paraffin over the plugs if longer periods of storage are contemplated.

It is necessary to burn the surface of the plug to destroy surface contamination, then to push the plug into the neck of the flask for a distance of 1 cm. to make room for the paraffin. Flasks hermetically sealed in this manner may remain visibly unchanged for weeks or even months. It is good practice to place a lead foil cap over the paraffin

Fig. 22.—Autoclave. (Park.)

plug and lead foil caps are better than paper caps as coverings for cotton plugs. Media in storage should be maintained at a temperature not exceeding 45° C., in a dry ice-box.

The Preparation of Nutrient Bouillon (Broth).—*Meat Infusion Broth.*—To 1000 c.c. of meat infusion (see page 189 for preparation), in a tared, agate-ware boiler, add 5 grams of common salt (NaCl) and heat to boiling. Dust 10 grams of Witte peptone over the surface and stir until it is thoroughly dissolved. Restore the loss by evaporation and adjust the reaction to the desired degree of acidity. Boil for five minutes, verify the reaction and filter through filter paper until the filtrate is perfectly clear. Sterilize in the autoclave.

Meat Extract Broth.[1]—To 1000 c.c. of meat extract (see page 190 for preparation) in a tared agate-ware boiler, add 10 grams of Witte peptone, dusting the peptone on the surface. Heat to boiling, restore loss by evaporation and adjust the reaction. Continue the boiling for five minutes, verify the reaction and *cool to room temperature.*[2] Filter cold through filter paper until perfectly clear and sterilize.

Nutrient Sugar-free Broth.—Meat infusion contains small amounts of muscle-sugar—dextrose—usually about 0.1 per cent. This sugar is present in nutrient meat infusion broth prepared as outlined above. It is frequently desirable to prepare meat infusion broth free from all sugars. The dextrose is readily removed by fermentation with Bacillus coli, adding a broth culture of this organism to the meat infusion before it is heated and maintaining the infusion at 37° C., for eighteen to twenty-four hours. The sugar which is attacked by Bacillus coli in preference to the protein constituents of the medium[3] is quantitatively removed. The organism must be killed as soon as the sugar is exhausted, otherwise the protein constituents will be attacked. The end of the fermentation may be judged with a fair degree of certainty if one removes some of the infusion seeded with Bacillus coli to a fermentation tube, kept at the same temperature, 37° C.; when gas is no longer evolved the sugar is exhausted. Sugar-free broth contains lactic acid, one of the products of fermentation of dextrose by Bacillus coli. After the sugar is removed the medium is sterilized in the usual manner, or made directly into sugar-free nutrient meat infusion broth as outlined above.

Nutrient Sugar Broth.—One per cent. of dextrose, lactose, saccharose, mannite, or other carbohydrate is added to nutrient sugar-free broth immediately before filtering. Media containing sugars are best sterilized in the Arnold sterilizer on three successive days; the high temperature of the autoclave tends to decompose carbohydrates.

[1] It is unnecessary to add salt to meat extract.

[2] A precipitate containing phosphates, soluble in the hot medium, settles out upon cooling. It must be removed before the medium is used.

[3] See chapter on Bacterial Metabolism.

Calcium Carbonate Nutrient Sugar Media.—Bacteria grown in sugar media frequently form acid products from the fermentation of the sugars—the amount of acid products may be sufficient to inhibit the development of the organisms even after one or two days' growth. The addition of insoluble carbonates—as calcium carbonate—neutralizes the acid as it is formed and thus maintains automatically a favorable reaction for prolonged development. Bolduan[1] has shown that pieces of marble about 0.5 centimeters square in 100 c.c. of broth not only restrain the development of free acid—the marble appears to create a somewhat more favorable medium, especially for the pneumococcus and streptococcus as well. The bits of marble should be sterilized in the hot-air sterilizer before they are introduced into the broth.

Nutrient Glycerin Broth.—To 1 liter of sugar-free broth add 3 to 5 per cent. pure, redistilled glycerin immediately before filtering. Sterilize in the autoclave fifteen minutes at fifteen pounds pressure. Glycerin broth is extensively used for the cultivation of the tubercle bacillus[2] and it is frequently employed in the culture of bacteria which are susceptible to desiccation—the glycerin conserves the moisture and retards evaporation.

The various sugar-broths may be prepared with meat extract as a basis; pathogenic bacteria develop less luxuriantly as a rule in extract media than in meat infusion media, however.

Dunham's Solution.—Five grams of common salt and 10 grams of Witte peptone are added to one liter of water and heated to boiling until the peptone is completely dissolved. Pass through filter paper until perfectly clear, tube, using 10 c.c. to each test tube, and sterilize in the autoclave. The reaction does not require adjustment.

This medium is frequently used to test the ability of bacteria to form indol. Indol is formed in the absence of utilizable sugars by Bacillus coli; members of the cholera group and other bacteria form tryptophan by the splitting off of alanin:

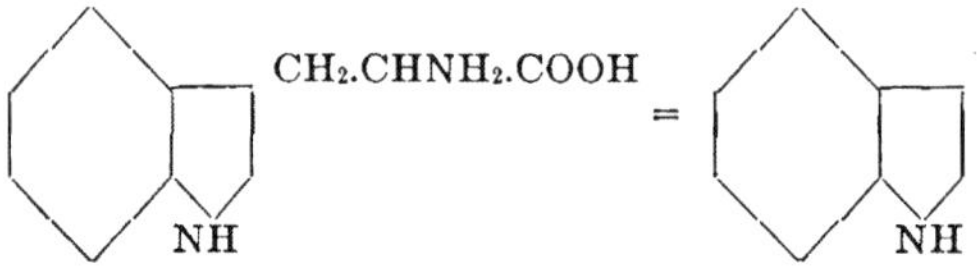

$CH_2.CHNH_2.COOH$ (Tryptophan, NH) = Indol (NH) + Alanin. The alanin is decomposed by the bacteria.

Tryptophan   Indol

[1] New York Medical Journal, May 13, 1905.

[2] The reaction of glycerin broth designed for the cultivation of tubercle bacilli should be +1.0 acid. The organism does not develop well in media neutral to phenolphthalein.

Samples of Witte peptone occasionally do not contain tryptophan, consequently each lot of peptone should be tested. When an especially favorable sample is found it should be reserved for this purpose. Plain neutral sugar-free broth is a better medium than Dunham's solution for the indol test, and it should be employed for this purpose whenever possible.

Nitrate Broth.—Add 10 grams of Witte peptone to one liter of water and dissolve by boiling. Then add 0.2 gram chemically pure potassium nitrate—free from nitrites—and filter. Sterilize in the autoclave. The reaction does not require adjustment.

Nutrient Gelatin Media.—Ten grams Witte peptone and 5 grams NaCl are added to 1 liter of sugar-free meat infusion[1] and dissolved by boiling. When the ingredients are in solution, 100[2] grams of "Gold Label" gelatin are added, a few leaves at a time, and stirred until dissolved. The reaction is then adjusted to the desired degree and verified after an additional five minutes' heating. The medium is cooled to 50° C., and clarified with eggs, using two eggs for each liter. Filter through a double layer of absorbent cotton in a large glass funnel until clear, and sterilize. When sterilization is accomplished, cool quickly and store in the ice-box.

Nutrient Agar.—(*a*) Dissolve 12 grams of powdered or shredded agar in one liter of meat infusion by the aid of heat and add 5 grams NaCl and 10 grams Witte peptone. Maintain a boiling temperature for at least thirty minutes, or until the ingredients are completely dissolved; restore the loss by evaporation, adjust the reaction, and filter through a double layer of absorbent cotton in a large glass funnel. Pass through filter until clear. It is frequently necessary to clarify agar with eggs. After the reaction is adjusted, cool to 50° C. add two eggs beaten up in water and mix thoroughly. Heat slowly to the boiling-point, boil ten minutes, and filter through absorbent cotton; sterilize in the autoclave.

(*b*) Prepare "double-strength" meat infusion; 1000 grams of finely comminuted lean meat are suspended in one liter of water; infuse in the ice-box for twenty-four hours; heat to boiling and filter through filter paper. Prepare nutrient meat infusion broth with this strong infusion as a basis and adjust the reaction to twice the desired acidity—thus, if +1.0 is to be the final reaction, make the infusion broth

[1] Meat extract may be used in place of meat infusion, but the medium is not as satisfactory for pathogenic bacteria.
[2] Use 120 grams gelatin during warm weather.

+2.0. Dissolve 24 grams agar in one liter of water, boiling steadily until complete solution is attained. Add to the meat infusion broth, boil for ten minutes and clarify with eggs in the usual manner. Filter and sterilize.

Meat Extract Agar.—Meat extract agar is made by substituting meat extract solution for meat infusion; otherwise the process is the same. The medium must be clarified with eggs.

Glycerin Agar.—Five per cent. of glycerin is added to meat infusion agar immediately before filtration. The reaction for cultivation of the tubercle bacillus should be +1.0 acid to phenolphthalein. Tubercle bacilli do not thrive in media neutral to phenolphthalein.

Löffler's Blood Serum.—Add one part by volume of 1 per cent. nutrient dextrose broth[1] to three parts of clear, hemoglobin-free beef or sheep serum, and distribute in test-tubes. The tubes are placed in a Koch's serum inspissator or in specially designed racks in an autoclave in an inclined position to produce a slanted surface, and slowly heated to 80° C. This temperature is maintained until the medium is firmly coagulated. The temperature is then raised to 95° or 100° C., and maintained for an hour on each of three successive days, or to 115° in the autoclave, and maintained for one hour. The medium is opaque and white and the surface is smooth and should be free from a metallic lustre when viewed by reflected light. The lustre indicates an accumulation of salts, which are inimical to the growth of many bacteria.

Coagulated Serum.[2]—Clear blood serum from the dog, sheep, cow, or other animal, preferably sterilized by filtration through Berkefeld filters, and with or without the addition of glycerin, is placed in test tubes and slanted and coagulated in a serum inspissator at a temperature of 75° to 80° C. An exposure of one hour to this temperature on each of six successive days is necessary to insure sterility. The medium should be translucent, free from bubbles, and firm.

Hiss Serum Water Media.—Hiss[3] has recommended a serum water medium for the cultivation of pneumococci and similar organisms. It is prepared in the following manner:

Sheep or beef serum,[4] clear and free from hemoglobin, is added to water in the proportion of one volume of serum to three of water.

[1] If the liquefaction of blood serum by bacteria is to be tested, sugar-free broth must be used in place of dextrose-broth.

[2] Theobald Smith, Tr. Am. Phys., 1898, xiii, 417.

[3] Jour. Exp. Med., 1905, vii, 223.

[4] It is advisable to sterilize the serum by passage through an unglazed porcelain filter.

Ten per cent. aqueous solutions of various sugars are prepared and sterilized, and a sufficient amount of the desired sugar to make a final concentration of 1 per cent. is added to the sterile serum solution. Sufficient sterile 5 per cent. litmus solution is added for an indicator. Fermentation of the carbohydrate is shown by the development of an acid reaction, and frequently by a well-defined coagulation of the medium as well.

Endo Medium for the Isolation of Typhoid, Paratyphoid, and Dysentery Bacilli.—I. *Preparation of Agar.*—(*a*) Prepare plain, sugar-free nutrient agar as described on page 197, using 15 grams of agar per liter.

(*b*) Adjust the reaction to a point just alkaline to litmus.

(*c*) Flask the agar, 100 c.c. to a flask, and sterilize in the autoclave.

II. *Preparation of Indicator.*—(*a*) Prepare a 10 per cent. solution of basic fuchsin in 96 per cent. alcohol. This solution is fairly stabile if kept away from the light.

(*b*) Prepare a 10 per cent. aqueous solution of chemically pure anhydrous sodium sulphite (1 gram in 10 c.c. water). This solution does not keep.

(*c*) Add 1 c.c. of "II, *a*" to 10 c.c. of "II, *b*" and heat in the Arnold sterilizer for twenty minutes. The color of the fuchsin is nearly discharged if the solutions are of proper strength. This solution must be prepared each day—it does not keep.

III. *Preparation and Use of Endomedium.*—(*a*) Add 1 gram of C. P. lactose (free from dextrose) to 100 c.c. of agar and place in the autoclave until melted and the lactose is thoroughly dissolved.

(*b*) Add a sufficient volume of "II, c" (about 1 c.c) to impart a faint pink color to the medium.

(c) Pour into sterile Petri dishes and allow to harden in a dark place with the covers partly removed. When cool the medium should be colorless when viewed from above and a very faint pink when viewed from the edge. The medium must be kept in a dark place because the color is restored by the action of daylight.

Those bacteria which ferment lactose—as Bacillus coli—form lactic acid which restores the color of the medium in the immediate neighborhood of the colony; the colony therefore is colored red. Some aldehydes also restore the color, but it is not very probable that aldehyde production is commonly observed among the lactose-fermenting organisms. Non-lactose fermenting bacteria grow as colorless colonies.

If the plates are to be incubated two or three days it may be advisable to increase the agar to 2.5 per cent. to limit the diffusion of color from the acid colonies. For rapid isolations the medium with the normal percentage of agar is preferable.[1]

The Technique of Inoculation of Endomedia is described on page 231.

Lactose Litmus Agar.—I. Prepare 1 per cent. lactose nutrient agar by adding 10 grams of C. P. lactose (free from dextrose) to one liter of plain nutrient agar. Adjust the reaction to a point slightly alkaline to litmus. Tube and sterilize in the Arnold sterilizer.

II. Prepare an aqueous solution of litmus—either a 5 per cent. solution of purified litmus (Merck) or a 1 per cent. solution of azolitmin (Kahlbaum) and sterilize.

To Use Lactose Litmus Agar.—Add about 1 c.c. of sterile litmus solution to a sterile Petri dish and pour over it the melted lactose agar, previously inoculated with the desired material. For water and milk, add 1 c.c. of water or milk (diluted to the proper degree) to the Petri dish before adding the lactose agar. Mix intimately by rotating gently, allow to harden, and incubate.

Those bacteria which ferment lactose with the production of acid appear as red colonies. Non-lactose-fermenting organisms appear as blue colonies.

Blood Agar.—Blood is drawn with aseptic precautions from the carotid or femoral artery of a dog or rabbit into a sterile flask containing beads. The blood is defibrinated by prolonged agitation and added to plain (not dextrose) nutrient agar previously melted and cooled to 45° C., in the proportion of 2 c.c. of blood to 10 c.c. of agar. Small amounts of blood may be withdrawn directly from the heart of an animal without difficulty, provided a small hypodermic needle is used. The blood may be injected directly into the melted agar without defibrination.

Occasionally human blood is added to agar; if a series of agar slants are prepared it is possible to convert them into blood agar with a small amount of blood, as follows:

Withdraw 10 c.c. of blood, using aseptic precautions, from the median basilic vein, in a large syringe. Inject the blood at once into four times the volume of plain nutrient agar melted and cooled to 45° C. Mix at once and run 2 c.c. over the slanted surface of each agar slant, and allow to harden in the inclined position in such a manner that a uniform layer of the blood-agar mixture is obtained. Incubate to prove sterility.

[1] Kendall and Day, Jour. Med. Res., 1911, xxv, 95.

Ascitic and Hydrocele Fluid Media.—*Ascitic and Hydrocele Agar.*[1]—Collect ascitic or hydrocele fluid in a sterile bottle, using aseptic precautions. Allow to stand in the ice-box until clear, and heat to 50° C. for half an hour to destroy enzymes. Two parts of hydrocele or ascitic fluid to eight or ten parts of plain nutrient agar previously melted and cooled to 45° C. make a medium especially adapted to the growth of many of the more fastidious pathogenic bacteria.[2]

Ascitic broth is prepared by adding 20 to 50 per cent. by volume of sterile ascitic fluid to plain nutrient broth. Incubate to prove sterility.

Egg Media.—Eggs are a very good substitute for blood serum in Löffler's medium. Eggs are carefully broken into a clean beaker stirred gently with a rod (avoiding the formation of air bubbles) until homogeneous, and mixed with dextrose broth in the proportion of one part by volume of broth to three volumes of egg. The medium is coagulated in a slanted position and sterilized precisely as Löffler's blood serum is coagulated and sterilized.

Egg Medium.—No. 1. Mix four to six volumes of thoroughly homogenized eggs with one volume of nutrient broth, and add sufficient glycerin to make the concentration of the latter 3 per cent. by weight. Coagulate and sterilize in the slanted position precisely as Löffler's blood serum is coagulated and sterilized. This medium is excellent for the cultivation of tubercle bacilli.

No. 2. Add one volume of physiological salt solution to ten volumes of egg which have been lightly stirred with a rod until the yolks and whites are intimately incorporated. Coagulate and sterilize in a slanted position in test tubes.

Milk and Litmus Milk.—One liter of fresh milk is thoroughly mixed and tubed in the ordinary manner. Litmus milk is prepared by adding sufficient litmus solution to impart a clear blue color. It is tubed, using 10 c.c. to each tube, and sterilized in the autoclave.

For some purposes it is desirable to remove the cream before tubing, but for cultural work the color of the cream ring in litmus milk is of some diagnostic importance. Thus, members of the paratyphoid group of bacilli almost invariably show a blue-green cream ring; the colon bacillus colors the cream ring red brown. It should be remembered that litmus milk does not coagulate as readily or as rapidly as plain milk.

[1] Ascitic and hydrocele fluids may be sterilized by passage through an unglazed porcelain filter.

[2] It should be remembered that ascitic and hydrocele fluids usually contain about 0.08 per cent. dextrose.

Potato.—New potatoes have an acid reaction, as a rule, and old potatoes are slightly alkaline.

Large potatoes are thoroughly scrubbed, the skin removed, and cut into cylinders with a cork-borer. The cylinders, which should be at least 1.5 cm. in diameter, are divided into equal parts by a diagonal cut. The pieces are placed in running water overnight so that they will not darken, and are inserted, base downward, in large test tubes. It is advisable to add about 1 c.c. of water to each tube to prevent drying. Sterilize in the autoclave.

Hiss's[1] Semisolid Medium.—

FORMULA.[1]

Water	1000 c.c.
Agar	8 grams
Peptone	10 "
Meat extract	3 "
NaCl	5 "
Gelatin[2]	40 "

When all the ingredients are dissolved, adjust the reaction to +0.5 (phenolphthalein), filter, and add sufficient litmus solution to impart a clear blue. Dissolve 1 per cent. of dextrose, lactose, saccharose, mannite, or other carbohydrate in the medium, and fill test-tubes with it. Sterilization of lactose and saccharose semisolid media is preferably carried out in the Arnold sterilizer. Dextrose and mannite media may be sterilized in the autoclave.

Semisolid media are inoculated by the stab method. A change in reaction is indicated by the litmus; gas-forming organisms form bubbles in the depth of the medium.

Russell Double Sugar Medium.—To 1 liter of nutrient agar, slightly alkaline to litmus, add sufficient sterile 5 per cent. litmus solution to impart a distinct clear blue color. Add 1 per cent. of C. P. lactose and 0.1 per cent. dextrose, and distribute in test-tubes.

Sterilize in the Arnold sterilizer for three successive days, and allow to harden in a slanted position.

Media for the Cultivation of Aciduric Bacteria.—*Acid Broth.*—Add sufficient glacial acetic acid to a liter of 2 per cent. dextrose broth to make the reaction equal to 50 c.c. of normal acid. A precipitate forms, which will settle out, leaving a clear supernatant fluid that may be removed to sterile test tubes with a sterile 10 c.c. pipette.

Oleate Agar.—The addition of 0.2 per cent. sodium oleate to dextrose agar makes a favorable medium for the cultivation of aciduric bacteria.

[1] Jour. Exp. Med., 1897, ii, 677.
[2] Add after the other ingredients are in solution.

V. **The Cultivation of Bacteria.—Inoculation of Culture Media.**—A platinum wire of 24-gauge is generally used to transfer bacteria from medium to medium. A piece of platinum wire[1] three inches in length is fused into the end of a glass rod 5 mm. in diameter and about 15 cm. in length. Metal handles are preferred by many; they possess the great advantage of not breaking, but become heated during the process of sterilization. The straight wire or "needle" is commonly used for the inoculation of slant and stab cultures in solid media; for the inoculation of fluid media a loop is formed on the end of the wire. The use of the loop permits of the transfer of a greater amount

Fig. 23.—Needle sterilizer. (A. de Khotinsky.)

of material. It is occasionally necessary to transfer more material than a drop or two obtained with a loop in order to insure growth, and for this purpose sterile capillary pipettes are very convenient. Many anerobic bacteria and organisms which grow poorly in artificial media must be transferred with the pipette.

The transfer of bacteria from media to media involves the following steps:

(*a*) Flame cotton plugs to destroy molds and spores of bacteria; extinguish flame.

(*b*) Twist cotton plugs to destroy adhesion to the neck of the tube. The plugs may then be removed intact.

(*c*) Sterilize platinum wire in Bunsen flame. Heat wire white hot and pass that portion of the handle adjoining the wire through the flame, rotating it between the fingers while doing so. Allow the wire to cool.

(*d*) Grasp the tubes in the left hand and remove plugs from the tubes, holding one between the third and fourth fingers of the right

[1] A cheap and efficient substitute for platinum wire is "Nichrome" wire. It is rather less durable than platinum, and melts at a lower temperature.

hand, the other between the second and third fingers, the plugs projecting outward. Flame the mouths of the tubes and test coolness of the platinum wire by plunging it for a distance of about a centimeter into the sterile medium.

(*e*) Remove some material from the infected tube by dipping the tip of the wire in it, and transfer to the sterile tube.

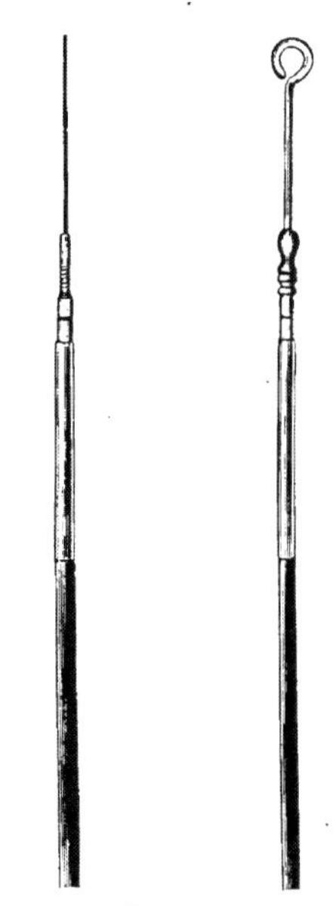

FIG. 24.—Platinum needle and platinum loop.

(*f*) Replace the plugs in their respective tubes and sterilize the wire before laying it down.

II. **The Isolation of Pure Cultures of Bacteria.**—*A. Aërobic Organisms.*—It is the exception rather than the rule that bacteria exist in nature or in many pathological processes in pure culture, that is, that a single kind of organism alone is present. From such mixtures of bacteria it is frequently necessary to isolate one or more organisms in a pure state, uncontaminated by other microörganisms. A common and efficient method of separating bacteria from mixtures is to distribute them in melted gelatin or agar,[1] in such a manner that individual cells are somewhat widely separated. The medium is then allowed to harden. The organisms are immobilized in or upon the medium and surrounded by nutrients; the descendants of each individual organism thus develop locally and apart from the descendants of other organisms. Under favorable conditions the descendants of individual cells become so numerous they may be seen with the unarmed eye as spots or colonies, each of which is made up of the progeny of a single organism. It is a simple matter to touch such a colony with a sterile, cool platinum needle, and infect sterile media with the adherent bacteria. In this manner pure cultures are obtained. The technic of the isolation of aërobic and facultatively anaërobic bacteria is technically termed plating, or streaking, depending upon the apparatus used.

1. *Plate Method.*—Three tubes of nutrient agar or gelatin are melted and cooled to 42° to 45° C. A platinum wire, previously sterilized

[1] Agar melts at about 95° *C.* and solidifies at about 40° *C.* It is necessary to work rapidly with melted and cooled agar, to carry out the technic of inoculation before solidification takes place.

and cooled, is dipped to a depth of about 0.5 c.c. in a mixture of bacteria in fluid media, or touched to a growth in solid media, and then rotated two or three times in a tube of the sterile melted medium. Without sterilizing the needle, the process is repeated in the second and third tubes. Each tube is then rotated between the palms of the hands, to distribute the organisms thoroughly, and poured individually into the sterile Petri dishes. The medium is flowed uniformly over the bottom of the dish and set aside to harden.

It will be seen that the first tube inoculated contains the greatest number of organisms, and that the third tube would theoretically contain but few. The colonies in one of the plates will be so widely separated that they can be "fished" with the platinum wire without the danger of touching other colonies, and transferred to fresh, sterile media. The success of this procedure depends largely upon a rigorous observance of details. The mouths of the tubes and the cotton plugs should be flamed thoroughly before inoculation is practiced, and the transfer of the contents of the tube to the Petri dish must be done carefully to prevent contamination. The cover of the Petri dish should be raised with the left hand, but directly over the bottom, to prevent the entrance of adventitious bacteria from the air. The mouth of the tube should not touch the bottom or edge of the Petri dish and, finally, the cover of the latter should be replaced at once.

After the medium has hardened the plates are incubated—gelatin plates at 20° C., agar plates at 37° C. It is customary to invert agar plates during incubation; when agar cools and becomes solid a contraction takes place which squeezes out some fluid. (This is well defined in slanted agar as the water of condensation.) If the fluid were allowed to remain on the surface of the agar plate it would convert the surface potentially into a broth culture, in which the various organisms would mix in hopeless confusion. Inversion of the plates prevents the accumulation of moisture on the surface to a large degree; the water of condensation collects on the cover instead. The porous tops recommended by Hill may advantageously be used—they absorb moisture as it is formed. Gelatin plates are not inverted; fluid is not expressed as the medium solidifies, and liquefied gelatin formed during the growth of actively proteolytic organisms would collect on the cover and probably contaminate the entire plate.

2. *Streak Method.*—The isolation of pure cultures of bacteria by the streak method differs from the plate method in that the medium (gelatin, agar, blood serum) is not inoculated in the fluid state; the

necessary dilution to secure isolated colonies is attained by drawing a platinum needle infected with bacteria several times across the surface of sterile, slanted gelatin, agar, blood serum, blood agar or other solid medium, each time covering an area not previously touched. Eventually a degree of dilution is reached where discrete colonies are discernible.

The plating method and streak method possess advantages and disadvantages. A considerable proportion of the growth in plates inoculated in the fluid state is beneath the surface, where it is less characteristic than surface colonies. The distribution of organisms, however, is more uniform, and small numbers of bacteria occurring in mixture with larger numbers of undesirable organisms are somewhat less likely to be overlooked. It is possible, moreover, to obtain a quantitative estimation of the numbers of bacteria in mixtures by the plate method. The streak method is advantageous both with respect to the economy of time necessary to inoculate the medium, and in that the colonies are wholly upon the surface of the medium. There is less danger of contamination when "fishing" from streak plates than from the regular method of plating, because there is no chance for submerged colonies to underlie those upon the surface.

The use of certain kinds of media, as that of Endo, of blood agar, and Löffler's blood serum, requires that surface inoculation shall be made. The possibility of missing or overlooking small numbers of the less hardy types of bacteria is greater with the streak method of isolation.

3. *The Barber Method for the Isolation of a Single Cell.*—It is occasionally necessary, in very refined bacteriological studies, to be absolutely certain that the starting point of a pure culture is a single organism. Theoretically, single cells are the progenitors of the colonies observed in media inoculated by the plate or the streak method, and such is usually the case. Undoubtedly it may happen that a chain of streptococci may remain adherent and their descendants appear as a single colony, and it is equally certain that two alien bacteria may occasionally become adherent by intertwining of flagella or adhesion of viscid capsular substance and develop into a mixed colony. The apparatus of Barber,[1] which consists essentially of a delicate capillary pipette mounted in the substage of the microscope, and capable of upward and downward motion in the optical axis of the instrument, is designed to circumvent this possibility. In practice a very thin

[1] Univ. Kansas Science Bull., No. 1, March, 1907.

emulsion of bacteria in a fluid medium is placed on the surface of a sterile thin plate of glass in such a manner that a drop of the contaminated fluid hangs in the opening in the stage ordinarily occupied by the condenser. The drop is manipulated by a mechanical stage, guided by direct observation with a one-sixth-inch lens until a single organism appears in the field of vision. The sterile capillary pipette is carefully brought upward until the tip engages the dependent drop; the organism will be seen to enter the pipette, which is then lowered and removed from its attachments to the microscope. The single cell is transferred to a suitable medium and incubated in the usual manner.

B. Anaërobic Bacteria.—1. *Plating Methods.*—The cultivation of anaërobic bacteria which do not grow in the presence of atmospheric (free) oxygen requires special apparatus and technic. The simplest method, and one which is successful if gas-forming bacteria are absent, is to make dilutions in dextrose agar precisely as described under Plating in the preceding paragraphs. The tubes should be filled to a depth of 10 cm. with the medium, and tubes of relatively large diameter—2 to 3 cm.—are preferable. The tubes, previously heated to the boiling point, and rapidly cooled to 43 to 45° C. to prevent reabsorption of oxygen, are inoculated by the dilution method, rotated between the hands to distribute the organisms uniformly, and cooled rapidly in an upright position.

Colonies appear within the depths of the media after incubation; in the thinly seeded tubes these colonies are discrete, and they may be removed without contamination, either in sterile capillary pipettes introduced through the surface of the medium, or after breaking the tubes from the side. It is, of course, necessary to sterilize the outside of the tube if it is to be broken. A greater degree of anaërobiosis may be obtained within the tubes if after solidifying they are placed neck downward with the cotton plugs removed, in a beaker containing freshly prepared alkaline pyrogallate solution.[1] Growth of anaërobic bacteria upon the surface of agar or blood serum may be obtained in this manner.[2] Those bacteria which produce gas during their growth cannot be isolated in pure culture in deep agar tubes; the liberation of gas bubbles fragments the medium and permits the various colonies to coalesce.

[1] Five grams of dry pyrogallic acid are placed in a beaker and covered with 15–25 c.c. of water: when dissolved a layer of kerosene or paraffin oil about 1 cm. in depth is added, and a 10 per cent. solution of sodium hydroxide is introduced below the oil layer with a pipette.

[2] See Rickards, Cent. f. Bakt., 1904, I Abt., xxxvi, 557.

The "bottle-plate" method of Simonds and Kendall[1] overcomes this difficulty to a considerable degree—through the use of simple appliances.

Sixteen-ounce French square tincture-mouth bottles are plugged with cotton and sterilized with dry heat. With the bottles lying on their sides, sufficient blood agar is poured in to form a layer 5 to 10 mm. deep, and allowed to harden. Dorset's egg medium, dextrose agar or other media may be substituted for the blood agar if desired.

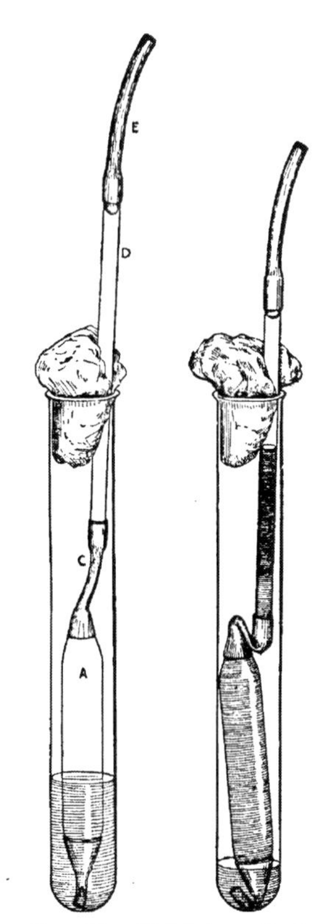

Fig. 25.—Wright's method of making anaërobic cultures in fluid media. (Mallory and Wright.)

As soon as the medium has hardened the bottles should be turned on the opposite side, thus bringing the medium uppermost and preventing condensation water from adhering to it. Inoculation is made with a bent glass rod infected with bacteria from a thin suspension in a liquid medium, and rubbed over the surface of the agar within the bottle. A partial vacuum is next created within the bottle, and residual oxygen dissolved in alkaline pyrogallate solution in the following manner: A closely fitting rubber stopper with one hole carrying a glass tube four inches in length is inserted in the bottle. The outer end of the glass tube projects three-quarters of an inch beyond the stopper and is fitted with a rubber tube three inches in length. That portion of the glass rod within the bottle is bent at an angle of 45° and the stopper is turned in such a manner that the end of the glass tube points toward the side of the bottle opposite the layer of agar. As much air as possible is aspirated from the bottle, and the rubber tube closed with a pinch-cock to prevent reëntrance of air.

The bottle is now placed on its side, with the medium uppermost, and with a pipette, 10 c.c. each of a 50 per cent. solution of pyrogallic acid and 10 per cent. sodium hydrate are run in through the rubber tube, avoiding the entrance of air. A few cubic centimeters of clean

[1] Jour. Inf. Dis., 1912, xi, 207.

water are also run in, to wash the rubber tube free from caustic alkali. The apparatus is working properly when the rubber tube between the pinch-cock and the bottle is collapsed, indicating a partial vacuum within the bottle itself. Residual oxygen is rapidly absorbed within the pyrogallate solution, leaving an inert atmosphere of nitrogen.

The bottle is incubated, medium uppermost, for the required time. Inspection of the surface of the medium will show the colonies. After incubation the pinch-cock is carefully opened, admitting air very gently to avoid spattering the medium, and the stopper is removed. The pyrogallate solution is poured out and residual traces removed with clean water. The bottle is drained standing upon end, mouth down, and then the colonies are ready for fishing. The colonies which develop are all surface growths: the isolation of gas-forming anaërobie bacteria is as readily accomplished as the isolation of non-aërogenic types.

Fig. 26.—Novy jar for anaërobic cultures. (*Park.*)

Pure cultures of anaërobic bacteria may be obtained in an atmosphere of hydrogen; plates prepared in the usual manner are placed on a rack in a Novy jar or other similar vessel provided with a tightly fitting stop-cock, through which hydrogen can be admitted in sufficient volume to displace the air. The stop-cock must be hydrogen-tight. The procedure is to place inoculated plates without covers on a rack within the jar in an inverted position, one above the other. A few grams of pyrogallic acid are placed on the bottom of the jar with a small piece of solid sodium hydroxide. At the last moment, when everything is in readiness, 20 to 30 c.c. of water are gently poured down the side of the jar to prevent spattering, and the cover quickly clamped down. A current of hydrogen gas, either from a cylinder or from a Kipp generator, is passed through the jar at a fairly rapid rate.

The hydrogen should enter at the top, and the outlet for the gas should be as near the bottom of the apparatus as possible. A sample of the escaping gas, collected in a test-tube by downward displacement, will ignite without an explosion when all oxygen is displaced. The inlet tubes are closed, and incubation practiced in the usual manner. An atmosphere of nitrogen is to be preferred to an atmosphere of hydrogen whenever it is practicable.

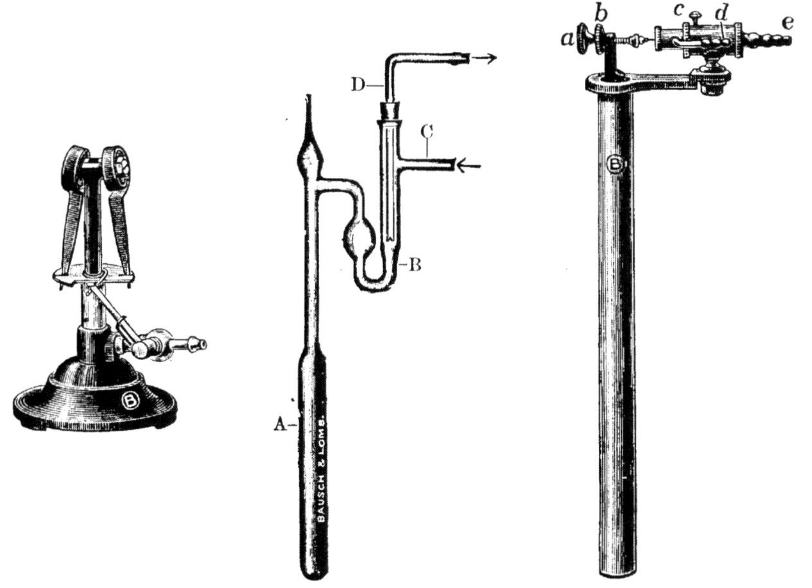

FIG. 27.—Koch safety burner. (*Park.*)

FIG. 28.—Dunham thermo-regulator. (*Park.*)

FIG. 29.—Roux Bimetallic regulator. (*Park.*)

2. *Anaërobic Cultures in Fluid Media.*—A simple method of maintaining anërobiosis in fluid media, sufficiently effective for ordinary usage, is to overlay a flask or test tube containing dextrose broth with a layer of albolene about 1 cm. in depth. Immediately before inoculation all residual oxygen in the medium should be removed by an exposure of half an hour in the Arnold sterilizer, or ten minutes in an autoclave. The liquid is cooled rapidly to minimize reabsorption of oxygen. Wright[1] has maintained anaërobie conditions in test-tube cultures with alkaline pyrogallate solution. Test tubes are prepared with absorbent cotton plugs, which are made tighter than ordinary usage demands. After the culture medium (freed from dissolved oxygen by heating and rapid cooling) is inoculated, the cotton

[1] Mallory and Wright, *Pathological Technic*, 4th ed., p. 126.

plug is pushed into the tube until the upper end is about 15 mm. below the top. The space above the cotton plug is filled loosely with dry pyrogallic acid and a strong solution of sodium hydroxide, 2 to 3 c.c., is added to dissolve the acid. Immediately a tightly fitting rubber stopper is inserted into the mouth of the tube. The alkaline pyrogallate solution absorbs the oxygen within the tube, leaving an atmosphere of nitrogen.

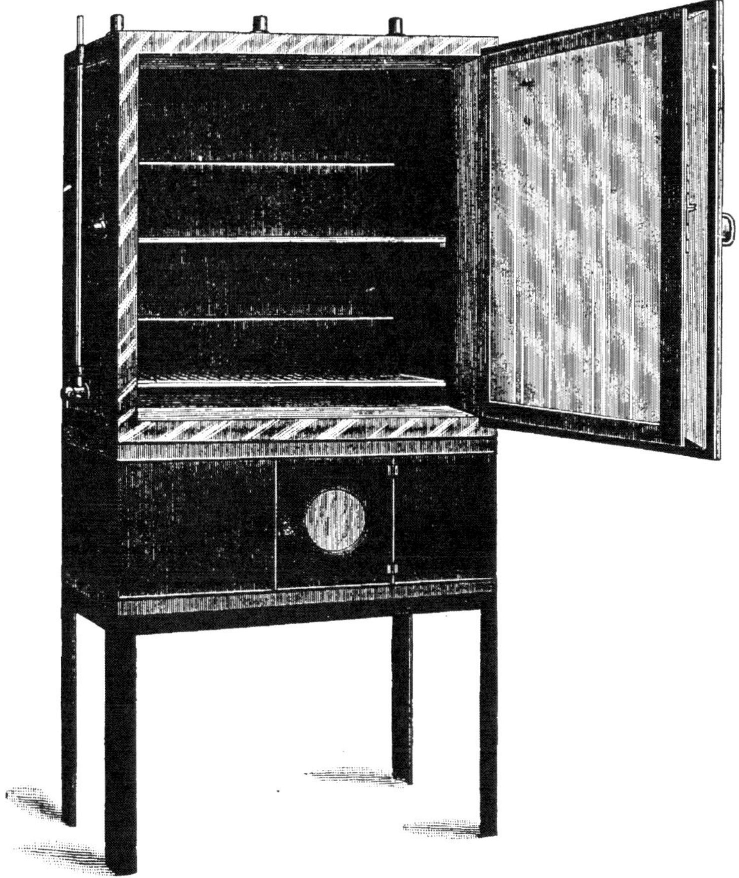

Fig. 30.—Incubator. (*Park.*)

The addition of bits of fresh, sterile tissue,[1] fresh, sterile defibrinated blood, or of the coagulum which is formed during the coagulation of meat infusion adds greatly to the nutritional value of cultures for the growth of anaërobic bacteria.

[1] Theobald Smith, *Cent. f. Bakt.*, 1890, vii, 502.

Special mention of the preparation of tissue media is made in the sections on Specific Anaërobic Organisms.

The Incubation of Bacterial Cultures.—The growth of bacteria in artificial media is markedly influenced by the temperature to which they are exposed. A majority of those organisms parasitic upon or pathogenic for man develop most luxuriantly at the temperature of the human body, 37° C. Exposure to temperatures but slightly above 37° C. leads to rapid death of these organisms, consequently incubators must be available within which cultures may be safely exposed to a uniform and constant degree of heat equal to that of the human body. Gelatin cultures must be maintained at temperatures not exceeding 22° C.

Incubators are single- or double-walled chambers of various sizes, heated directly by gas or electricity, or indirectly through a water jacket. The latter run more uniformly, because water receives and imparts heat more slowly than air. On the other hand, large incubators cannot be surrounded with water jackets because of mechanical difficulties. The regulation of temperature within incubators is controlled by bimetallic regulators which actuate valves or electromagnets controlling the supply of gas or electricity which heats the chamber, or by mercurial thermoregulators working upon the principle of the mercury thermometer. Bimetallic regulators, in which the movement imparted to the regulator of the source of supply of heat is due to the differential expansion or contraction of two dissimilar metals, are more sensitive to slight variations in heat and they possess the additional advantage of being less fragile than mercurial regulators. Various patterns of thermoregulators of tried efficiency are on the market and a selection between them is largely a matter of mechanical adaptability to local needs.

VI. **The Study of Bacterial Cultures.**—I. **Growth in Solid Media.**—(*a*) *Colonies.*—The macroscopic appearance of bacterial colonies upon solid media is of considerable value for the differentiation and recognition of the various types; in a similar manner their microscopic appearance, stained or unstained, permits of some differentiation.

The aspect of a colony is influenced.

1. By the kind of organism—Streptococcus colonies, for example, are habitually small and nearly transparent; anthrax colonies are habitually larger and opaque.

2. By the consistency of the medium—in firm, dense media the growth of bacteria is limited and relatively dry; in moist, semisolid

media the growth of the same organism is usually luxuriant, moist, and spreading.

3. By the composition and reaction of the medium—the addition of specifically nutritive substances, as of fresh sterile tissue to media for the cultivation of anaërobes; of utilizable carbohydrates to media for the cultivation of carbohydrophilic bacteria; of fresh, defibrinated blood to media for the cultivation of hemoglobinophilic organisms; these may improve conditions otherwise unfavorable for bacterial development.

The reaction of the medium, furthermore, is important; many bacteria are extremely sensitive to slightly acid media; the aciduric bacteria thrive in media too acid for the existence of other organisms. Even the ordinary laboratory media, made according to a definite formula, vary sufficiently in chemical and physical properties to influence materially the appearance of bacterial colonies. The degree of influence is more pronounced in the feebly growing forms, but it may affect the appearance of colonies of the more hardy types as well.

4. The rate of growth of bacteria also affects the appearance of colonies.

It is useless, as a scientific procedure, to attempt to recognize differences of greater refinement than the accuracy of the method permits of, and for this reason the descriptions of bacterial colonies should not be carried to extremes. In general, bacterial growths on solid media are described as solids in space—the average size, form, color, lustre and texture. This applies equally well to colonies, slant and stab cultures. The really valuable information gleaned from a study of bacterial growths is the recognition of types of growth. For example, spore-forming bacteria (aërobic) produce rather heavy, opaque, flocculent colonies; members of the Alcaligenes—dysentery, typhoid, paratyphoid group—grow characteristically as rather small, round, transparent colonies.

(*b*) *The Enumeration of Bacteria.*—A very practical application of the plating method for the isolation of bacteria is the enumeration of bacteria in water, milk and other similar substances. The principle involved depends upon the development of colonies of bacteria from single cells. If a definite volume of water, 1 c.c. for example, is distributed in melted agar, thoroughly mixed in the tube by rotation between the hands, and poured carefully into a sterile Petri dish, the number of colonies which develop within a definite period of incubation may be regarded as a measure of the number of living bacterial cells in a cubic centimeter of water. Experience has shown that the accuracy of the method is influenced somewhat by the number of organisms in the sample. A large number of bacteria, by mutual antagonism, will fail to develop into a proportionate number of colonies. The most accurate results are obtained when the bacterial content of the sample as plated lies between fifty and two hundred individual organisms. If more than two hundred bacteria are probably present, a dilution of the sample with sterile water is made before plating, to reduce this source of error. It is convenient in making dilutions to use a multiple of ten, because the subsequent calculation is much simplified. A dilution of 1 to 10 is made by adding 1 c.c. of the sample to 9 c.c. of sterile water, shaking thoroughly and plating 1 c.c. If the technic is all right, each colony on the plate represents one-tenth the number of living bacteria in the original sample. The total number of colonies multiplied by ten gives the theoretical bacterial count of the sample. A dilution 1 to 100 is made by adding 1 c.c. of the sample to 99 c.c. of sterile water. The plating method is inexact, partly because an unknown proportion of organisms in the original sample will fail to develop for various reasons in the cultural medium; furthermore, certain types of organisms, as streptococci, may remain adherent in chains of greater or lesser length and develop as a single colony. Anaërobic bacteria do not develop under aërobie conditions.

A template of paper or glass ruled in square centimeters is used to facilitate the enumeration of colonies; for densely colonized plates, each centimeter square of the template is subdivided into smaller units, usually one-ninth of a square centimeter. The Petri dish containing colonies is placed upon the template in such a manner that the colonies appear superimposed upon the rulings. It is a simple matter, with the lines as a guide, to count either the entire number of colonies in the Petri dish, or a few representative areas, which can be multiplied by a factor. (The average Petri dish contains about 63 square centimeters.)

Example.—A sample of milk diluted 1 to 100 shows a large number of colonies after forty-eight hours' incubation. The total count of nine squares (each a square centimeter) is 180 colonies, an average of twenty colonies per square centimeter. The colonies upon the entire plate (63 square centimeters) is 63 x 20, or 1260. The number of living bacteria in 1 c.c. of the sample of milk would be 1260 x 100 or 126,000, because the number of colonies upon the plate is $\frac{1}{100}$ the entire number in 1 cm.

The value of the method as a convenient means of comparison of the bacterial content of various samples of milk, water, sewage, and the like depends largely upon the supposition that the same types of bacteria present in different samples will grow quantitatively under like conditions. The comparison of bacterial counts is therefore a comparison of a section of the total bacterial flora, not an absolute measure of the number of living organisms. The method of counting bacterial colonies has been highly developed for the regulation of water and milk supplies of cities. (See section Water and Milk.)

(c) *Growth of Bacteria in Gelatin.*—Gelatin is added to cultural media both to confer upon the media the property of solidifying, and to enrich the content in nitrogenous substances.

Pure gelatin does not contain tyrosine and it is relatively rich in diamino acids; according to Hausmann,[1] nearly 36 per cent. of the nitrogen in gelatin is diamino nitrogen—about 63 per cent. in the form of mono-amino acids. Chemically, gelatin media are convenient for the demonstration of soluble, proteolytic enzymes.[2] In the absence of utilizable carbohydrate, several types of bacteria "liquefy" gelatin, that is, through the activity of their proteolytic enzymes the gelatin molecule is split by hydrolytic cleavage to molecules so simple in their state of aggregation that they can no longer produce a "gel." The presence of utilizable carbohydrate prevents the liquefaction of gelatin by many bacteria.[3]

Formerly the morphology of the liquefied zone in gelatin stab cultures was regarded as distinctive for individual organisms; thus, the napiform liquefaction produced by cholera vibrios was supposed to be sufficiently constant to possess diagnostic value. It is now generally conceded that this morphological characteristic is of comparatively little importance for the identification of the organism. On the other hand, the liquefaction of gelatin and of coagulated blood serum and casein as well is important from a chemical viewpoint, because it indicates the activity of a soluble proteolytic enzyme.[4]

II. **Growth of Bacteria in Fluid Media.**—(a) *Plain Broth.*—Plain broth, prepared from meat infusion and peptone in the usual manner,

[1] Zeit. f. physiol. Chem., 1899, xxvii, 95.
[2] Kendall, Boston Med. and Surg. Jour., 1913, clxviii, 825.
[3] Kendall and Walker, Jour. Inf. Dis., 1915, xvii, 442.
[4] The enzyme may be obtained sterile and in an active state in the filtrates of liquefied gelatin, blood serum, casein, and from plain broth cultures as well, if the bacteria are removed by filtration through unglazed porcelain: Auerbach, Arch. f. Hyg., 1897, xxxi, 311; Berghaus, ibid., 1906, lxiv, 1; Kendall and Walker, Jour. Inf. Dis., 1915, xvii, 442.

and freed from sugar with Bacillus coli,[1] contains, on the average, about 300 milligrams of nitrogen per 100 c.c. A small but variable amount of the nitrogen exists as free ammonia.[2] Less than 10 per cent. of the total nitrogen, as a rule, exists as aminonitrogen (determined by the method of Van Slyke).[3]

The visible changes in the appearance of broth cultures incidental to the development of bacteria are not of great importance; they consist essentially of turbidity, sediment, and occasionally a ring or pellicle. The development of a pellicle is of importance in the production of toxin by the diphtheria bacillus, however, because it indicates the maximum oxygenation of the bacteria. The character of the turbidity and sediment—the viscidity and color—may afford some information of the character of the organism. Products of importance are frequently detected by chemical or physiological examination in plain broth cultures of bacteria. Thus, in the absence of utilizable carbohydrate, diphtheria and tetanus bacilli produce their very potent toxins;[4] proteolytic bacteria elaborate soluble enzymes;[5] Bacillus coli, Bacillus proteus and other bacteria form indol and phenolic bodies from tryptophan and tyrosine respectively; the cholera vibrios form nitroso indol,[6] and in sugar-free broth containing freshly drawn, sterile, defibrinated blood, various bacteria produce hemolysis.

The rate of decomposition of the protein constituents of the broth may be measured by the Folin microscopic method for ammonia; the increase in ammonia indicates the extent of deaminization.[7] The rate of hydrolysis of protein is conveniently estimated with the Van Slyke[8] amino-acid apparatus, after removal of ammonia from the medium.[9] A combination of these methods affords an approximate analysis of plain broth media before and after bacterial growth. Undoubtedly the application of the Emil Fischer esterification method of aminonitrogen determination will throw much light upon the utilization of various amino acids by specific bacteria during their growth in artificial media. Amino acids containing aromatic nuclei—as tryp-

[1] Theobald Smith, Cent. f. Bakt., 1897, xxii, 45.
[2] Determined by the Folin Method, Jour. Biol. Chem., 1912, xi, 523.
[3] Jour. Biol. Chem., 1912, xii, 275; 1913, xvi, 121.
[4] Theobald Smith, Tr. Assn. Am. Phys., 1896; Jour. Exp. Med., 1899, iv, 373.
[5] Kendall and Walker, Loc. cit.
[6] Kendall, Jour. Med. Res., 1911, xxv, 117.
[7] Kendall and Farmer, Jour. Biol. Chem., 1912, xii, 13, 215, 219, 465; xiii, 63; Kendall, Day and Walker, Jour. Am. Chem. Soc., 1913, xxxv, 1201; 1914, xxxvi, 1937.
[8] Van Slyke, Loc. cit.
[9] The rate of hydrolysis may also be estimated by Sörenson's formol titration method, but this is less accurate for small amounts than Van Slyke's method.

tophan, tyrosine, histidin—give colored compounds with various reagents because they contain the chromophoric group, C = C. The formation of indol from tryptophan (see page 74 for chemistry) has long been used as a diagnostic test for Bacillus coli and other bacteria. The test depends upon the removal of alanin from the tryptophan molecule by the activity of the organism, and the addition of an auxochromic group, NO_2 in the beta position of the pyrrol ring, previously occupied by alanin. In an acid medium the compound, betanitroso-indol, is brownish red.

Procedure, Indol Test.—To a three-day plain broth culture of Bacillus coli (or other organism) add 1 c.c. of concentrated hydrochloric acid.[1] Mix thoroughly and overlay the acid broth with 1 to 2 c.c. of a 0.1 per cent. solution of sodium nitrite.[2] At the junction of the two solutions a brownish-red ring of nitroso indol develops.

(*b*) *Carbohydrate Broths.*—The addition of sugars, as dextrose, lactose, saccharose, or of alcohols, as glycerol, to plain broth media, greatly enriches the medium in non-nitrogenous substances which may be readily utilizable sources of energy for bacteria. It is hardly necessary to emphasize the importance of purity in all sugars and other carbohydrates intended for bacterial purposes, nor the fallacy of attempting to determine the action of bacteria upon specific carbohydrates in media not freed from muscle sugar (dextrose). The use of serum as a basis for fermentation media frequently introduces a source of error, because blood serum normally contains about 0.08 per cent. of dextrose, an amount quite sufficient to give rise to considerable amounts of acid.[3]

The observations made in carbohydrate media are usually restricted to:

(*a*) Change in reaction.

(*b*) Gas formation, and in fermentation tubes, to growth in the closed arm as well.

[1] Any strong mineral acid will answer the purpose.

[2] Best accomplished by running the nitrite solution carefully down the side of the tube held in a slanting position; a stratification of the two liquids should be obtained.

[3] The significance of fermentation media for the classification and identification of bacteria depends upon their content both of protein and carbohydrate. Bacteria derive their energy requirements from carbohydrate, if it is utilizable, but of course they must obtain their "Bausteine" from the nitrogenous constituents. If the carbohydrate cannot be utilized, both structural and energy requirements are derived from the protein constituents. Bacteria vary greatly in their ability to ferment carbohydrates; some types, as Bacillus alcaligenes, do not appear to ferment even dextrose. Bacillus lactis aërogenes, on the contrary, can ferment hexoses, bioses, and even starches. The fermentability of a carbohydrate depends apparently upon its stereo-isomeric configuration, and relatively slight differences in the configuration of similar carbohydrates may determine their value for specific organisms as sources of energy. This point is discussed somewhat later in this section.

Bacteria which can utilize carbohydrates for their energy requirements usually produce acid; many types produce gas as well. The acid, which is commonly lactic, together with small amounts of acetic and other fatty acids may be estimated by titration with standard alkali. A more accurate estimation is based upon the determination of the hydrogen ion concentration.[1] The gases formed are usually carbon dioxide and hydrogen. An approximate ratio of the proportion H/CO_2 is conveniently made in the Smith Fermentation Tube,[2] in the following manner:

The level of the gas in the closed arm is marked with a wax pencil. The bulb of the fermentation tube is then completely filled with a 1 per cent. solution of sodium solution, and the gas brought into contact with the alkaline solution by inverting the tube several times. The gas is then entirely run back into the closed arm, and the volume again measured. The volume is diminished proportionately to the absorption of CO_2 by the caustic alkali.

Smith[3] has determined the "gas ratio" for the principal aërogenic bacteria as follows

Organism.	Dextrose.		Lactose.		Saccharose	
	H	CO_2	H	CO_2	H	CO_2
B. coli	63	37	63	37	63	37
Hog cholera	66	34	—	—	—	—
B. lactis aërogenes	65	35	62	38	80	20
Friedlander bacillus	67	33	86	14	67	33
B. edematis maligni	67	33	?	..	?	
B. proteus	72	28	—	—	67	33
B. cloacæ	70	30	37	63	58	42

Bacteria which ferment sugars grow in the closed arm of the fermentation tube; those organisms, with very few exceptions, which cannot utilize the carbohydrate of a fermentation medium fail to grow in the closed arm where free oxygen is not available; growth appears only in the open arm.

Occasionally a very slight change in the stereo-isomeric formula of a carbohydrate, or a very small change in its terminal groups will determine its fermentability by various organisms. Thus dextrose, mannose, and their respective alcohols, sorbite and mannite, according to Emil Fischer,[4] have the following stereo-isomeric formulæ:

[1] Clark, Jour. Inf. Dis., 1915, xvii, 109.
[2] Theobald Smith, The Fermentation Tube, Wilder Quarter Century Book, 1895, 187 et seq.
[3] Loc. cit.
[4] Untersuchungen über Kohlenhydrate und Fermente, 1884–1908, Berlin, 1902.

		H	H
		|	|
H—C=O	H—C=O	H—C—OH	H—C—OH
|	|	|	|
H—C—OH	HO—C—H	H—C—OH	HO—C—H
|	|	|	|
HO—C—H	HO—C—H	HO—C—H	HO—C—H
|	|	|	|
H—C—OH	H—C—OH	H—C—OH	H—C—OH
|	|	|	|
H—C—OH	H—C—OH	H—C—OH	H—C—OH
|	|	|	|
H—C—H	H—C—H	H—C—H	H—C—H
|	|	|	|
O	O	O	O
H	H	H	H
D. Glucose.	D. Mannose.	D. Sorbite.	D. Mannite

The fermentation of these hexoses and their respective alcohols by certain bacteria is shown in the accompanying table:

Organism.	D. Dextrose	D. Mannose.	D. Sorbite.	D. Mannite
B. dysenteriæ Shiga . .	+	+	—	—
B. dysenteriæ Flexner .	+	+	—	+
B. Morgan No. 1 . . .	+	+	—	—
B. paratyphosus Beta .	+	+	+	+

An explanation for the phenomenon set forth in the table does not readily suggest itself. Somewhat similar selective action upon specific amino acids by other bacteria is known, qualitatively at least. Thus, members of the Hemorrhagic Septicemia Group, particularly those derived from animal sources, produce indol in plain broth media. Typhoid bacilli, diphtheria bacilli and many other pathogenic organisms usually fail to produce indol in ordinary media under similar conditions. It is possible that this noteworthy action of members of the Hemorrhagic Septicemia Group upon tryptophan may be related to the fact that this amino acid is an important constituent of the hemoglobin, the coloring matter of the blood; the Hemorrhagic Septicemia Bacteria are particularly likely to grow in the blood stream of infected animals.

Fermentation reactions of bacteria in varied carbohydrate media are of importance in their cultural identification. The table on page 316 illustrates the separation of members of the Intestinal Group of Bacteria by their fermentation reaction in various carbohydrates.

Milk.—Milk is an important natural medium for bacterial growth. It contains protein, carbohydrate and fat, together with inorganic salts. A variety of reactions and changes in milk are produced by bacterial development.

(*a*) *Change in Reaction.*—Milk contains, in addition to protein, two carbohydrates, which play a prominent part in determining the reaction of the medium. The principal carbohydrate is lactose, but fresh milk contains in addition, a small amount—about 0.08 per cent.—of a sugar reacting like dextrose.[1] Changes in the reaction of milk caused by bacterial activity, therefore, may be of several types. An initial acidity followed by an alkaline reaction, as exhibited by the dysentery bacilli and other organisms, is probably due to the initial fermentation of the small amount of dextrose, which results in the formation of acid—and then the production of alkaline products from the decomposition of protein when the dextrose is exhausted. These organisms do not ferment lactose.

A permanent acid reaction is induced either by bacteria which ferment lactose, or less commonly, by the decomposition of fat with the liberation of fatty acids. Bacillus typhosus and Bacillus paratyphosus alpha produce a permanently acid reaction in milk, but do not ferment lactose. The exact chemistry of their activity in the medium is not known. An alkaline reaction in milk is usually an indication of proteolytic action with the formation of basic products of protein decomposition.

The accumulation of acid incidental to the fermentation of lactose, as, for example, by B. coli, may be sufficient in amount to cause an acid coagulation of the casein.[2] An acid coagulation can be distinguished from an enzyme (lab or rennin) coagulation; the acid coagulum will redissolve in alkali, but an enzyme coagulum fails to redissolve by merely neutralizing the reaction.

Some types of bacteria, as Bacillus aërogenes capsulatus, ferment lactose energetically, liberating a large amount of gas, and forming butyric acid as well. For some unknown reason, Bacillus coli and allied organisms, which ferment lactose in fermentation tubes with the liberation of considerable amounts of gas, fail to produce gas from the lactose as it exists in milk. It has been shown, however,[3] that the colon bacillus will liberate gas from lactose if the milk is first acted upon by a strongly proteolytic organism, as B. mesentericus.

Proteolytic bacteria, which are unable to utilize either the small amount of dextrose, the lactose or the fats of milk, usually produce

[1] Theobald Smith, Jour. Boston Soc. Med. Sci., 1897, ii, 236; Jones, Jour. Inf. Dis., 1914, xv, 357.

[2] It must be remembered that bacteria grown in litmus milk frequently fail to cause coagulation unless the medium is heated to boiling.

[3] Kendall, Boston Med. and Surg. Jour., 1910, clxiii, 322.

an alkaline reaction which may be a simple alkalinity without obvious change in the appearance of the medium (as, for example, B. alkaligenes) or a deep peptonization of the casein, as illustrated by B. pyocyaneus. B. mesentericus peptonizes casein energetically, but the reaction of the medium is persistently acid. The initial acidity is probably due to the formation of acid from the dextrose of the milk; the residual acid may be associated with the activity of an esterase which certain strains of this organism appear to elaborate. Fatty acids are formed by hydrolysis of the glycerides of the cream by the soluble esterase, while the metabolic activities of the organism appear to be largely directed to the proteins of the medium.[1]

It is apparent, therefore, that the chemical and physical changes induced in milk incidental to bacterial development in the medium are, or may be, complex in their origin. A knowledge of the proteolytic and fermentative activities of bacteria in the simpler media, however, will frequently furnish an explanation for the more involved reactions in the highly complex medium, milk.

[1] Kendall, Day and Walker, Jour. Am. Chem. Soc., 1914, xxxvi, 1937.

CHAPTER X.

BACTERIOLOGICAL EXAMINATION OF MATERIAL FROM THE PATIENT AND THE CADAVER.

THE successful outcome of a bacteriological examination of material from a patient or a cadaver depends to a large degree upon the application of proper technic at the time of collection. Naturally this is varied according to the nature of the case:

Postmortem cultures are taken from organs or tissues usually indicated by the nature of the infection, and a choice of media for the isolation of a specific bacteria, or types of bacteria, is made with this information in view. The value of a postmortem bacteriological examination is frequently measured by the promptness with which it is made after death; postmortem invasion of tissues, organs, and even the heart and larger bloodvessels by bacteria from the mouth and gastro-intestinal tract takes place very quickly. Even if the cadaver is placed at once in a cold room, some time must elapse before the internal organs are cooled sufficiently to arrest bacterial growth.

The spleen, liver, kidneys, and bloodvessels are more commonly examined for evidence of pathogenic microörganisms. The surface of the undisturbed organ is first seared with a hot iron, then incised through the sterile area, and some of the contents withdrawn in a platinum loop or with a sterile capillary pipette and introduced at once into suitable media. (The kind of media to be used is clearly set forth for each organism, in succeeding chapters.) Blood may be obtained from the heart, after searing the surface of the organ, or from the larger veins of the extremities. Exudates from the pleural, peritoneal or pericardial cavities may be removed with sterile pipettes and transferred temporarily to sterile test-tubes or flasks. Purulent discharges are, if small in amount aspirated directly into sterile capillary pipettes; if in considerable quantity, removed to test tubes or flasks, and inoculated as soon as practicable into suitable media.

MATERIAL FROM THE LIVING SUBJECT.

Blood Cultures.—The organisms of septicemia may be numerous or few in number in the blood stream—furthermore, they may be associated with specific lysins and agglutinins, as occasionally happens in typhoid fever. For these various reasons, experience has shown that from 5 to 15 c.c. of blood, drawn aseptically, should be discharged at once with aseptic precautions, into at least 100 c.c. of 0.1 per cent. meat infusion dextrose broth,[1] and evenly distributed by careful agitation. The degree of dilution attained practically renders lytic action and agglutination ineffective; the enrichment of the medium by the relatively large proportion of blood creates a very favorable medium for the development of the organisms.

Technic of Blood Cultures.—1. *Apparatus.*—An all-glass syringe with a platinum-iridium needle of moderately large bore is sterilized in the autoclave, preferably enclosed in a large test tube. A syringe cannot be sterilized for bacterial purposes by boiling in water.

As an alternate apparatus, a 250 c.c. Ehrlenmeyer flask fitted with a rubber stopper containing two glass tubes bent at right angles may be used. The flask contains 100 c.c. of 0.1 dextrose meat infusion broth. One tube is connected with a platinum-iridium needle by a short length of rubber tubing, and the needle is protected during sterilization by a small test tube slipped over it and extending its full length. The test tube is removed when the blood is to be taken. The other tube is protected by a short length of rubber tubing containing a small filter of absorbent cotton. Suction is applied through the latter tube. It will be seen that blood may be drawn directly into the broth in this apparatus, and in practice it has been found convenient to replace the rubber stopper with a sterile cotton plug after the blood is mixed with the media.

2. *Collection of Blood.*—The skin over the median basilic vein is cleansed with green soap and alcohol, dried, and sterilized by the application of tincture of iodine, which is allowed to act for two to

[1] See Media.

three minutes. Then the point of the needle is gently inserted into the vein (which may be made prominent by gentle pressure with a tourniquet applied to the arm above the elbow), and from 5 to 20 c.c. of blood withdrawn. This is introduced at once into broth, as outlined above.[1]

It may be desirable to estimate the number of bacteria in the blood: 1 c.c. of blood is mixed at once with 10 c.c. of agar previously melted and cooled to 42° C., and plated in a Petri dish. If desired, dilution may be made 1 to 10, 1 to 100 in succeeding tubes of agar.

Typhoid and paratyphoid bacilli grow readily in the broth cultures. They may be identified by their cultural and agglutination reactions with highly potent specific sera. Streptococcus and pneumococcus cultures are obtained in a similar manner from the blood stream in blood bouillon. The organisms are isolated in pure culture by smearing the broth, after incubation, upon the surface of freshly prepared blood agar plates. The Streptococcus colonies usually exhibit a wide clear zone of hemolysis. Pneumococcus colonies are characterized by a narrower green zone of altered blood pigment around them. Plague bacilli and Micrococcus melitensis are frequently detected in the blood stream; occasionally the organisms are present in sufficient numbers to develop in blood agar plates inoculated with 1 to 2 c.c. of blood. The former produces characteristic lesions in guinea-pigs; the latter develops very slowly, frequently becoming visible only after five to seven days' incubation.

Bacteriological Examination of Cerebrospinal Fluid.—Spinal fluid for bacteriological examination is obtained by lumbar puncture with a sterile hypodermic needle, or fine trochar about 8 cm. long and 1 mm. in bore. The needle is introduced preferably in the fourth intravertebral space; the fasciculi of the cauda equinum are not tense at this level and are readily pushed aside by the needle without injury. An imaginary line touching the crests of the ilia intersects the spinous process of the fourth lumbar vertebra; the sterile needle is inserted through the previously sterilized skin at a point 1 cm. to the right (or left) of the lower rim of the spinous process, and directed obliquely upward and inward to such a degree that the point of the needle will reach the median line at a depth of 5 to 6 cm. The subarachnoid space is reached at this level and resistance to the passage of the needle

[1] Occasionally circumstances arise which make it necessary to send the blood to a distance for examination; mixing the blood with an equal volume of glycerine bile (one part glycerin, ten parts ox bile; sterilize in autoclave) is said to be an efficient method for preserving the bacterial content of blood practically unchanged for several hours.

ceases, and spinal fluid should flow at once. The fluid should be collected in a sterile test tube. Usually from 20 to 30 c.c. of fluid flow spontaneously; the flow may be much greater, 75 c.c. or even more. Rarely but a few drops, or even none at all may be obtained. Normal spinal fluid is clear and practically colorless. Only a few cells, chiefly lymphocytes, may be found in the sediment obtained by centrifugalization. Pathologically the fluid may contain numerous cellular elements. A blood-stained spinal fluid may be due to injury to bloodvessels during the passage of the needle, or to blood from hemorrhage in the brain or upper levels of the cord. In the former case the blood will clot if the spinal fluid is allowed to stand; in the latter case the blood settles to the bottom, but fails to clot. A turbid spinal fluid is indicative of an inflammatory process in the cerebrospinal axis. If the turbidity is uniform, pus cells are almost invariably present. Occasionally the fluid appears clear, but upon standing, solitary, cobweb-like coagula appear, which enmesh cellular elements and bacteria that may be present. Sometimes an artificial stimulus to coagulation is produced by adding a fibre or two of sterile cotton.

The spinal fluid should be centrifugalized and some of the sediment stained with Wright's stain to determine the types of leukocytes and organisms present. Polymorphonuclear leukocytes indicate an infection with meningococcus, parameningococcus, streptococcus, staphylococcus, typhoid, colon, influenza or plague bacilli. The fluid is usually more or less turbid. Tubercular infection, which, next to meningococcus infection, is the most common, is usually accompanied by a clear spinal fluid from which the cobweb coagula mentioned above may be obtained upon standing. About 75 per cent. of cases of tubercular meningitis may be diagnosed through the recognition of acid-fast bacilli in the stained smears from these coagula. It is essential, in doubtful cases, to inject 1 to 2 c.c. of spinal fluid subcutaneously into guinea-pigs. If the inguinal glands are injured mechanically by squeezing them between thumb and index finger before the injection is made, and the material is introduced as near the glands as possible, a definite diagnosis of tuberculosis may frequently be made within two weeks; ordinarily four to six weeks are required for the development of tuberculosis in the guinea-pig.

For the diagnosis of acute infections of the cerebrospinal axis, about 10 c.c. of spinal fluid should be withdrawn with aseptic precautions into a sterile test tube. If this fluid is visibly turbid, direct smears stained by Gram's stain and with Wright's method will furnish valuable evidence of the etiological organism, and will indicate the medium to use for its isolation and identification. Blood agar is best suited for the meningococcus, parameningococcus, streptococcus and influenza bacillus. The staphylococcus, typhoid, colon and plague bacilli are less fastidious in their requirements. Less commonly, bacteria other than those described above are found in the cerebrospinal fluid following infection of the sinuses, otitis media, mastoid infection or septicemia. The virus of anterior poliomyelitis is also found in the spinal fluid. The most practical method of diagnosis for the latter is to filter the clear spinal fluid through a Berkefeld filter to eliminate all bacteria, and to inject 5 to 10 c.c. of the filtrate intraspinously into monkeys. The animal usually will exhibit symptoms within two weeks if the virus is present.

The Examination of Peritoneal, Pleural and Pericardial Fluids.—Fluids or exudations from the peritoneum, pericardium or pleuræ should be stained by Gram's method to determine the type of organism, and by Wright's method to distinguish the types of cellular elements and their relation to the microörganisms. If the fluid is clear, or if lymphoid cells predominate, an infection with the tubercle bacillus is immediately suggested. Sediment from such a fluid should be injected into a guinea-pig, using the method outlined for suspected spinal fluid. A turbid fluid usually indicates an infection with the streptococcus, pneumococcus, staphylococcus or pneumobacillus, if the material is from the pleuræ or pericardium; an infection with the streptococcus or members of the intestinal group if the source is the peritoneal cavity. Rarely the gonococcus has been found. An examination of the Gram-stained smear will indicate the proper medium to use for the isolation of the organisms in pure culture.

Pus.—A Gram stain of pus will indicate, as a rule, the proper medium to use for the isolation and identification of the organisms. Pus from "cold" abscesses frequently contains no organisms recognizable either by Gram or acid-fast stains; experience has clearly demonstrated, however, that a small amount of the material injected subcutaneously into guinea-pigs will cause their death, frequently within three weeks. At autopsy, tubercles and tubercle bacilli are found in abundance. Much and others believe that tubercle bacilli found in the pus from cold abscesses do not exist in their normal form, but appear as granules—the so-called Much granules—which are, however, viable and virulent for guinea-pigs. In this animal the organisms regain their normal morphology and staining reactions. The possibility of Hyphomycetes in the pus from old cavities in the lungs should be borne in mind. Actinomyces are usually visible to the naked eye as minute, yellowish granules which exhibit the characteristic club when viewed under the microscope in properly stained specimens. Pus from abscesses in the cervical region may contain spiral organisms. The occurrence of these organisms should suggest the possibility of a sinus connecting the abscess with the mouth. Frequently such a sinus originates at the base of a carious tooth.

Examination of Urine.—A bacteriological examination of the urine is of value not only in the diagnosis of infection of the genito-urinary system; it may afford information of the causative organisms in septicemia, and occasionally those concerned in the more chronic heart or joint lesions as well.

The external genitalia are usually contaminated with B. smegmatis, which resembles the tubercle bacillus, and with various adventitious organisms as well. Prominent among the latter is Bacillus coli. A satisfactory sample of urine for bacteriological examination may be obtained from males if the glans and meatus are thoroughly cleansed with soap and water. The greater amount of urine passed should be rejected, and the last portion should be collected in a sterile, wide-mouthed bottle. It is necessary to catheterize females after a preliminary cleansing with soap and water, to obtain a satisfactory specimen for bacteriological examination. A sterile catheter must be used, and the first portion of the urine should be discarded. Under ordinary conditions, except in tubercle infections the causative organisms will be present in sufficient numbers so that a direct smear of the sediment, stained by Gram's method, will furnish a valuable clue to the method and media to be used for the isolation and identification of the organism.

Blood agar is a favorable medium for the isolation of the streptococcus, pneumococcus, gonococcus and staphylococcus. The gonococcus is usually recognized by a Gram-stained smear without further attempt at isolation. It is a Gram-negative diplococcus which, in acute infection, usually appears both intra- and extracellularly among polymorphonuclear leukocytes. Micrococcus catarrhalis, which might easily be confused with the gonococcus, occurs very rarely in genito-urinary infections; ordinarily it may be disregarded. Micrococcus melitensis grows very slowly upon ordinary media. Its very small size together with the deliberateness of growth usually suffice to attract attention to its presence. An agglutination with a specific serum completes the diagnosis. Streptococci and pneumococci produce distinctive changes in the hemoglobin of blood agar plates. Their final identification is discussed in the section devoted to these organisms. Bacillus coli and Bacillus proteus are common incitants of cystitis; they grow readily upon ordinary media and their recognition depends upon the changes pure cultures induce in artificial media. (See table, page 316.)

Bacillus typhosus and members of the Paratyphoid Group are occasionally found in the urine of patients and convalescents. The organisms are readily obtained in pure culture by plating upon nutrient agar, or, better, upon Endo medium (see page 201). Their cultural characteristics and agglutination with specific sera establish their identity. Tubercle bacilli may be found in the urine; the only satisfactory and trustworthy diagnosis is made by injecting the sediment of a twenty-four-hour sample of urine subcutaneously into a guinea-pig. The animal will succumb to infection if tubercle bacilli are present, but will fail to react to smegma bacilli, which are acid-fast and resemble tubercle bacilli morphologically.

Examination of Feces.—(See also Special Section, *Bacteriology of the Feces.*)—The isolation and identification of pathogenic microörganisms from the feces is frequently a difficult task because the normal intestinal bacteria preponderate even in severe infections. Nevertheless, the use of special media has greatly reduced the difficulties and a search for specific microörganisms is now possible with a very favorable outlook for success.

For convenience, intestinal infections may be divided into those caused by cocci, by bacilli, and spiral organisms. Of the spherical organisms or cocci, the streptococcus is by far the most common pathogenic organism encountered in intestinal infections, although an overgrowth of Micrococcus ovalis may be associated with a distinct symptomatology. The streptococcus is a common inhabitant of the intestinal tract, and for this reason streptococcus infection of the alimentary canal is denied by many observers. The streptococcus is frequently an important secondary invader of the intestinal mucosa in bacillary dysentery, and possibly in typhoid and paratyphoid infections as well. It is also frequently associated with an overgrowth of the "gas bacillus" (Bacillus aërogenes capsulatus) in intestinal infection with the latter organism. The occasional acute enteritis observed both sporadically and epidemically among young children is also incited by streptococci. The distinction, if any exists, between the intestinal streptococcus and Streptococcus pyogenes is not clearly established. The isolation of streptococci from intestinal contents is made either by direct plating upon dextrose agar, or by inoculation of feces into dextrose broth. The streptococcus, as a general rule, produces enough acid in the medium after one or two days' growth at body temperature to seriously restrain the development of the intestinal bacteria. A Gram stain prepared from the sediment of the fermentation tube will frequently reveal a nearly pure culture of the organism. A direct smear from the feces, stained by Gram's method, also will indicate the unusual preponderance of streptococci in acute streptococcus enteritis.

The members of the alcaligenes, dysentery, typhoid, paratyphoid group—comprise the more important bacilli ordinarily sought for in the intestinal contents. Their isolation upon ordinary media is difficult because Bacillus coli, the most important of the intestinal organisms, greatly outnumbers the more delicate pathogenic bacteria; its colonies on ordinary media are not readily distinguished from typhoid colonies. The Endo medium (see page 201) however, affords a ready means of identification between the pathogenic bacteria and Bacillus coli. The Endo medium is essentially lactose agar containing a small amount of basic fuchsin decolorized with sodium sulphite. Organic acids including lactic acid restore the color to fuchsin. None of the members of the Alcaligenes-typhoid Group ferment lactose, therefore no lactic acid is formed in and around colonies of these bacilli. Bacillus coli, on the other hand, ferments lactose, and consequently the colonies of this organism are colored red. The lactic acid resulting from the fermentation of the lactose locally restores the color to the fuchsin.

Procedure.—A thin suspension in plain broth, prepared from a freshly passed specimen of feces, is incubated if possible, for an hour at 37° C., then rubbed gently over the surface of an Endo plate with a sterile bent-glass rod or platinum needle. At the end of eighteen to twenty-four hours, small colorless transparent colonies are removed to 0.1 per cent. dextrose meat infusion broth for further development. Inasmuch as colonies of B. alcaligenes, dysenteriæ (Flexner, Shiga and other strains) typhosus, paratyphosus alpha and beta, and the Morgan bacillus are practically identical in appearance, a final identification must depend upon their cultural characteristics (see page 316 for table) and their agglutination with specific sera of high potency.

Members of the Mucosus Capsulatus Group are occasionally found in acute and subacute diarrheas. They grow readily upon the surface of Endo plates as very viscid, slimy colonies which are readily recognized by their macroscopic appearance. Bacillus pyocyaneus is an occasional incitant of intestinal disturbance. Its colonies upon ordinary agar are surrounded by a yellowish or greenish halo. The same general appearance characterizes its growth upon Endo medium. Among the anaërobie bacilli, the "gas bacillus" (Bacillus aërogenes capsulatus) is the most important. The organism is present in variable but small numbers in the feces of healthy adults, and occasionally in young children as well. It may occasionally become a very prominent organism among the fecal flora. The isolation and recognition of the gas bacillus from the intestinal contents depends primarily upon the energetic fermentation in milk cultures inoculated with feces and heated to 80° C. for twenty minutes prior to incubation. (See Chapter XXV for details.) Members of the spiral group, including the highly pathogenic cholera vibrio, are readily isolated and identified by the procedure described in the section on Vibrio Choleræ (Chapter XXVI).

Tubercle bacilli are not infrequently found in the feces of individuals who have advanced pulmonary tuberculosis. It is almost certain that the organisms have been swallowed in a majority of such cases. Occasionally a diagnosis of tuberculosis may be made thus in young children from whom it is difficult or impossible to obtain a satisfactory specimen of sputum. Tubercle bacilli are also found in the feces, derived from tuberculous ulcerations. A diagnosis of tubercle bacillus cannot safely be made from a demonstration of acid-fast organisms in the fecal contents, because acid-fast bacteria other than tubercle bacilli may be present. A guinea-pig furnishes the only reliable method of distinguishing tubercle bacilli from adventitious non-pathogenic acid-fast organisms.

Examination of Sputum, of Buccal and Pharyngeal Material.[1]—A sample of sputum suitable for bacteriological examination should be collected with care. The mouth should be clean, the receptacle should be sterile, and the material should be raised by a deep pulmonary cough, not by a superficial effort. Buccal and pharyngeal material for bacteriological examination is usually obtained upon sterile cotton swabs. Bits of membrane may be removed with sterile forceps.

Examination by Staining.—A Gram-stained preparation of sputum, buccal or pharyngeal material usually contains a variety of microorganisms comprising cocci, spiral forms, and even fungi and yeasts. Many of the organisms may be normal inhabitants of the buccal

[1] An excellent discussion of Infections of the Respiratory Tract and of Sputum as a Means of Diagnosis is that of Leutscher, Arch. Int. Med., 1915, xvi, 657.

cavity, and of the pathogenic organisms, pneumococci, streptococci, and occasionally diphtheria bacilli are found. Usually clinical signs or an abnormal appearance of the sputum, mouth, or throat lead to a microscopic examination of the material from this region and, as a rule, the nature of the symptomatology is a reliable guide to the stain to be used. Among the organisms which stain by Gram's method, pneumococci, streptococci, staphylococci, Micrococcus tetragenus, and occasionally Diplococcus crassus are the more common spherical organisms. Micrococcus catarrhalis, the meningococcus and parameningococcus are the only Gram-negative cocci, so far as is known.

Of the Gram-staining bacilli, the diphtheria and pseudodiphtheria bacilli together with Bacillus subtilis and rarely Bacillus anthracis may be found. The bacillus of Friedlander, typhoid, influenza, pertussis, plague and glanders bacilli are Gram-negative. Bacillus fusiformis and Vincent's spirillum are Gram-negative as well. They color somewhat indistinctly with Löffler's methylene blue and very distinctly with Wright's or Giemsa's stain. Mouth spirals and Treponema pallidum are best stained with the latter method. Tubercle, leprosy and nasal secretion bacilli (Karlinski) stain with the acid-fast stain.

Higher bacteria and moulds are occasionally identified in material from the buccal cavity. Actinomyces, Oïdium albicans, aspergillus, mucor, streptothrix, and yeasts have been detected. The virus of poliomyelitis has also been demonstrated in material from the nasopharynx which has been freed from bacteria by passage through a Berkefeld filter and injected into a monkey.

For the routine examination of sputum, three stains are ordinarily employed—Ziehl-Neelsen for tubercle bacilli, Löffler's alkaline methylene blue for diphtheria, pseudodiphtheria, and fusiform bacilli (and Vincent's spirillum), and the Gram stain, using dilute carbol fuchsin as a counterstain for pneumococci, streptococci, influenza, and pertussis bacilli principally. Smith's stain for sputum (see page 186) is advantageous for pneumonic sputum.

The organisms mentioned previously but not detailed in the routine examination of sputum are of comparatively rare occurrence. They must be studied by purely cultural methods.

Cultural Methods.—Antiseptic gargles should not be used before collecting sputum or material from the mouth or pharynx for cultural examination. Sputum or exudate, obtained in a suitable manner, is first washed through six or seven portions of sterile salt solution, if its cohesiveness permits, to remove or diminish surface contamination.

For a majority of bacteria, freshly prepared blood agar plates are the most satisfactory media to employ.[1] Hemolytic streptococci, pneumococci, Pneumococcus mucosus and influenza bacilli grow upon this medium.

Diphtheria bacilli are grown upon Löffler's blood serum, as described in the section on diphtheria.

Tubercle bacilli can be readily distinguished from lepra bacilli, nasal secretion bacilli and adventitious acid-fast organisms by the injection of washed, cheesy particles from sputum into guinea-pigs.

The organism commonly found in Vincent's angina (Bacillus fusiformis) is not readily cultivated upon ordinary media. Its recognition usually depends upon its demonstration in smears prepared directly from the lesions.

Bacteriological Examination of the Eye.—The normal conjunctival sac frequently contains Staphylococcus albus and Bacillus xerosis; indeed these organisms are so commonly found in this region that they are regarded as normal inhabitants. Abnormally a variety of bacteria may develop on the conjunctiva, frequently causing a violent inflammation. Material for bacteriological examination is best obtained after gently flooding the conjunctival sac with a few drops of sterile salt solution, which are removed with a sterile cotton swab. Then a small sterile cotton swab is gently rubbed over the conjunctival surface and inoculated into suitable media after a Gram-stained smear has been examined.

The gonococcus, Koch-Weeks bacillus, and the pneumococcus are more commonly the incitants of acute inflammation of the conjunctiva; less frequently hemoglobinophilic bacilli (B. influenzæ particularly) or Bacillus pyocyaneus may be found. An examination of Gram-stained smears will indicate the media to be employed if isolation of the organisms in pure culture is desired. The meningococcus is occasionally found in conjunctival inflammations in cases of cerebrospinal meningitis; it must not be confused with the gonococcus. Micrococcus catarrhalis, which resembles both the gonococcus and meningococcus in its morphology and staining reactions, does not produce an acute conjunctival inflammation with a profuse purulent discharge—rather, this organism usually gives rise to a slight reaction, even though the

[1] Several drops of sterile blood, obtained from the finger or the lobe of the ear after a preliminary sterilization, are placed in the centre of an agar plate. The material to be studied is streaked out radially from the blood. Enough blood can be moved with the organisms by this method to insure growth.

organisms are numerous.[1] Blood agar plates are preferable for the cultivation of bacteria from the eye. Not only do the hemoglobinophilic organisms and the gonococcus grow in this medium—the less fastidious forms also develop rapidly.

Subacute Conjunctivitis.—The Morax-Axenfeld bacillus is a common excitant of subacute conjunctivitis, particularly when the internal angle is involved. The secretion is meagre and best obtained in the morning. The bacilli are short and thick, Gram negative, and occur singly and in pairs, both free and in leukocytes. They must be distinguished from members of the Mucosus Capsulatus Group, which are comparatively common in ozena which involves the nasal ducts. The latter are capsulated, which distinguishes them from the Morax-Axenfeld organism.

Corneal ulcerations may be caused by pneumococci, streptococci, leprosy bacilli, and rarely by tubercle bacilli. The latter organism is best detected by animal inoculation.

Pseudomembranous conjunctivitis is frequently the result of a localization and development of diphtheria bacilli, less commonly of streptococci. The etiology of phlyctenular conjunctivitis is still unknown.

Bacteriological Examination of the Ear and Nose.—The middle ear normally is sterile, but bacteria may reach it either by extension of growth from the nasopharynx through the Eustachian tube, or directly from the blood and lymph channels. By far the most common incitant of infection of the middle ear is the streptococcus alone or less frequently in association with other organisms. This organism is also commonly isolated from thrombosed sinuses. The pneumococcus and Pneumococcus mucosus are also frequently isolated from otitis media. Bacillus pyocyaneus or Bacillus proteus are not uncommonly found in middle ear infections, particularly those containing fetid pus. Bacillus coli has also been detected in foul-smelling pus from the middle ear. Staphylococci, Micrococcus catarrhalis, Micrococcus tetragenus, influenza bacilli, members of the Mucosus Capsulatus Group of bacilli, typhoid and diphtheria bacilli have also been isolated from otitis media.

Infection of the external auditory meatus, which contains cerumen, is frequently the result of an overgrowth of various moulds, particularly Aspergillus and Mucor.

[1] For a discussion of Gram-negative diplococci found in the eye, see Blue, Arch. Ophthal., 1915, xliv, No. 6.

The normal nasal cavity, although freely exposed to the exterior and theoretically, at least, continually contaminated with bacteria both from the inspired air and the microörganisms washed from the eyes in the lachrymal secretions, is relatively free from microörganisms. Staphylococcus albus, non-hemolytic short-chain streptococci and pseudodiphtheria bacilli appear to be the more common organisms isolated from the healthy nasal cavity. Material for examination is obtained after cleaning the external nares with sterile salt solution upon swabs of sterile cotton.

Diphtheria, leprosy, ozena, rhinoscleroma and various coryzas are the common types of nasal infection, but a variety of organisms may be present there either transiently, or somewhat more permanently during bronchial infections. Thus pneumococci, influenza and pertussis bacilli have occasionally been isolated from the nasal secretion during pneumonia, influenza or whooping cough respectively. Meningococci and parameningococci have been demonstrated both in patients and carriers during epidemics of cerebrospinal meningitis. It is not unlikely that Micrococcus catarrhalis has been incorrectly diagnosed as the meningococcus in the past, because both organisms are Gram-negative diplococci. Microcococcus catarrhalis is occasionally found in large numbers in the nasal secretion of acute coryza.

The bacteriology of ozena is a subject of controversy. Bacillus ozaenæ and Bacillus rhinoscleromatis, both members of the Mucosus Capsulatus Group of bacteria, have been regarded as the etiological agents in the past.

The earliest lesion of leprosy appears to be a nasal ulcer, more frequently located at the junction of the bony and cartilaginous septum, hence an examination of the nasal cavity is of paramount importance for the early diagnosis of this disease.

Tuberculous ulcerations of the nose are comparatively infrequent; the tubercle bacillus is readily distinguished from the lepra bacillus by injection of suspected material into a guinea-pig. The animal is very susceptible to infection with the tubercle bacillus, but refractory to lepra bacilli. Occasionally acid-fast bacilli, which are neither lepra nor tubercle bacilli, have been reported as occurring in the nasal secretion. Karlinski's nasal secretion bacillus is the best known of the Saprophytic Acid-fast Group. It grows promptly and with considerable luxuriance upon glycerin agar, which at once distinguishes it from the pathogenic acid-fast bacilli.

Nasal diphtheria is not an uncommon type of infection with the diphtheria bacillus. The organism is readily distinguished by its morphology with the methylene blue stain both from the nasal secretion and from cultures upon Löffler's blood serum. When the nasal secretion is profuse, as, for example, in acute or subacute coryza, saprophytic bacteria, as Bacillus proteus, may develop in the nasal secretion, causing extremely offensive odors. There is little evidence that the organism is exciting inflammation, however; it would appear that the secretion is a favorable medium for the development of the organism.

The virus of poliomyelitis may be found in the nasal secretion. Its identification has been discussed above.

THE UTILIZATION OF ANIMALS FOR BACTERIAL DIAGNOSIS AND EXPERIMENTATION.

Pasteur's brilliant animal experiments led Koch to formulate his Postulates for the etiological relationship of bacteria to disease. A rigorous demonstration of the etiological relationship of bacteria to specific disease, said Koch, must fulfill the following conditions:

1. A specific microörganism must be constantly associated with the disease.
2. The organism must be isolated from the lesion and cultivated outside the body of the host.
3. A pure culture of the organism must incite the disease when introduced into a normal animal.
4. The organism must be isolated from the experimental animal again in pure culture.

Experience has shown that many diseases of man cannot be exactly reproduced in experimental animals and Koch's Postulates, therefore, cannot be fulfilled with exactitude in these instances. Nevertheless, experimental animals are indispensable both in diagnostic and experimental bacteriological laboratories. They are used:

1. As culture media for certain types of bacteria which grow slowly or feebly upon artificial media, particularly when the number of such organisms is too small to permit of cultivation under artificial conditions. The isolation of tubercle bacilli from urine, of glanders bacilli from the lesions of glanders are illustrative.
2. To obtain pure cultures of bacteria from mixtures, as the inoculation of white mice with pneumonic sputum for the pneumococcus, or rubbing mixtures containing plague bacilli upon the shaved abdomen of a guinea-pig to obtain pure cultures of B. pestis.

3. To study experimentally the lesions incited by specific micro-organisms.

4. To distinguish sharply between closely related bacteria, as for example, between bovine and human tubercle bacilli. Thus, rabbits

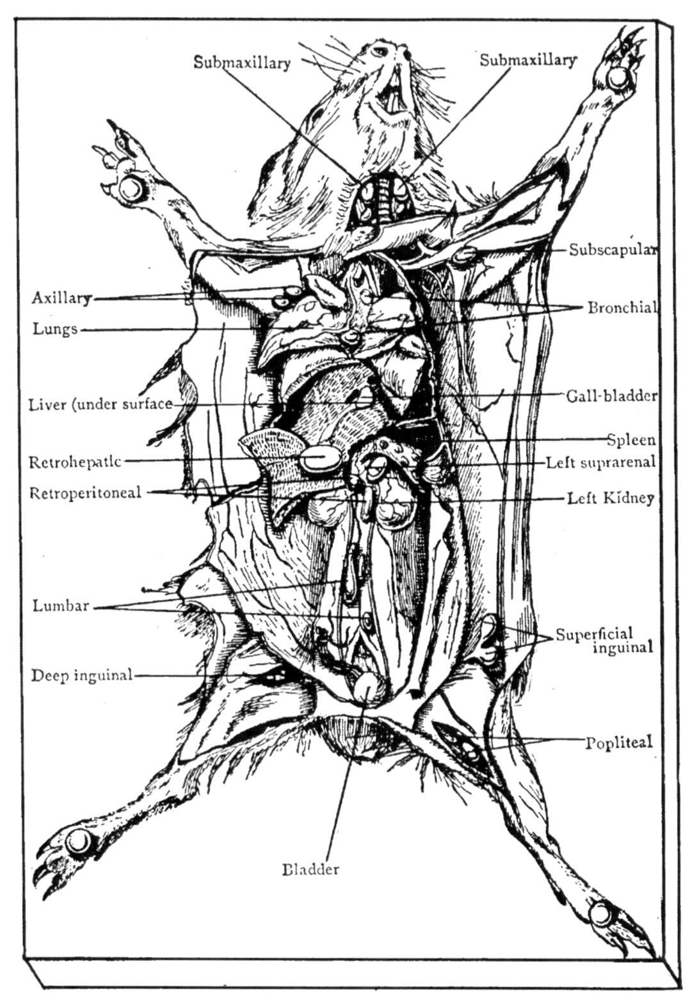

Fig. 31.—Guinea-pig dissection to show anatomical relations of internal organs and important lymph glands. (From Eyre, Bacteriological Technique, Saunders & Co.)

are susceptible to infection with bovine, but not with human tubercle bacilli. Guinea-pigs are susceptible to infection with both types.

5. To study the virulence of various microörganisms.

6. To test the toxicity of bacterial toxins and other products, and to measure the potency of curative sera.

7. For the production of various antibodies, as antitoxins, agglutinins, precipitins and lysins.

The choice of animals depends chiefly upon the nature of the observation to be made. Rabbits, guinea-pigs, white rats and mice, dogs and cats are more commonly made use of for these various examinations. The method and site of inoculation, as well as the dosage, may influence the course of the infection.

The Inoculation of Animals.—Animals may be inoculated through natural channels, as by inhalation into the respiratory tract, or ingestion into the alimentary tract. More frequently, however, material is introduced parenterally into the tissues direct. The site of inoculation is usually the skin, the body fluids or body cavities. The skin must necessarily be entered to reach the deeper tissues. For this reason the site of injection should be shaved and sterilized with tincture of iodin.[1]

Cutaneous Inoculation.—(*a*) Cutaneous: Material is rubbed upon a shaved area of skin.

(*b*) Intracutaneous: Injection is made directly into the skin.

(*c*) Subcutaneous: Material is introduced beneath the skin. A pocket is sometimes made by separating the skin from the cellular subcutaneous tissue, into which solid fragments of tissue are placed. The skin over the abdomen is a common site for inoculation with fluid cultures; the hypodermic needle is introduced at one side of the median line and forced through the subcutaneous tissue in a transverse direction, to a point well beyond the median line on the opposite side. The abdominal wall becomes somewhat tense and does not permit leakage to the outside if this procedure is followed.

Intravenous inoculations are made either into the blood stream through a vein, or directly into the heart. Rabbits are readily injected through the marginal ear veins; the vein is pinched close to the head of the animal and gently massaged; this causes distention and makes the vein prominent. A hypodermic needle will then readily enter the vein; it should be gently forced along its course for a centimeter or two before injection.

Body Cavities.—The peritoneal cavity is commonly selected, but intrapleural injections are readily made. Before introducing a hypodermic needle into the peritoneal cavity, the animal, guinea-pig or

[1] Tincture of iodin should be freshly prepared and painted upon the dry surface it is desired to sterilize. Sterilization is usually accomplished after two or three minute's exposure to the iodin solution.

rabbit, is held head downward to permit the intestines to pass anteriorly as far as possible. The needle is first introduced somewhat obliquely through the abdominal skin posteriorly, then directly into a fold of the abdominal wall pinched between the fingers. The needle should be pressed in until resistance to its passage has ceased. Unless the precaution is taken to dip the point of the needle in sterile vaseline, some of the contents will be introduced into the cutaneous or subcutaneous tissues as well as the peritoneal cavity. The "Hitchens" syringe with its side-arm containing salt solution to rinse the entire charge from the needle before withdrawal from the animal is highly recommended for this purpose.

Intracerebral injections are made either through the optic foramen, or through the dura after trephining the skull.

Intratracheal injections are occasionally made, but more commonly the material is introduced deep into the bronchi through a flexible rubber cannula. The animal should be anesthetized for this operation.

White mice and rats are usually inoculated in the loose subcutaneous tissue at the base of the tail. The needle should pass somewhat obliquely to avoid the spinal cord.

Care of Animals.—Guinea-pigs and rabbits are very susceptible to "snuffles" and frequently perish from contagious pneumonia and other epizootics of the respiratory tract.[1] The first symptoms are usually nasal discharge and a mucopurulent exudation from the eyes. Such animals should be killed at once and their cages thoroughly sterilized. Animals in adjacent cages should be quarantined.

Inoculated animals are best kept in separate cages apart from the healthy stock. If they become moribund it is better to chloroform them and perform the autopsy at once; fresh, uncontaminated cultures may be obtained only at this time. If animals are permitted to die, frequently several hours intervene before an autopsy is performed, and postmortem bacterial invasion of the tissues and blood stream is usually a disturbing factor. Infected material is obtained from animals with the same precautions and technic as those for a human autopsy.

[1] Theobald Smith, Jour. Med. Research, xxix, 291, for discussion.

CHAPTER XI.

PRACTICAL STERILIZATION, ANTISEPSIS AND DISINFECTION.

THE terms sterilization, disinfection, antisepsis and deodorization are frequently used indiscriminately, but it is important to distinguish between them. Sterilization and disinfection imply the destruction of microörganisms, the latter being restricted largely to hygienic procedure, as the disinfection of excreta, etc. A restriction of bacterial growth not necessarily involving the death of microörganisms is properly termed antisepsis. Deodorants, as the term signifies, are those substances which destroy or mask odors; deodorants may or may not destroy bacteria.

LABORATORY STERILIZATION.

The many kinds of apparatus and media used in the study of bacteria must be freed from adventitious organisms before they are applicable to bacteriological investigation. Physical and chemical agents are commonly made use of for this purpose.

Physical Agents.—1. **Heat.**—(*a*) *Incineration.*—Incineration is a most efficient method of sterilizing articles of little value. The free flame is commonly used for sterilizing platinum needles and platinum loops. If the latter are charged with pathogenic bacteria, and particularly bacteria which contain fats, as the tubercle bacillus, it is necessary to dry the material by holding the loop near the flame before incineration to prevent "spattering." The "bacteria incinerator" made by de Khotinsky is particularly to be recommended for this purpose.[1]

(*b*) *Dry Heat.*—Test tubes, flasks, Petri dishes, pipettes and other laboratory glassware are sterilized in the hot-air sterilizer—an oven heated with a gas flame. An exposure of one and a half hours at 160° C. or one hour at 180° C. will effectually kill all spores. The heat should be applied gradually and reduced gradually to diminish the danger of cracking. Dry heat has but little power of penetration. Glassware is conveniently wrapped in paper before sterilization to protect it from dust prior to its use. The cotton plugs of flasks and beakers are also covered with paper before sterilization, for the same reason.

(*c*) *Moist Heat.*—1. The most satisfactory agent for the sterilization of articles uninjured by moisture is steam under pressure. Many kinds of media and laboratory apparatus, and fomites as well are quickly and completely sterilized by steam. The autoclave is commonly used for laboratory purposes. It consists essentially of a double-walled chamber with close-fitting cover, into which steam may be introduced. There are many patterns, but the essential features are—the steam should enter the chamber from the top, and the bottom of the chamber should be provided with a stop-cock, through which the residual air and condensation can escape.

Operation.—A single layer of apparatus should be sterilized at one time. If several layers are introduced, condensation water from the upper layer may collect on the lower layers, permitting of subsequent contamination. Steam is admitted to the chamber to displace the air, and the air-cock should remain open until live steam flows freely from it, because hot air is far less efficient than steam for sterilization. Also, the condensed steam escapes through the same orifice. When all the air is replaced by dry steam the pressure is gradually increased until fifteen pounds are recorded on the pressure gauge. This pressure is maintained from ten to twenty minutes, depending upon the nature of the material to be sterilized. In general, media

[1] It consists essentially of a tube about 12 cm. in length and 1 cm. in diameter, of fire clay surrounded by a resistance coil of sufficiently fine wire and numerous layers to heat the interior of the tube to a white heat. The charged platinum wire is placed in the tube, and within a few seconds it becomes white-hot. There is absolutely no danger from "spattering," because the extruded organisms fall upon the hot walls of the tube (see Fig. 23, page 205).

in test-tubes is more quickly sterilized than media in flasks. At the end of the allotted time, the pressure is *gradually* reduced until equilibrium is reached with the atmospheric pressure; a sudden release of pressure would cause violent ebullition of fluid, and a wetting or even expulsion of cotton plugs from test-tubes or flasks.

TABLE OF PRESSURE AND TEMPERATURE.

Pressure, pounds.	Temperature, Centigrade.
0	100.0°
5	107.7°
10	115.5°
15	121.5°
20	126.5°

2. **Live Steam.**—Many solutions are injured by temperatures above 100° C. Media containing sugars (particularly bioses) milk and gelatin are partly decomposed by prolonged sterilization in the autoclave. An exposure to live steam at 100° C. for thirty minutes on each of three successive days usually suffices to effect sterilization of these media without injury to the constituents of the medium. This method of *fractional sterilization* depends upon the destruction of all vegetative cells during the heating process, and the germination of spores into vegetative organisms between heatings. It is assumed that all viable spores will have germinated before the third exposure to heat, but Theobald Smith[1] has shown that spores of anaërobic bacteria may not vegetate within the specified time. A fourth exposure to heat after two or three days may be required to insure sterilization. The Arnold sterilizer is widely used for fractional sterilization with live steam. It consists essentially of a double-walled copper chamber surmounting a double-bottomed water reservoir, the lower compartment of which is shallow and contains but little water. A flame applied to this shallow reservoir soon generates steam, which rises through a central passage to the chamber in which the material to be sterilized is placed. Condensed steam flows by gravity to the upper water compartment, and from thence to the lower heated reservoir to replace the evaporation. It takes but a few minutes to generate sufficient steam to fill the sterilizing chamber. The sterilizing process begins when the contents of the sterilizing chamber have reached 100° C.

3. **Fractional Sterilization** at temperatures from 60° to 80° C. is frequently made use of for materials such as blood serum, which would be injured by exposure to 100° C. The sterilizing process is repeated

[1] Jour. Exp. Med., 1898, iii, 647.

for an hour daily over a period of five to seven days. The sterilization of Löffler's blood serum in a Koch inspissator is carried out at this lower temperature.

4. **Boiling Water.**—Petri dishes, culture tubes and other apparatus containing pathogenic bacteria may be freed from bacteria by boiling in water for five minutes. Practically no pathogenic bacteria form spores. If tetanus, anthrax or gas bacillus cultures are to be destroyed, the autoclave is necessary.

Chemical Solutions.—Chemical disinfectants are most efficient in aqueous solutions, and they must therefore be soluble in water. Moisture is also essential for gaseous disinfectants.

The theory of the germicidal action of disinfectants is not well understood; apparently the efficiency of salts of heavy metals is associated with their noteworthy affinity for proteins, with which they form firm combinations. It must be remembered that these salts react more quickly with animal proteins than bacterial proteins, therefore greater concentrations of metallic salts are required to kill bacteria suspended in protein solutions than to destroy the same organisms in aqueous suspension. Thus, typhoid bacilli may be killed by 1 to 500,000 bichloride of mercury if they are suspended in water, but a concentration of at least 1 to 1500 is required to sterilize the same organism in blood serum. Absolute alcohol does not appear to be a very powerful germicide; possibly its rather limited germicidal value is associated with its dehydrating properties. Dilute solutions of alcohol, 20 to 30 per cent., are practically as destructive of bacteria as absolute alcohol is. Phenols are excellent germicides in aqueous solutions, but their tendency to go into solution in oils (which do not readily penetrate the ectoplasm of cellular structures) makes them unreliable germicides in oily menstrua.

Salts of Heavy Metals.—1. *Mercuric Chloride,* $HgCl_2$.—Mercuric chloride or bichloride of mercury is a powerful germicide, very soluble in hot water, less soluble in cold water.[1] It is usually dispensed in tablet form mixed with NaCl, which increases its solubility and also prevents somewhat its marked tendency to unite with proteins. This is of importance in the treatment of wounds and secretions of wounds with this germicide. A 1 to 1000 solution of bichloride in water is the dilution commonly used for practical purposes. This strength

[1] One part of bichloride will dissolve in 3 to 4 parts of boiling, distilled water; upon cooling, much of the bichloride becomes insoluble; one part of the salt will dissolve in 16 to 18 parts of water at room temperature.

of solution will kill all pathogenic bacteria in a very short time; a solution of 1 to 500 strength will even kill anthrax spores within a few hours.

The advantage of bichloride of mercury as a germicide resides in its great bactericidal powers. Its disadvantages are: its marked affinity for protein which, in the case of wounds, may lead to local necrosis of tissue, or in greater concentrations, by absorption, to toxic action on the kidneys, intestinal tract, and even the central nervous system. It is unreliable for the disinfection of sputum, feces, urine, purulent discharges, and other excreta, and it should never be employed in the sterilization of instruments or eating utensils. Linen soiled with blood or stained in any way should not be immersed in bichloride, for it acts as a mordant and "sets" the stain.

2. *Silver Salts.*—Silver nitrate is a much less efficient germicide than mercuric chloride, but it is quite extensively used upon mucous membranes. The soluble organic compounds of silver, as Protargol, are less irritating than the inorganic salts and apparently nearly as efficient.

Oxidizing Solutions.—1. *Potassium Permanganate,* $KMnO_4$.—Potassium permanganate is a strong disinfecting agent, but it is almost instantly reduced and rendered inert by organic substances. This greatly impairs its practical value. Nevertheless, it is used in surgical asepsis and also in wells and cisterns which are to be freed from pathogenic bacteria. A strong solution is thrown into the well or cistern, enough to impart a very pronounced pink color to the water, and left for several hours. The water is fit for use when the last traces of color are removed by dilution or emptying and washing out the reservoir. This process is spoken of as "pinking" a well.

2. *Hydrogen Peroxide,* H_2O_2—Hydrogen peroxide is a valuable germicide, applicable to the cleansing of mucous surfaces and wounds. It is readily reduced to H_2O and nascent oxygen in contact with organic substances, and its efficiency is attributable to the latter element. It is essential that the peroxide actually reach the organism to be destroyed in order to be effective. Usually hydrogen peroxide is quite acid in reaction and irritating for this reason.

3. *Chlorinated Lime or "Bleach."*—Chlorinated lime is an excellent deodorant and germicide when it is fresh, but it soon loses chlorine when exposed to the air. Nascent chlorine is liberated from aqueous solutions, and reacts with water to form nascent oxygen and hydrochloric acid, according to the equation $2Cl + H_2O = 2HCl + O$.

One part of nascent chlorine to 1,000,000 parts of water—a milligram to a litre in other words—will kill moderate numbers of bacteria within a few minutes. For this reason, chlorinated lime is extensively used in the treatment of swimming pools to reduce the bacterial count. It is also used for the practical sterilization of urine, bath water, feces, and in the solid state, in privies, cellars, and stables.

Phenols, Cresols.—Phenol, popularly known as carbolic acid, and cresols, of which three are known—ortho, meta, and para—are powerful germicides:

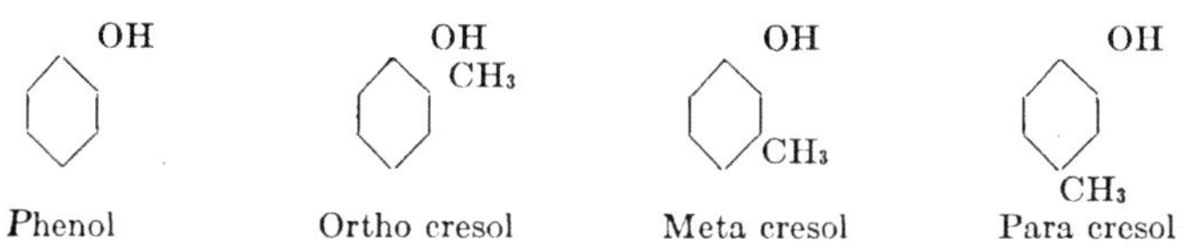

Phenol　Ortho cresol　Meta cresol　Para cresol

Phenol and the cresols are somewhat sparingly soluble in water. A 6 per cent. aqueous solution of carbolic acid, and 5 per cent. solutions of the cresols are about the limits of solubility; 3 to 5 per cent. solutions are used for most practical purposes. Phenol and cresols are not only very toxic for bacteria, they are caustic and poisonous for human tissue as well. Stronger solutions are anesthetic, suggestive of a definite action upon nervous tissue. These substances appear to be readily absorbed from mucous surfaces, the skin, and wounds. They are excreted, in part at least, through the kidneys. "Smoky urine," indicating an irritation of the kidney tissue, is a not uncommon sequel of carbolic acid poisoning.

A 3 per cent. solution of phenol is approximately equivalent in its disinfectant value to a 1 to 1000 solution of bichloride of mercury, but it does not unite readily with proteins to form insoluble, inert compounds, and it is not destructive of fabrics, metals and articles of every-day use.[1] For sputum, urine, feces, purulent discharges, and for stained and soiled linen, a 5 per cent. solution, equal in volume to the bulk of the material to be disinfected, is used and allowed to remain at least one hour before being disturbed.

Cresols form soaps with caustic solutions, which are strongly germicidal. An excellent cresol soap may be made by adding one part by volume of cresols to an equal amount of soft soap (potash soap). This is stirred thoroughly and allowed to stand twenty-four hours. A 5 per cent. aqueous solution of this preparation is nearly three times as efficient in its disinfectant value as a 5 per cent. solution of carbolic acid.

[1] Hamilton, Therapeutic Gazette, 1914, xxxviii, 311.

Tincture of Iodin.—*In vitro,* tincture of iodin is of little value as a germicide, but freshly prepared tincture of iodin applied to the skin appears to possess very considerable germicidal value. This solution seems to be most effective when it is freshly prepared and works most satisfactorily when the part upon which it is to be used has been cleaned with alcohol and allowed to dry. Nascent iodin is liberated, and it is stated that iodin *in statu nascendi* is the active germicidal factor. Tincture of iodin is rather widely used as a skin disinfectant for minor operations, for sterilizing the epidermis prior to spinal puncture, collecting blood for cultural purposes, and for operations upon laboratory animals. Iodin is absorbed through the skin, and in large amounts it is toxic.

Boric Acid.—Boric acid is frequently used upon mucous surfaces and other exposed parts when a very mild antiseptic solution is required. Boric acid is rather an antiseptic than a germicide: its chief advantage lies in the fact that 1 to 3 per cent. aqueous solutions have but little action on the tissues.

All disinfectants appear to be cellular poisons to a greater or lesser degree; in lesser concentrations they are without marked effect upon microörganisms; in effective concentrations they appear to form combinations with tissues if they are used in or on man.

Disinfection of the tissues has been attempted with specific bactericidal sera, which are without noteworthy harmful effects upon the patient. At the present time immune sera are not wholly satisfactory for this purpose, but sufficiently encouraging results have been obtained to justify their present use and to afford promise of their improvement in the future.

A majority of chemical disinfectants are, to use Ehrlich's terminology, organotrophic rather than parasitotrophic, that is, they have a greater affinity for the tissues of the host than for the parasite. Quinine, on the contrary, appears to be parasitotrophic—it is almost a specific for malarial parasites. Ehrlich's brilliant researches in chemotherapy have added organic compounds containing arsenic to the list of parasitotrophic substances; they have a very direct and inimical action upon trypanosomes and the Treponemata, and but minimal action upon the tissues of the host.

Formaldehyde.—A solution of formaldehyde gas in water, commercially known as formalin, is a powerful disinfectant; it does not react as strongly as mercuric chloride with protein solutions;[1] it does not

[1] Formaldehyde unites with ammonia and with the amino-nitrogen of amino acids to form stable compounds; there is relatively little action upon native proteins, however.

injure metals or ordinary fabrics. The commercial solution contains about 35 per cent. of formaldehyde, hence a 10 per cent. solution of "formalin" will contain but 3.5 per cent. of "formaldehyde," which is, of course, the reactive substance. Formaldehyde is an excellent disinfectant for sputum, urine and feces, and other excretions; a 5 per cent. solution of formalin (corresponding to about 2 per cent. formaldehyde) in the proportion of two volumes of the disinfectant to one of the excretion will effect practical sterilization of feces within an hour. Fomites are sterilized in the same manner. The fumes are irritating, and disinfection should not be practiced in the sick-room.

Essential Oils.—Essential oils have been used extensively in the past, particularly in the treatment of nasal and pharyngeal infections, and for mouth-washes. Menthol, thymol and eucalyptol, the active principles of oil of peppermint, thyme and eucalyptus respectively, undoubtedly possess antiseptic and feebly germicidal properties. Cloves, cinnamon and other spices have been used for the preservation of certain types of foods; their efficiency probably depends largely upon their content of essential oils. The expense of these substances compared with their efficiency as antiseptics makes their use practically prohibitive.

Soaps.—Cleanliness is a very important barrier to the spread of disease. Very few pathogenic bacteria upon exposed surfaces of rooms can survive an application of hot soap suds applied with a vigorous arm and a scrubbing brush. A 5 per cent. solution of washing soda (commercial sodium carbonate) is even more efficient if applied hot, but there are limitations to its use. Fine furnishings and hangings, wall paper and similar objects cannot ordinarily be treated with liquid disinfectants.

Testing and Standardizing Liquid Disinfectants.—The first satisfactory method of comparing the disinfectant value of chemical disinfectants was that of Rideal and Walker,[1] widely known as the "Carbolic Coefficient" method. A modification of this method, proposed by Anderson and McClintic,[2] is widely used in the United States. Briefly, the method as modified by Anderson and McClintic consists in comparing the activity of the unknown disinfectant solution in various dilutions with a standard solution of carbolic acid; Bacillus

[1] Jour. Sanitary Institute, London, xxiv.

[2] Bull. Hyg. Lab., Washington, D. C., April, 1912, No. 82. Full details of method and the disinfectant value of a large number of substances are given.

typhosus is the organism selected for the purpose, and the strength of solution of both the unknown and known solutions are carefully measured. The time and temperature of exposure of the organism to the disinfectant solutions and the nature of the medium in which the exposure is made are carefully controlled. Even with the most rigorous attention to details, the carbolic coefficient of the same disinfectant determined by this method varies nearly 50 per cent. in the hands of different observers;[1] for the present, the standards of the Public Health and Marine Hospital Service[2] are regarded as official for the United States.

Gaseous Disinfectants.—Pathogenic bacteria which are known or suspected to be present upon fabrics or furnishings injured by chemical disinfectant solutions, as well as bacteria promiscuously distributed in rooms through droplet infection and by dust may be killed by gaseous disinfectants, of which several are available.

1. **Formaldehyde.**—Formaldehyde is the most efficient of the gaseous disinfectants for superficial disinfection, but its limited power of penetration must be borne in mind. Formaldehyde is dispensed commercially under the name "formalin," which signifies a 40 per cent. volume solution of the gas (formaldehyde) in water. Commercial formalin rarely contains more than 36 per cent. of formaldehyde by volume, however, and in practice it is well to estimate 35 per cent. as a working basis. Commercial solutions, it must be remembered, are always acid, and the gas itself in small amounts is irritating to mucous membranes. Prolonged exposure to concentrations of the gas sufficient to kill bacteria may be fatal to animals. The gas has practically no insecticidal value. In sufficient concentration the gas is inflammable and may be ignited by any free flame.

In the past formaldehyde was liberated from its aqueous solution in the gaseous state in complicated retorts, autoclaves or lamps of special design. Much simpler methods have been evolved, which are now used almost exclusively in practical gaseous disinfection. Of these the permanganate method and the "sheet-volatilization method" are the most widely used; the former possesses the dual advantage of a quick liberation of the entire available amount of disinfectant, and very simple apparatus; the latter is advantageous when a gradual evolution of gas and a prolonged exposure to its action are desired.

The Permanganate Method.—When formalin is poured upon crystals of potassium permanganate, an energetic reaction with the evolution

[1] Hamilton and Ohno, Jour. Pub. Health, 1913; ibid., 1914, iv, 163.
[2] Bull. Hyg. Lab., Washington, D. C., April, 1912, No. 82.

of sufficient heat to boil the liquid takes place. Formaldehyde gas and heated water vapor are evolved. The entire process requires but a few minutes, and when two parts of formalin to one part of permanganate are used the residue is small in amount and practically dry and free from reactive substances.

Ten ounces of formalin and five ounces of permanganate of potash crystals are required for each thousand cubic feet of space to be disinfected. The temperature must be not less than 60° F., and the humidity must be at least 60 per cent. for successful results. It is convenient to place the permanganate in a three-gallon, galvanized-iron pail with flaring sides, because the reaction between permanganate and formalin is attended with considerable spattering. It is also advisable to place two or three layers of heavy paper under the pail, of sufficient size to project two feet at least in all directions, or better, to place a galvanized-iron plate of similar dimensions under the pail to catch all the liquid which is ejected from the pail during the process of evolution of the gas. For successful disinfection, all closets, drawers and alcoves should be opened as freely as possible; doors, windows and fireplaces leading to the exterior should be tightly closed. The room should be left closed and undisturbed for at least four hours.

The Sheet Volatilization Method.—This method requires no apparatus except sheets, and some mechanical device for spraying formalin upon the sheets. The conditions of moisture and humidity and the same general preparation of the room as for the potassium permanganate formalin method must prevail.

Sheets are hung upon tightly stretched cords or other similar support, in such a manner that they rest at an angle of about 45° with the perpendicular. They are wet with warm water, are "wrung out" to remove the excess, and sprayed with formalin in the proportion of ten ounces to each thirty square feet of surface. One sheet (thirty feet square) is sufficient for each thousand cubic feet of room space.

The evolution of formaldehyde is slower with the sheet method than with the permanganate method, but equally efficient disinfection is obtained if the room is kept closed eight hours.

2. **Paraform.**—Paraform is a polymer of formaldehyde; it is a white solid which is readily ignited, and burns with a bluish flame. It offers no advantages over formaldehyde, except that it occupies much less space. Special lamps have been devised to liberate formaldehyde from it in the gaseous state, but the efficiency of the method is not greater than the permanganate method, and the apparatus is some-

what more expensive, and bulky to transport. Paraform dissolved in warm water, in the proportion of two ounces of the former to half a pint of the latter may be used in place of formalin either in the permanganate method or the volatilization method described in the foregoing.

Attempts have been made to combine paraform and sulphur in the form of candles or pastilles for purposes of disinfection. Such preparations are valueless so far as the generation of formaldehyde is concerned, because the products of combustion of this substance are carbon dioxide and water.

3. **Sulphur.**—Sulphur was formerly highly regarded as a gaseous disinfectant, but it is now used chiefly for insecticidal fumigation. The products of combustion are SO_2 and SO_3, both gases; in the presence of moisture they have considerable germicidal activity, but little penetrating power.

Sulphur dioxide and trioxide are vigorous bleaching agents; they destroy fabrics, fine furnishings, and are injurious to painted or varnished surfaces. Consequently, the usefulness of sulphur as a germicide is restricted to the holds of ships, to warehouses and similar structures, where the destruction of vermin is an important factor in the disinfecting process.

At least 5 pounds of sulphur for each 1000 cubic feet of space to be disinfected are placed in a broad, shallow iron pot, preferably from one to two feet in diameter and from three to six inches high. These are placed in pans containing about two inches of water, both to prevent damage if the pot cracks during the burning process, and to supply moisture essential to the success of the disinfection. The sulphur should be not more than three inches deep in the pot and should slope gently from the edges of the pot to the center, where a crater is hollowed out and filled with an ounce of alcohol to start combustion. The sulphur burns slowly, and all cracks, doors and windows should be sealed with paper and paste to prevent escape of the fumes. At least twelve hours should be allowed before the room is opened.

Liquid sulphur dioxide is sometimes used in place of burning sulphur; the cost is several times that of burning sulphur, and for the practical disinfection of rooms it is rarely used.

4. **Chlorine Gas.**—Chlorine gas, particularly in humid atmospheres, possesses considerable germicidal power, but its extremely corrosive action upon fabrics and furnishings has materially restricted its field of usefulness for practical disinfection.

5. **Ozone.**—Nascent oxygen in actual contact with bacteria is a powerful germicide, and aside from the cost of production, it is of value for the purification of water for domestic purposes. As an aërial disinfectant, however, it has been disappointing.

PRACTICAL DISINFECTION.

Sputum.—The bacteria and other microörganisms which incite disease of the mouth, nose and respiratory tract leave the patient chiefly in the nasal secretion and sputum. They are eliminated in "droplets" of sputum during violent expulsion of the expired air, as in coughing and sneezing. The patient, therefore, should be instructed to cough or sneeze into paper or cloth napkins, to prevent the escape of infected droplets of sputum, and to expectorate into a sputum box provided with a cover. The paper napkins should be placed in a covered receptacle and eventually burned. Cloth napkins may be satisfactorily treated by complete immersion in *boiling* water for at least fifteen minutes.

Sputum may be disinfected with 5 per cent. carbolic or cresol solution, or with a 5 per cent. solution of formalin. At least one hour's exposure to the disinfectant is required.

Vomitus.—An elimination of pathogenic bacteria from the body in vomitus is by no means impossible, although relatively little attention has been paid to this subject in the past. The cholera vibrio s probably the most formidable organism to be reckoned with, but the possibility of typhoid bacilli must be borne in mind. Vomitus should be handled with the same precautions as infected feces.

Feces and Urine.—Those organisms which are the etiological agents of infections involving the gastro-intestinal tract, as typhoid, dysentery, paratyphoid bacilli and cholera vibrios, amoebæ, and probably the unknown excitants of the intestinal disorders escape from the diseased host chiefly in the feces, and occasionally in the urine.

The feces and urine should be received in porcelain or metal containers of appropriate pattern to prevent mechanical loss of material and immediately mixed with twice the volume of carbolic acid or cresol solution, an equal volume of 5 per cent. formalin solution, or with chloride of lime in the proportion of 10 per cent. of the total volume of feces and urine. The fecal mass, unless completely fluid, should be intimately mixed with the disinfectant solution and allowed to remain in contact with it at least an hour. The soiled parts of the patient should be wiped with a cloth dipped in 2 per cent. carbolic acid or cresol solution, then with water to remove the disinfectant. The cloths should be either placed at once in briskly boiling water, or in the bedpan, and treated with the feces.

Fomites.—Soiled linen, clothing and bedding should be immersed in a liberal amount of 2 or 3 per cent. carbolic acid solution and left at least two hours. An exposure of fifteen minutes in briskly boiling water, provided a considerable volume is used, is also sufficient to disinfect soiled fomites.

Bath Water.—The water in which patients suffering from intestinal infections have bathed should be disinfected before it is discharged into a drain. An ounce of chlorinated lime thoroughly mixed with the bath water will disinfect it within an hour. The sides of the bath-tub above the level of the water must be disinfected as well as the water itself.

Skin and Hands.—Infection of the skin and hands, both of the patient and attendants, is frequently unavoidable in intestinal diseases. A vigorous application of a scrubbing brush and green soap and a thorough cleansing of the nails frequently suffices for the hands. An application of 2 to 3 per cent. carbolic acid, or 1 to 1000 bichloride of mercury for several minutes will remove all danger of infection.

Sterilization of the hands for surgical operations is still a subject of debate; there is little uniformity in the methods advocated by leading surgeons. Wearing sterilized rubber gloves during operations is a common practice.

Instruments.—The preparation of instruments for surgical use, often erroneously called "sterilization," must be sharply distinguished from true sterilization in the bacteriological sense. Simple boiling of surgical appliances in soda solution does not necessarily render them free from bacterial spores, although the method is efficient for surgical technic because the residual bacteria which may survive this treatment do not germinate in the tissues. It is frequently deemed sufficient to boil syringes and other appliances used for removing blood or other material for bacteriological study; the only trustworthy method for this purpose is the autoclave or the hot-air sterilizer, depending upon the nature of the appliance.

The use of carbolic acid is not recommended for bacteriological syringes and other apparatus used in collecting material for bacteriological examination; it is difficult to remove the last traces of the disinfectant without contaminating the instrument itself.

Clinical Thermometers, Dental Instruments.—Clinical thermometers and dental instruments are ethically on a par with the common drinking cup and the common towel. Barbers' razors and brushes also belong to this group. The cost of such instruments is prohibitive for individual use, however, and their disinfection appears to be the practical solution of the problem. In hospitals the thermometers can be sterilized readily, first, by wiping them carefully to remove adherent mucus, then immersing them in 5 per cent. formalin solution,[1] where they remain until wanted again. A thorough rinsing in water will remove the formalin. The clinician who has an extensive visiting practice cannot afford individual thermometers; for practical purposes his thermometer can be kept free from bacteria if it is washed each time in running water until free from mucus, and kept in a metallic case containing 10 per cent. formalin solution prepared daily. Running water will remove all traces of formalin before use. At least two hours should be allowed before sterilization is regarded as complete. Several thermometers may be required to permit of this period of sterilization for each individual instrument.

Dentists' instruments almost without exception can be safely sterilized in a boiling 5 per cent. solution of washing soda (sodium carbonate) within five minutes' exposure. If they are then wiped dry there is little danger of rusting. The sterilization of dental mouth mirrors is a problem which would appear to require special investigation.

[1] A covered container is required; the fumes of formaldehyde are very irritating to the patient.

SECTION II.

PATHOGENIC BACTERIA.

CHAPTER XII.

THE PYOGENIC COCCI.

THE BACTERIA OF INFLAMMATION.

THERE is a group of bacteria which possesses in common the ability to incite that type of infection which is commonly spoken of as inflammation. A majority of these organisms are habitual parasites of man living upon the exposed surfaces of the body, the skin and mucous membranes chiefly: with respect to their pathogenic properties they may be regarded as "opportunists," not as a rule requiring a well-defined portal of entry through definite tissues to become invasive. Any break in the continuity of the skin or a weakening or change in the physiological state of a mucous membrane (frequently caused by intracurrent infection) provides the necessary atrium for invasion of the underlying tissues.

Not only are these bacteria ordinarily unable of themselves to locate and force an entrance to the tissues of their host; after invasion is accomplished they are unable to escape from the tissues in sufficient numbers to cause progressive disease of like nature in other hosts. They are locked up in the body, as it were, and eventually perish. They have not perfected their pathogenic mechanism. (See chapter on Parasitism.)

Bacteria of the "opportunist" type may be raised to very considerable pathogenic powers if artificially created atria of entrance to and escape from the tissues are provided, as for example, by passage through suitable animals, but they soon tend to lose their artificially acquired pathogenic properties under ordinary conditions and return again to a parasitic existence.

Prominent among these habitually parasitic bacteria which occur on the skin and mucous membranes of man are the various members of the Staphylococcus and Streptococcus Groups.

THE STAPHYLOCOCCUS GROUP.

Micrococcus Aureus.—Synonyms.—*Staphylococcus pyogenes aureus; Staphylococcus aureus; Micrococcus pyogenes aureus; Micrococcus salivarius aureus.*

Historical.—Staphylococci probably were first seen by Klebs, somewhat later by Billroth, in unstained pus. Pasteur[1] repeatedly isolated them from the pus of furuncles, and in one case of osteomyelitis, and suggested their etiological relationship to these lesions, but to Rosenbach[2] belongs the priority of growing them in cultures of undoubted purity.

Morphology.—The organisms in the free state are spherical, measuring from 0.7 to 0.9 micron in diameter. Those just about to divide are frequently oval. They occur singly, in pairs, or in irregular masses, both in culture and in pus; rarely chains of four to six cocci are found.

Staphylococci are non-motile and possess no flagella; they do not form capsules, and spores have not been observed. They stain readily with ordinary anilin dyes, some individuals more intensely than their fellows. They are Gram-positive.

Isolation and Culture.—Staphylococci are readily obtained in pure culture by plating or streaking the suspected material directly upon agar or gelatin. The colonies on gelatin after thirty to forty-eight hours' incubation at room temperature become visible as gray, glistening growths 0.5 to 1 mm. in diameter; somewhat later the colonies sink into saucer-shaped depressions of liquefied gelatin, the bacteria collect at the bottom of the depression and soon become golden-yellow in color. The growth upon agar plates at 37° C. is more rapid: at the end of forty-eight hours' incubation the colonies are golden-yellow and have attained a diameter of 1 to 3 mm.

Staphylococci grow readily in the ordinary cultural media. Gelatin, coagulated blood serum (sugar-free) and casein are liquefied.

[1] Compt. rend. Acad. Sci., 1880, xc, 1035.
[2] Mikroörganismen bei den Wundinfektionskrankheiten des Menschen, Wiesbaden, 1884, 19.

Acid is produced in dextrose, lactose, saccharose and mannite broths. Milk is coagulated, usually within three days at 37°.C.; many strains subsequently partially digest the coagulum. In plain and dextrose broths a turbidity is produced after twelve to fourteen hours' incubation at 37° C.; after forty-eight hours' growth a golden-yellow sediment collects in the bottom of the tubes. Growth on slanted agar is golden-yellow in color, moist and spreading. Pigment production is especially luxuriant on slanted potato.

The organisms are aërobic, facultatively anaërobic. The optimum temperature of growth lies between 28° and 38° C.; growth ceases below 8° C. and above 43° C.

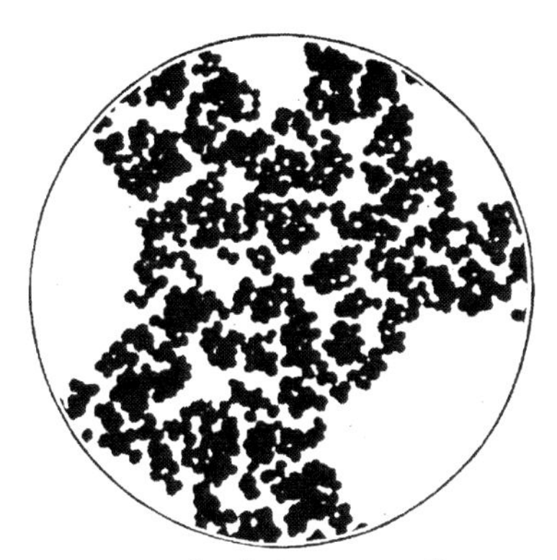

Fig. 32.—Staphylococcus. X 1000.

Staphylococci are among the most resistant of the non-spore-forming bacteria to physical agents. An exposure of one hour at 80° C or two hours at 70° C. moist heat is usually fatal. Several minutes' exposure at 100° C. (flowing steam) or twelve hours' exposure to direct sunlight may fail to kill them. Indirect daylight may fail to destroy their vitality even after two weeks; three months' continuous drying (on cloth or paper) is equally ineffective; 0.001 per cent. mercuric chloride and 5 per cent. carbolic acid usually kill the naked germs in about ten minutes.

Products of Growth.—Acids, chiefly lactic, but with demonstrable amounts of propionic, butyric, and valerianic, are formed during the fermentation of ordinary sugars. No gas is produced. The pus of staphylococcus abscesses is usually acid in reaction; the organisms appear to form limited amounts of acid from protein.[1] Emmering[2]

[1] Kendall, Day and Walker, Jour. Am. Chem. Assn., 1913, xxxv, 1246.
[2] Berlin. deut. chem. Gesellsch., 1896, 2721.

has identified indol, phenol, skatol, and trimethylamine among the decomposition products of staphylococci grown anaërobically in protein media. Cacace[1] has shown that the earlier decomposition products produced from gelatin and coagulated blood serum are chiefly proteoses and peptones; as proteolysis proceeds, these products are degraded to simpler amino acid compounds.

Pigment.—Staphylococci isolated directly from severe inflammations usually produce a golden-yellow pigment, but prolonged cultivation upon artificial media may result in a partial or complete loss of chromogenesis. Armand[2] has isolated non-chromogenic strains of staphylococci directly from typically chromogenic cultures by the plate method. The yellow pigment, which is produced most abundantly in media containing carbohydrates (particularly on potato) in the presence of free oxygen, appears to lie between the individual organisms, not within their substance. It is insoluble, or nearly so, in water, readily soluble in alcohol. It is related to the lipochromes. The pigment can be saponified readily, and it evolves an odor of acrolein when it is dry-heated. Strong acids, notably sulphuric, change the yellow color to a green-blue (lipocyanin). Lugol's solution (iodin-potassium iodide) turns it green.

Enzymes.—1. *Proteolytic.*—Old sugar-free broth and gelatin cultures of staphylococcus contain a proteolytic enzyme which will liquefy gelatin—a gelatinase. This enzyme may be obtained in an active state free from bacteria by filtering either broth or liquefied gelatin cultures of the organism through unglazed porcelain.[3] An enzyme which liquefies casein is demonstrable in milk cultures; whether the latter enzyme is identical with the gelatinase has not been determined.

2. *Amylolytic.*—According to Buxton,[4] staphylococci produce a maltase which hydrolyzes maltose; no other inverting enzymes have been observed.

3. *Lipolytic.*—Wells and Corper[5] have demonstrated a lipase of moderate activity in autolyzed agar slant cultures of staphylococci.

4. *Hemolytic.*—Neisser[6] and Wechsberg[7] have shown that old (7- to 14-day) broth cultures of staphylococci, particularly the more virulent strains, contain a soluble enzyme which hemolyzes blood

[1] Cent. f. Bakt., 1901, xxx, 244.
[2] Quoted by Lehmann and Neumann, Bacteriology, 1904, 3d ed., 193.
[3] Loeb, Cent. f. Bakt., 1902, xxxii, 471.
[4] Am. Med., 1903, vi, 137.
[5] Jour. Inf. Dis., 1912, xi, 388.
[6] Zeit. f. Hyg., 1901, xxxvi, 299.
[7] Cent. f. Bakt., Orig., 1903, xxxiv, 857.

both *in vivo* and *in vitro*. *In vitro* this enzyme, staphylolysin, appears to digest the stroma of red blood cells, liberating hemoglobin from them. A quantitative measure of the activity of this hemolysin can be made by adding gradually decreasing amounts of broth culture (filtered through unglazed porcelain) to well-washed red blood cells suspended in salt solution; the mixtures are incubated at 37° C. for one hour, then kept in the ice box twenty-four hours. The greatest dilution of broth showing hemolysis is considered the unit.[1] This enzyme is destroyed or inactivated at a temperature of 60° C. in twenty minutes. Whether this hemolysin is identical with or produced parallel to the proteolytic enzyme of the staphylococcus has not been determined. Burckhardt[2] believes the staphylolysin is a true hemolytic bacterial toxin; from his observations it appears to be non-protein in nature, not giving the biuret reaction. It is soluble in ether.

Leucocidin.—Van de Velde[3] has obtained an enzyme which destroys leukocytes by injecting virulent staphylococci into the pleural cavities of rabbits; the exudate, freed from cellular detritus by filtration through unglazed porcelain, rapidly kills and even dissolves fresh leukocytes. Neisser has shown that fresh leukocytes will reduce the color of dilute methylene blue solutions to the point of extinction; if dilute methylene blue is added to tubes containing leukocytes and leukocidin, no reduction occurs, thus indicating that the leukocytes are injured or destroyed. Leukocidin solutions alone fail to remove the color.

Thrombokinase.—Loeb's observation[4] that the products of growth of staphylococci cause blood to coagulate more rapidly than normal has been interpreted by Much[5] to be due to a substance reacting like a thrombokinase.

Distribution in Nature.—Staphylococci are found widely distributed in nature, but associated rather closely with man and the higher domestic animals. These organisms do not appear to be adapted to a purely saprophytic existence. They are found in dust, particularly that of stables, houses, and hospitals; they are common on the skin, the mucous membranes of the nose, mouth, and to a lesser extent in the gastro-intestinal tract,[6] the eye, the external ear, and nearly always

[1] It must be remembered that the sera of normal men and of animals frequently exhibit antihemolytic powers, hence the necessity of washing red blood cells thoroughly before testing the activity of staphylolysin upon them.

[2] Arch. exp. Path. u. Pharm., 1910, lxiii, 107.

[3] Ann. Inst. Past., 1896.

[4] Jour. Med. Res., 1903, x, 407.

[5] Biochem. Zeit., 1908, xiv, 143.

[6] Moro, Jahrb. f. Kinderheilk., lii, 530; Streit, Inaug. Diss., Bonn, 1897.

under the finger nails and in the hair follicles in man, which makes sterilization of the skin and hands difficult.

Chemotaxis.—The bodies of staphylococci appear to contain substances of unknown composition which attract leukocytes; the cell substance of killed cocci injected in the cornea frequently causes an accumulation of leukocytes in the anterior chamber of the eye—hypopyon.

Pathogenesis.—*Man.*—Ordinarily the organisms exist on the intact surfaces of man as "opportunists," occasionally gaining entrance to the underlying tissues through abrasions, chiefly in the skin, causing localized abscesses, furuncles, or metastatic inflammations. Of the metastatic inflammations, acute osteomyelitis and endocarditis are the more common; less commonly generalized purulent pyemias develop. It is assumed that metastatic pyemias are caused either by direct invasion of the blood stream or less commonly by transmission of staphylococci in leukocytes to remote parts of the body; there they escape from the leukocytes and set up new foci of infection. Suppurative pleurisy and pericarditis are not uncommon. The occurrence of furunculosis in diabetics is so frequent as to lead to the supposition that not only is the general average resistance to invasion by staphylococci reduced in this disease, there may be a peculiar local lack of resistance in the skin itself. Occasional individuals exhibit a certain vulnerability to infection in particular regions; the neck and buttocks are more frequently affected. One invasion appears to predispose to subsequent infection. Staphylococci frequently are secondary invaders in pulmonary tuberculosis, diphtheria and other severe infections. Generally speaking, staphylococci cause acute focal inflammations. Generalized infections of staphylococcus causation are relatively uncommon. Prolonged infections frequently result in profound generalized symptoms; chills with intermittent fever are the more common clinical signs. Parenchymatous or even amyloid degeneration of certain glandular organs, notably the kidneys, is the more common pathological lesion in such cases.

Experimental Reproduction of Lesions.—A satisfactory explanation of the pathogenesis of staphylococci for man is not available. Neither the staphylolysin nor the leukocidin appears to play a prominent part in the morbid process. There is little definite evidence that the cell substance of the organisms themselves is the important factor. Nevertheless, the etiological relationship of staphylococci to furunculosis has been definitely established by the experiments of Carré[1] and Engels,[2] both of whom rubbed virulent cultures of these organisms upon their skin, producing there typical furuncles.

Animals.—Rabbits are the best of the laboratory animals for experimental inoculation. Subcutaneous inoculations of virulent strains frequently result in abscess formation and the development of a febrile reaction. These abscesses commonly ulcerate, discharge and heal spontaneously. By no means do all virulent strains induce lesions, however; there is great difference between them in this respect. Intraperitoneal injections frequently cause a rapidly fatal peritonitis with or without septicemia. The intravenous injection of 0.25 to 1 c.c. of an eighteen-hour broth culture usually causes a generalized pyemia with septic foci, particularly frequent in the kidneys and liver. Orth[3] and Wyssokowitsch[4] have shown that mechanical injury to the heart valves prior to the intravenous injection of staphylococci usually causes a localization of the organisms there, producing an endocarditis. If a bone is injured prior to an intravenous injection, a typical osteomyelitis frequently results. It should be remembered that the pus produced by staphylococci in rabbits is more dry than that produced in man. Guinea-pigs are less susceptible than rabbits to infection with the staphylococcus.

Immunity and Immunization.—Staphylococci do not ordinarily exhibit invasive powers for man or animals; they are usually parasitic. Whenever the continuity of the skin is destroyed, as by abrasion, or weakened, as in diabetes, the organisms reach the underlying tissues and induce inflammatory reactions. Repeated injections first of killed then live staphylococci will frequently raise the threshold of infection in experimental animals to a very considerable degree, but the process of immunization can not be always relied upon—many animals die rather abruptly with rather extensive amyloid degeneration, particularly of the kidneys. Leukocytes, particularly the polymorphonuclear leukocytes, appear to play a prominent part in the immunity against staphylococci; it can be shown by experiment that the leukocytes are more active phagocytically in immunized than in non-immunized animals.

Similarly, the resistance to staphylococcus infection, which appears

[1] Fortschritt d. Med., 1885, 170.
[2] Cent. f. Bakt., Orig., 1903, xxxiv, 96.
[3] Cent. f. d. med. Wissensch., 1905, No. 33.
[4] Virchow's Arch., 1886, ciii.

to be rather marked in the average normal man, seems to depend largely on the phagocytic activity of leukocytes in the last analysis; and the efficiency of vaccines, particularly the autogenous vaccines, in the treatment of furunculosis has focused attention sharply upon the part played by opsonins in these infections. Generally speaking, injections of killed cultures of staphylococci in graduated doses beginning with one hundred millions and increasing to a thousand millions or more at appropriate intervals exert a favorable influence on the course of the infection. The efficiency of this vaccination (active immunization) is attributed to the gradual development of specific opsonins (bacteriotropins) which reënforce the action of normal opsonins, whose activity is somewhat below normal. In practice this is accomplished in the following manner: the organism is isolated on agar slants in pure culture, washed off, after twenty-four hours' incubation, in normal salt solution, thoroughly emulsified, and standardized so that each cubic centimeter contains the requisite number of bacteria. They are killed either by heating to 80° C. for one hour, or, better, by the addition of 0.5 per cent. carbolic acid, and incubation at 37° C. for twenty-four hours. The sterility of the culture must be demonstrated before it is used. This vaccine is inoculated subcutaneiously, with surgical precautions, using the dosage mentioned above as a routine. The inoculations are repeated at intervals of from five to eight days. The duration of the immunity induced by vaccination is not known. Vaccines are less effective in pyemia and metastatic staphylococcus infections than in the localized infections.

The lessened lipase activity of the blood, manifested by a decreased splitting of ethyl butyrate, is a frequent result of staphylococcus invasion, according to Clerc;[1] according to V. Dungern,[2] the blood serum from cases of extensive osteomyelitis is several times as inhibitory to the staphylococcus enzymes as is that of normal individuals.

Antibodies.—The cell substance of staphylococci does not appear to be very poisonous to experimental animals,[3] and although an antistaphylolysin and an antileukocidin are relatively easily produced in experimental animals, they do not appear to confer any considerable degree of immunity. Agglutinins do not appear to have been demonstrated in the blood serum of man and animals suffering from staphylococcal infections, but Kolb and Otto, and Proscher[4] claim

[1] Compt. rend. Soc. de biol., 1901, liii, 1131.
[2] München. med. Wchnschr., 1898, xlv, 1040.
[3] Kruse, Allgemeine Mikrobiologie, Leipzig, 1910, p. 968.
[4] Cent. f. Bakt., 1903, xxxiv: quoted by Besson, Practical Bacteriology, 1913.

to have prepared sera of marked agglutinating value, which clump virulent strains in higher dilution than non-virulent strains.

Precipitins.—Specific precipitin reactions appear to have been demonstrated in animals infected with staphylococci.

Bacteriological Diagnosis.—(*a*) *Microscopic.*—A Gram stain of the suspected material usually suffices to establish a diagnosis. It must be remembered, however, that staphylococci from pus and exudates may occur in pairs and even in short chains; they may, therefore, be mistaken for streptococci. An absolute diagnosis can be made only by the identification of pure cultures.

(*b*) *Cultural.*—Pure cultures of staphylococci are usually obtained readily by "streaking out" or plating the organisms on agar. Blood agar is preferable, if streptococci or pneumococci are also suspected

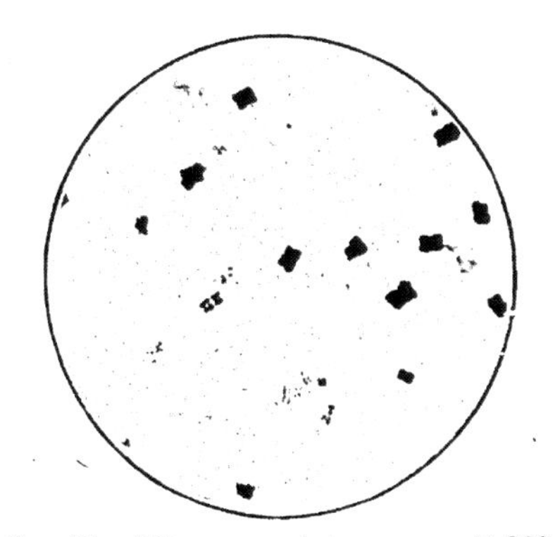

Fig. 33.—Micrococcus tetragenus. × 800.

to be present, otherwise the latter may be overlooked. The identification of the colonies on agar usually can be made by the examination of a Gram-stained preparation. Staphylococci are common on the skin, and precautions must be taken to eliminate this source of error before making cultures.

(*c*) *Animal Inoculation.*—The virulence exhibited by staphylococci for animals is not a reliable index of their virulence for man.

Dissemination and Prophylaxis.—The wide distribution of staphylococci on the mucous membranes, particularly on the skin and in the hair follicles, makes the prevention of their introduction to underlying tissues through cuts and abrasions difficult. The customary procedures of aseptic surgery are the best preventatives of infection. The skin may be sterilized for operation (after thorough cleansing and drying, which is imperative) by painting with freshly prepared tincture of iodin or iodoform. Sterilization is usually accomplished within ten minutes after the iodin is applied.

Staphylococcus Pyogenes Citreus.—This organism differs from Staphylococcus aureus chiefly in the color of the pigment it produces, a lemon yellow, and a lessened ability to liquefy gelatin.

Staphylococcus Pyogenes Albus.—In many instances this organism is an achromogenic variant of Staphylococcus aureus: it produces white colonies on agar and gelatin, it liquefies gelatin slowly, and it is somewhat less pathogenic for rabbits; many strains do not ferment mannite.

Staphylococcus Epidermidis Albus.—Welch first described this organism, which appears to be a degenerate Staphylococcus albus; it does not liquefy gelatin and its pathogenic powers are practically *nil.* It frequently causes the troublesome but relatively benign "stitch abscesses." It appears to be a very constant parasite on the skin.

Micrococcus Tetragenus.—Micrococcus tetragenus was first described by Gaffky;[1] he found it in cavities of the lung in pulmonary tuberculosis. It occurs but rarely in pure culture in abscesses either in man or animals,[2] but it is often present in the saliva; occasionally it has been recovered from dento-alveolar abscesses.[3]

Morphology.—The organism occurs typically in tetrads, enclosed in transparent gelatinous capsules which require special staining methods for their demonstration. The individual cells are about 1 micron in diameter. In artificial media the tetrad arrangement may disappear and the cocci occur chiefly in pairs and groups of three or four pairs. The tetrad arrangement and the capsule are restored by passage through animals. The organism is non-motile, and possesses no flagella. It forms no spores and stains readily with ordinary anilin dyes. It is Gram-positive.

Isolation and Culture.—Micrococcus tetragenus grows rather slowly in all ordinary media, particularly the first transfers from the tissues to artificial media. It can be isolated readily in pure culture in gelatin or agar plates; the colonies are small, round and grayish, 0.5 to 0.75 mm. in diameter.

Growth in Media.—The organism does not liquefy gelatin, casein, or blood serum. Acid is produced in dextrose, lactose, saccharose, and

[1] Mitt. a. d. kais. Gesamte, i, p. 1.
[2] Müller, Wien. klin. Wchnschr., 1904, xvii, 1815.
[3] Goadby, Mycology of the Mouth, 1903, p. 101.

mannite broths. A uniform turbidity is produced in plain and sugar broths; the growth is more luxuriant in the latter. Milk is slightly acidulated, but no coagulation or peptonization takes place. Micrococcus tetragenus is aërobic, facultatively anaërobie. The optimum temperature of growth is 37° C., the maximum about 44° C., the minimum about 12° C. The resistance to physical and chemical agents is undetermined.

Products of Growth.—Unknown: no toxin has been described.

Pathogenesis.—The frequent occurrence of the organism in the sputum of the tuberculous and its occasional isolation from tuberculous cavities has led to the theory that Micrococcus tetragenus may play a secondary part in the destruction of lung tissue. This is not definitely determined, however. It is also found in the saliva of healthy individuals. Less commonly it has been found in the pus of empyemas which follow pneumonia; but the organism can hardly be regarded as a human pathogen.

Injected subcutaneously into white mice, Micrococcus tetragenus usually causes a fatal septicemia; the organism may be recovered from the heart blood, spleen and liver. House and field mice appear to be relatively immune. Intraperitoneal injection into guinea-pigs may cause a fatal peritonitis with much pus in which typical tetrads are found. Rabbits and dogs are not susceptible. Infections with the organism in man are so uncommon that nothing is definitely known of human susceptibility and immunity. Vaccines have been tried in a very few cases with somewhat promising results.

Bacteriological Diagnosis.—The finding of Gram-positive cocci about 1 micron in diameter in pus, which occur habitually in tetrads, usually suffices to establish a satisfactory bacteriological diagnosis. The saliva occasionally contains tetracocci which resemble Micrococcus tetragenus very closely, but it is claimed by many that these organisms are not necessarily Micrococcus tetragenus. Isolation and identification by cultural methods must be resorted to in suspected cases.

Micrococcus Ovalis.—**Synonym.**—Enterococcus.[1]

Historical.—Micrococcus ovalis was described by Escherich,[2] who found it very commonly in the intestinal tracts of nurslings and bottle-fed infants.

Morphology.—The organism is oval in outline, measuring 0.6 to 0.9 microns in the lesser diameter, and it occurs habitually in pairs,

[1] Thiercelin, Thèse de Paris, 1894.
[2] Darmbakterien des Säuglings, Stuttgart, 1886, p. 89.

with a tendency for the proximal ends to be slightly flattened and the distal ends to be somewhat pointed. In this respect Micrococcus ovalis resembles the pneumococcus very closely. In fluid media, particularly sugar broths, the pairs of organisms remain adherent in chains of greater or lesser length giving rise to a diplostreptococcus arrangement which is precisely like that exhibited by the pneumococcus under the same conditions.

Micrococcus ovalis is non-motile and possesses no flagella. It forms no spores. According to Lewkowicz[1] and others, capsules are produced when the organism is isolated directly from lesions. The organism stains readily with ordinary anilin dyes, and it is Gram positive.

Isolation and Culture.—Micrococcus ovalis grows with moderate vigor on agar plates, better in dextrose or lactose agar. The colonies after forty-eight hours' incubation at 37° C. are round, translucent, colorless, and measure about 1 to 2.5 microns in diameter. They are not distinctive. Colonies on gelatin plates are very small and develop slowly. The medium is not liquefied. Blood agar appears to be a better medium for isolation of Micrococcus ovalis than any other; the colonies are 1 to 3 mm. in diameter even after eighteen hours' incubation, grayish and succulent. No hemolysis takes place. A slight turbidity, which soon settles, forms in plain broth; the addition of dextrose or lactose greatly enriches the growth. Milk is usually coagulated in one to three days (acid coagulation), but the coagulum does not become digested.

Micrococcus ovalis is an aërobic, facultatively anaërobic organism. The lower limit of growth is about 8° C., the optimum from 37° to 39° C., and the maximum about 45° C. Its resistance to chemical and physical agents is about the same as that of the streptococcus.

Products of Growth.—*Chemical.*—The organism exhibits no evidence of proteolysins; it is relatively inert in protein media. No indol, skatol or volatile sulphur compounds are produced. Acid is produced in dextrose and lactose broths; the action on other sugars is yet to be determined.

Enzymes.—No enzymes are known.

Toxins.—No toxic products have been detected in cultures of Micrococcus ovalis.

Distribution.—The normal habitat of Micrococcus ovalis appears to be the intestinal tract of man; it occurs in the meconium frequently,[2]

[1] Cent. f. Bakt., 1901, xxix, 635.
[2] Escherich, loc. cit.

and it is a constant inhabitant of the intestinal flora of artificially fed infants; it also occurs commonly, but in lesser numbers, in the intestinal flora of the normal nursling. The organism has been repeatedly isolated from the feces of adults, and it has also been isolated from the intestinal tract of cattle.[1]

Pathogenesis.—*Man.*—Micrococcus ovalis is ordinarily a harmless parasite of the intestinal tract; occasionally it becomes invasive (usually secondarily) and produces various inflammations, according to the tissues invaded and its association with other bacteria. Lewkowicz[2] isolated Micrococcus ovalis in nearly pure culture from three cases of severe dysentery; the organisms were found to be capsulated and resembled pneumococci in a striking manner. Jouhaud,[3] Thiercelin,[4] Ramonovitsch,[5] and Gilbert and Lippman[6] have isolated the organism either in pure culture or in association with other bacteria from cases of cholecystitis, puerperal fever, appendicitis, various intestinal inflammations, and even from the cerebrospinal canal in cases of meningitis. The close resemblance of the organism to the pneumococcus, which has been observed by Kruse,[7] Sittler[8] and others, has doubtless led to confusion; many cases of "pneumococcus" infection of the stomach, gall-bladder, appendix and other intestinal adnexa are probably infections with Micrococcus ovalis, and *vice versa.*

Animal.—Wilhelmi[9] has isolated Micrococcus ovalis from enteritides of young cattle; Lewkowicz[10] has found the organism isolated directly from human inflammations to be pathogenic for white mice. It exhibits no pathogenicity as it occurs normally in the intestinal tract.[11]

Bacteriological Diagnosis.—1. *Microscopical.*—The presence of considerable numbers of diplococci in the feces with their approximated ends slightly flattened, their distal ends somewhat pointed, staining intensely with the Gram stain, is frequently sufficient evidence to establish a tentative diagnosis of Micrococcus ovalis.

2. *Cultural.*—Various dilutions of feces or products of inflammation are plated either on dextrose agar or "streaked out" on blood agar.

[1] Wilhelmi, Landwirthschaft. Jahrb. f. Schweiz., 1899, xiii.
[2] Cent. f. Bakt., 1901, xxix, 635.
[3] Thèse de Paris, 1903.
[4] Comp. rend. Soc. de biol., 1902, No. 27; 1908, lxiv, 76.
[5] Ibid., 1911, lxx, 122.
[6] Ibid., 1902, No. 30.
[7] Cent. f. Bakt., Orig., 1903, xxxiv, 737.
[8] Die wichtigsten Bakterientypen der Darmflora beim Säuglinge, u. s. w., Wurzburg, 1909.
[9] Landwirthschaftl. Jahrb. f. Schweiz., 1899, xiii.
[10] Loc. cit.
[11] Thiercelin, Thèse de Paris, 1894; Compt. rend. Soc. de biol., April 15, 1899. Jouhaud, Thèse de Paris, 1903.

The morphology and cultural reactions outlined above suffice to establish a diagnosis. The absence of hemolysis or of green discoloration of the hemoglobin separates the streptococcus and pneumococcus from Micrococcus ovalis.

3. *Serological.*—Not practicable.

Dissemination and Prophylaxis.—Micrococcus ovalis does not cause progressive disease from man to man; it is an intestinal parasite habitually and only occasionally becomes invasive. No precautions other than the careful sterilization of dejecta are necessary. The hands of attendants should be kept surgically clean when caring for intestinal disturbances incited by Micrococcus ovalis, or, indeed, by any microörganism.

CHAPTER XIII.

THE STREPTOCOCCUS-PNEUMOCOCCUS GROUP.

THE STREPTOCOCCUS GROUP.

THE Streptococcus Group comprises those spherical bacteria in which as multiplication proceeds the successive planes of division are parallel and the individual cells remain adherent in longer or shorter chains. The limits of the group are poorly defined, both morphologically and pathogenically. It includes organisms which occur habitually in chains, both in culture and in the animal body, and its limits have been extended to enclose types which exhibit chain formation only in fluid media. The latter, of which Micrococcus ovalis and the pneumococcus are examples, occur in the animal body as diplococci, and grow thus on solid media; in fluid media they grow habitually in chains of greater or lesser length, in which, however, the typical diplococcal arrangement persists. The term streptococcus, therefore, is a purely morphological one; it includes organisms which excite various types of inflammation in man and in animals, together with those which are ordinarily saprophytic.

The most important members of the group exist on the skin, and particularly on the mucous membranes of man, as habitual parasites or "opportunists." Streptococcus pyogenes and its variants are the most common of these and the most versatile in their pathogenesis.

Streptococcus Pyogenes.—Synonyms.—Streptococcus erysipelatos; Streptococcus scarlatinosus; Streptococcus septicus.

Historical.—Streptococci were seen in unstained pus by Klebs in 1872. Several years later Koch[1] demonstrated them in stained sections and in inflammatory exudates. Pasteur[2] appears to have been the first to cultivate streptococci from cases of puerperal fever and to differentiate them from staphylococci, both morphologically and by

[1] Untersuchungen über Wundinfektion, 1878.
[2] Compt. rend. Acad. sci., 1880, xc, 1035.

the character of the lesions which they excite. Ogsten[1] independently confirmed Pasteur's observations. Fehleisen,[2] using more exact cultural methods, isolated streptococci from a case of erysipelas; Rosenbach[3] studied the organism in great detail and introduced the name, Streptococcus pyogenes.

Morphology.—The individual cells are spherical, less commonly oval, measuring from 0.5 to 1 micron in diameter. The size of individual cells varies somewhat even in the same culture. The organisms remain adherent in chains which vary in length from four to twenty or more elements, in which a definite association of cocci in pairs with their proximate sides flattened is occasionally observed. The number of elements in the chain varies somewhat according to the origin of the culture; it has been observed that streptococci freshly isolated from lesions tend to occur in longer chains, while those organisms which grow habitually upon the normal surfaces and mucous membranes of the body appear more frequently in shorter chains. V. Lingelsheim[4] has designated those strains which form chains of eight or more cocci, Streptococcus longus; the short-chain types are called Streptococcus brevis. Notwithstanding the frequent parallelism of pathogenesis and development of long chains of cocci in artificial media, in contradistinction to the lesser virulence of the short-chain types, experience has shown that the length of the chains may also be influenced directly by variations in the culture media.[5] This distinction, therefore, is untenable. Streptococci grown on solid media are prone to group themselves in pairs, or even irregular masses, resembling staphylococci. Similarly, the typical streptococcal arrangement is frequently lacking in purulent inflammations of streptococcal causation. Occasional cells in a chain of streptococci, especially in old cultures, are met with which are distinctly larger than their fellows; they color somewhat differently and were formerly regarded as spores —arthrospores.[6] It is now known that they are not noticeably more resistant than the more typical cells, and they are probably to be regarded as involution forms.

Streptococcus pyogenes is non-motile, non-flagellated, and does not produce true endospores. Occasional strains, isolated directly

[1] Brit. Med. Jour., 1881.
[2] Aetiol. d. Erysipelas, Berlin, 1883.
[3] Mikroörganismen bei Wundinfektions-Krankh. des Menschen, Wiesbaden, 1884.
[4] Zeit. f. Hyg., 1891, x, 331.
[5] Hueppe, Die Methoden der Bakterien-Forschung, Wiesbaden, 1889, 24, 130.
[6] See Aronson (Berl. klin. Wchnschr., 1896, No. 32; 1902, No. 42) and Vincent (Arch. de méd. exp., etc., 1902) for details.

from lesions or from animals, exhibit a delicate stainable zone around individual organisms or pairs of organisms, which suggests capsules. Howard and Perkins[1] have isolated such an organism which exhibited a very definite capsule. It grew habitually in short chains in fluid media, the individuals occurring typically in pairs. The organism is closely related to the pneumococcus, and Dochez and Gillespie[2] have named it Pneumococcus mucosus.

Streptococcus pyogenes stains readily with ordinary anilin dyes. It is typically Gram-positive, although old cultures may fail to retain the Gram stain. The saprophytic types frequently are Gram-negative.

Isolation and Culture.—Streptococci may be isolated directly from inflamed areas and from pus upon agar plates, better upon dextrose agar plates. The colonies are minute, gray and transparent, and may be readily overlooked; if they occur in association with staphylococci or other rapidly growing organisms, they are readily overgrown. The more virulent varieties develop less readily, and require the addition of blood or ascitic fluid to ordinary media for their initial growth outside the body. On blood agar plates (one part human blood, two parts of nutrient, *sugar-free* agar) the majority of virulent streptococci produce a wide, clear zone of hemolysis 4 to 8 mm. in diameter around each colony. This medium is particularly valuable for the isolation of streptococci.[3] On Löffler's blood serum growth is moderately luxuriant; typical chains are found in the condensation water of solid media, but not as a rule upon the surface. The organisms grow feebly in gelatin stab cultures producing a few small discrete gray colonies along the line of inoculation. Little or no growth is found on the surface of the medium. Liquefaction does not take place.

A slightly alkaline reaction (neutral to phenolphthalein) is most favorable for the growth of streptococci; the addition of sugars, particularly dextrose, to ordinary media (but not blood agar) increases the rate and extent of development, which, however, are soon limited by the accumulation of acid products of fermentation. The addition

[1] Jour. Med. Research, 1901, vi, 163.

[2] Jour. Am. Med. Assn., 1913, lxi, 727.

[3] Schottmüller (Müncb. med. Wchnschr., 1903, xx, 849) has classified streptococci according to the changes they produce in blood agar as follows:

I. Streptococcus longus pyogenes seu erysipelatis (Streptococcus pyogenes) produces a wide, clear zone of hemolysis around the colony; in blood broth the color changes to a burgundy red. Long-chained streptococci.

II. Streptococcus mitior seu viridans (Streptococcus viridans) produces a greenish area around the colony; a brownish color in blood broth. Short-chained streptococci.

III. Streptococcus mucosus. No hemolysis on blood agar. Colonies viscid. Organisms distinctly encapsulated.

of solid calcium carbonate (marble) to sugar media is important; it neutralizes the excess of acid, and also appears to add somewhat to the nutritive value of the medium.[1]

Streptococci grow slowly in plain broth, producing a sediment after twenty-four to forty-eight hours' incubation. A flocculent sediment consisting of long chains of organisms is characteristic but not distinctive of many virulent strains (Streptococcus conglomeratus); a granular sediment usually contains short-chain streptococci almost exclusively.

Streptococcus pyogenes ferments dextrose, lactose, maltose and saccharose and sorbite with the formation of considerable amounts of acid. Mannite is not as a rule attacked. Milk is coagulated in from three to five days, the coagulum resulting from the accumulation of the acid fermentation of the lactose. The coagulum is never dissolved. Andrewes and Horder[2] state that Streptococcus pyogenes does not coagulate milk, although the organism produces a considerable amount of acid in this medium. Smith and Brown[3] have shown that boiling the milk may be necessary to make the coagulum visible.

Streptococcus pyogenes is an aërobic, facultatively anaërobic organism. Pathogenic strains do not as a rule grow below 16 to 18° C. The optimum temperature lies between 35° and 39° C., the maximum about 44° C. The parasitic types are not long-lived away from the human body. Exposure to 60° C. for one hour will kill most streptococci; a longer time is required if the organisms are exposed in albuminous media. Five per cent. carbolic acid and 1 to 1000 mercuric chloride will kill the naked germs in from five to ten minutes. Streptococci dried in sputum will resist a temperature of 100° C. (in flowing steam) for several minutes, and drying at ordinary temperatures in the dark for several weeks. Direct sunlight kills them in about ten hours. The organisms survive and retain their virulence if they are suspended in sterile, defibrinated blood and kept in the ice box for several weeks.

Products of Growth.—*Chemical.*—Streptococci exhibit but little evidence of proteolytic activity. No indol, skatol, phenol or other aromatic derivatives of amino acids have been detected in cultures; gelatin is not liquefied and casein and coagulated blood serum are not visibly changed. Emmerling[4] found peptone, leucin, tyrosin,

[1] Bolduan, New York Med. Jour., 1905, May 13.
[2] Lancet, 1906, ii, 708.
[3] Jour. Med. Research, 1914, xxxi, 455.
[4] Berl. chem. Gesell., 1897, 1863.

ammonia, methylamine, propyl pyridin, succinic acid, butyric acid and other volatile acids among the anaërobie decomposition products of fibrin by this organism, but no aromatic derivatives.

Toxin.—A soluble toxin has not been demonstrated in cultures of streptococci, although substances have been isolated by Marmorek[1] and others which will kill guinea-pigs. These substances do not exhibit sufficient potency to warrant the assumption that they are important factors in the production of the grave symptoms characteristic of severe streptococcus infections. Attempts to demonstrate endotoxin have also been unsuccessful; the bodies of the organisms are but slightly toxic to experimental animals. The manifestations of toxemia in streptococcal infections, however, are too striking to

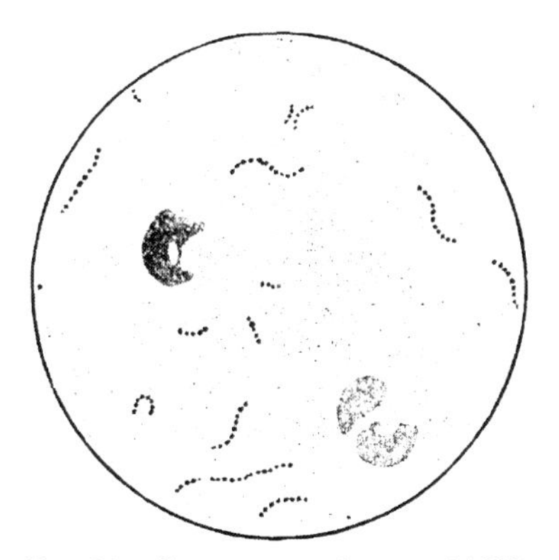

FIG. 34.—Streptococcus in pus. × 800.

be reconciled with the negative results of these investigations; the nature of the mechanism of streptococcus infection remains to be elucidated.

Hemolysin—Streptocolysin.—Bordet[2] and Besredka[3] have shown that filtered broth cultures of streptococci will dissolve red blood corpuscles, liberating hemoglobin, and that this hemolytic substance —streptocolysin—is active both *in vivo* and *in vitro*. Frequently the blood of rabbits injected with streptocolysin was found to be "laked" just before death. Besredka's observations would indicate that the substance is rather firmly bound to the organisms and does not appear in the medium to any considerable degree. M'Leod,[4] M'Leod and

[1] Berl. klin. Wchnschr., 1902, xiv, 253.
[2] Ann. Inst. Past., 1897, xi, 177.
[3] Ibid., 1901, xv, 880.
[4] Jour. Path. and Bact., 1912, xvi, 321.

M'Nee,[1] and Lyall[2] have studied the conditions favoring the formation of the hemolysin and find that sugar-free ascitic broth is suitable for this purpose. The substance is thermolabile and is found in an active state only during the first twelve to twenty-four hours of culture, at which time small amounts of sterile (filtered) broth, 0.01 to 0.10 c.c., are strongly hemolytic. The hemolysin does not induce antibody formation when it is injected into susceptible animals. Hemoglobinemia and hemoglobinurea are produced in rabbits that are very susceptible to the hemolysin; less susceptible rabbits react but slightly. There is no definite evidence that streptocolysin plays a prominent part in the streptococcus infections of man. Virulence and hemolytic activity are frequently, but by no means necessarily, parallel phenomena.

Distribution in Nature.—Streptococci are widely distributed in nature, always, however, in rather intimate association with man or the higher animals. They are found in the soil, water, milk, and they exist as "opportunists" on the exposed surfaces and mucous membranes of man. They are common in the mouth, nose and throat, the intestinal tract, and rare in the normal vagina.

Pathogenesis.—*Human.*—Streptococci excite both local inflammatory and suppurative processes and generalized septicemic infections, the latter being the more common and characteristic. Superficial lesions may be mild in character, resembling those caused by staphylococci. The organisms may, and frequently do, enter the blood or lymph channels, and spread rapidly through the body, inciting the most severe generalized infections. Streptococci are the etiological agents of erysipelas, frequently of general and puerperal sepsis and phlebitis, and inflammations of the internal organs; of these, the middle ear, the endocardium, the peritoneum, the meninges or joints are more commonly involved.[3] Escherich[4] and others have described a severe type of enteritis, particularly of young children—streptococcus enteritis—which occasionally exhibits an epidemic tendency in the summer months.[5] Attention has been directed in recent years to severe epidemics of septic sore throat in which the

[1] Ibid., 1913, xvii, 524.
[2] Jour. Med. Research, 1914, xxx, 487.
[3] Menzer, Deut. med. Wchnschr., 1901, 97. Meyer, Zeit. f. klin. Med., 1902, xlvi, 311; Internat. Beiträge zur inn. Med., 1902, ii, 443. Philipp, Deut. Arch. f. klin. Med., 1903, lxxvi, 150. Poynton and Payne, Cent. f. Bakt., Orig., 1902, xxxi, 502. Cole, Jour. Inf. Dis., 1904, i, 714. Rosenow, Jour. Inf. Dis., 1910, vii, 411; ibid., 1912, xi, 210; Jour. Am. Med. Assn., 1913, lx, 1223.
[4] Jahrb. f. Kinderheilk., 1899, xlix, 137.
[5] Kendall, Day and Bagg, Boston Med. and Surg. Jour., 1913, clxix, 741.

evidence points to streptococci transmitted through milk as the etiological agent. The type of streptococcus involved has been a subject of controversy, but the extensive studies of Smith and Brown[1] show clearly that Streptococcus pyogenes is by far the most common organism found. They demonstrated that the streptococcus which is isolated from bovine mastitis is not, except possibly in rare instances, a causative factor in epidemic sore throat.

Streptococci occur frequently as secondary invaders in diphtheria, many gastro-intestinal diseases, and diseases of the lungs, where they may be at times even more formidable than the primary infecting organism. As Theobald Smith has admirably expressed it, they are "organisms of the diseased state." The virulence exhibited by streptococci varies considerably, as does the type of lesions they excite. This variation in virulence is not at all well understood at the present time, but experiments indicate that the site of infection and the past history of the organism exercise some influence. Rosenow[2] has isolated streptococci, using special methods, from the regional glands in arthritis, gall-bladders, and gastric ulcers. He states that the freshly-isolated strains exhibit rather marked tendencies to localize in the homologous tissues of experimental animals. This specific tissue affinity is rapidly lost during cultivation of the organisms in artificial media, however.

Animal.—Fränkel,[3] Petruschky,[4] and Koch and Petruschky[5] showed that the virulence of the same strain of streptococcus varied materially according to the conditions of culture, and that the lesions produced in rabbits varied likewise; thus the descendants of the same culture would produce variously a rapidly fatal septicemia, erysipelas, arthritis, endocarditis or peritonitis. Marmorek has shown that the virulence of streptococci for animals may be greatly increased by repeated passage; after a series of passages an incredibly small amount of culture, even one one-hundred-millionth of a cubic centimeter of a forty-eight-hour broth culture introduced intraperitoneally may cause death within two days. Streptococci which are virulent for man frequently exhibit but little virulence for animals; it is essential, therefore, that large amounts of material be injected into experimental animals to obtain infection. Rabbits are more susceptible than other labora-

[1] Jour. Med. Research, 1914, xxxi, 455.
[2] Jour. Am. Med. Assn., 1913, lx, 1223; lxi, 1947; 1914, lxiii, 1835. Jour. Inf. Dis., 1915, xvi, No. 2.
[3] Cent. f. Bakt., 1889, vi, 671.
[4] Zeit. f. Hyg., 1896, xxiii, 144.
[5] Ibid., p. 478.

tory animals. Subcutaneous injections of morbid material into rabbits result variously, depending upon the virulence of the strain for this animal (not necessarily upon its virulence for man); a localized abscess may form or an erysipelatoid inflammation may occur, which is usually somewhat localized, but may develop into a wide-spread cellulitis. Intraperitoneal injections are usually followed by rapidly-fatal peritonitis. Death may occur within twenty-four hours. Intravenous injections may cause a rapidly fatal generalized septicemia, or, if the strain is less virulent and death does not occur during the first three to four days, the serous surfaces may be violently inflamed. Less virulent strains which do not cause acute death usually lead to endocardial or joint involvement. Mice are nearly as susceptible to strepto-

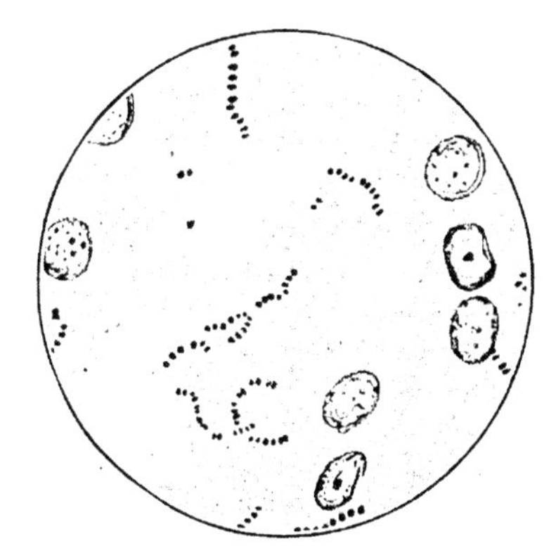

FIG. 35.—Streptococci in liver, section stained by Gram's method. X 800. (Kolle and Hetsch.)

coccus infection as rabbits. Guinea-pigs are less susceptible; subcutaneous inoculations usually lead to abscess formation, which soon heals, but intraperitoneal injections may result in peritonitis and death. Horses are quite susceptible to infection with streptococci, particularly with Streptococcus equi (Streptococcus coryzæ contagiosæ equorum), which causes equine distemper or strangles. The udders of milch cattle occasionally become infected with streptococci resulting in a severe inflammation, mastitis or garget, which may lead to loss of function of one or more quarters of the udder. It is probable from the investigations of Smith and Brown[1] that streptococci of bovine origin are not commonly the etiological agents of septic sore throat in man.

[1] Loc. cit.

Immunity and Immunization.—Streptococcus infections, mild or severe, do not appear to induce any considerable degree of active immunity. Not infrequently recovery is a matter of some time; the acute symptoms may abate and the organisms disappear from the blood stream, only to localize in some internal organ, a structure as for example, a joint, where they may cause a chronic, obstinate arthritis. It is possible that various strains of streptococci which can not be differentiated by our somewhat artificial cultural criteria may exist, and that subsequent infection may be with another strain. A similar condition exists in lobar pneumonia. Van de Velde[1] has stated that the serum of an animal immunized against one strain of streptococcus will protect against the homologous strain, but not against heterologous strains of streptococci, a somewhat parallel situation. On the other hand, experiments are recorded which are not in accord with this hypothesis. A patient suffering from an inoperable tumor was inoculated subcutaneously with a culture of streptococcus; the inoculation resulted in a moderately severe erysipelas which persisted for about ten days; when the inflammation had subsided a second reinoculation was made in the same place, and a secondary erysipelatoid inflammation spread over the same area. A third inoculation resulted similarly. These experiments indicate that this patient did not develop immunity at the site of infection.[2]

Rabbits have been actively immunized to streptococci through repeated vaccination, first with killed cultures, then gradually increasing doses of living, virulent organisms; eventually the animals will resist successfully several times the original fatal dose of the homologous strain. Active immunization with polyvalent vaccines containing many strains of streptococci from lesions is considerably more efficient in protecting the animal against subsequent infection with heterogeneous strains. The sera of such actively immunized animals do not possess noteworthy antihemolytic properties; their antitoxic content, if indeed there be any, is unknown. The chief demonstrable change in the serum appears to be an increased phagocytic power, causing leukocytes *in vitro* to take up more streptococci than they would normally. The injection of sera of actively immunized animals appears to increase the resistance of non-immunized animals to otherwise fatal amounts of streptococci.

[1] *Cent.* f. Bakt., 1898, xxiv, 688.

[2] Coley has injected streptococci into malignant tumors with occasional beneficial results; the observations are too few to warrant any definite statement of the efficiency of the procedure.

Marmorek,[1] Tavel[2] and others have prepared antistreptococcic immune sera on a large scale by immunizing horses first with killed cultures, then with increasing amounts of living cultures. Marmorek, a staunch supporter of the "Einheit" theory that all streptococci were identical, used a single strain of organism, whose virulence was greatly increased for rabbits prior to injection into horses. Immunization requires several months. He found that for some days following each injection the horse exhibited a febrile reaction, and during that period the serum was toxic for rabbits; streptococci may be found in the blood stream during this period. After the temperature has reached normal—three weeks or more after the injection—the toxic properties disappear and the serum exhibits protective powers when it is introduced into rabbits with a lethal dose of streptococci. This serum has been used extensively in the treatment of erysipelas, puerperal fever, and scarlet fever, but its curative value is still a matter of discussion.

Tavel's serum is essentially like that of Marmorek, except that a polyvalent vaccine is used for immunization. Besredka also uses a polyvalent vaccine for immunizing horses, but the organisms are not exalted in virulence for rabbits by passage through a series of them before inoculating horses. Besredka believes that passage through rabbits may modify the virulence of the streptococci for man, from whom the organisms are obtained for immunizing the horses, and for whom the serum is to be used. Streptococcal sera are as yet of debatable value; in localized lesions they have frequently exhibited some therapeutic value; in the severe generalized infections in man they are usually either irregular in their action or inactive. Somewhat more encouraging results have been reported where the specific immune serum is used in connection with autogenous vaccines of streptococci.

Antibodies.—Agglutinins are present in the sera of animals immunized with streptococcus vaccines, and the degree of agglutinating power may be very considerable for homologous strains. The results are usually less definite with heterologous strains, and agglutinins developed during immunization with streptococci are of no considerable value in prognosis. The part they may play in immunization is problematical.

Complement fixation has not been found a satisfactory method for

[1] Ann. Inst. Past., 1895, ix, 593.
[2] Loc. cit.

identifying streptococci; the results are occasionally variable without apparent cause.

Bacteriological Diagnosis.—1. *Microscopical Examination.*—Smears from abscesses or inflammatory areas usually exhibit pairs and short chains of cocci which retain the Gram stain. Occasionally the organisms can not be distinguished with certainty from staphylococci. Frequently, when microscopic examination fails (and this is usually the case when blood is examined), streptococci are found by cultural methods.

2. *Cultural Examination.*—If the material is purulent, it may be streaked or plated out on 0.1 per cent. dextrose agar; the colonies are small and transparent, and may be easily overlooked. Blood, lymph or serum should be plated on blood agar. If the material is blood, one part may be added to two parts of melted plain agar, and the whole, after thorough mixing, may be poured into sterile Petri dishes. Usually small, gray colonies with relatively broad, clear areas of hemolysis appear within forty-eight hours. If lymph and serum be the suspected material, blood agar should be used for plating out. Hemolytic colonies, as above, appear usually within two days. It is always well to inoculate 1 or 2 c.c. of blood serum or lymph into broth and maintain it at 37° C. for twenty-four hours to enrich the culture, then plate on blood agar; also inoculate a like amount into a rabbit.

3. *Animal Inoculation.*—The intraperitoneal injection of suspected fluids into rabbits frequently results in a fatal peritonitis, from which the organism may be recovered from the blood stream. Relatively large amounts should be used.

The detection of streptococci in the blood of a patient is frequently an unfavorable clinical sign; it does not necessarily, however, justify a grave prognosis. Cases are met with which present symptoms of septicemia, yet the organisms may not be obtained from the blood. Occasionally the patient dies from toxemia, due apparently to the absorption of toxic substances from the local infection. Streptococci from erysipelas, septicemia, scarlet fever, and even from articular rheumatism are so similar culturally and morphologically that the various strains can not be differentiated with certainty; slight variations in cultural reactions are exhibited by all these strains. Neither does animal experimentation afford definite criteria for the establishment of types. Even one passage through an animal may modify the pathogenicity greatly.

In the light of our present knowledge the resistance of different tissues and the portal of entry play a prominent part in determining both the type of lesion which will result from invasion of the body by streptococci, and the modification in virulence they may undergo in man or animal as the struggle between host and invader is extended in time.

Prophylaxis.—General surgical aseptic methods. Autogenous vaccines have been extensively used in streptococcus infections, but with less favorable results than autogenous staphylococcus vaccines.

The Streptococcus Einheit or Vielheit.—Considerable discussion has arisen concerning the unity or the plurality of types included within the organism known as Streptococcus pyogenes. Marmorek[1] and others have stoutly maintained the Einheit theory. Considerable

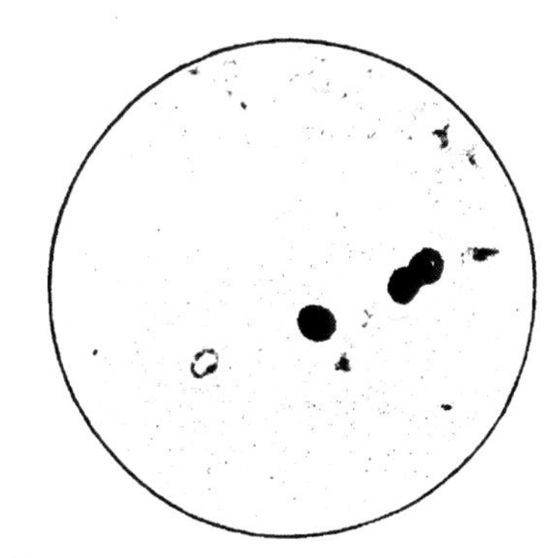

FIG. 36.—*Pneumococcus mucosus* showing capsule. X 1000.

evidence in favor of this view was advanced by Koch and Petruschky,[2] who showed that a streptococcus obtained from a fatal puerperal sepsis caused erysipelas in a rabbit when it was injected subcutaneously, peritonitis when injected intraperitoneally, and septicemia when introduced intravenously. The organisms freshly isolated caused a rapidly fatal septicemia when introduced through the blood stream, but the virulence was gradually lost following cultivation on artificial media; the septicemic phenomena diminished in intensity and there was evidence of a localization of the organisms. Their conclusions were that the type of lesion produced by Streptococcus pyogenes depended largely upon the virulence of the culture, the tissue invaded, and the number of organisms. With a comparatively slight loss in virulence the endocardium appeared to be somewhat more frequently

[1] Berl. klin. Wchnschr., 1902, xxxix, 299.
[2] Loc. cit.

the site of the focal infection; with a greater loss of virulence, the joints. It must be remembered in this connection that the virulence of a streptococcus for man does not necessarily determine the virulence for animals.

It is possible to raise the virulence of streptococci very materially by artificially creating portals of entry and of escape which are not usually available to the streptococcus. This is accomplished by passage through experimental animals. By passage it is possible to reproduce with considerable accuracy the various reactions mentioned above, depending upon the virulence of the organism, the tissue into which the injection is made, and the number of organisms introduced. It is also important to remember than an increase in virulence for one animal, attained by frequent passages, frequently results in a loss, partial or complete, of the virulence of the organism for another animal. Too little is known of the mechanism of virulence, however, to place a final interpretation upon the biological significance of changes in pathogenic powers.

Additional evidence of the Einheit of streptococci has been brought forward by Rosenow,[1] who states that he has changed streptococci to pneumococci and back again by special methods of culture and animal inoculation. Two possibilities present themselves to explain this phenomenon, if Rosenow's claims are substantiated. First, the streptococcus-pneumococcus complex is a single organism which exhibits nodes of relative cultural stability (assuming that present-day methods for the recognition of bacterial types are fundamentally sound), and the organism may pass from one node to another under the stress of environmental stimuli. The second possibility is that the streptococcus and pneumococcus are in reality distinct biological entities and that an actual discontinuous mutation has occurred. The many variables to be considered in this connection—variations in virulence, adaptability to various hosts, and changes in appearance in different media, all of which may change independently of or parallel to each other—complicate the problem to a considerable degree; final judgment must await the establishment of authoritative standards for bacterial diagnosis of unquestioned fundamental stability.

Neufeld, and Cole and his associates have presented a new aspect of the problem. They found that the older conception of the unity of the pneumococcus type was untenable. They found there were four distinct types of pneumococcus which were recognizable both

[1] Loc. cit.

by serological and pathological methods, and that these types were mutually stable, for long-continued passage through animals failed to alter or modify their general cultural and agglutinating properties, although the virulence of the respective types for one or another animal could be increased or decreased. It is not improbable that a thorough study of the streptococcus group may reveal similar serological variance and that in the type now designated Streptococcus pyogenes several individual types parallel to those of the pneumococcus may be demonstrated.

The important question for the moment is, do these changes of virulence, *et cetera*, exhibited by the streptococcus influence the diagnostic aspect of the question? Theobald Smith has admirably summed up the present status of the subject in the following words: "Spontaneous changes in the cultural characters of the streptococcus do not proceed rapidly enough, if they go on at all, to interfere with current bacteriological methods. Tendencies toward slow changes may be used as further valuable distinguishing characters."[1]

THE PNEUMOCOCCUS.

Synonyms.—Micrococcus pasteuri, Diplococcus pneumoniæ, Diplococcus lanceolatus, Streptococcus lanceolatus.

Historical.—Although the pneumococcus was observed by Sternberg[2] and independently by Pasteur[3] in the blood of rabbits inoculated with sputum, the etiological relationship of the organism to lobar pneumonia was not established until 1886, when Fränkel[4] and Weichselbaum[5] published their respective studies upon lobar pneumonia.

Morphology.—Viewed under the microscope, the pneumococcus presents two distinct appearances, depending upon the source of the culture. Observed in human or animal tissues, exudates or body fluids, or in media containing non-coagulated albuminous fluids, as blood serum, ascitic or hydrocele fluids, the organisms occur typically in pairs surrounded by a definite capsule, or less commonly in short chains enclosed in a capsule. The individual cells are typically lanceolate in shape with the apposed surfaces of each pair flattened, and the distal ends somewhat pointed. Less commonly the organisms are oval, or nearly spherical. The paired arrangement is maintained when the organisms remain adherent to form short chains. Cultures

[1] Smith and Brown, Jour. Med. Research, 1914, xxxi, 501.
[2] National Bureau of Health, 1881.
[3] Compt. rend. Acad. Sci., 1881, xcii, 159.
[4] Zeit. f. klin. Med., 1886, x, 401. Ibid, xi, 437.
[5]

in artificial media which do not contain albuminous fluids are not encapsulated, and the distinctive lanceolate shape is frequently lost; the organisms become more nearly oval or spherical in outline, but the tendency to remain adherent in pairs is usually maintained. Chains of from four to eight elements are developed in broth cultures, which has led many observers to include the pneumococcus in the streptococcus group. The size of the organisms varies considerably; ordinarily the lesser diameter measures from 0.5 to 0.8 microns, and the longer diameter from 1 to 1.3 microns.

The pneumococcus is non-motile and possesses no flagella. The capsule, which surrounds pairs of organisms derived from sputum, tissue, body fluids and exudates of man and animals, as well as those

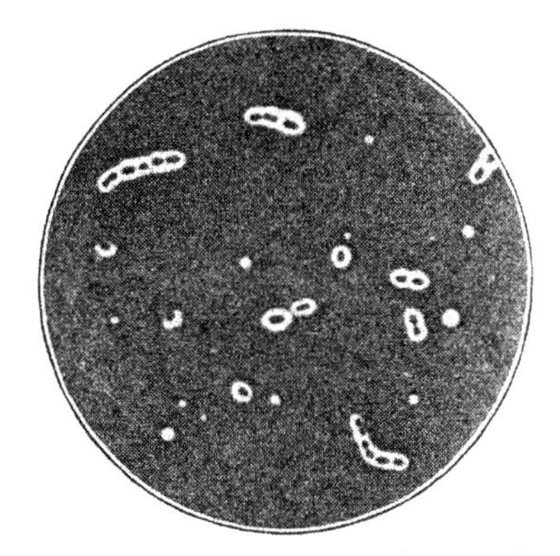

Fig. 37.—Pneumococcus showing capsules.

cultivated in milk or media containing uncoagulated albuminous substances, is readily demonstrated by the methods of Welch,[1] Hiss[2] and Rosenow.[3] The capsule is poorly formed or absent from pneumococci derived from chronic processes or from mucous surfaces where the organisms are growing as parasites or "opportunists."

The ordinary anilin dyes stain pneumococci readily, and they are Gram-positive when freshly isolated, but tend to become Gram-negative during cultivation in artificial media.

Isolation and Culture.—Pneumococci grow slowly and feebly upon ordinary laboratory media, and they soon perish. Cultures may be obtained from the blood stream in a large percentage of cases from the fifth day of the disease to the crisis[4] by inoculating 5 to 10 c.c. of

[1] Bull. Johns Hopkins Hospital, 1892, xiii, 128.
[2] Cent. f. Bakt., Ref., 1902, xxxi, 302.
[3] Jour. Infec. Dis., 1911, ix, 1.
[4] Rosenow, Jour. Inf. Dis., 1904, i, 280.

blood into 100 to 150 c.c. of 0.1 per cent. dextrose broth, and incubating for twenty-four hours at 37° C. Isolation of pneumococci from sputum by cultural methods is practically hopeless; but pure cultures may be obtained from the heart blood of white mice inoculated subcutaneously with sputum.

The organisms may be obtained from inflammatory exudates and pus either by inoculation of the material into white mice or infecting the surface of blood agar, serum, ascitic or hydrocele agar plates. Colonies on blood agar plates are minute, gray, and surrounded by a greenish halo which Butterfield and Peabody[1] and Cole[2] have shown to be methemoglobin. Colonies on ascitic agar are small, transparent and colorless. The growth upon plain nutrient agar or gelatin is very scanty. Gelatin is not liquefied. The addition of dextrose to agar increases the nutritive value of the medium, but the acid formed by the fermentation of the dextrose soon kills the bacteria unless calcium carbonate is added to neutralize the acid. Many strains of pneumococci grow in milk, producing as a rule sufficient acid to cause coagulation. The coagulum is never liquefied. Growth upon Löffler's blood serum is moderately luxuriant, particularly for subcultures; initial development of the organisms directly from human or animal sources is not extensive upon this medium. The colonies are small, clear and colorless, and not distinctive. Growth is more rapid in fluid than in solid media. Secondary inoculations into plain broth or broth containing utilizable carbohydrates result in a clouding of the medium and extensive development, more luxuriant in the latter than the former. The addition of blood, blood serum or ascitic fluid to media increases the nutritive value greatly. The organisms die within a few days, and even after twenty-four hours' incubation degenerative forms appear, and they become Gram-negative. Transfer at frequent intervals to fresh media is essential to maintain viable cultures of the pneumococcus.

The pneumococcus is an aërobic, facultatively anaërobic organism whose limits of growth lie between 25° C., below which development ceases, and about 42° to 43° C.; the optimum temperature of growth is 37° C. The organisms are not resistant to heat, being killed by an exposure of ten to fifteen minutes to 55° C.[3] Chemical disinfectants, as 5 per cent. carbolic acid or 1 to 1000 bichloride of mercury, destroy pneumococci readily. Dried rapidly in sputum, they retain their

[1] Jour. Exp. Med., 1913, xvii, 587.
[2] Ibid., 1914, xx, 363.
[3] See Wood, Jour. Exp. Med., 1905, vii, 592, for literature.

viability for nearly two weeks, but sunlight is rapidly fatal. The virulence is rapidly lost during cultivation in artificial media, but it may be retained practically unimpaired for weeks if the organisms suspended in blood are sealed in glass tubes and maintained in the dark at ice-box temperature. Pneumococci obtained from sputum, either of healthy individuals or from the "rusty sputum" characteristic of the earlier stages of lobar pneumonia, possess sufficient virulence to kill white mice.

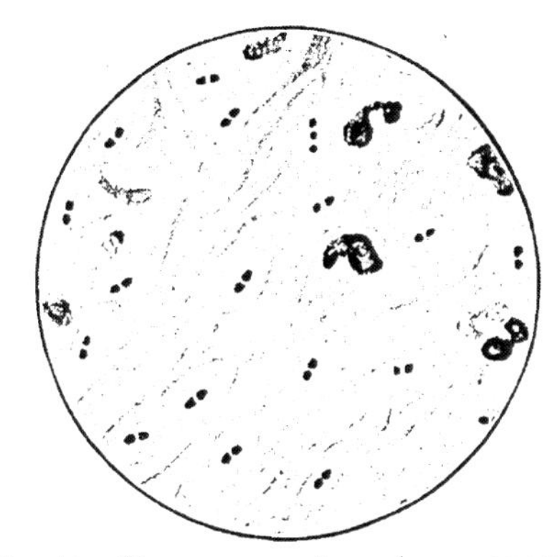

FIG. 38.—*Pneumococcus in sputum.* × 1000.

The original virulence may frequently be restored to cultures on artificial media by passage through white mice, provided large doses are administered at the start. Repeated, rapid inoculations of virulent pneumococci frequently lead to a decided increase of virulence above that originally exhibited by the organisms.

Products of Growth.—*Chemical.*—The pneumococcus produces acids, chiefly lactic, but smaller amounts of formic acid, in hexoses, bioses, and many starches. Hiss[1] has shown that the fermentation of inulin by the pneumococcus is a very constant cultural differentiation of the organism from the streptococcus, which is unable to ferment this starch. Another important method of distinguishing between pneumococci and streptococci is the solubility of the former in bile or a freshly prepared solution of sodium chlorate.[2] [3] Colonies of the pneumococcus on blood agar are surrounded by a greenish zone of methemoglobin.[4]

[1] Jour. Exp. Med., August, 1905, vii., 547.

[2] Neufeld, Zeit. f. Hyg., 1900, xxxiv, 454. Wadsworth, Jour. Med. Research, 1904, x, 228.

[3] The test is made as follows: 1 c.c. of a twenty-four-hour broth culture of the suspected organism is mixed with 0.1 c.c. of a freshly-prepared 2 per cent. solution of sodium chlorate and maintained at 37° *C.* Clearing of the solution indicating solution of the organisms does not take place uniformly; some cultures dissolve more rapidly than others. Cole, Jour. Exp. Med., 1912, xvi, 658. Acids interfere with the success of the test.

[4] Butterfield and Peabody, loc. cit. Cole, loc. cit.

Enzymes have not been demonstrated in cultures of the pneumococcus.

Toxins.—Soluble toxins have not been detected in cultures of pneumococci, although the filtrates obtained by Klemperer,[1] Washbourn,[2] and Isaeff[3] were toxic for small laboratory animals. The toxicity observed in these preparations was probably due to the liberation of endotoxins as the result of autolysis of pneumococci in the medium. Macfadyen[4] has obtained toxic substances from two- to three-day agar cultures of virulent pneumococci, which were ground finely after freezing with liquid air (method of Macfadyen and Roland), then extracted with 1 to 1000 potassium hydrate, centrifugalized to remove fragments of the organisms and filtered. A small amount of the filtrate, 0.5 to 1 c.c. in rabbits, 0.1 to 1 c.c. in guinea-pigs, produced death when injected intravenously or intraperitoneally. The toxicity of the filtrate was roughly proportional to the virulence of the organisms for rabbits. Heating the filtrate to 55° C. for an hour, or exposure to chloroform vapor for the same time reduced the toxicity of the preparation very considerably. Neufeld and Dold[5] and Rosenow[6] obtained toxic substances from pneumococci, the former by extraction of the organisms in 0.1 per cent. lecithin in physiological salt solution, the latter by simple autolysis, which induced symptoms in guinea-pigs suggesting acute anaphylaxis. Cole[7] has repeated these experiments with results that were irregular: thus, of 213 guinea-pigs injected with extracts of pneumococci in salt solution, 8 died acutely with symptoms resembling acute anaphylactic shock, 83 died within twelve hours, the remainder were negative. Cole concludes that extracts of pneumococci in salt solution may be toxic, but not uniformly so. The exact conditions under which these solutions become toxic are unknown. Solutions of pneumococci dissolved in dilute solutions of bile salts were found to be very constantly toxic.[8] The intravenous injection of these solutions into rabbits and guinea-pigs elicits symptoms resembling closely those of acute anaphylaxis. Many of the animals die acutely.

[1] Zeit. klin. Med., 1891, xx, 165.
[2] Jour. Path. and Bact., 1897, iii, 214.
[3] Ann. Inst. Past., 1892, vii, 259.
[4] Cent. f. Bakt., Orig., 1907, xliii, 30.
[5] Berl. klin. Wchnschr., 1911, xlviii, 1069.
[6] Jour. Infec. Dis., 1911, ix, 190.
[7] Jour. Exp. Med., 1912, xvi, 644.
[8] Casagrandi (quoted by Pribram: Kolle and Wassermann Handb., 2 ed., 1913, ii$_2$, 1350) states that normal rabbit blood contains antihemolysins.

Hemotoxin.—Recently Cole[1] has shown that solutions obtained by dissolving pneumococci in dilute solutions of bile salts, or by trituration, are hemolytic for rabbits, guinea-pigs, sheep and human red blood cells, and that their activity is inhibited by minute amounts of cholesterin. The injection of these solutions in gradually increasing amount leads to an inhibition of their action; in other words, this "hemolytic endotoxin" appears to act as an antigen.

Pathogenesis.—*Human.*—At least 90 per cent. of all cases of lobar pneumonia, one of the most prevalent and fatal of human diseases, is caused by the pneumococcus, but this disease is by no means the only one in which the organism is an etiological factor. Many bronchopneumonias which follow acute infections, as typhoid, diphtheria, so-called "aspiration pneumonia," are also of pneumococcic causation. Pleurisy, a frequent complication of both types of pneumonia, is quite commonly a pneumococcus infection, and a majority of sporadic cases of meningitis, particularly in children, are also caused by the organism. Indeed, in children the pneumococcus is rather more commonly isolated from suppurative processes than any other organism; in adults the incidence of pneumococci in suppurations is on the whole considerably less. Middle ear involvement, inflamed mastoids, endo- and pericarditis are all frequently caused by the pneumococcus. The channel of infection appears to be through the blood stream, and pneumococci have been isolated from the blood stream in a very large percentage of all cases of lobar pneumonia.[2] Less commonly the organisms become localized in joints, causing arthritis, and around the shafts of bones, causing osteomyelitis. Conjunctival inflammation of varying degrees of severity which occasionally leads to ulcer formation is frequently a pneumococcus infection.

It was formerly stated that virulent pneumococci could be obtained from the sputum of fully 30 per cent. of normal individuals. The supposition was that the patient became the victim of his own organisms. Recent studies by Dochez and Avery[3] suggest strongly that the pneumococci found in the sputum during pneumonia are commonly replaced by pneumococci of a less virulent type soon after convalescence. Their observations, furthermore, make it justifiable to consider those patients who harbor the more virulent types after

[1] Jour. Exp. Med., 1914, xx, 346.
[2] Rosenow, loc. cit.
[3] Quoted by Cole, New York Med. Jour., January 2 and 9, 1915.

recovery as carriers, precisely as typhoid carriers harbor typhoid bacilli after recovery from typhoid fever.

Animal.—Mice are the most susceptible of laboratory animals to infection with the pneumococcus. Small amounts of pneumonic sputum, exudate or pus injected subcutaneously lead to a rapidly fatal septicemia. Encapsulated pneumococci are found in the blood and visceral organs, particularly the spleen, which is enlarged, and the peritoneal fluid. Rabbits are somewhat less susceptible and the results of inoculation of pneumococcic exudates or cultures depend upon the virulence of the organisms, the size of the dose, and the method of inoculation.[1] The intravenous or subcutaneous inoculation of virulent cultures leads to a fatal septicemia, death occurring within five days as a rule. The less virulent organisms, which do not kill the animal within a few days after inoculation, frequently cause localized abscess formation with a fibrinous exudate. The nature and extent of the lesions induced depend largely upon the time which elapses between inoculation and the death of the animal. In general it may be stated that localized lesions appear when less virulent organisms are injected. Intravenous injections are more effective than subcutaneous inoculations of the same amount of organisms. Guinea-pigs are relatively non-susceptible to pneumococcus infection.

Many attempts have been made to reproduce the typical pathological lesions of lobar pneumonia in experimental animals. Wadsworth[2] succeeded in reproducing typical lobar pneumonia in rabbits by first partially immunizing them to the organism in order to localize the lesions in the lungs. Lamar and Meltzer,[3] and Wollstein and Meltzer[4] produced lobar pneumonia in dogs by the method of tracheal insufflation devised by Meltzer; and Winternitz and Hirschfelder[5] have been equally successful in producing lobar pneumonia in rabbits. The method consists essentially in forcing suspensions of pneumococci deep into the terminal bronchioles and their alveoli. Cole[6] has shown that the strain of organism influences the results; organisms of slight virulence give negative results, and organisms possessing too great virulence cause a generalized septicemia with congestion and edema of the lungs as the only local pulmonary manifestations.

[1] Kruse and Pansini, Zeit. f. Hyg., 1892, xi, 279 *et seq.*
[2] Am. Jour. Med. Sc., 1904, cxxvii, 851.
[3] Jour. Exp. Med., 1912, xv, 133.
[4] Ibid., 1913, xvii, 353, 424.
[5] Ibid., 1912, xvii, 657.
[6] Arch. Int. Med., 1914, xiv, 56.

Types of Pneumococci.—Kruse and Pansini[1] as early as 1891 called attention to the differences, both cultural, morphological and in virulence, which they observed in studying eighty-four strains of pneumococci isolated from many cases of pneumonia. They believe that there was no sharp line of demarcation between the pneumococcus and Streptococcus pyogenes, because their various strains included all variants between the two types of organisms. Recently Rosenow[2] has reported the transmutation of typical pneumococci to Streptococcus pyogenes by a series of animal passages and cultural manipulations. Cole[3] has been unable to confirm this observation in any one of several hundred strains, but it should be stated that he has not employed Rosenow's procedure in detail.

Much light has been shed upon the apparent variability of strains of pneumococci by the observations of Neufeld and Händel,[4] and Dochez,[5] and Dochez and Gillespie.[6] These observers have shown by serological reactions that pneumococci may be divided into four groups or types, each of which fails to agglutinate with sera other than the homologous serum. These groups have been tentatively designated I to IV inclusive. Groups I and II are typical virulent pneumococci. Group III comprises the organism formerly known as Streptococcus mucosus, now called Pneumococcus mucosus; and Group IV includes relatively avirulent strains which are commonly found in the mouths of healthy persons. Group IV is somewhat more heterogenous, judging from agglutination reactions, than Groups I to III. Group III contains the most virulent organisms. A study of the distribution of the various types in seventy-two cases of pneumonia illustrates this point.[7]

Infection type.	No. cases.	No. deaths.	Per cent.
1	34	8	24
2	13	8	61
3	10	6	60
4	15	1	7
—	—	—	—
Total	72	23	32

It is possible that "mixed infections" will be found when more cases are carefully studied. The same general types have since been reported in Europe and in Philadelphia.[8]

[1] Loc. cit. [2] Jour. Am. Med. Assn., 1913, lxi, 2007. [3] Loc. cit.
[4] Zeit. f. Immunitätsforsch., 1909, iii, 159; Berl. klin. Woch., 1912, xlix, 680.
[5] Jour. Exp. Med., 1912, xvi, 680. [6] Jour. Am. Med. Assn., 1913, lxi, 727.
[7] Cole, Arch. Int. Med., 1914, xiv, 33.
[8] Cole, New York Med. Jour., January 2 and 9, 1915.

Immunity and Immunization.—Relatively little is known of the nature and extent of immunity following recovery from an attack of pneumonia. One attack appears to predispose somewhat to a subsequent attack, which was explained formerly on the basis that little or no immunity was conferred on the patient. The extensive work of Cole and his associates suggests that a second attack of the disease may be caused by a different type of pneumococcus; their experiments indicate that antibodies specific for one type are not protective against infection with the other types.

The serum of convalescent pneumonia patients exhibits relatively feeble bactericidal activity, even upon the homologous strain of the pneumococcus, and the mechanism which leads to recovery is not definitely known. Neufeld[1] and others have advanced the hypothesis, based upon careful observation, that the crisis in pneumonia, which usually marks the end of the prominent clinical symptoms, is associated with a somewhat abrupt increase in the amount of specific opsonin of the blood—an increase in bacteriotropins in Neufeld's terminology. This theory assumes that leukocytes play a prominent part in the healing process, and that phagocytic activity becomes efficient at or about the time of the crisis.

Neufeld and Händel,[2] and Cole and his associates[3] have produced a serum which protects susceptible animals, as mice, against many times the fatal dose of the homologous strain of organism by injecting gradually increasing doses of very virulent pneumococci into horses. Cole has used these sera clinically in the treatment of pneumonia with promising results in infections caused by Types I and II of the pneumococcus. The serum appears to destroy or greatly reduce the number of pneumococci in the blood, and to be of material benefit in reducing the severity of the infection. At present a satisfactory serum for infection with Type III, Pneumococcus mucosus, has not been prepared. Cole specifically directs attention to the necessity of identifying the type of infecting organism (by agglutination reactions) before administering the serum. It is imperative that the homologous serum be used.

Bacteriological Diagnosis.—Pneumococci are found in the healthy throats of a very considerable percentage of adults, consequently the identification of pneumococci in the sputum is of little clinical signifi-

[1] Zeit. f. Immunitätsforsch., 1909, iii, 159.
[2] Arb. a. d. kais. Gesamte, 1910, xxxiv, 169.
[3] Jour. Am. Med. Assn., 1913, lxi, 663; New York Med. Jour., January 2 and 9, 1915.

cance unless the type of the organism is determined. Dochez and Avery[1] have found that the common mouth pneumococcus is usually the avirulent type (Type IV); convalescents from pneumonia usually exhibit the virulent types, I to III, as a rule. These types can be identified by agglutination reactions with the specific sera prepared by Cole.

Pneumococci isolated from pleural and pericardial exudates, middle-ear infection, empyema and pneumococcic cerebrospinal meningitis can be identified morphologically by their lanceolate shape and Gram-positiveness; the type of organism, however, must be determined by serological reactions.

They are best obtained in pure culture, if they are mixed with other bacteria, upon blood agar plates. A green halo surrounds the typical pneumococcus colony.

The prophylaxis is the same as for any acute respiratory disease.

[1] Quoted by *Cole*, loc. cit.

CHAPTER XIV.

THE MENINGOCOCCUS—GONOCOCCUS GROUP.

THE MENINGOCOCCUS GROUP.
Micrococcus Meningitidis.
Parameningococcus.

THE GONOCOCCUS GROUP.
Micrococcus Gonorrheæ.
Micrococcus Catarrhalis.

THE MENINGOCOCCUS GROUP.

Micrococcus Meningitidis.—Synonyms.—Diplococcus intracellularis meningitidis; Diplococcus weichselbaumii, Meningococcus.

Historical.—Micrococcus meningitidis was isolated in pure culture by Weichselbaum[1] from purulent cerebrospinal fluids of several typical cases of cerebrospinal meningitis. The injection of pure cultures of the organisms directly into the meninges of dogs resulted in well-marked meningeal inflammation and encephalitis. Other organisms, pneumococci, streptococci, Bacillus influenzæ, for example, may incite inflammations of the cerebrospinal membranes, but these bacteria do not ordinarily cause epidemics of the disease. The meningococcus frequently causes wide-spread infection, and, unlike the organisms just mentioned (except the pneumococcus occasionally) the typical lesions are primarily of the cerebrospinal axis.

Morphology.—Meningococci obtained directly from the cerebrospinal fluid or from meningeal exudates occur characteristically in pairs with their apposed sides flattened and somewhat elongated. They measure about one micron in diameter, although the size varies even in the same culture. The individuals are fairly uniform in size and shape in very young, fresh cultures, but in older cultures considerable variations in size are met with. Examined directly in inflammatory exudates from the spinal fluid or meninges during the acute stages of the disease, the organisms occur typically and characteristically as intra- and extracellular diplococci and tetrads. They are found in polymorphonuclear leukocytes, but never in lymphocytes or other body cells.[2] They are intracellular but never intranuclear, according to Councilman, Mallory and Wright.

[1] Fortschr. d. Med., 1887, Nos. 18 and 19.

[2] Councilman, Mallory and Wright, Epidemic Cerebrospinal Meningitis. A Report to the Mass. St. Bd. of Health, 1898, p. 75.

The organisms are non-motile and possess no flagella. No spores are formed and no capsules have been demonstrated. (Jaeger[1] believed that the organisms produced capsules, but his observations are unconfirmed.) Ordinary anilin dyes stain meningococci, but quite irregularly. Occasionally one element of a pair stains intensely while its fellow stains faintly or not at all. Relatively large oval or round forms are frequently seen in cultures and in purulent exudates as well, which exhibit a brightly staining point in the centre of the organism; the remainder of the cell is scarcely colored.[2] Carbol-thionin is one of the best stains for the organism. The meningococcus

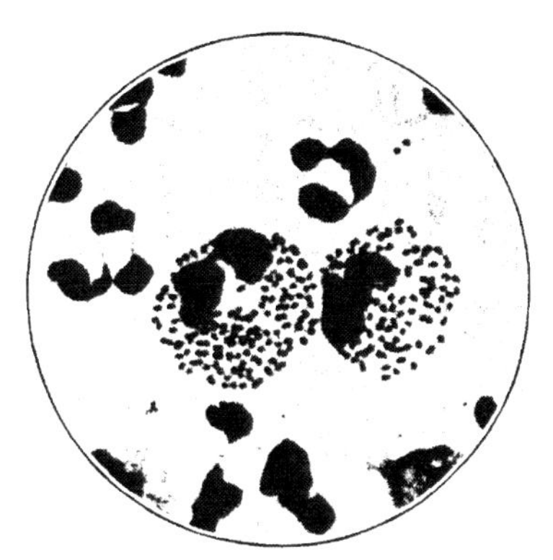

Fig. 39.—Meningococci in pus. × 1000.

is Gram-negative. Meningococci obtained from purulent exudates or from cultures on artificial media can not be definitely differentiated from gonococci or even from Micrococcus catarrhalis by any known staining methods. The source of the material should be known before even a tentative morphological diagnosis is attempted.

Isolation and Culture.—The meningococcus grows feebly or not at all upon ordinary artificial media. Growths may be obtained upon agar containing animal protein, as defibrinated blood or ascitic fluid, or upon Löffler's blood serum by smearing cerebrospinal fluid (drawn with aseptic precautions by lumbar puncture) in liberal amounts upon the surface of these media.[3] The addition of 1 per cent. of dextrose to the media favors development of the cocci. If the fluid obtained is not turbid, centrifugalization should be resorted to and

[1] Zeit. f. Hyg., 1895, xix, 358.
[2] Councilman, Mallory and Wright, loc. cit., p. 74.
[3] For technic of lumbar puncture, see page 226.

the sediment distributed as densely as possible in the manner indicated. A few small, transparent, round colonies are usually obtained when relatively large amounts of material are inoculated. The first growth upon artificial media is difficult to obtain; secondary transfers, if made within three days from initial cultivations, are usually successful and development is somewhat more vigorous. It should be emphasized that relatively large amounts of cocci must be inoculated to insure growth in artificial media.[1] Little or no growth occurs in plain broth; the addition of calcium carbonate[2] to dextrose broth makes a favorable medium for the development of the organism. Ascitic and serum broths are suitable media for the meningococcus. A coherent sediment gradually accumulates in these media and a delicate pellicle usually forms on the surface after a few days. Secondary transfers in milk usually grow, but there is little or no detectable change in the physical properties of the medium.

The meningococcus is essentially an aërobic organism, at least in its development outside the human body. The optimum temperature of growth is 37° C., and growth ceases when the temperature exceeds 42° C. or falls below 25° C. The organism is soon killed by low temperatures. Stock cultures can not be maintained at the temperature of the ice-box; they should be kept at temperatures between 32° and 38° C. Frequent transfers (every two or three days) must be made to maintain the viability of the organism; exceptionally strains are met with which become acclimatized to the conditions obtaining in artificial media to such a degree that transfers made at less frequent intervals suffice to maintain the viability of the culture.

The meningococcus exhibits little resistance to heat, drying or the action of chemical agents. Five minutes' exposure to 65° C. or two minutes' exposure at 80° C. suffices to sterilize the culture. Drying for a few hours at 20° C. is likewise fatal to the organism. Exposure of the organism to carbolic acid broth (1 to 800) inhibits development, and drying in the dark for seventy-two hours is fatal; sixty hours' exposure to drying is insufficient to kill the organisms.[3]

Products of Growth.—Meningococci are culturally very inert. No proteolytic enzymes have been demonstrated; gelatin and blood serum are not liquefied, and no coagulation or peptonization of milk occurs. Indol, skatol, phenol or other products of similar nature are

[1] The organisms, like gonococci, degenerate rapidly in artificial media. This may explain the necessity of transferring the organisms at frequent intervals.

[2] Bolduan, New York Med. Jour., 1905, May 13.

[3] Councilman, Mallory and Wright, loc. cit., p. 78.

not demonstrable in cultures of the organism. Acid, but no gas, is produced with considerable regularity in dextrose and maltose broths;[1] other ordinary carbohydrates are unattacked. These fermentation reactions are of considerable value in the cultural differentiation of meningococci from other organisms which may readily be confused with them.

Toxins.—Soluble exotoxins have never been demonstrated among the products produced by the meningococcus; killed cultures of the organism appear to be as fatal for ordinary experimental animals as the living organisms. This would suggest that the toxic phenomenon may be attributable to the liberation of endotoxins rather than to a soluble toxin.

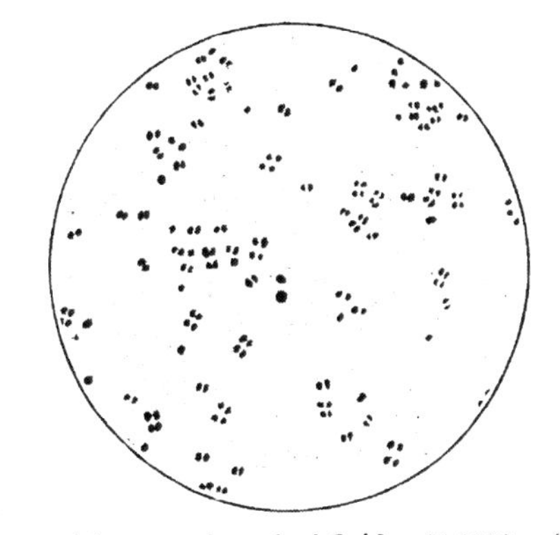

Fig. 40.—Meningococci from cerebrospinal fluid. × 1200. (Kolle and Hetsch.)

Pathogenesis.—The meningococcus possesses but feeble pathogenic powers for guinea-pigs; all attempts to induce infection by subcutaneous injections, according to Councilman, Mallory and Wright,[2] were negative. Occasionally successful results were obtained from intraperitoneal and intrapleural inoculation. A slight fibrinopurulent exudate was found postmortem in the peritoneal or pleural cavities in the fatal cases. Intracranial inoculations were uniformly negative. One successful infection of a goat by spinal canal inoculation was obtained by these observers; the animal died within twenty-four hours, and autopsy revealed intense congestion of the meninges of the cord and brain. A small amount of purulent spinal fluid was

[1] Kopetsky, Meningitis, The Laryngoscope, 1912, xxii, 797, has called attention to the early disappearance of the reducible substance (dextrose?) normally present in the spinal fluid in cerebrospinal meningitis. It is possible that the action of the organism upon this substance explains the phenomenon.

[2] Loc. cit., p. 76.

obtained containing but little fibrin. Small numbers of cocci were found within the polymorphonuclear leukocytes. Flexner[1] and Von Lingelsheim and Leuchs[2] have reproduced the essential lesions of cerebrospinal meningitis in monkeys by the subdural injection of suspensions of the organisms. The organisms were recovered in pure culture at autopsy.

The evidence of the etiological relation of the meningococcus to cerebrospinal meningitis in man is essentially the common, almost constant demonstration of meningococci in the cerebrospinal fluid and exudates antemortem, and from the tissues of the brain and cord postmortem. It must be remembered that other organisms can produce essentially the same lesions, however. The nature and extent of the lesions observed in fatal cases varies somewhat with the time which elapses between the onset of symptoms and death. The rapidly fatal cases frequently exhibit intense congestion of the membranes of the cord and brain; usually a fibrinopurulent exudate forms, more extensive as a rule at the base of the brain but readily demonstrable in the spinal fluid obtained by lumbar puncture. According to Westenhöffer,[3] there is commonly a swelling of the tonsils and pharynx in the early stages of the disease; middle ear involvement is comparatively frequent. It is probable that the organism passes from the nose and nasopharynx through the lymphatics to the base of the brain. The accessory sinuses of the nasal cavity appear to be inflamed in a majority of cases, particularly during the initial clinical period of the disease. There is a thickening of the meninges in those cases which run a more chronic course, frequently with considerable distention of the ventricles. Intracranial pressure is usually a prominent symptom. The organism has been isolated from the blood by Jacobitz,[4] Dieudonné,[5] Elser,[6] Elser and Huntoon,[7] the latter in 25 per cent. of their large series of cases.

Immunity and Immunization.—Little is definitely known of man's immunity to the meningococcus. One of the surprising results of the intensive study of the epidemic disease is the occurrence of the organism in the nasopharynx in a very considerable number of apparently healthy individuals, chiefly among those in actual contact with patients, less commonly among those not intimately in association with cases but in regions where the disease is epidemic, and rarely

[1] Cent. f. Bakt., 1907, xliii, 99. [2] Klin. Jarhb., 1906, xv, 489.
[3] Berl. med. Gesellsch., 1905, May 17; abstr. Cent. f. Bakt., Ref., 1905, xxxvi, 754.
[4] München. med. Wchnschr., 1905. [5] Cent. f. Bakt., Orig., 1906, xli, 420.
[6] Jour. Med. Research, 1906, xiv, 89. [7] Jour. Med. Research, 1909, xx, 371.

among individuals residing in areas where but few sporadic cases have been reported. The percentage of positive examinations varies considerably. Dieudonné[1] found about 12 per cent. of normal soldiers in a garrison at Munich, where an outbreak occurred, gave positive cultures from the nasopharynx. Bruns and Hohn[2] found 465 carriers among 3154 healthy individuals in a community where the disease was epidemic. They also found the percentage of carriers was greatest when the epidemic was at its height. Usually these carriers are temporary carriers; smaller numbers become permanent carriers or periodic carriers.[3]

Serum Therapy.—Many attempts have been made to prepare sera for the treatment of epidemic cerebrospinal meningitis, and two preparations have stood the test of actual practice, Kolle and Wassermann's[4] serum and the serum prepared by Flexner and Jobling. The method of immunization adopted by Flexner and Jobling appears from available data to be essentially that of Wassermann. It is as follows: horses are injected subcutaneously, first with dead cultures of meningococci, secondly with live cultures, and finally with autolysates of cultures. The latter are prepared by suspending virulent meningococci in sterile water for two days at 37° C. and injecting the supernatant fluid. The serum thus produced appears to combine phagocytic properties, increasing the destruction of the organisms by leukocytes; bacteriolytic properties, killing and dissolving the cocci, and possibly some antitoxic properties as well. It is essential, as Flexner has pointed out, to inject the serum directly into the spinal canal. This is accomplished by lumbar puncture. The turbid spinal fluid is allowed to escape through the needle with which the puncture is made until symptoms of intercranial pressure are reduced. An additional amount of fluid is then withdrawn to make way for the serum which is injected directly, 15 to 20 c.c. for young children and 20 to 40 c.c. for adults. The treatment is repeated from two to several times, until the spinal fluid is clear and has a normal appearance and cellular content. The serum must be used early in the disease to obtain the best results. Flexner and Jobling[5] have analyzed 328 cases with the following mortality:

	Per cent.
Injection during first to third day of disease	mortality 19.9
Injection during fourth to seventh day of disease	mortality 22.0
Injection after seventh day of disease	mortality 36.4

[1] Loc. cit. [2] Klin. Jahrb., 1908, xviii, 285.
[3] Mayer and Waldmann, Münch. med. Wchnschr., 1910, 475. Mayer, Waldmann, Fürst and Gruber, München. med. Wchnschr., 1910, 1584.
[4] Deut. med. Wchnschr., 1906. [5] Jour. Am. Med. Assn., 1908, li, No. 4.

Similar results have been obtained in Germany with Wassermann's serum.[1] Later observations by Flexner[2] confirm these results. The mortality has been reduced from about 70 per cent. to about 20 to 25 per cent.

Bacteriological Diagnosis.—(*a*) *Morphological.*—The demonstration of Gram-negative, biscuit-shaped diplococci in purulent spinal fluid from patients exhibiting the characteristic clinical symptoms is sufficient to establish a diagnosis of the meningococcus. It is to be remembered that the spinal fluid is clear for the first twenty-four hours of the disease, and usually clear after the tenth day to the fourteenth day even in untreated cases. Centrifugalization in sterile tubes must be resorted to in such cases; the sediment is examined as above. Smears from the nasopharynx, from middle-ear infections, and from suspected carriers can not be definitely diagnosed upon morphological characters alone. Cultural characteristics must be studied as well.

Cultural Characters.—Spinal fluid removed aseptically (and centrifugalized if the fluid is clear) and material from the nasopharynx, nasal cavity, or accessory nasal sinuses[3] is spread upon Löffler's blood serum and incubated at 37° C. After twenty-four to forty-eight hours' incubation, small, clear, round colonies develop in the majority of cases in which meningococci are present. These should be transferred to ascitic broth (preferably containing 1 per cent. of dextrose and a small piece of calcium carbonate) and examined after twenty-four hours' incubation at body temperature. If growth occurs, inoculation should be made in ascitic fluid dextrose and ascitic fluid maltose broths to determine if acid is produced. Several diplococci have been found which resemble the meningococcus microscopically but which differ from it in their fermentation reactions. A negative result does not exclude the possibility of an infection with the meningococcus; negative cultures occur quite frequently. Von Lingelsheim[4] and Elser and Huntoon[5] have studied these organisms carefully and give the following differential table:

[1] Wassermann, Deut. med. Wchnschr., 1907, 1585; Wassermann and Leuchs, Klin. Jahrb., 1908, xix, Heft 3.
[2] Jour. Am. Med. Assn., 1909, liii, 1443.
[3] Material for examination from the nasopharynx is best obtained upon sterile swabs; the infected swab should be immediately rubbed over the surface of a series of blood serum tubes or ascitic agar plates. This method is particularly adapted for the examination of suspected carriers.
[4] Klin. Jahrb., 1906, xv, Heft 2.
[5] Jour. Med. Research, 1909, xx, 377.

	Gram.[1]	Dextrose.[2]	Levulose.[2]	Galactose.[2]	Maltose.[2]	Lactose.[2]	Saccharose.[2]
Meningococcus	—	+	—	—	+	—	—
Pseudomeningococcus	—	+	—	—	+	—	—
Gonococcus	—	+	—	—	—	—	—
Micrococcus catarrhalis	—	—	—	—	—	—	—
Diplococcus crassus[3]	+	+	+	+	+	+	+
Diplococcus flavus	—	+	+	—	+	—	—
Micrococcus pharyngis siccus	—	+	+	—	+	—	—
Pigmented coccus I.	—	+	+	—	+	—	+
" " II.	—	+	+	—	+	—	—
" " III.	—	+	—	—	+	—	—
Micrococcus cinereus[4]	—	—	—	—	—	—	—

It will be seen that the meningococcus produces acid in dextrose and maltose. A differentiation between the gonococcus, Micrococcus catarrhalis and the meningococcus can frequently be made by their growths upon cultural media. The gonococcus grows poorly or not at all upon blood serum (Löffler's), the meningococcus grows with moderate rapidity upon it, and Micrococcus catarrhalis grows even upon plain agar.

The final diagnosis of the meningococcus depends upon its agglutination with specific sera. Positive agglutination will take place in dilutions of 1 to 500, even in 1 to 2000. Kutscher[5] has isolated strains of the organism which failed to agglutinate (macroscopic method) at 37° C., but agglutinated typically at 55° C. This should be tried in doubtful cases.

Serological Diagnosis.—Bettencourt[6] and Franca,[7] von Lingelsheim, Elser and Huntoon[8] and others have shown that the sera of convalescent cases of cerebrospinal meningitis very frequently exhibit specific agglutinins for the meningococcus. Of 593 tests, von Lingelsheim found 24.1 per cent. positive during the first five days of the disease, 56.7 per cent. positive from the sixth to the tenth day. Normal sera did not agglutinate with the organism in dilutions greater than 1 to 25; the sera of patients agglutinated in dilutions as high as 1 to 200. Elser and Huntoon have obtained agglutination in dilutions as high as 1 to 400.

The method of complement-fixation has not been satisfactory in the diagnosis of cerebrospinal meningitis.[9]

[1] + = Gram-positive
— = Gram-negative

[2] + = acid produced
— = no acid produced.

[3] Jaeger's meningococcus.

[4] Micrococcus catarrhalis?

[5] Kolle and Wassermann, Handb. d. path. Mikroörganismen, I. Erganzbd., 1907, 518.

[6] Zeit. f. Hyg., 1904, xlvi, 463.

[7] Klin. Jahrb., 1906, xv, Heft 2.

[8] Loc. cit.

[9] Von Lingelsheim XIV Cong. for Demog. and Hyg., Berlin, September, 1907.

Dissemination and Prophylaxis.—The disease is usually more frequent in children and young adults, usually in the winter and spring months. Frequently a nasal inflammation is prevalent before the disease begins to spread. The disease spreads by contact, as the organisms die out rapidly away from the human body. Many cases do not progress beyond the stage of nasal pharyngitis and sore throat, and it is probable that these cases are potentially carriers. According to Bruns and Hohn,[1] there may be from ten to twenty times as many carriers as cases. The disease is very likely to occur in barracks and boarding houses. Many people may be exposed to infection but comparatively few acquire the disease, suggesting a rather high natural resistance to the organism. The meningococcus may remain for months in the nasal passages of carriers, although ordinarily they remain less than a week.

Ward attendants should be segregated and quarantined, and nasal sprays used on the patients and attendants. It is quite probable that infected handkerchiefs or inhalation of infectious droplets are important in spreading the organism. It should be treated like any other acute infectious disease of the respiratory tract.

Parameningococcus.—In a critical discussion of the treatment of epidemic cerebrospinal meningitis with a specific antimeningococcus serum, Flexner[2] had directed attention to a relatively small group of cases which either failed to respond favorably to the serum, or reacted for a short time and later failed to improve. The spinal fluid of these cases contained organisms microscopically indistinguishable from typical meningococci. It was assumed tentatively that there might be two types of meningococcus, one of which was naturally "serum-fast," the other acquired "serum-fastness" during the course of the treatment with the serum. Dopter[3] has described an organism—the parameningococcus—apparently identical with the typical meningococcus in its morphological and cultural characteristics, but specifically different in its serological reactions. The parameningococcus, like the meningococcus, has been isolated from the nasal and oral cavities of man, and, in a few cases, from the blood stream and the meninges as well. The clinical manifestations incited by the parameningococcus are indistinguishable from those of epidemic cerebrospinal meningitis, but they fail to respond favorably to the administration of meningococcus serum. Dopter[4] has prepared a specific

[1] Loc. cit.
[2] Jour. Exp. Med., 1913, xvii, 553.
[3] *Compt. rend.*, Soc. de Biol., 1909, lxvii, 74.
[4] Semaine méd., 1912, xxxii, 298.

parameningococcic serum which is stated to have effected rapid improvement in the few cases of parameningococcus infection in which it was tried. These cases failed to respond to injections of meningococcus serum.

Wollstein[1] has made careful comparative studies of the morphological, cultural and serological reactions exhibited by a series of meningococci and parameningococci; her conclusions, which follow, summarize the available information of the relationship between these two organisms:

"The parameningococci of Dopter are culturally indistinguishable from true or normal meningococci, but serologically they exhibit differences as regards agglutination, opsonization, and complement deviation.

"Because of the variations and irregularities of serum reactions existing among otherwise normal strains of meningococci, it does not seem either possible or desirable to separate the parameningococci into a strictly definite class. It appears desirable to consider them as constituting a special strain among meningococci, not, however, wholly consistent in itself.

"The distinctions in serum reactions between normal and parameningococci are supported by the differences in protective effects of the monovalent immune sera upon infection in guinea-pigs and monkeys.

"It is therefore concluded that it is highly desirable to employ strains of parameningococcus in the preparation of the usual polyvalent antimeningococcus serum. It remains to be determined where it is better to employ the parameningococci along with normal meningococci in immunizing horses, or to employ normal and para strains separately in the immunization process and to combine afterward, in certain proportions, the sera from the two kinds of immunized horses."

THE GONOCOCCUS GROUP.

Micrococcus Gonorrheæ.—**Synonyms.**—Diplococcus gonorrheæ, gonococcus.

Historical.—The gonococcus was first observed by Neisser[2] in purulent urethral and vaginal discharges. Some years later Bumm[3] grew the organism in pure culture upon coagulated human blood serum and reproduced acute gonorrhea in men by urethral injections.

[1] Jour. Exp. Med., 1914, xx, 201.

[2] *Cent.* f. d. med. Wiss., 1879, No. 28.

[3] Die Mickroörganismen des gonorrhoischen Schleimhauterkrankungen Gonococcus, Neisser, Wiesbaden, 1885, No. 28.

Morphology.—The gonococcus occurs typically as a diplococcus, the proximated surfaces of pairs of cocci being flattened and elongated; they resemble coffee beans in shape. The longer diameter measures about 1.5 microns, the shorter diameter about 0.8 micron. The polymorphonuclear leukocytes of pus from cases of acute gonorrhea usually contain from one to several pairs of gonococci which are within the cytoplasm of the leukocyte but rarely or never within the nuclei. The organisms are also found within desquamated epithelial cells and occur free in pus as well. Gonococci are less numerous in the subacute and chronic stages of the disease, and they occur chiefly extracellu-

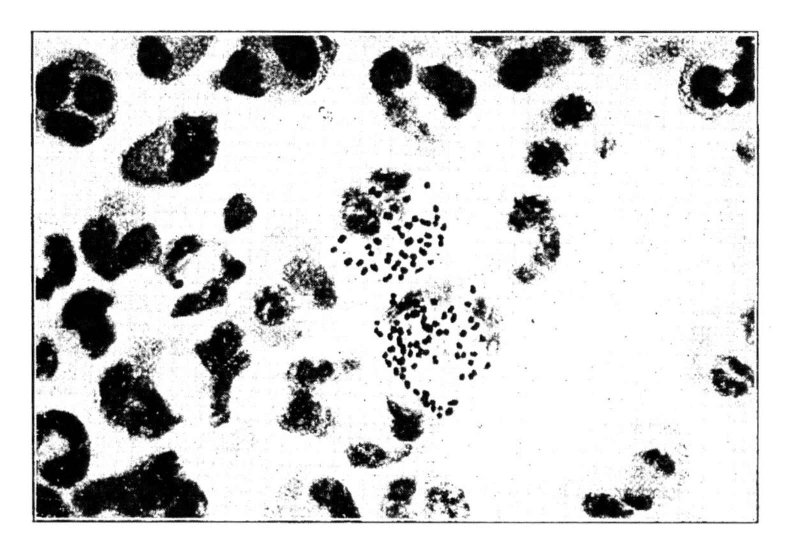

FIG. 41.—Gonococcus smear of pus from acute case. Methylene blue stain. (Warden.)

larly, with occasional pairs or clusters of gonococci in epithelial cells, less commonly in polymorphonuclear leukocytes. The organisms undergo degeneration rapidly, and even in pus from the more acute cases many large faintly staining cocci are found in association with those which are more typical in morphology and staining. In the chronic stage of the disease degenerated forms are very common.

The gonococcus is non-motile, and possesses no flagella; it forms no spores and capsules have not been detected. It stains with ordinary anilin dyes, but with some difficulty. It is Gram-negative.

Isolation and Culture.—The organism does not grow upon ordinary media; for the first growths outside the human body media containing uncoagulated protein, preferably that of human origin, is required. Agar[1] smeared with sterile defibrinated blood,[2] or agar mixed with hydrocele or ascitic fluid (one part fluid, two parts agar) furnishes a satisfactory nutrient substrate. Pus from acute cases (after preliminary cleaning and sterilization of the external genitalia) spread upon one of the media described above, should exhibit colonies after twenty-four hours' incubation at 37° C. The colonies are minute, clear and colorless; they resemble small dewdrops and exhibit a tendency to coalesce. Organisms stained from these colonies remain typical in morphology only for one or two days. Very soon degeneration (autolysis) commences, and in a very short time the organisms are dead[3] and partially dissolved. Secondary growths may be obtained from colonies, provided the inoculations are made within twenty-four to forty-eight hours from the time of plating. Ascitic broth is an especially favorable medium for this purpose.

The gonococcus is markedly aërobic; little or no growth occurs in media from which oxygen is excluded. The temperature limits are very restricted; growth ceases below 30° C. and above 40° to 42° C. The optimum temperature is 37° C. The organism is extremely sensitive to desiccation, and cultures die spontaneously within six to eight days. Repeated transfers of the cocci at intervals of two to three days will prolong the life of the culture almost indefinitely, provided they are maintained at 37° C. The organisms are very readily killed (outside the body) by the usual disinfectants. Gonococci in the urethra can not be killed readily by chemical disinfectants; the organisms penetrate rather deeply into the walls and the disinfectant can not reach them in sufficient concentration to be effective. This is particularly true of the subacute and chronic stages of the disease.

Products of Growth.—No enzymes have been detected in cultures of gonococci. Culturally the organism is inert; no development occurs in ordinary media. Acid is produced in dextrose-ascitic broth, but no other sugars are fermented. (See page 299 for comparison of cultural characters of gonococcus and similar Gram-negative diplococci.)

Toxins.—No soluble (exo-) toxin has been demonstrated in cultures of gonococci.

Finger, Ghon and Schlagenhaufer,[4] Nicolaysen,[5] Wassermann[6] and de Christmas[7] have shown that the cell substance itself is toxic.

[1] Glycerin agar is better than ordinary agar for this purpose.

[2] The blood agar should be heated to 56° *C*. for thirty minutes to destroy its bactericidal properties before use.

[3] Warden, Jour. Infec. Dis., 1913, xii, 93.

[4] Arch. f. Derm. u. Syph., 1894, xxviii, Nos. 1 and 2; Cent. f. Bakt., 1894, xvi, 350.

[5] Cent. f. Bakt., 1897, xxii, 305.

[6] Zeit. f. Hyg., 1898, xxvii, 307.

[7] Ann. Inst. Past., 1900, 349.

De Christmas has shown that the poisonous substance (endotoxin) diffuses readily into the culture medium, probably because of the rapid autolysis which is a noteworthy feature of the organism. The endotoxin is fairly resistant to heat; a brief exposure to 120° C. fails to entirely destroy its potency.

Pathogenesis.—*Experimental.*—Bumm[1] and Finger, Ghon and Schlagenhaufer[2] have reproduced typical urethritis in man with pure cultures of the gonococcus. The latter successfully infected the urethras of six healthy men with the organism (serum agar culture). The incubation period was from two to three days, and the clinical picture was typical in each instance. The organism was recovered in pure culture from each patient.

Animal.—Laboratory animals are not susceptible to urethral infection with the gonococcus. Intraperitoneal injections of cultures into white mice produce a purulent peritonitis, but there is little evidence that the organisms multiply there. Acute joint inflammations with purulent exudation follows the inoculation of the cocci into the joints of rabbits, and purulent conjunctivitis can be produced in young rabbits by rubbing gonococci on the conjunctiva. There is no evidence that the organisms multiply in these sites; the reverse appears to be the case for the cocci disappear rather rapidly. The endotoxins are responsible for the local reactions.

Human.—Man is very susceptible to infection with the gonococcus. The usual portals of entry are the mucous membranes of the urethra, vagina, and the conjunctiva. The urethral mucous membrane is particularly susceptible and it is commonly the primary site of invasion. The uterine mucosa and adnexa are also readily infected in adults; in young children the cervix is closed and infection of the uterus by continuity of growth from the vagina is rare in them, but vulvovaginitis is common, especially in hospital wards where infection is readily transmitted by thermometers, hands of ward attendants, and by direct contact.

The initial development of the organisms is upon the surface of the mucosa, then they penetrate to the deeper layers, infecting the prostate, and by continuity the epididymis in the male. Infection may spread from the vagina to the uterus in the female, then by continuity of growth to the Fallopian tubes, the ovaries, and the peritoneum, causing endometritis, salpingitis, oöphoritis, and peritonitis. Sterility is usually the result. Cystitis and arthritis are not uncommon sequelæ

[1] Loc. cit. [2] Loc. cit

of infection with the gonococcus, and the mucosa of the rectum is occasionally involved. Serous surfaces are rarely involved. Occasionally a generalized invasion takes place frequently resulting in septicemia with endocarditis. Ophthalmia neonatorum is a particularly common infection of the newborn of infected mothers. The conjunctivæ become contaminated with gonococci as the child passes through the vagina. A large percentage of the blind have lost their eyesight in this manner at birth. The instillation of silver preparations, required by law in many States, has greatly reduced this form of infection.

Immunity.—Man exhibits little or no resistance to infection with the gonococcus and the mucous membranes may actually be more susceptible to reinfection than they were originally.[1] In the chronic cases, where the organisms lie dormant for months, even years, the tissues appear to be somewhat less suited for growth of the organisms, but the patient can infect others even at this stage of the disease. Various attempts to prepare sera for curative purposes have not been generally successful, although Rogers[2] has reported cures in cases of gonorrheal rheumatism and chronic gonorrheal urethritis by the injection of the serum prepared by Torrey.[3]

Vaccines have been used with variable success. The injection of an autogenous vaccine, containing from five to ten million gonococci from a twenty-four-hour ascitic fluid agar culture, appears to give the best clinical results. Probably the extremely rapid autolysis of the gonococci plays a prominent part in the ineffectual attempts to induce improvement by the use of vaccines.[4]

Bacteriological Diagnosis.—(*a*) *Microscopical.*—Pus from the urethra of acute cases of gonorrhea should be dropped upon a cover glass or slide and spread by gently pressing a second cover glass or slide upon the first, then sliding them apart. By so doing the organisms remain in the polymorphonuclear leukocytes and epithelial cells, a very important diagnostic point. A Gram stain and a methylene-blue stain should be made. The former reveals intracellular and intercellular bean-shaped diplococci which are Gram-negative. Occasionally leukocytes contain as many as twenty pairs of the cocci. Dilute methylene blue 1 to 10 (Löffler's) usually stains gonococci intensely; the remainder of the cellular elements are faintly colored. The morphology of the organisms is clearly shown by this procedure.

[1] It is uncommon, however, to find auto eye infections from venereal lesions; even in cases of gonorrheal vulvovaginitis the eyes are rarely infected with the gonococcus.
[2] Am. Jour. Med. Sc., 1906, xlvi, No. 4.
[3] Ibid.
[4] Lespinasse; Illinois Med. Jour., April, 1912.

In chronic cases the discharge is scanty and it is better to receive the morning urine in a sedimentation glass containing a crystal or two of thymol. After a short time threads of mucus separate out; these should be removed with a capillary pipette and examined as above. The pus from old cases of gonorrhea frequently contains but few gonococci, which are difficult to find. It has been found that the local injection of silver nitrate (properly diluted) will usually cause an elimination of pus which frequently contains the organisms in somewhat larger numbers. Drinking beer is said to produce the same result. Vaginal smears may be obtained from swabs which are introduced into the vagina, or by means of long pipettes with rounded ends, containing a few drops of 1 to 1000 mercuric chloride, which are expressed and drawn up into the pipette several times deep in the vagina. The material thus removed is stained in the usual manner. Smears from the conjunctiva should be diagnosed very conservatively; Micrococcus catarrhalis and other Gram-negative diplococci which may occur within polymorphonuclear leukocytes are occasionally associated with an inflamed conjunctiva. The clinical picture should be considered in making the final diagnosis in such cases, and whenever possible cultures should be made to confirm the results.

(*b*) *Cultural.*—Cultures of the gonococcus are best obtained early in the disease, when secondary infection with staphylococci or other organisms has not taken place. The external genitalia should be carefully cleaned as for a surgical operation, and pus collected on a sterile swab which is rubbed over the surface of blood- or ascitic agar. The isolation of gonococci from pus of the subacute and chronic stages of the disease is extremely difficult; indeed, it is practically a matter of chance if pure cultures are obtained at this time. Vaginal cultures may be obtained upon sterile swabs which are inoculated in the same manner.

The gonococcus does not grow upon ordinary media, not even Löffler's blood serum, which distinguishes it from the meningococcus and from other Gram-negative cocci, including Micrococcus catarrhalis. (For fermentation reaction of the gonococcus, see page 299).

Serological Diagnosis.—*Agglutinins.*—The diagnosis of gonorrhea by agglutination of the gonococcus with the serum of the patient has not been successful.[1]

[1] Torrey (Journ. Med. Res., 1908, xx, 771) has isolated ten strains of gonococcus identical morphologically and culturally, but distinct serologically. This may explain in part the irregularity of the reaction of agglutination provided but a single strain of organism is used.

Complement-fixation Reaction.—Diagnosis of gonococcus infection by the method of complement fixation has been shown to be of considerable value, particularly in the more chronic cases, provided an homologous strain of the organism is used for the antigen. A mixed antigen composed of several strains is frequently employed in practice.[1] Much additional work is required, however, to determine the limits of variability of the various strains of the organism before the method is placed upon a thoroughly satisfactory basis for routine work.

Shattuck and Whittemore[2] have prepared concentrated polyvalent glycerin extracts and autolysates of gonococci to test the value of the skin reactions in gonococcus infections. The tests were made intradermally and by the von Pirquet method. Their results were unsatisfactory diagnostically.

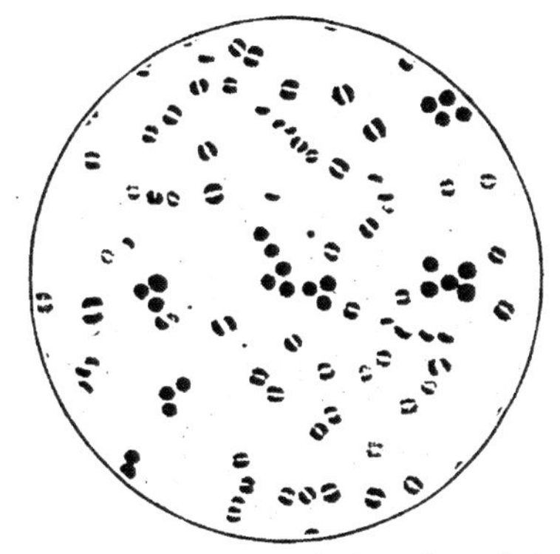

FIG. 42.—Micrococcus catarrhalis and staphylococcus.

The medicolegal aspects of gonorrheal infections make it incumbent upon the examiner to be very cautious in diagnosing the organism.

Dissemination and Prophylaxis.—The common towel has in the past been responsible for many cases of gonorrheal ophthalmia, but laws forbidding its use have largely removed this danger. It is certain that ordinary care will prevent infection of the innocent with the organism. Ophthalmia neonatorum is prevented by the instillation of silver salts in the manner indicated above.

Micrococcus Catarrhalis.—**Historical.**—Micrococcus catarrhalis appears to have been described first by Seifert[3] and by Kirchner;[4] the

[1] Lespinasse and Wolff, Illinois Med. Jour., January, 1913. Torrey's ten strains should be used in preparing the gonococcus antigen.

[2] Boston Med. and Surg. Jour., 1913, xlxix, 373.

[3] Volkmann's Sammlung klinischer Vorträge, No. 240.

[4] Zeit. f. Hyg., 1890, ix, 528.

name first appears in *Die Mikroörganismen* (Flügge), 3d edition, in 1896, credited to R. Pfeiffer.

Morphology.—Micrococcus catarrhalis occurs typically as a diplococcus with the apposed surfaces of adjacent cocci flattened and somewhat elongated. It measures about one micron in diameter. Occasionally the organisms are arranged in tetrads, particularly in young, active cultures in artificial media; in older cultures a tendency toward short chain formation is frequently observed. Degenerated cocci occur in older cultures. In sputum, bronchial secretions and other material from inflammation of the upper respiratory tract, in which Micrococcus catarrhalis is a primary or accessory factor, the organisms occur both within and without the pus cells. In the acute stages they are usually extracellular.[1] The organism is non-motile, and it has no flagella. It forms neither spores nor capsules. It colors readily with ordinary anilin dyes, some cells more intensely than their fellows, and it is Gram-negative.

Isolation and Culture.—The organism grows with moderate vigor upon agar; after twenty-four hours' incubation the colonies are small, translucent and gray. After three to four days the colonies are larger with an opaque centre, the periphery being translucent. Old colonies tend to become somewhat brownish. Development is more vigorous in media containing blood, blood serum, or ascitic fluid. Hemolysis of the blood does not occur. The growth in gelatin is slow, and usually feeble. A slight turbidity develops in broth. Moderate development occurs in milk. Micrococcus catarrhalis grows best at 37° C.; restricted development takes place at 16° C.; no growth can be detected at 43° C.

Products of Growth.—The organism is culturally inert. It does not produce any demonstrable proteolytic enzymes, and it produces no acid in any sugar. No toxic products are known. Filtrates of broth cultures have no apparent action upon white mice. No pathogenesis for laboratory animals has been detected.

Human Pathogenesis.—Micrococcus catarrhalis has occasionally been reported as a causative factor in catarrhal inflammations of the upper respiratory tract, and even in atypical pneumonia[2] and in bronchitis.[3] Ordinarily it is an opportunist found in the upper respiratory tract.

Bacteriological Diagnosis.—The organism is of importance chiefly through its striking resemblance to the meningococcus and the gono-

[1] Ghon, Pfeiffer and Sederl, Zeit. f. klin. Med., 1902, xliv, 262.
[2] Bernheim, Deut. med. Wchnschr., 1900. [3] Ritchie, Jour. Path. and Bact., 1900

coccus. It differs from these diplococci both in its relatively luxuriant growth upon artificial media and its ability to grow at room temperature. It resembles them in its intracellular disposition and in its staining reactions.

Droplet infection and transmission by contact are possible means of dissemination, and appropriate precautions should be taken to prevent this.

CHAPTER XV.

MICROCOCCUS MELITENSIS.

Historical.—The organism was discovered by Bruce.[1]

Morphology.—Micrococcus melitensis is a very small oval coccus, occurring singly or in pairs, rarely in short chains; the individual cells measure about 0.3 to 0.4 micron in diameter. Some observers declare the organism to be a short bacillus, a view which is perhaps based upon its appearance in old cultures, where various involution forms are readily observed. The coccus form almost invariably predominates in fresh material. The organism is non-motile, possesses no flagella and forms no capsule. Spore formation has never been observed. It stains readily with ordinary anilin dyes, and is Gram-negative.

Isolation and Culture.—One of the noteworthy cultural characters of Micrococcus melitensis is its slow growth on artificial media, even at 37° C. Suspected material, either blood, urine, milk, or material from splenic puncture, should be spread upon the surface of slightly acid agar and examined after three or four days' incubation for very minute white colonies which have a darker center. The organism grows slowly in gelatin, without producing liquefaction, and it produces a slight turbidity in broth. Milk appears to be a good medium for its development, goats' milk being better than cows' milk for this purpose.

The coccus is aërobic, facultatively anaërobic. The minimum temperature of growth is about 8° C., the optimum 37° C., and the maximum about 44° C. Direct sunlight kills it in a few hours; an exposure to 55° C. is usually fatal within an hour; 1 per cent. carbolic acid kills it in ten to fifteen minutes.[2] It resists drying in the cold and in the dark for several weeks.

Products of Growth.—Micrococcus melitensis is culturally inert;[3] it produces no proteolytic enzymes and it produces neither acid nor gas in any sugars. Milk, particularly goats' milk, becomes progressively alkaline in reaction. No toxins have been demonstrated.

[1] Practitioner, September, 1887, xxxix, 161.
[2] Mohler and Eichhorn, Bureau of Animal Industry, 1911, xxviii, 125.
[3] Kendall, Day and Walker, Jour. Am. Chem. Soc., 1913, xxxv, 1247.

Pathogenesis.—*Animal.*—Apes are susceptible to Micrococcus melitensis; the subcutaneous inoculation of cultures of the organism leads to definite clinical symptoms parallel to those observed in man. The disease usually runs a prolonged course and is often fatal. Monkeys are somewhat less favorable subjects than apes. (Goats, sheep, cattle, and horses are also susceptible to infection, although the disease is rarely generalized; the presence of the virus in the urine of the males, the milk and urine of the females of these species is the principal indication of infection.) The incubation period is from five to fourteen days. Eyre[1] states that rabbits and guinea-pigs may be infected, but not rats and mice.

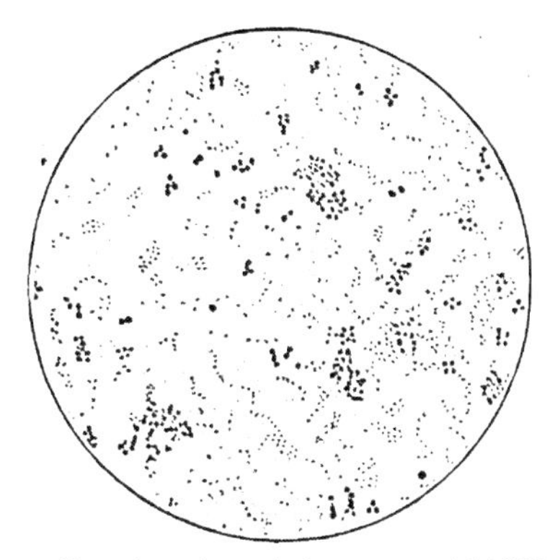

FIG. 43.—Micrococcus melitensis and staphylococcus. × 1000. (Kolle and Hetsch.)

Milk appears to be the chief source of infection; on the Island of Malta, where Malta fever was first described, fully 10 per cent. of the female goats contained the organism in their milk. Monkeys readily contracted the disease by drinking this milk. The urine of both male and female goats was shown to be infected as well.

Immunity and Immunization.—The blood and urine of infected individuals contain the virus of the disease and specific agglutinins are present in the blood even early in the disease. The agglutinins may persist for years after convalescence. Dilutions of $\frac{1}{100}$ to $\frac{1}{10,000}$ are made from the blood serum with suitable controls. A small amount of growth from a three-day agar culture of the organism is thoroughly emulsified in each dilution of serum and in the controls; the emulsions are incubated at 37° C. for two hours, then placed in the ice-box for twenty-four hours before the readings are made. A control with a

[1] Kolle u. Wasserman, Handb. d. Path. Mikroörganismen, I. Erganzband.

non-specific serum ($\frac{1}{50}$) and the organism should be made at the same time and incubated in the same manner, for experience has shown that the serum of normal individuals may agglutinate Micrococcus melitensis in moderately high dilution. Wright has immunized horses with repeated injections of Micrococcus melitensis. The blood serum agglutinated the organism in high dilution; it was claimed by him that the serum possessed curative value, the chief phenomena following its administration being a fall in temperature and a shortening of the course of the disease. This is still debatable.

Bacteriological Diagnosis.—A. Blood,[1] urine, milk, or material from splenic puncture is plated out as outlined above. The organisms are agglutinated with a serum of high potency.

B. The blood of the patient should be examined in high dilution ($\frac{1}{100}$) for specific agglutinins.

Dissemination and Prophylaxis.—The organisms leave the body through the milk or urine. Pasteurization of the milk and disinfection of the urine of infected animals is the best prophylaxis. It should be remembered that the organisms can enter the body through cutaneous wounds.

[1] The organisms are not always present in the blood of patients in demonstrable numbers; a negative culture is not conclusive.

CHAPTER XVI.

THE ALCALIGENES—DYSENTERY—TYPHOID—PARA-TYPHOID GROUP.

BACILLUS ALCALIGENES.

Bacillus alcaligenes was first isolated by Petruschky[1] from the feces of a patient presenting the clinical symptoms of typhoid fever. The serum did not agglutinate the typhoid bacillus and no typhoid bacilli were recovered from the blood or dejecta. Several similar cases are now on record in which Bacillus alcaligenes has been isolated both from the blood stream and the intestinal contents; the sera of these cases agglutinated the specific organism in dilutions of 1 to 50 or even higher, and Bacillus typhosus was not found. Bacillus alcaligenes occurs occasionally in acute intestinal disturbances of young children, not infrequently in association with organisms of the dysentery and paratyphoid groups.[2] Less commonly it is found in the dejecta of normal children, adults[3] and in water.

Morphology.—The organism both in size and shape resembles the typhoid bacillus very closely. It is actively motile and has peritrichic flagella. It does not form spores, and so far as is known, does not exhibit a capsule. Ordinary anilin dyes color it readily and it fails to retain the Gram stain.

Isolation and Cultures.—The organism grows readily in ordinary media. On agar the colonies are transparent, colorless, and round, and after eighteen hours' incubation at 37° C. attain a diameter of from 1 to 3 mm. The organism grows with moderate luxuriance on gelatin, but produces no liquefaction. In broth there is a uniform clouding, and after a few days a delicate pellicle usually forms. Bacillus alcaligenes grows fairly readily in milk; the reaction becomes progressively alkaline. In sugars no acid or gas is developed.

The organism is aërobic, facultatively anaërobic. The minimum temperature of growth is about 6° C., the optimum 37° C., and the

[1] Cent. f. Bakt., 1896, xix, 187.
[2] Kendall, Day and Bagg, Boston Med. and Surg. Jour., 1913, clxix, 741.
[3] Ford, Studies from the Royal Victoria Hospital. Montreal, 1903, i, No. 5.

maximum about 44° C. The resistance of Bacillus alcaligenes to physical and chemical reagents is similar to that of the typhoid bacillus.

Products of Growth.—Bacillus alcaligenes is characterized culturally by its inertness. Neither acid nor gas is produced from any known sugar. A moderate amount of proteolysis similar in degree to that of the typhoid bacillus in sugar-free broth is characteristic of the development of this organism in all the ordinary media.[1] Milk is not coagulated nor peptonized, but a progressive alkalinity develops, associated with the liberation of small amounts of ammonia.[2] No enzymes have been detected, and no toxins have been demonstrated in cultures of the organism.

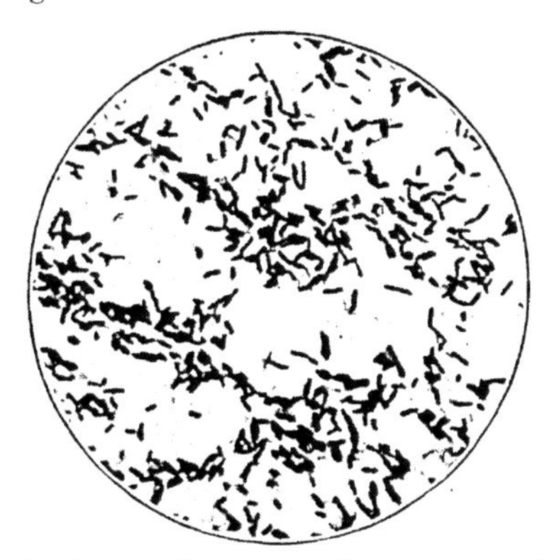

FIG. 44. Bacillus alcaligenes; bouillon culture. × 1000.

Pathogenesis.—The comparatively few cases of infection with Bacillus alcaligenes have not been studied in sufficient detail to throw any light upon the character of the lesions produced by the organism. The disease resembles typhoid fever clinically, and it is possible that in the past occasional typhoidal fevers have been incorrectly diagnosed. Animal experimentation has been uniformly negative.

Immunity.—Nothing definite is known of the immunological relations of Bacillus alcaligenes. Specific agglutinins have been demonstrated in a few instances where infection with the organism has been confirmed bacteriologically.

Bacteriological Diagnosis.—The organism may be isolated occasionally from the blood; ordinarily, however, the diagnosis is made by the isolation of the bacilli from the feces. Upon the Endo medium the organism grows precisely like the typhoid bacillus. It is readily dif-

[1] Kendall, Day and Walker, Jour. Am. Chem. Assn., 1913, xxxv, 1216.
[2] Ibid., 1914, xxxvi, 1940.

ferentiated from the typhoid bacillus by cultural reactions, Bacillus alcaligenes forming neither acid nor gas in dextrose, lactose, saccharose, or mannite. It does not liquefy gelatin, and it produces a permanent alkalinity in milk. The differential cultural reactions are shown in the table (page 316).

Dissemination and Prophylaxis.—Nothing is known of the method of dissemination of Bacillus alcaligenes. It appears to be an organism whose portal of entry is the gastro-intestinal tract. Carriers have never been satisfactorily demonstrated. Prophylaxis is precisely the same as that for other intestinal organisms.

THE GROUP OF THE DYSENTERY BACILLI.

The term dysentery as it is used in the clinical way includes at least two entirely distinct entities: amebic dysentery, a semi-acute or chronic infection caused by an ameba, which is usually restricted to the tropics and subtropics; and an acute type caused by members of the dysentery bacillus group, more frequently encountered in temperate zones. The latter type not uncommonly assumes epidemic proportions, but occurs sporadically as well. Japan has suffered greatly in the past from the ravages of bacillary dysentery. Ogata and Eldridge[1] state that 1,136,067 cases with 257,289 deaths occurred in that country during the period between 1878 and 1899 inclusive. The mortality, which varied markedly from year to year, averaged 22.6 per cent. of all cases. The disease appears to be rare in England, but it has been reported in Germany.[2] The Atlantic seacoast cities of the United States have experienced epidemics of the disease, but the inland cities appear to have been relatively free from it. During inter-epidemic years mild, atypical, sporadic cases and moderate numbers of bacilli carriers (both of the Shiga and Flexner types of organisms) have been discovered.[3]

The most virulent of the dysentery bacilli was isolated and described by Shiga[4] during the great epidemic of 1897 to 1898 in Japan. Flexner[5] recovered an organism which he believed was identical with the Shiga bacillus from cases of dysentery in the Philippines. Later studies of this organism by Martini and Lentz[6] revealed specific differences in

[1] Quoted in Public Health Reports, 1900, xv, 1.
[2] Kruse, Deut. med. Wchnschr., 1900, vol. xxvi.
[3] Kendall, Boston Med. and Surg. Jour., 1913, clxix, 754; May 20, 1915.
[4] Cent. f. Bakt., 1898, xxiii, 599; xxiv, 817, 870, 913.
[5] Ibid., 1900, xxviii, 625.
[6] Zeit. f. Hyg., 1902, xli, 540.

agglutinins from the Shiga bacillus, and Lentz[1] showed that the Shiga bacillus did not ferment mannite; the Flexner bacillus ferments this alcohol with the production of acid. Later intensive studies of bacillary dysentery bacilli by Park and Dunham, Hiss and Russell, and others confirmed the work of the earlier observers and added several strains to the group, which differ from the Shiga and Flexner strains both with respect to their specific agglutinating powers and their cultural reactions. The principal cultural reactions of the more prominent Gram-negative intestinal bacteria, including not only the pathogenic organisms but the habitually parasitic organisms as well, follow:

	Gram.	Motility.	Dextrose.	Lactose.	Saccharose.	Mannite.	Levulose.	Galactose.	Maltose.	Gelatin liquefaction.	Milk.
B. alcaligenes	−	+	−	−	−	−	−	−	−	−	−
B. dysenteriæ Shiga	−	−	+	−	−	−	−	−	−	−	±
B. dysenteriæ Flexner	−	−	+	−	−	+	+	+	−	−	±
B. dysenteriæ Hiss-Russell	−	−	+	−	+	+	+	+	+	−	±
B. dysenteriæ Rosen	−	+	+	+	−	+	+	+	+	−	+
B. pyogenes fœtidus	−	+	+	+	+	+	+	+	+	−	+
B. typhosus a	−	+	+	−	−	+	+	+	+	−	±
B. typhosus b	−	+	+	−	−	+	+	+	+	−	+
B. paratyphosus alpha	−	+	g	−	−	g	g	g	g	−	+
B. paratyphosus beta	−	+	g	−	−	g	g	g	g	−	±
B. Morgan No. 1	−	+	g	−	−	−	?	?	?	−	±
B. coli a	−	+	g	g	−	g	g	g	g	−	c
B. coli b	−	+	g	g	g	g	g	g	g	−	c
B. proteus	−	+	g	−	g	−	g	g	g	+	p
B. cloacæ	−	+	g	g	g	g	g	g	g	+	c/p

Legend: carbohydrate solutions: − = no fermentation, + = acid produced, g = gas produced.
milk: − = no fermentation, alkaline reaction, ± = initial acidity, terminal alkalinity, + = acid, c = coagulation, p = peptonization.

Morphology.—The morphology of the members of the dysentery group of bacilli is practically identical; they are medium-sized, rod-shaped organisms, measuring from 0.8 to 1 micron in diameter, and from 1.5 to 3 microns in length. They have rounded ends and occur singly or in pairs, rarely in short chains. Frequently elongated somewhat irregular involution forms are found in old broth cultures. The bacilli are non-motile (except the "Rosen" strain, which is sluggishly motile), possess no flagella, form no capsules and produce no spores. They stain fairly readily with ordinary anilin dyes; frequently,

[1] Zeit. f. Hyg., 1902, p. 559.

the ends of the organisms stain somewhat more heavily than the centre. All the organisms comprising this group are Gram-negative.

Isolation and Culture.—The dysentery bacilli grow well on ordinary laboratory media. Colonies on agar, after eighteen to twenty-four hours' incubation at the body temperature, are round, transparent and colorless; frequently they attain a diameter of from 1 to 3 mm. The colonies are indistinguishable from those produced by bacilli of the typhoid and paratyphoid groups. There is moderate growth along the line of inoculation in gelatin, but no liquefaction. In broth after eighteen to twenty-four hours' growth a uniform turbidity develops, somewhat more luxuriant in dextrose than in plain broth. After several days' growth in plain broth a delicate pellicle frequently

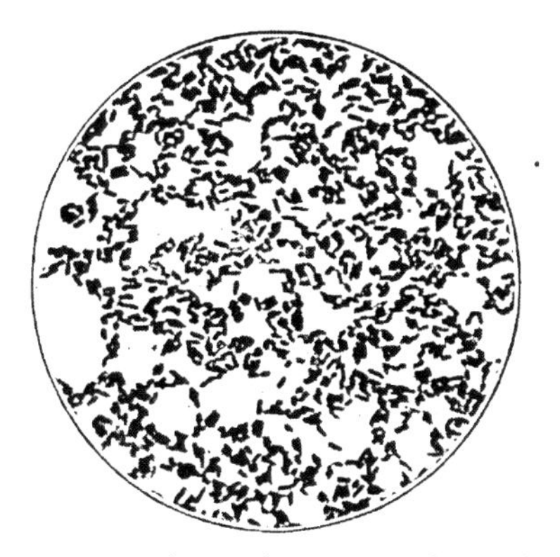

FIG. 45.—Bacillus dysenteriæ. Shiga type, bouillon culture. × 1000.

appears on the surface of the latter medium. In milk moderate development takes place with no coagulation. There is an initial acidity followed after from two to five days by an alkaline reaction, which increases somewhat in intensity with the age of the culture. On potato the growth is very similar to that of the typhoid bacillus; on acid potato the growth is almost invisible; on alkaline potato the growth is brownish and of moderate luxuriance.

The dysentery bacilli are aërobic, facultatively anaërobic bacilli whose limits are approximately the following; minimum temperature of growth 8° C.; maximum 42° to 44° C.; optimum 37° C.

Cultures of dysentery bacilli vary somewhat in their resistance to heat. The majority of cultures are killed by an exposure of ten minutes at 65° C. Some strains, however, are only killed by an exposure of ten minutes at 70° C. The organisms are moderately resistant to cold. Cultures may retain their viability in the ice-box, 6° to 10° C., for nearly two months. In sterile water the organisms at ordinary temperatures do not as a rule survive more than a week. Pfuhl[1] has found that dysentery bacilli may remain alive for 101 days in moist soil protected from sunlight; in dry soil under otherwise the same conditions they do not survive more than thirty days. In cheese and in butter they remain alive for at least nine days, and in sterile milk for about three weeks. Dried on linen, they also survive about three weeks.

Products of Growth.—*Chemical Products.*—Plain broth cultures of Shiga and Flexner bacilli do not contain indol or phenols, even after prolonged incubation. The statements with reference to indol production in the group, however, are somewhat conflicting, particularly with reference to the Flexner type of organism. Morgan and others[2] have stated that Flexner bacilli produce indol; on the other hand, Kendall, Bagg, Day and Walker[3] have isolated over 200 strains of Flexner bacilli from dysenteric cases and have found almost without exception that indol is not formed. These strains were identified by their cultural reactions and by agglutination with specific Flexner serum of high potency. Dopter[4] has found that strains of Flexner bacilli obtained from different sources, which were identical culturally and agglutinated the same with specific sera, vary in indol production some producing indol, others not producing it.

Acid Production in Carbohydrate Media.—All members of the dysentery group agree in two important characteristics: they do not form gas in carbohydrate media, and form acid in dextrose. Lentz[5] has called attention to an important cultural differentiation of the Flexner and Shiga bacillus, the former producing acid in mannite, the latter not fermenting this alcohol. Further study has shown that the fermentation of various carbohydrates is important in the recognition of the various types. The fermentation and other cultural reactions of members of the dysentery bacillus group are shown in the table on page 316. The members of the dysentery group produce an initial acidity in milk; fermentation of the small amount of dextrose, amounting to about 0.1 per cent., which is found in fresh milk (Theobald Smith[6]) followed by an alkaline reaction (action of the organisms upon protein when the utilizable carbohydrate is exhausted).[7]

[1] Ztschr. f. Hyg., 1902, xl, 555.

[2] Brit. Med. Jour., April 6, 1907, 908; July 6, 16.

[3] Boston Med. and Surg. Jour., 1911, clxiv, 301; 1913, clxix, 741, 753; Jour. Am. Chem. Soc., 1913, xxxv, 1211.

[4] Les Dysenteries, Paris, 1909, 36.

[5] Boston Jour. Med. Sci., 1897, ii, 236; Jones, Jour. Inf. Dis., 1914, xv, 357.

[6] Ztschr. f. Hyg., 1902, xli, 559.

[7] See Kendall, Day and Walker for essential analytical details, Jour. Am. Chem. Assn., 1914, xxxvi, 1940.

Enzymes.—Dysentery bacilli do not appear to produce extracellular proteolytic enzymes. They do not liquefy gelatin, blood serum or fibrin, and do not coagulate milk. Wells and Corper[1] have demonstrated a lipase of moderate activity in the autolysates of dysentery bacilli.

Toxins.—(*a*) *Exotoxin.*—The nature of the poison produced by the Shiga bacillus, the most virulent of the dysentery bacilli, is a matter of debate. Todd,[2] Ludke,[3] Doerr,[4] and Kraus and Doerr[5] state that the organism produces a soluble (exo-) toxin which stimulates antibody formation in suitable animals; the sera are specifically antitoxic and protect laboratory animals against several times the fatal dose of the toxin. According to Kraus and Doerr,[6] this toxin acts somewhat like that of the diphtheria bacillus; the lesions observed in the large intestine are comparable to the lesions of the diphtheria bacillus on the tonsils and pharynx. The nervous lesions are somewhat like those of poliomyelitis. Intravenous injection of large doses in rabbits causes death in from six to eight hours; smaller doses cause paresis, diarrhea, which is frequently bloody, paralysis of the bladder, hypothermia and death in one to four weeks. Postmortem there is a mucohemorrhagic enteritis, usually localized in the cecum. It is stated that the entire intestinal tract is involved in dogs, with the duodenum particularly affected. Intraperitoneal and subcutaneous injections give a much milder reaction with a prolonged incubation period. The toxin is inactivated by acids, but its potency may be partially restored when the acid is neutralized with alkali. Conradi[7] and others find dead cultures almost as toxic as the living bacilli; they call attention to the toxic properties of autolysates (in sterile water) of the Shiga bacillus, a fact which was pointed out by Gay[8] some time before. It is probable that both soluble and autolytic poisons are concerned in the toxicity of filtrates of broth cultures of the organism. The toxic substances may be obtained in dry form by saturating the broth (freed from bacilli by filtration through unglazed porcelain) with ammonium sulphate, dialyzing the precipitate to remove the ammonium salts, and evaporation of the

[1] Jour. Infec. Dis., 1912, xi, 388.
[2] Brit. Med. Jour., December 5, 1902, ii; October 4, 1903, ii.
[3] Jour. Path. and Bact., 1905, x, 328.
[4] Cent. f. Bakt., Orig., 1905, xxxviii, 420, 511.
[5] Ztschr. f. Hyg., 1906, lv, 1.
[6] Loc. cit.
[7] Deutsch. med. Wchnschr., 1903, xxix, 26.
[8] Penna. Med. Bull., 1902.

dialyzed solution to dryness *in vacuo*. The dried residue is very toxic for rabbits; 0.002 to 0.005 grams dissolved in a small amount of sterile salt solution will usually kill these animals when injected intravenously. Smaller amounts gradually increased stimulate antibody formation.[1] The antitoxin, however, has little curative value, for the toxin appears to have a greater affinity for the epithelium of the intestinal mucosa and central nervous system than it has for the antitoxin. The other members of the dysentery group do not produce soluble toxic substances in demonstrable amounts.

(*b*) *Endotoxin.*—Neisser and Shiga[2] have found that autolysates of Shiga bacilli produce a mucohemorrhagic enteritis in rabbits. Besredka,[3] Conradi[4] and others have also extracted substances from the organisms by grinding them with sand, by alternate freezing and thawing (method of MacFadyen and Roland), or by autolysis, which in small amounts will kill experimental animals when injected intravenously, intraperitoneally, or subcutaneously. Administration by mouth is without noteworthy effect. The potency of the endotoxin is not appreciably impaired by an exposure to 70° C. for an hour; an exposure to 80° C. renders it inactive. Conradi[5] has shown that occasional strains of dysentery bacilli (Shiga type) produce small amounts of soluble hemotoxin.

Pathogenesis.—*Experimental.*—Direct experimental evidence of the etiological relationship of the dysentery bacillus to bacillary dysentery is afforded by a few laboratory accidents in which the clinical disease has followed the accidental ingestion of cultures of dysentery bacilli. The most conclusive experiment, however, is that reported by Strong and Musgrave.[6] A forty-eight-hour broth culture of B. dysenteriæ (Shiga type) was swallowed by a condemned criminal after a dose of sodium hydrogen carbonate was given to neutralize the gastric acidity. The initial symptoms of a typical attack of bacillary dysentery followed after an incubation period of thirty-six hours. The organisms were isolated from the mucopurulent, bloody feces. Ravant and Dopter[7] produced clinical dysentery in an ape by feeding it Shiga bacilli.

Human.—Infection with the Shiga bacillus is somewhat less common in the United States than infection with the Flexner and other

[1] Todd, loc. cit.; Kraus and Doerr, loc. cit.
[2] Deutsch. med. Wchnschr., 1903, No. 4.
[3] Ann. Inst. Past., April, 1906, vol. xxv.
[4] Loc. cit.
[5] Loc. cit.
[6] Report of the Surgeon-General, United States Army, 1900.
[7] Quoted by Kolle and Hetsch, Die experimentelle Bakt., II. Aufl., i, 304.

types of the dysentery group, but far more fatal. Mixed infections in which both Shiga and Flexner bacilli are present are occasionally seen.[1] Among adults infection with the Flexner type of organism tends to be sporadic in distribution and less severe than infections with the Shiga type which more commonly assume epidemic tendencies.

The incubation period of bacillary dysentery may be as brief as forty-eight hours, or even less, and as a rule there are no distinctive prodromal symptoms. The feces, at first watery, may be very frequent, as many as twenty to thirty per diem, and become mucopurulent with considerable amounts of fresh blood mixed in them. The organisms are present in variable numbers. Dysentery bacilli do not as a rule appear to invade the blood stream, but at least three instances are on record where pure cultures of the Shiga bacillus have been isolated antemortem from the general circulation;[2] occasionally pure cultures of dysentery bacilli may be obtained from mesenteric lymph nodes postmortem.

Lesions.—The lesions, which are found chiefly in the large intestine, vary with the severity and duration of the disease. In the early stages of the disease there is a severe catarrhal inflammation of the mucous membrane of the large intestine with some necrosis of the epithelium, associated with hyperemia of the mucosa of the small intestine as well. The mesenteric glands are usually swollen and hyperemic. Later the inflammation may become very severe; a pseudomembrane may form in the large intestine with extensive superficial ulceration of the mucosa. The ulcers do not extend as a rule to the submucosa; consequently, perforation is rare in uncomplicated cases. The submucosa, however, may be swollen and somewhat edematous.

The nervous symptoms which are a feature of severe dysentery infections would suggest that in addition to the intestinal lesions there may be involvement of the nervous system. Southard, McGaffin and Richards[3] have shown that in addition to lesions of the intestinal tract, the Shiga toxin has a special affinity for the anterior horn ganglion cells, thus explaining on a definite anatomical basis the nervous symptoms which are a feature of fatal cases of bacillary dysentery. Dopter[4] has expressed the same opinion. He believes the toxin of the

[1] Kendall, Bagg, Day and Walker, loc. cit.

[2] Rosenthal, Deutsch. med. Wchnschr., 1903, No. 6; Kendall, Bagg and Day, Boston Med. and Surg. Jour., 1913, clxix, 741; Darling and Bates, Am. Jour. Med. Sc., 1912, clxiii, No. 1.

[3] Boston Med. and Surg. Jour., 1909, clxi, 65, 108.

[4] Loc. cit., p. 77.

Shiga bacillus has an elective affinity for the intestinal mucosa of the large intestine, and it is the toxin secreted by the dysentery bacilli during their multiplication in the intestinal mucous membrane which induces the anatomical and nervous lesions characteristic of the disease.

Animals.—Typical bacillary dysentery has not been produced in laboratory animals by feeding the organisms. The intravenous inoculation or intraperitoneal injection of living or killed broth cultures of Shiga or Flexner bacilli, however, are usually fatal, particularly to rabbits. Vaillard and Dopter,[1] and Flexner[2] have shown that small amounts of forty-eight-hour broth cultures of Shiga bacilli introduced intravenously into young rabbits frequently lead to diarrhea, which at first is mucous in character; later it becomes mucosanguineous. After two or three days symptoms of paraplegia develop. At autopsy the large intestine is swollen and frequently edematous. The mesentery is hyperemic with enlarged glands. The intestinal contents are mucosanguineous in character and the intestinal wall is considerably thickened. If the animal survives for several days, more advanced lesions are sometimes seen, particularly beginning ulceration and necrosis. Flexner states that the intestinal lesions of bacillary dysentery in man and in animals are probably due, in part at least, to the direct action of the dysentery toxin.

Immunity and Immunization.—Shiga[3] and others have succeeded in immunizing laboratory animals, particularly rabbits, guinea-pigs, and horses, with dysentery bacilli, beginning by injecting killed cultures of these organisms, first with very small amounts which are slowly and cautiously increased, finally with living bacilli. It is difficult to immunize animals because of the toxicity of the organism. The sera of these animals contain specific agglutinins, lysins, precipitins, and opsonins, frequently of high potency. According to Todd, and Kraus and Doerr,[4] specific antitoxins are also demonstrable in the sera of these animals, particularly in animals immunized to the Shiga bacillus. The agglutinins which are specific for the type of organism used in immunization are, according to Dopter,[5] as a rule of greater potency when killed cultures exclusively are used for immunizing. In thoroughly immunized animals the agglutinins may be active even in dilutions of 1 to 5000.

[1] Ann. Inst. Past., 1903, p. 472.
[2] Jour. Exp. Med., 1906, vol. viii.
[3] Ztschr. f. Hyg., 1902, xli, 355.
[4] Loc. cit.
[5] Loc. cit., p. 84.

Specific bacteriolysins have been demonstrated in immune sera *in vitro* by Shiga[1] and *in vivo* by Kruse.[2] Specific precipitins, which in dilutions of 1 to 10 or greater will produce a precipitate in broth filtrates of the homologous strain, but not, as a rule, for other types of the dysentery bacilli are also found. The sera of patients who have recovered from attacks of bacillary dysentery usually contain specific agglutinins which are active even in dilutions of 1 to 50. Specific precipitins, lysins, and opsonins are also demonstrable in the sera of these patients.

Therapy.—Attempts to immunize man with vaccines, both mono- and polyvalent,[3] sensitized vaccines (bacteria which have been in contact with antidysentery serum, then centrifugalized, washed, and suspended in salt solution, according to the method of Besredka and of Gay), and the use of antisera, usually derived from immunized horses, have not been generally successful, although a few favorable results have been recorded.

Bacteriological Diagnosis.—(*a*) *Agglutinin Reaction.*—The sera of normal individuals rarely agglutinate dysentery bacilli in dilutions greater than 1 to 10, although Dopter[4] states that the Flexner organism may be clumped with the serum of apparently normal individuals in a dilution greater than 1 to 10. For this reason agglutination tests should be made in a dilution of 1 to 20 to 1 to 30 with the Shiga organism, and 1 to 80 to 1 to 100 with the Flexner strain in each case examined, since one or the other organism, or both, may be present in typical cases of bacillary dysentery. Agglutinins do not as a rule appear in mild cases, and in severe cases they are not demonstrable until from the seventh to the tenth day on the average.

The serum of dysentery carriers, both those giving a history of a previous attack and those with the negative dysentery history, frequently agglutinates either with Shiga or Flexner bacilli. The agglutination reaction, therefore, is not conclusive for clinical diagnosis unless a negative reaction is obtained early in the disease followed by a positive reaction on or after the seventh to the tenth day.

(*b*) *Isolation of Dysentery Bacilli from the Feces.*—Dysentery bacilli do not invade the blood stream as a rule, and they are not found in the urine. The bacteriological diagnosis, therefore, depends upon

[1] Loc. cit.
[2] Deutsch. med. Wchnschr., 1902.
[3] Shiga, Deutsch. med. Wchnschr., 1901, Nos. 43 and 45; Kruse, ibid., 1903, Nos. 1 and 3.
[4] Loc. cit., p. 91.

the isolation of the organisms from the feces and their identification by cultural and serological reactions.

A bit of blood-stained mucus offers the best material for isolation of the organisms: it should be washed two or three times in sterile salt solution to remove extraneous organisms as far as possible, for experience has shown that dysentery bacilli are frequently enclosed in mucus. The mucus is then macerated in sterile broth, and if possible incubated for one or two hours at 37° C. It is then spread upon the surface of Endo-plates and incubated for eighteen to twenty-four hours at 37° C. The colonies are precisely similar to those of typhoid and paratyphoid bacilli; the final identification of the dysentery bacilli is made by their cultural reactions (see page 316) and by agglutination with specific sera of high potency. The rapid method of isolating and identifying typhoid bacilli described on page 338 is equally applicable to dysentery bacilli. The possibility of carriers should be borne in mind when mild and atypical cases are under consideration.

Dissemination and Prophylaxis.—Dysentery bacilli appear to be widely distributed in certain areas of the temperate zone, and outbreaks occur at varying intervals. Interepidemic years are occasionally characterized by considerable numbers of atypical, mild cases, and carriers are not uncommon.[1]

The organisms enter the body through the mouth and intestinal tract, and leave it in the feces; consequently the method of transmission of the disease is similar to that of typhoid and other excrementitious disorders. There is some evidence that the disease may be milk-borne; exclusively breast-fed infants are rarely or never infected; bottle-fed babies of the same age may be infected in relatively large numbers during years which exhibit an epidemic tendency of bacillary dysentery. Zinsser[2] has produced evidence in favor of the occasional milk transmission. The organism may also reach the body by direct transmission through carriers, in hospitals, and through contaminated water and food. Flies may also play a part in the spread of the disease.

The precautions to be observed are those for any intestinal infection.

[1] Kendall, Boston Med. and Surg. Jour., 1913, clxix, 7493; ibid., May 20, 1915.
[2] Proc. New York Path. Soc., 1907.

TYPHOID BACILLUS.

Historical.—Typhoid bacilli were first seen in sections of tissue from autopsies by Klebs in 1876. Somewhat later Eberth[1] successfully demonstrated them in sections of mesenteric glands, lymph nodes and the spleen by the use of the recently introduced tissue stains. Gaffky[2] first isolated the organisms in pure culture and established their probable etiological relationship to typhoid fever. Later investigations with more refined methods have completely substantiated Gaffky's observations.

Morphology—Typhoid bacilli are rod-shaped organisms of moderate size, measuring from 0.5 to 0.8 microns in diameter and from 1 to 3 microns in length. The dimensions vary within the limits given upon different media, the organisms being as a rule somewhat longer in fluid media than upon solid media. Elongated rods and even filaments are occasionally found in old gelatin and potato cultures. The bacilli have rounded ends and occur as a rule singly or in pairs. They are actively motile, particularly in young cultures grown in 0.1 per cent. dextrose broth; plain broth cultures are usually more sluggish. Each organism possesses characteristically from eight to ten peritrichic flagella; rarely as many as twenty may be attached to a single organism. The flagella are somewhat wavy in outline and measure from 6 to 8 microns in length. No spores are produced It was formerly held that typhoid bacilli formed no capsules. Carpano,[3] and Gay and Claypole,[4] however, have demonstrated capsules around typhoid bacilli grown in blood media.

The organisms stain readily with ordinary anilin dyes and they are Gram-negative.

Isolation and Culture.—The typhoid bacillus grows readily upon the ordinary media. Colonies on agar plates are round, colorless, flat, and nearly transparent; they attain a diameter of from 0.5 to 1.5 mm. after eighteen to twenty-four hours' incubation at 37° C. Development in gelatin is less rapid, and the colonies after two to three days' incubation at 20° C. are somewhat brownish in color. A uniform turbidity is produced in plain broth after eighteen hours' growth at 37° C.; development in dextrose broth is more intense, but after five to seven days it ceases and the organisms die, due to the accumula-

[1] Virchows Arch.. 1880, lxxxi, 58; 1881, lxxxiii, 486.
[2] Mitt. a. d. kais. Gesamte, 1884, ii, 370.
[3] Cent. f. Bakt., Orig., 1913, lxx, 42.
[4] Arch. Int. Med., 1913, xii, 624.

tion of acid. Growth is luxuriant in milk, but there is little chemical change in the composition of the medium as the result of the growth.[1] Two types of reaction are observed in litmus milk: (*a*) The reaction becomes slightly acid, turning the litmus to a lilac color which persists. This is much more common than (*b*); the milk becomes slightly acid, as in "*a*," then it becomes slowly but progressively alkaline. Relatively few authentic strains of typhoid bacilli appear to produce the transient acidity in this medium. At one time potato was regarded as an important differential medium for the recognition of the typhoid bacillus. The "invisible growth" described by Gaffky[2] is now known to be dependent largely upon the reaction; potatoes having an acid reaction give this invisible growth; old potatoes which usually have

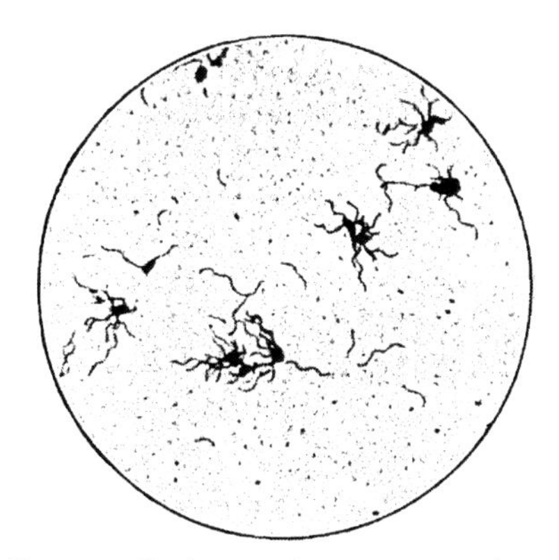

FIG. 46.—Bacillus typhosus, flagella stain.

a slightly alkaline reaction give a heavy, brownish growth much like that of the colon bacillus. The addition of small amounts of alkali, as sodium carbonate, to potato prior to inoculation makes the growth visible and brown; the addition of a small amount of organic acid to the medium usually results in the development of the invisible type of growth.

The typhoid bacillus is an aërobic, facultatively anaërobic organism, whose minimal temperature of growth is about 8° C.; development is maximal at 37° C., and ceases when the culture is exposed to temperatures above 43° to 44° C. An exposure of ten to twenty minutes at 60° C. will kill the naked organisms; a longer exposure at a higher temperature is required to kill them when they are suspended in organic

[1] Kendall, Day and Walker, Jour. Am. Chem. Assn., 1914, xxxvi, 1958.
[2] Loc. cit.

matter, as feces. Cultures exposed to temperatures from 0° C. to −10° C. for three months occasionally contain viable organisms. Alternate freezing and thawing is more fatal than simple freezing. The typhoid bacillus dies out rather rapidly in potable water, less rapidly in sterilized potable water. The addition of organic matter, particularly of fecal origin, appears to promote longevity somewhat. The observations of Jordan, Russell and Zeit[1] would indicate that a large percentage of organisms exposed in potable water die within three days. Kersten[2] has shown that typhoid bacilli will develop with considerable rapidity in raw milk. The bacilli may remain alive in soil for several months, provided they are shielded from direct sunlight, and they may resist drying under similar conditions for

Fig. 47.—Bacillus typhosus, bouillon culture. × 1000.

several weeks. A maximum exposure of from four to eight hours to direct sunlight in the months of June, July and August (Northern Hemisphere) usually kills the organisms. Mercuric chloride 1 to 1000 kills the naked germs in about ten minutes; 5 per cent. carbolic acid kills them in from five to ten minutes, as a rule.

Products of Growth.—The typhoid bacillus liberates ammonia from protein in sugar-free media, and forms small amounts of non-volatile alkaline products as well. The reaction, therefore, becomes progressively alkaline. A radical change in the nature of the products of metabolism occurs when the bacilli are grown in protein media containing utilizable carbohydrates, as dextrose or mannite. The reaction becomes strongly acid, due to the fermentation of the sugar. The

[1] Jour. Infec. Dis., 1904, i, 641.
[2] Arb. a. d. kais. Gesamt., 1909, xxx, 341.

protein under these conditions is left unattacked except for minute amounts necessary to supply the nitrogenous requirements of the organism. The acids formed are chiefly lactic acid, together with smaller amounts of formic acid.[1] Indol or phenols are not formed in ordinary media, but Peckham[2] has shown that indol may be produced in protein media of special composition.

The essential cultural characters of B. typhosus are indicated in the table on page 316. Culturally Bacillus typhosus is relatively inert; it does not produce proteolytic enzymes which liquefy gelatin, blood serum or fibrin. A fat-splitting ferment has been demonstrated in autolyzed typhoid bacilli by Wells and Corper.[3] An esterase which liberates butyric acid from ethyl butyrate is detectable in sterile filtrates of plain and dextrose broth cultures of the organism.[4]

Typhohemolysin (typholysin). Castellani,[5] and E. Levy and P. Levy[6] have found that filtrates of (sugar-free) broth cultures of typhoid bacilli are hemolytic. They appear to have demonstrated specific antihemolytic properties in the blood of animals injected with hemolytic filtrates, thus meeting the objection that the hemolysis might be due to the alkalinity of the medium itself. There is no evidence at present which would suggest that this hemolysin plays any important part in typhoid infections of man. The typholysin is relatively thermostabile.

Toxins.—A soluble toxin has never been satisfactorily demonstrated among the products of growth of the typhoid bacillus, and the consensus of opinion at the present time is in favor of the view that the principal toxic substance of the organism is an endotoxin. The endotoxin has been studied with special thoroughness by MacFadyen and Roland,[7] and Besredka.[8] It has been obtained in various ways: by grinding the organisms with sand, by freezing in liquid air and triturating, or by autolysis of the bacilli in sterile distilled water. Relatively small amounts of endotoxin obtained by any of these methods will usually kill guinea-pigs. No antitoxin has been produced in the sera of animals inoculated with gradually increasing amounts of this endotoxin.

[1] Kendall, Jour. Med. Research, 1911, xxiv, 411; 1912, xxv, 117. Boston Med. and Surg. Jour., 1911, lxiv, 288. Kendall, Day and Walker, Jour. Am. Chem. Assn., 1913, xxxv, 1214.

[2] Jour. Exper. Med., 1897, ii, 549.

[3] Jour. Infec. Dis., 1912, xi, 388.

[4] Kendall and Simonds, Jour. Infec. Dis., 1914, xv, 354.

[5] Lancet, February 15, 1902.

[6] Cent. f. Bakt., 1901, xxx, 405.

[7] Cent. f. Bakt., Orig., 1903, xxxiv, 618, 765; MacFadyen, ibid., 1903, xxxv, 415.

[8] Ann. Inst. Past., 1905-1906.

Typhoid Fever.—**Pathogenesis.**—*Experimental.*—Typhoid fever is a disease of man only, and until recently rigorous experimental proof that the typhoid bacillus is the specific cause of this infection has been lacking. The evidence of the etiological relationship of the typhoid bacillus is of two kinds: (1) a few cases where laboratory attendants have accidentally or purposely swallowed cultures of typhoid fever and have developed the disease; (2) experiments of Metchnikoff and Besredka.[1]

The experiments of Metchnikoff and Besredka appear to be conclusive. They produced typhoid fever in anthropoid apes by feeding the animal food infected with fecal material containing typhoid bacilli. The animals (fifteen in all) developed fever and diarrhea after eight days, and typhoid bacilli were isolated from the blood stream on the tenth day. Three died. Specific agglutinins were demonstrable in the blood serum, and the clinical picture was essentially that of typical typhoid fever. These observers ruled out the possibility of a filterable virus.

Pathogenesis in Man.—*Portal of Entry.*—Typhoid bacilli enter the body through the mouth and pass through the gastro-intestinal tract. They lodge in lymphatic tissue of the intestines, particularly Peyer's patches, then invade the general lymphatic system and spleen, and are found in the blood, especially during the first week of the clinical disease. Typhoid fever, therefore, is a bacteremia. Rose spots, which are frequently found on the abdomen during the first week of the clinical disease, contain colonies of typhoid bacilli which are localized in the subcutaneous tissue.[2] Characteristic lesions are found in Peyer's patches which at first are swollen and hyperemic. After a few days the glands become rather pale, caused, in part at least, by hyperplasia of the lymphoid and endothelioid cells, which cuts off the blood supply in whole or in part, leaving these areas even more prominent (medullary swelling).[3] Necrosis then commences and the glands gradually become yellowish in color and softer in consistency. Soon the necrosis ceases rather abruptly as immunity checks the process and the necrotic tissue then sloughs away, leaving a somewhat irregular elongated ulcer which usually extends to or through the muscular layer of the intestine. About the end of the third week scar tissue begins to appear in these ulcers, which in time practically fills up the original area, leaving the

[1] Ann. Inst. Past., March 25, 1911; xxv, 193, 865.
[2] Richardson, Philadelphia Med. Jour., March, 1900. (Special Typhoid Fever Number.)
[3] Mallory, Jour. Exp. Med., 1898, iii, No. 6, p. 611.

site of the ulcer marked by a somewhat depressed cicatrix. Occasionally secondary infection of the ulcers results in perforation or hemorrhage, and sometimes an uninfected ulcer may erode through a blood vessel, causing hemorrhage. It should be remembered that typhoid ulcers tend to run along the long axis of the intestine, whereas tuberculous ulcers, on the contrary, run transversely, following the course of the lymphatics.

In addition to the intestinal lesions, there is in typhoid fever an acute splenic tumor with a great proliferation of typhoid bacilli in this organ. Foci of typhoid bacilli are commonly found also in the kidneys and the liver, mesenteric lymph nodes, less commonlv in lungs, meninges, bone marrow, certain muscles and the tonsils. Parenchymatous degeneration of the heart, liver and kidneys is common, as is a catarrhal inflammation of the respiratory tract and a severe inflammation of the entire intestinal mucous membrane. Somewhat uncommonly, typhoid cases have been recorded in which there are no intestinal lesions. In these cases it would appear that the disease is septicemic in character.[1] In typhoid fever there is leucopenia, due apparently to some interference with the activity of the bone marrow. The febrile reaction is usually attributed to the liberation of endotoxin from typhoid bacilli, which are dissolved in the blood stream by specific lysins. This toxin exhibits both a general and local reaction. The general reaction is characterized chiefly by fever and symptoms of generalized toxemia; the local reaction is particularly marked in those areas where typhoid bacilli undergo solution, as in the spleen and Peyer's patches.

Various complications of typhoid fever are occasionally reported, caused by the localization of typhoid bacilli either alone or in association with other bacteria, as the streptococcus, staphylococcus, or pneumococcus, in various organs. Peritonitis, usually following perforation of an ulcer in the intestinal wall, is one of the most severe of these complications. Abscess formation in various deep-seated organs, as the spleen and psoas muscle, is not uncommon. Bronchopneumonia, pleurisy, pericarditis, osteitis, and inflammation of the membranes of the cord (meningitis) and brain have also been attributed to the typhoid bacillus.

Carriers.—Typhoid bacilli can not be isolated from the majority of typhoid patients after the fifth week of the disease. In a small

[1] Possett, Atypische Typhusinfektion. Lubarsch and Ostertag, Ergebn. d. allgem. Pathol., 1912, xvi, 184.

percentage of cases, however, the organisms may be excreted in the urine, or more commonly in the feces, for months or even years after recovery. Thus, Philipowicz[1] isolated typhoid bacilli from a case of cholecystitis who had had typhoid fever thirty-eight years previous to the operation. In this case very few typhoid bacilli were present in the feces, and it is probable that the few organisms were overwhelmed by the intestinal bacteria during their passage through the intestinal tract. From 1 to 4 per cent. of all typhoid cases which recover appear to become fecal typhoid carriers; a smaller percentage become urinary carriers. No history of typhoid fever can be elicited from some of these carriers, and the supposition is that either the carrier had in the past a mild unrecognized case, or less commonly that the organism had become acclimatized in the intestinal tract without inducing disease. Many carriers give a positive Widal reaction.

The residual focus of typhoid bacilli in carriers is usually the gall-bladder and the ducts of the gall-bladder, less commonly the urinary bladder. From the gall-bladder the organisms pass in irregular numbers into the intestinal tract; occasionally in sufficient numbers to be demonstrable in the feces. A considerable proportion of operations for cholecystitis and gall-stones—the greater majority being among women—give positive typhoid cultures when the contents are examined bacteriologically.

Pathogenesis in Animals.—All animals, except possibly anthropoid apes, are naturally immune to typhoid fever, and inoculation of old laboratory cultures of typhoid bacilli into laboratory animals is usually without noteworthy effect; virulent cultures of typhoid bacilli, particularly those produced by repeated passage through laboratory animals, may produce peritonitis and death when they are introduced into the animals by the intraperitoneal route. The infection, however, does not resemble typhoid fever. The lesions observed post-mortem are marked congestion of the abdominal organs, particularly the spleen, kidneys and liver, as well as involvement of the intestinal lymph apparatus; the thoracic organs are less involved as a rule.

The organisms may be recovered from the peritoneal fluid, the blood stream, and from various abdominal organs. Gay and Claypole[2] have succeeded in inducing with great regularity the carrier state in rabbits by injecting into them typhoid bacilli which have been grown for several successive transfers on agar overlaid with fresh defibrinated

[1] Wien. klin. Wchnschr., 1911, 1802.
[2] Arch. Int. Med., December, 1913.

rabbit's blood. They found that the typhoid bacilli localize themselves in the gall-bladders of the rabbits, and that they may from time to time invade the blood stream. In a more recent communication[1] they have shown that the carrier state occurs much less frequently if the animals are immunized with their dried sensitized vaccine.

Antibody Production.—Animals may be immunized by repeated injections of typhoid bacilli to such a degree that they will successfully resist several times the original fatal dose of these organisms.[2] Successive injections of typhoid bacilli stimulate antibody formation in horses, rabbits, guinea-pigs, and other animals. Of these antibodies, the lysins and agglutinins may be produced in high potency if the injections are continued long enough. Other antibodies, opsonins and precipitins particularly, are also produced. Gay and Claypole[3] have produced experimental evidence indicating that the titre of the specific agglutinins which develop during the process of immunization of rabbits affords no indication of the degree of protection attained by the immunizing process.

Protective Immunization.—As a rule, one attack of typhoid fever confers immunity; subsequent attacks are unusual.

During the last few years definite progress has been made in the protective immunization of human beings, both by the use of killed cultures of typhoid bacilli and by live cultures. The vaccine treatment for typhoid fever is the best known and the most widely practiced. The procedure is to grow typhoid bacilli on agar slants, wash them off with sterile physiological salt solution, kill them by heating to 60° C. for an hour, standardizing the suspension of typhoid bacilli, and injecting as a first dose five hundred million killed typhoid organisms. After an interval of seven to ten days a second injection of a billion killed typhoid bacilli is made, and after an equal interval a third and last injection of a billion killed typhoid bacilli is made. In about 20 per cent. of the cases injected general symptoms which consist of a febrile reaction and malaise develop, accompanied by local symptoms of pain, redness, and swelling at the site of inoculation. These symptoms may appear after the second or even after the third injection. It is customary to make the inoculation about four o'clock in the afternoon, so that the patient in the majority of cases sleeps through the general symptoms.

[1] Arch. Int. Med., 1914, xiv, 671.
[2] See Gay and Claypole, Arch. Int. Med., 1914, xiv, 671, for essential details.
[3] Loc. cit.

The immunity produced is generally considered to be relatively complete for from six months to a year. It must be remembered that for at least three weeks following the vaccination there is a diminution in the resistance of the individual to typhoid fever; consequently, typhoid vaccination should not be undertaken if there is a possibility of exposure to typhoid during this period. Vaccination is also very undesirable if it is performed during the incubation period of typhoid fever. It should be practiced only on perfectly healthy subjects free from all general and local organic defects or infections, particularly tuberculosis. Nurses, ward orderlies, doctors, and those engaged in the care of typhoid patients are particularly likely to benefit by these inoculations. Gay and Claypole[1] have demonstrated experimentally that a satisfactory degree of protection may be attained in animals by three injections, at intervals of two days each, of a dried sensitized vaccine. Observations upon man immunized with this vaccine indicate that the reactions are milder and the whole process can be completed within a week, thus diminishing very materially the time element which has been an important factor in the past. It is very probable that the period of increased susceptibility to infection may be decidedly shortened as well.

Vaccination with Living Cultures.—Metchnikoff and Besredka[2] found that the subcutaneous injection of living sensitized cultures produced an immunity in anthropoid apes which was apparently as definite as that produced by an actual attack of typhoid fever. The organisms were shown not to appear in the urine or feces or blood when introduced subcutaneously. They were unable to induce immunity in the chimpanzee with killed cultures of typhoid bacilli or with autolysates of killed cultures. Having in mind the efficiency of living cultures, they[3] attempted the vaccination of man with living cultures of the typhoid bacillus. They used sensitized cultures which appeared to cause only a feeble local reaction and no general reaction in the chimpanzee, in preference to non-sensitized living cultures, which they found produced rather intense local and general reactions. The vaccine was prepared by emulsifying agar cultures of typhoid bacilli in normal salt solution and permitting the organisms to remain in contact with antityphoid serum for twenty-four hours at 37° C. The organisms are then removed by centrifuging, washed

[1] Loc. cit.
[2] Ann. Inst. Past., 1913, xxvii, 597. Besredka, Ann. Inst. Past., 1913, xxvii, 607.
[3] Semaine Méd., July 24, 1912, 355.

repeatedly, then re-emulsified in normal saline solution and heated to 50° C. for thirty minutes, then standardized in the usual manner. Nearly eight hundred people have been vaccinated with these sensitized living cultures; the local reaction was slight in each instance, and only exceptionally was there any general reaction. A careful examination of the blood, urine and feces of sixty-four of these cases failed to show typhoid bacilli, which would suggest that individuals vaccinated with living typhoid bacilli neither develop typhoid fever nor become carriers. The cases are too few in number to compare statistically with the cases vaccinated with killed cultures. Gay and Claypole[1] have taken issue with Metchnikoff upon this point and their experiments indicate that their sensitized dried vaccine may be equally or more efficient without the theoretical dangers which attend the use of living bacilli.

Various attempts have been made to induce passive immunity to typhoid infection by the injection of sera obtained from horses which have received numerous injections of typhoid bacilli or their soluble products. The results have on the whole not been encouraging. Gay and Force[2] have applied a preparation of typhoid bacilli ("typhoidin") made like Koch's old tuberculin, by the von Pirquet method, to patients that have recovered from typhoid fever and to those who have been vaccinated with typhoid bacilli. They find that 95 per cent. of recovered cases from typhoid (20 cases out of 21 examined) gave a clear-cut cutaneous reaction. One case had typhoid forty-one years previously. The reaction was negative in 85 per cent. of individuals not giving a history of typhoid (and presumably not vaccinated)—41 cases tested. The 9 cases (15 per cent.) that gave a positive reaction were suspected to have had a mild undiagnosed attack. Several, but not all, of those vaccinated within four years (9 out of 15) gave a positive reaction. Gay and Force suggest that the test is of presumptive value as an index of protection against typhoid by vaccination. Later observations by them confirm this view.

Diagnosis.—The diagnosis of typhoid fever in the living subject may be made either by the isolation and identification of the specific organism, Bacillus typhosus, or by the demonstration of antibodies specific for this organism in the body fluids of the patient.

(*a*) Bacteriological Diagnosis.—1. Isolation of typhoid bacilli from the blood stream and from rose spots.

[1] Loc. cit.

[2] University of California Publications in Pathology, 1913, ii, No. 14; Arch. Int. Med., 1914, xiii, 471.

Typhoid bacilli are found in the peripheral blood of a large percentage of typical cases of typhoid fever during the first week of the clinical disease. The organisms are found less frequently in the later stages. The statistics reported by Coleman and Buxton,[1] covering 1137 cases, show this clearly.

	Cases.	Positive, per cent.
First week of clinical disease	224	89
Second week of clinical disease	484	73
Third week of clinical disease	268	60
Fourth week of clinical disease	103	38
Fifth week of clinical disease	58	26

The organisms have also been isolated from rose spots (which appear as a rule early in the clinical course of the disease) by Richardson and others. From these observations typhoid fever may be regarded primarily as a bacteremia.[2] It should be remembered, however, that the organisms are destroyed in the blood stream by specific lysins, and that their presence in the circulating fluids of the body are partly caused by an overflow of organisms from foci in the spleen and other organisms.

Method of Collecting Blood.—The skin of the elbow is thoroughly cleansed as for a surgical operation, a tourniquet is applied, and a large hypodermic needle is introduced into a vein, preferably the median basilic. From 5 to 15 c.c. of blood are removed, discharged at once into a flask containing 150 to 250 c.c. of dextrose broth (0.1 per cent.), and mixed thoroughly before clotting takes place. This considerable dilution of the blood is important, partly because clotting takes place more slowly and thus favors the escape of the organisms into the broth, and also because it dilutes the lysins which are usually present in the blood of typhoid patients. It is necessary to reduce the concentration of lysins, for lysins dissolve typhoid bacilli. Incubation of the culture at 37° C. for twenty-four hours usually results in a growth of bacteria in which the specific organisms are present, either alone or mixed with skin cocci.

Coleman and Buxton[3] recommend an ox bile glycerin peptone medium for the isolation of typhoid bacilli. The medium as prepared by them has the following composition: Ox bile, 900 c.c.; glycerin, 100 c.c.; peptone, 20 grams. This is sterilized in the autoclave and

[1] Am. Jour. Med. Sc., 1907, cxxxiii.

[2] Brion and Kayser, Deut. Arch. f. klin. Med., 1906, lxxxv, 552. Coleman and Buxton, Jour. Med. Research, 1909, xxi, 83. Kolle and Hetsch, Experimentelle Bakt. und. Infektionskrank., 1911, 3ed., i, 250.

[3] Loc. cit.

distributed in flasks, 25 c.c. to a flask. The ox bile prevents the coagulation of the blood. Three c.c. of blood, according to the Coleman technic, are added to the flask of this medium, incubated for eighteen to twenty-four hours, then plated out on agar. Experience has shown that larger amounts of blood are more satisfactory, for it has been found that not infrequently 5 c.c. of blood will not give a growth of typhoid bacilli, whereas 10 c.c. or, better, 15 c.c. will give a growth. The organisms obtained in pure culture are identified by agglutination with a known specific typhoid serum of high potency. Such a serum used in high dilution reduces the possibility of "group agglutinins" which might otherwise give an erroneous diagnosis. It must be remembered that occasional strains of typhoid bacilli are isolated from the body which are typical culturally, but which are non-agglutinable. Frequently a few successive transfers of these organisms on artificial media will restore their agglutinating properties; occasionally, however, a strain is met with which will not agglutinate with specific typhoid serum even after long-continued transfer on artificial media. McIntosh and McQueen[1] have found that at least certain strains of these non-agglutinable typhoid bacilli will stimulate the production of typical typhoid agglutinins if they are injected into animals. The agglutinins developed in these animals will promptly clump agglutinable typhoid bacilli, but will not agglutinate the non-agglutinable strains which incited the production of these agglutinins. These non-agglutinable strains, however, will absorb the agglutinins apparently as readily as the agglutinating strairs. Gay and Claypole[2] have found similarly that occasional strains of typhoid bacilli isolated from "typhoid carrier" rabbits may be non-agglutinable. They absorb agglutinin, however. They suggest the use of sera obtained from animals immunized with cultures of typhoid bacilli grown upon agar containing the blood of man. The isolation of typhoid bacilli from the blood stream and their identification establishes the diagnosis of typhoid fever beyond question of doubt.

The isolation of typhoid bacilli from rose spots is performed in essentially the same manner, except that fluid is expressed from the rose spot after the skin is sterilized over it, and the expressed fluid is grown either in the dextrose broth or in the bile medium. Neufeld[3] and Richardson[4] have successfully isolated typhoid bacilli from the roseola

[1] Jour. Hyg., 1914, xiii, 409.
[2] Jour. Am. Med. Assn., 1913, lx, 1141; Arch. Int. Med., 1913, xii, 613.
[3] Ztschr. f. Hyg., 1899, xxx, 498.
[4] Philadelphia Med. Jour., March 3, 1900.

of typhoid fever in a considerable number of cases. Thus, Neufeld[1] obtained cultures in 13 of 14 cases examined, and Richardson obtained them in 5 out of 6 cases. Both Neufeld and Richardson emphasize the importance of incising several spots. The technic developed by Richardson is as follows: the skin over several rose spots is cleaned as for a surgical operation and then frozen by a spray of ethyl chloride. This procedure drives out most of the blood, as well as making the operation practically painless. A small incision is then made with a sterile knife and the substance of the rose spot is removed with a small skin curette and at once placed in 0.1 per cent. dextrose broth, and incubated for eighteen to twenty-four hours. The identification of the bacilli which develop in the broth is made by the usual cultural and agglutination reactions.

2. *Isolation of Typhoid Bacilli from the Urine.*—Typhoid bacilli have been found in the urine in from 25 to 35 per cent. of the cases examined. Such urines frequently contain albumin. The organisms do not as a rule appear until the third week of the disease, consequently their isolation is of comparatively little value diagnostically, although their recognition is of great importance for the prevention of secondary cases. The organisms may exist in the urine for a few weeks after recovery. Rarely they persist for months or very rarely for years after recovery. Frequently their presence is not manifested by clinical symptoms, but occasionally persistent cystitis may be caused by their continued growth in the urinary bladder. Usually the bacilli present in the urine are found in pure culture. Occasionally colon bacilli are found either in association with typhoid bacilli or even in pure culture after the typhoid bacilli have disappeared.

3. *Isolation of Typhoid Bacilli from Feces.*—Typhoid bacilli are usually found in pure culture or nearly pure culture in the blood, and, if the proper precautions are observed, in the urine as well. In the feces, on the contrary, they are usually in the minority and their isolation presents certain difficulties. It has been claimed by many authorities that typhoid bacilli are not found in the feces in demonstrable numbers, at least until about the middle of the second week. Klinger[2] has collected statistics from 812 contact cases which indicate the danger of infection from feces even before the development of clinical symptoms.

[1] Loc. cit.
[2] Public Health Reports, 1911, xxvi, 319.

SECONDARY CASES INFECTED FROM PRIMARY CASES.

First week of incubation period	33
Second week of incubation period	150
First week of disease	187
Second week of disease	158
Third week of disease	116
Fourth week of disease	59
Fifth week of disease	34
Sixth week of disease	22
Seventh week of disease	14
Eighth week of disease	16
Ninth week of disease	15

The isolation and identification of typhoid bacilli from the feces is by no means proof that the case under consideration is typhoid fever; the patient may be a carrier.

Technic of Isolation of Typhoid Bacilli from Feces.—A *thin* uniform emulsion of feces suspected to contain typhoid bacilli is made in 0.1 per cent. dextrose broth and incubated, if time permits, for one hour at 37° C.

The emulsion is best made by repeatedly running a rather heavy platinum needle through the fecal mass to insure a representative sample. The process is continued until the desired density of bacteria in the broth tube is attained. Incubation of one hour permits of a slight development of all the organisms; it particularly acclimatizes the typhoid bacilli to artificial media. The emulsion is then spread with a bent sterile glass rod on the surface of Endo medium previously prepared in large Petri dishes.[1] The Petri dishes after inoculation are inverted and placed in the incubator at 37° C. and examined eighteen to twenty-four hours later for clear, colorless, transparent colonies which rarely attain a diameter exceeding 2 mm. These colonies are transferred to 0.1 per cent. dextrose broth and after incubation for eighteen to twenty-four hours at 37° C. are mixed with a high potency antityphoid serum and examined for agglutination.

Rapid Method of Isolating Typhoid Bacilli.[2]—It is frequently possible to identify typhoid bacilli (and paratyphoid and dysentery bacilli as well) in feces within twenty-four hours by taking advantage of the microscopic agglutination method with a high potency serum in the following manner: Endo plates are inoculated as indicated above and incubated at 37° C. for fifteen to eighteen hours. Typical colonies are removed entire to small test-tubes containing 1 c.c. of 0.1 per cent. dextrose broth which have been kept at incubator temperature.

[1] For preparation and use of the Endo medium, see page 201.

[2] Kendall and Day, Jour. Med. Research, 1911, xx, 95.

Incubation of these infected tubes for one to two hours almost invariably gives sufficient numbers of organisms to make a microscopic agglutination. A confirmatory cultural diagnosis may be obtained by the inoculation of small tubes of semi-solid media and milk with the remainder of the broth culture. This method differs from the one usually employed merely in the small amount of broth used, which requires less bacteria to produce turbidity, and in the fact that the growth is practically continuous from the Endo medium to the tube, the broth being warmed to the body temperature at the start. Taking advantage of these factors cuts down the time required for diagnosis nearly twenty-four hours.

(*b*) Serological Diagnosis.—The blood serum of patients who have recovered from a typical attack of typhoid fever contains elements which give specific reactions with the typhoid bacillus or its products; of these, lysins, agglutinins, opsonins and precipitins have been carefully studied. The method of fixation of complement and the ophthalmo reaction have received less attention.

The lysins, which appear early in the course of the disease, dissolve typhoid bacilli, but not other bacteria, at least in the dilutions ordinarily used. It is probable that the lysins not only dissolve typhoid bacilli *in vitro*, they destroy the organisms in the blood stream as well,[1] liberating endotoxins which play a prominent part in the production of the febrile reaction.

Agglutinins are formed in the majority of cases, which will clump typhoid bacilli. The significance of agglutinins in the typhoid complex is not definitely established.

The opsonic index of the serum of immunized animals and of clinical cases of typhoid fever in man appears to be increased, but available methods of measuring the opsonic index do not furnish information consistent enough to warrant definite conclusions.

The reaction of fixation of complement has been used diagnostically in a limited number of cases. The technical skill required to elicit satisfactory results has doubtless interfered with its general application.

The agglutination reaction is by far the most commonly used antibody reaction employed in the diagnosis of typhoid fever.

The Widal Reaction.—Historical.—Gruber and Durham appear to have first demonstrated that the sera of animals immunized to typhoid bacilli would agglutinate the typhoid bacilli, even if the

[1] Coleman and Buxton, Medical and Surgical Report of Bellevue and Allied Hospitals, 1909–10, iv, 46.

serum were diluted many times. Grünbaum and later Widal applied this principle in the diagnosis of typhoid fever. It is now recognized that the principle involved is a general one for certain kinds of bacteria, and the Gruber-Durham-Grünbaum-Widal reaction is used practically in the diagnosis of several diseases. The sera of such animals frequently contain agglutinins which are active even in dilutions of $\frac{1}{10,000}$ or even higher. Specific lysins are also produced, which in dilutions of $\frac{1}{100}$ to $\frac{1}{1000}$ will dissolve (and kill) typhoid bacilli.

1. *Collection of Blood for the Agglutination Test.*—Dried blood, blood serum, blister fluid, or whole blood may be used for this reaction.

(*a*) *Dried Blood.*—A generous drop of blood is dropped upon a thin sheet of aluminum or upon clean, glazed paper, and allowed to dry. The advantages of dried blood are: (1) it is easily obtained by making a puncture in the ear of the patient and collecting a drop of blood; (2) it does not lose its agglutinating properties readily; (3) it is not readily contaminated; and (4) the blood may be removed quantitatively after it is dried (scaled off), weighed and then diluted to the desired degree as accurately as blood serum. The disadvantages are: (1) flies will readily remove a film of dried blood; and (2) typhoid bacilli are rarely found in blood clots. There is, however, very little danger of spreading typhoid in this way. In practice dried blood is diluted with physiological normal saline solution to a pale rose color, which corresponds to a dilution of 1 to 20. This dilution is somewhat inaccurate and anemic bloods introduce a disturbing factor. This method of dilution, however, is sufficiently accurate for all except unusual cases, and it is a method generally used in routine board of health examinations.

(*b*) *Blood Serum.*—A few drops of blood are collected in a capillary pipette or small test-tube and allowed to clot. The serum is removed and diluted accurately with salt solution. The advantages are: (1) the accuracy with which dilution may be made; and (2) the ease with which serum is obtained. The disadvantages are: (1) that blood serum is readily contaminated; and (2) it does not keep well, it deteriorates. Blood serum is the best for accurate work.

(*c*) *Blister Fluid.*—This possesses no advantages over blood serum. It is somewhat more difficult to obtain and probably somewhat less accurate than blood serum.

(*d*) *Whole Blood.*—Aside from clotting, whole blood is as reliable as blood serum, so far as accuracy of dilution and potency of agglutinins is concerned. It must be remembered, however, that the red blood cells appear in the field viewed under the microscope. Fresh whole blood presents one great disadvantage—the fibrin in it may cause a pseudoagglutination, for the fibrin network that forms as coagulation proceeds entangles typhoid bacilli in its meshes, giving the appearance of a true agglutination. Whole blood can be conveniently drawn into a blood-counting pipette and diluted accurately and immediately.

The Culture to be Used.—Old stock cultures of typhoid bacilli usually give the best results. Freshly isolated cultures not infrequently agglutinate less readily than those which have been on artificial media for some time. The organisms should be grown in 0.1 per cent. dextrose broth for eighteen hours at 30° to 32° C. It has been found that typhoid bacilli grown at this temperature agglutinate somewhat better than those grown at 37° C. Killed cultures are frequently used, but the results obtained are somewhat less accurate than those with living cultures. In rare instances it has been found that killed cultures will agglutinate with typhoid sera at 45° C. when living cultures fail to agglutinate. Controls must always be made: the typhoid culture is diluted with an equal volume of salt solution. Spontaneous agglutination sometimes takes place when no serum is present. This is shown in the control and at once invalidates the agglutination which may be obtained with the serum.

Technic of Test.—*(A) Microscopic Method.*—Dried blood, blood serum, blister fluid, or whole blood is diluted 1 to 20 with physiological salt solution. A loopful of this diluted fluid is mixed intimately with a loopful of typhoid broth culture on a coverglass and suspended in a hanging drop slide ringed with vaseline to prevent evaporation. The final dilution of the blood is 1 to 40 by this procedure. A control is made using a loopful of salt solution and a loopful of typhoid culture prepared in the same manner. Both the serum and the control are kept at room temperature. A preliminary examination should show actively motile bacteria in the control preparation and usually actively motile bacteria in the serum preparation. It sometimes happens that agglutination takes place in the serum preparation almost immediately. If the preliminary examination is satisfactory, the final examination is made at the end of an hour. Both preparations are examined and the controls should show actively motile unclumped organisms. A positive agglutination is recorded if the control is as stated and the organisms in the serum preparation are non-motile and gathered together in clumps with few or no free-swimming bacteria between the clumps.

(*B*) *Macroscopic Method.*—Various dilutions of serum are placed in small sterile test-tubes, 1 c.c. in each test-tube. As a routine, a dilution of 1 to 20 is used, but a series of dilutions up to the limits of the serum are frequently made. To each tube is added 1 c.c. of a broth culture of typhoid bacilli. A control is made by adding 1 c.c. of a broth culture of typhoid bacilli to 1 c.c. of salt solution. These mixtures are respectively shaken and incubated together with the control at 37° C. for two hours, then they are placed in the ice-box, and examined eighteen to twenty-four hours later. A positive agglutination is indicated when the supernatant fluid of the serum typhoid mixtures is clear, while the control containing no serum remains uniformly cloudy.

The microscopic method is much more rapid than the macroscopic method and is sufficiently accurate for ordinary purposes. The macroscopic method requires a much longer time, but it is more accurate, for the dilutions can be made carefully with graduated pipettes.

Discussion.—Available statistics show that about 20 per cent. of typhoid patients exhibit a positive agglutination reaction at the end of the first week; 60 per cent. at the end of the second week; 80 per cent. at the end of the third week; and 90 per cent. at the end of the fourth week. These agglutinins persist; about 75 per cent. of all patients exhibit a positive agglutination after two months. Occasionally agglutinins may persist for several years.[1] The amount of agglutination present, as indicated by the degree of dilution which will still clump typhoid bacilli, has no known relationship to the severity of the attack. An occasional mild case of typhoid may be accompanied by the appearance of agglutinins of great potency; severe attacks may exhibit little or no agglutinin in the blood. Occasionally, agglutinins are not demonstrable in the blood serum of undoubted cases of typhoid fever. This has been found to be the case by Moreschi[2] in several cases of chronic leukemia. Moreschi[3] has made the interesting observation that even the vaccination of these leukemics with killed cultures of typhoid bacilli may not lead to the development of agglutinins. In icterus an agglutination is not infrequently encountered even if the serum is highly diluted. It is very probable that at least some of these cases are typhoid carriers, having typhoid bacilli in the gall-bladder. They may be ambulatory cases. It has been claimed

[1] An initial negative reaction (first week) followed by a positive reaction is conclusive. It rules out the possibility of persistent agglutinins from previous cases, and those due to protective vaccination.

[2] Ztschr. f. Immunitätsforsch., 1914, xxi, 410.

[3] Loc. cit.

by some observers that the agglutination seen in icteric patients is due to bile in the blood stream. This, however, has not been proven. A negative agglutination, when the clinical symptoms suggest typhoid fever, should suggest the possibility of a paratyphoid infection.

Ophthalmo Reaction.—Chantemesse[1] has found that an ophthalmo reaction may be elicited in typhoid patients similar to that produced by the introduction of tuberculin in the eye of the tuberculous patient. Broth cultures of typhoid bacilli are precipitated with alcohol; the precipitate is dried and pulverized; $\frac{1}{50}$ milligram of the powder is dissolved in a few drops of sterile saline solution and introduced into the eye. A transient redness with a flow of tears occurs in normal individuals; a severe reaction (even accompanied by a serofibrinous exudate in unusual cases), which reaches its maximum intensity within twelve hours, is elicited in typhoid patients, and, occasionally, in individuals who have recovered from the disease. The diagnostic value of the reaction is as yet undetermined.

Dissemination and Prophylaxis.—The disease typhoid fever occurs only by transmission of typhoid bacilli directly or indirectly from preëxisting cases. The disease is acquired only by the ingestion of the specific organisms, and infection by any other channel than the alimentary canal has not so far been satisfactorily demonstrated.

Prophylactic measures, therefore, should begin with the isolation of the patient and disinfection of all excreta and all utensils which have been in contact with the patient. The organism may occur in the fecal discharges of patients before clinical symptoms develop, in patients recently recovered from the disease, in carriers (which number about 2 per cent. of all cases diagnosed), and probably in a relatively few individuals in whom the organism may gain a temporary foothold without producing symptoms. The bacilli may be transmitted to others by the hands of those who care for the patients, and the hands of carriers. Fecal matter containing typhoid bacilli may be transferred by flies, by water, through milk, and perhaps by vegetables which are eaten uncooked. The water in which typhoid patients have bathed is frequently grossly contaminated with the organisms. Rarely, wells and water supplies are contaminated by urinary typhoid carriers, in which event the colon bacillus, which is ordinarily relied upon for evidence of contamination, may be absent. A thorough disinfection of excreta including urine will prevent spread of the disease from known cases.

[1] IV. International Cong. of Demog. and Hyg., Berlin, September 26, 1907.

THE PARATYPHOID GROUP.

There is a group of closely related bacilli which exhibit cultural and pathogenic characters intermediate between those of the typhoid, dysentery and colon groups of bacteria, respectively. These organisms are variously known as the hog cholera, Salmonella, Gärtner, enteritidis, intermediate, paracolon or paratyphoid group.

Smith and Salmon[1] isolated the type organism of the group from the intestinal contents of swine infected with hog cholera. They named their organism the hog cholera bacillus.[2] Three years later Gärtner[3] described an organism, B. enteritidis, recovered by him both from the spleen and blood of a fatal case of meat-poisoning, and from the suspected meat (beef) itself. Numerous epidemics of meat poisoning[4] have been studied bacteriologically during the years following Gärtner's discovery, and very similar, if not identical, bacilli have been recovered from many of the patients.

In 1893 Smith and Moore[5] made the important observation that organisms culturally indistinguishable from the hog cholera bacillus could be isolated not infrequently from the intestinal contents of normal cattle, swine, sheep, cats and dogs. The significance of this discovery from the view-point of meat poisoning was not understood at that time.

In 1896 Achard and Bensaude[6] described paratyphoid fever and outlined the essential clinical and bacteriological diagnostic differences between this disease and typhoid fever. They obtained paratyphoid bacilli from the urine and blood stream of several cases, and recovered the organism from a secondary purulent arthritis in one of them as as well. Schottmüller[7] also obtained cultures of paratyphoid bacilli both from the feces and the blood stream of several cases of paratyphoid fever. Brion and Kayser[8] separated these organisms into two types: B. paratyphosus alpha, which produced a slight permanent acidity in litmus milk and gave an "invisible" growth on potato

[1] Ann. Rep. United States Bur. Animal Ind., 1885, vol. ii.

[2] A year earlier Klein (Virchows Arch., 1884, xcv, 468) obtained a bacillus from diseased swine which he regarded as the causative factor of hog cholera, but his organism produced spores, which at once distinguished it from the paratyphoid type. Neither the Klein bacillus nor the Smith-Salmon bacillus cause hog cholera; a filterable virus is the probable infecting agent.

[3] Correspondz.-Blatt des allgem. ärztl. Vereins von Thüringen, 1888, No. 9.

[4] Not to be confused with botulismus (see B. botulinus).

[5] Additional investigations concerning swine diseases, Washington, D. C., 1893.

[6] Soc. Méd. des Hôp. de Paris, 1896, 3d Série, xiii, 679.

[7] Deutsch. med. Wchnschr., 1900, p. 511.

[8] München. med. Wchnschr., 1902, p. 611.

like the typhoid bacillus; and B. paratyphosus beta, which produced an initial acidity in litmus milk followed by a progressively alkaline reaction. These observations, both clinical and bacteriological, have been confirmed by later investigations.

Morphology.—The members of the intermediate group are indistinguishable morphologically. They are rod-shaped bacilli with rounded ends, measuring from 0.8 to 1 micron in diameter, and 1.5 to 3.5 microns in length, occurring singly or in pairs, seldom in chains. In actively-growing cultures the organisms may be short, almost ovoid. In old cultures the organisms may be elongated; filamentous forms are more commonly seen in old gelatin cultures. The members of the group are actively motile and possess from four to twelve peritrichic flagella. Motility is greater in dextrose broth than in plain broth; this is particularly the case in young cultures. The organisms form no spores and appear to possess no capsules. They stain readily with ordinary anilin dyes; occasionally organisms from cultures several days old exhibit a tendency toward bipolar staining. They are Gram-negative.

Isolation and Culture.—The organisms of the paratyphoid group grow readily upon ordinary artificial media, B. paratyphosus alpha somewhat less luxuriantly than the remaining members. The colonies produced on agar after eighteen hours' incubation at 37° C. resemble those of the typhoid-dysentery group—small, round, and transparent—measuring from 1 to 3 mm. in diameter. On Endo medium the colonies, like those of B. typhosus and the dysentery bacilli, are clear and colorless and somewhat smaller than those developing upon plain agar. They usually measure from 0.75 to 2 mm. in diameter. The organisms grow well in gelatin, but do not cause liquefaction. They produce acid and gas in dextrose and mannite; lactose and saccharose are not fermented.

Milk.—Plain milk is not coagulated. All the members of the group except B. paratyphosus alpha cause a slow change in this medium, which becomes thin, brownish, and almost opalescent after two or more weeks' incubation. In litmus milk the cream ring is colored a deep blue-green, which is so constant as to be suggestive diagnostically. B. paratyphosus alpha produces a slight acidity which is permanent; the milk assumes a lilac color. B. paratyphosus beta and other members of the group produce a transient acidity[1] which

[1] For an explanation of the phenomenon, see page 222.

is followed by a progressive alkalinity, associated with the liberation of small amounts of ammonia.[1]

All members of the intermediate group produce considerable turbidity in plain and sugar broths. A pellicle may develop in plain broth after several days' incubation. Potato: B. paratyphosus alpha grows much like the typhoid bacillus on potato; the growth is nearly invisible on acid potato, but comparatively luxuriant. On alkaline potato the growth is brownish. B. paratyphosus beta produces a brownish growth even on slightly acid potato, which resembles that characteristic of B. coli.

The members of the intermediate group are all aërobic, facultatively anaërobic. The minimum temperature of growth is about 6° to 8° C., the optimum 37° C., and growth ceases at approximately 44° C. The resistance of the members of the intermediate group to environmental conditions, drying and to chemicals is similar to that of the typhoid bacillus. They are, however, somewhat more resistant to heat; an exposure of fifteen minutes at 70° C. or of five minutes at 75° C., kills the bacilli. This is a point of importance in meats infected with the organisms; temperatures lower than 75° C. *in the centre of the meat* can not be relied upon to remove danger of infection. Higher temperatures, 100° C., are preferable to remove all danger from the poisonous substances of the bacilli, which are not destroyed by gastro-intestinal digestion.

Products of Growth.—(*a*) *Chemical.*—Paratyphoid bacilli are rather more active proteolytically than typhoid and dysentery bacilli, but they produce neither phenols nor indol.[2] Dextrose and mannite are fermented with the formation of carbon dioxide and hydrogen, lactic acid, and smaller amounts of acetic and formic acids. Lactose and saccharose are not fermented. Numerous attempts have been made to classify the paratyphoid bacilli into several varieties upon the basis of the fermentation of carbohydrates other than those mentioned above, but the lack of agreement has proved an insurmountable obstacle to their general acceptance.

(*b*) *Enzymes.*—The members of the paratyphoid group do not produce soluble proteolytic ferments, and they do not liquefy coagulated blood serum, gelatin, fibrin or egg albumen. Neither lipolytic nor amylolytic enzymes have been demonstrated in cultures of these organisms.

[1] Kendall, Day and Walker, Jour. Am. Chem. Soc., 1914, xxxvi, 1943.
[2] Ibid., 1913, xxxv, 1221.

(*c*) *Toxins.*—Soluble toxins have not been demonstrated in cultures of paratyphoid bacilli. Cathcart[1] and Franchetti[2] have shown that minute amounts of autolysates of the organisms are rapidly fatal to rabbits and other small laboratory animals. According to Cathcart,[3] the poisonous substance (endotoxin) liberated from the organisms during autolysis is relatively thermostabile; a brief exposure of it to 100° C. does not completely destroy its potency.

Classification and Identification of the Paratyphoid Group.—It is possible to divide the Paratyphoid Group into two distinct types by their reaction in milk: the alpha type, of which several strains have been described, differing somewhat in their serological reactions; and the beta type. The former appears to be limited to man, but the latter comprises organisms which are rather widely distributed not only in man but in the lower animals as well. The better known strains of the beta type comprise not only B. paratyphosus beta, B. enteritidis and the hog cholera bacillus (B. choleræ suis, B. suipestifer) mentioned above, but B. psittacosis, obtained from infectious enteritis of parrots, which produces a pneumonic infection in man.[4] B. icteroides, Sanarelli, originally supposed to cause yellow fever, but now known to be indistinguishable from the hog cholera bacillus, the Danysz bacillus of rat plague, and B. typhi murium, Löffler, obtained from epizoötics of rodents, B. ærtrycke, de Nobele, and B. moorseele, van Ermengem, from epidemics of meat poisoning, and B. morbificans bovis, Basenau, isolated from a diseased cow, all belong to the same group. They possess in common cultural characteristics which differ somewhat quantitatively, but not qualitatively. Bainbridge and O'Brien[5] have attempted to classify the organisms by agglutination and absorption tests; they recognize four groups as follows: (1) B. paratyphosus alpha; (2) B. paratyphosus beta; (3) B. suipestifer (hog cholera bacillus), including B. psittacosis, B. ærtrycke and some strains of B. typhi murium; (4) B. enteritidis, including the Danysz bacillus, B. morbificans bovis, and some strains of B. typhi murium. This classification, if substantiated, possesses the advantage of separating those organisms which cause paratyphoid fever, the alpha and beta types, from the bacilli more commonly associated with the lower

[1] Jour. Hyg., 1906, vi, 112.
[2] Ztschr. f. Hyg., 1908, lx, 127.
[3] Loc. cit.
[4] Nocard, Conseil d'hygiene pub. et Salubrité du Dept. du Seine, Séance, March 24, 1893.
[5] Jour. Hyg., 1911, xi, 68.

animals, of which the hog cholera bacillus and B. enteritidis are the types. This classification has not been universally accepted, however. Doubtless the multiplicity of strains which have received the same name has led to confusion in standard type organisms which are especially essential in this line of investigation. It is not an assured fact that the paratyphoid bacilli, *alpha* and *beta*, are restricted to the production of paratyphoid fever in man; nor can it be stated definitely that B. enteritidis and the hog cholera bacillus consistently cause meat poisoning. Available information suggests that occasionally the choleraic symptoms of meat poisoning may be elicited by paratyphoid bacilli, and that the symptoms of paratyphoid fever may follow infection with B. enteritidis or B. suipestifer.

Pathogenesis.—*Animal.*—The members of the Paratyphoid Group are, as a rule, very pathogenic for small laboratory animals. The intraperitoneal injection of very minute amounts of bacilli usually causes acute death in guinea-pigs and mice. Rats are somewhat more resistant. B. typhi murium and other "rat viruses" produce a fatal enteritis in mice and rats; the bacilli are present not only in the intestinal contents, they may be obtained from the tissues and organs postmortem as well. Bacilli belonging to the Paratyphoid Group have been isolated from epizoötics and sporadic cases of enteritis in cattle, parrots, and rodents. The organisms appear to be widely distributed among the lower animals.

Human.—Three types of disease are produced in man by the bacteria of the paratyphoid group: (*a*) meat poisoning: the symptoms are choleraic in character, and they may be severe enough to be confused with true cholera;[1] infection usually follows the ingestion of imperfectly cooked beef or pork contaminated with B. enteritidis or the hog cholera bacillus. Somewhat similar symptoms have resulted from the accidental ingestion of the "rat virus" of Danysz and others;[2] (*b*) paratyphoid fever, a disease clinically resembling mild typhoid fever, usually caused by B. paratyphosus alpha or B. paratyphosus beta; (*c*) a rare type of disease, pneumonic in character, produced by B. psittacosis, which produces an epizoötic disease among parrots.

(*a*) *Meat Poisoning.*—The disease is more prevalent in summer and fall than it is in winter and spring, probably due in part to decreased

[1] Hetsch, Klin. Jahrb., 1907, xvi, 267.

[2] Mayer, München. med. Wchnschr., 1906, No. 47; Shibayama, München. med. Wchnschr., 1907, 979.

efficiency of refrigeration of meats in the warmer months. The incubation period may be as brief as four to six hours, or as long as twenty-four to seventy-two hours after ingestion of the infected food. The initial symptoms are usually a severe headache and chill, rapidly followed by acute gastro-intestinal disturbances, dizziness, nausea and vomiting, abdominal pain and diarrhea. Nervous symptoms and marked restlessness are characteristic of the severe and fatal cases. Usually the symptoms and fever abate within a week; they may persist for several weeks. The mortality is, as a rule, low, averaging from 1 to 2 per cent. The conspicuous lesion observed at autopsy is an intense hyperemia of the gastro-intestinal mucosa, usually without noteworthy involvement of Peyer's patches. Fatty degeneration of the liver is common. Bacilli (usually B. enteritidis or B. choleræ suis,[1] less commonly B. paratyphosus beta) may be isolated from the feces and blood stream in many of the acute cases during the first few days of the disease. They are almost invariably recovered from the heart blood and spleen at autopsy. Serum reactions, especially specific agglutinins, may be demonstrated at the end of the first week in many but not all cases.

An epidemic of meat poisoning is characterized by the sudden, practically simultaneous onset of symptoms in those who have eaten the contaminated food, and the limitation of the disease to the primary cases. Secondary infection is uncommon. It should be remembered that not all epidemics of meat poisoning are caused by members of the paratyphoid group of bacteria.

Distribution of Organisms.—The hog cholera bacillus (B. choleræ suis, B. suipestifer) is frequently found in the intestinal tracts of swine, rats and mice; probably somewhat less commonly in cattle. B. enteritidis is a frequent inhabitant of the intestinal contents of rats and mice, and relatively uncommon in healthy cattle.[2] It is suspected that a postmortem infection of beef is more common than an antemortem invasion; this is reasonably suggested by the wide distribution of rats and mice in slaughter houses. The organisms possess the somewhat unusual property of rapidly diffusing themselves through the substance of meat after they have been distributed on the surface of it by careless handling. Unless infected meat is thoroughly cooked, the organisms are not killed, and they may not be even weakened if the degree of heat and time of exposure is insuffi-

[1] Bainbridge, Lancet, March 16, 23, 30, 1912.
[2] Ibid.

cient. The endotoxins of the bacilli, furthermore, are relatively thermostabile. Thorough cooking of such meat is essential to insure safety.

(*b*) *Paratyphoid Fever.*—Bacteriologically, paratyphoid fever may be caused either by B. paratyphosus alpha or B. paratyphosus beta. Clinically there is little or no difference between the two infections. According to Bainbridge,[1] paratyphoid fever in Asia, particularly in India, is more frequently an infection with the *alpha* organism; in Europe the *beta* organism is much more frequently reported. Both types are found in the United States.[2] The organisms are occasionally found in the intestinal contents and feces of young children and adults who give no history of infection.

The incubation period of paratyphoid fever varies from eight to twenty days; the average is about two weeks. The onset is gradual; the usual prodromal symptoms are severe head- and backache, malaise and anorexia. Bronchitis and sore throat are common. There may be an initial chill, then the temperature rises rather rapidly to a maximum of 103° to 105° C.; after the fifth to the seventh day it falls slowly; it is normal by the end of the second week. Rose spots are occasionally seen early in the disease. Less commonly acute gastroenteric symptoms, resembling those of meat poisoning, complicate the clinical picture. Paratyphoid fever is a bacteremia, very similar to typhoid fever in this respect. The mortality is low, averaging from 1 to 2 per cent. of all cases. The lesions observed postmortem are intense hyperemia of the gastro-intestinal tract, usually with superficial ulcerations in the ileum and cecum, not necessarily, however, involving Peyer's patches. Acute splenic tumor is usually not a feature of paratyphoid infections. The bacilli may be isolated from the heart blood and visceral organs.

Bacterial Diagnosis.—(*a*) *Isolation of Bacilli.*—Blood cultures made during the first week are frequently positive. The organisms are usually present in the feces, occasionally in the urine. The identification of the bacilli depends upon the cultural characters outlined above; gas production in dextrose and mannite, no liquefaction of gelatin, and a permanent acidity in litmus milk (alpha type) or a transient acidity followed by a progressively alkaline reaction in this

[1] Loc. cit.

[2] Gwyn, Bull. Johns Hopkins Hospital, 1898, vol. ix. Cushing, ibid., 1900, vol. xi; Buxton and Coleman, Proc. Path. Soc. New York, February, 1902; Proescher and Roddy, Jour. Am. Med. Assn., 1909, lii, No. 6; Kendall, Bagg and Day, Boston Med. and Surg. Jour., 1913, clxix, 741; Kendall and Day, ibid., 1913, clxix, 753.

medium (beta type). Isolation from the feces is made upon Endo-plates in the same manner that dysentery and typhoid bacilli are obtained. The final diagnosis depends upon the agglutination of the bacilli with specific agglutinating sera of high potency.[1]

(*b*) *Serological.*—As a routine measure the diagnosis of paratyphoid fever by the agglutination test is unreliable. Not infrequently the blood serum of a patient agglutinates typhoid bacilli in dilutions approaching those ultimate for the homologous organism. The paratyphoid bacilli and B. typhosus possess in common group agglutinins which greatly vitiate the value of the test. The same objection does not hold for the diagnosis of typhoid fever by the agglutination reaction, however.

The isolation of B. paratyphosus (alpha or beta) from the blood stream during life, or from the internal organs at autopsy is the only reliable method of diagnosis. Carriers are not uncommon, and like typhoid bacillus carriers the organisms frequently remain in the gall-bladder, consequently isolation of the bacilli from feces does not necessarily establish a correct clinical diagnosis. Paratyphoid bacilli have been isolated occasionally from gall-stones and from cases of cholecystitis, particularly in women.

SUMMARY.

THE MORE IMPORTANT DIFFERENTIAL DETAILS OF PARATYPHOID FEVER AND OF MEAT POISONING.

	Meat Poisoning.	Paratyphoid Fever.
Organism	Hog cholera bacillus. B. enteritidis.	B. paratyphosus alpha. B. paratyphosus beta.
Habitat of organism	Intestinal canal of lower animals chiefly: hog cholera in swine, enteriditis common in rodents.	Chiefly intestinal tract of man.
Mode of infection	Usually contaminated meat (human carriers rare).	Usually human bacilli carriers.
Incubation period	Six to forty-eight hours.	Eight to twenty days.
Symptoms	Choleraic.	Typhoidal.

Pneumonic Infection with B. Psittacosis.—B. psittacosis causes a fatal enteritis in parrots, and it has been noticed, particularly in France, that coincidently with enteric disease in parrots a pneumonic infection has appeared in those associated with them. The disease in man presents no definite clinical features which would differentiate it from typhoid fever complicated by pneumonia. The incubation

[1] Sera that will agglutinate homologous strains in dilutions of 1 to 40,000 are readily prepared; such sera in dilutions of 1 to 10,000 may be regarded as specific for the identification of members of the group, if typical agglutination occurs.

period varies from five days to three weeks, usually, however, less than ten days. The onset is gradual in some cases, like typhoid, but it may be abrupt with an initial chill, as in pneumonia. The spleen is enlarged, but rose spots are rarely found. The mortality varies; it may be as high as 30 per cent. The postmortem lesions have not been established. In one case the bacillus was isolated from the heart's blood postmortem. Specific agglutinins in the patient's blood serum have not been satisfactorily studied, and the disease as a clinical entity is yet to be defined. The principal evidence of the causative relationship of B. psittacosis to the disease rests at present upon the occasional household epidemics following closely upon the presence of a diseased parrot.

Immunity and Immunization to Paratyphoid Infection.—The duration of immunity following recovery from an attack of paratyphoid fever or of meat poisoning is as yet undetermined. The brilliant results of protective immunization against typhoid fever with vaccines or residues of the typhoid bacillus have led to similar vaccination against paratyphoid infection with polyvalent vaccines composed of the principal strains of the paratyphoid group. Combined protective vaccination against typhoid and paratyphoid by the use of compound vaccines has also been attempted. The efficiency of the immunization can not be stated at the present time because statistics are unavailable.

Dissemination and Prophylaxis.—Paratyphoid fever appears to be spread by mild unrecognized cases, by carriers, and by the occasional transmission of bacilli through food, water or milk. Flies may also be a factor in the dissemination of the organisms. Meat poisoning is chiefly disseminated by infected meats, more frequently that of cattle or swine. The customary precautions appropriate for excrementitious diseases, including the restriction of carriers, may be confidently relied upon to prevent the spread of paratyphoid fever. Thorough cooking will largely reduce the occasional danger from contaminated meats.

CHAPTER XVII.

THE COLI—CLOACÆ—PROTEUS GROUP.

BACILLUS COLI.

Historical.—Bacillus coli was isolated in pure culture from the feces of infants, and its important cultural characters determined by Escherich in 1886.[1] It is very probable, as Escherich suggested,[2] that Emmerich's B. neapolitanus, Brieger's "propionic acid bacillus," and Fränkel's bacilli[3] are identical with the colon bacillus.

Morphology.—Bacillus coli is a rod-shaped organism which varies in shape from oval organisms resembling cocci to bacilli of moderate length. The organism varies in size from 0.5 to 0.8 micron in diameter and from 1 to 3 microns in length. The bacilli occur singly and in pairs; in older cultures short chains and elongated organisms are frequently observed. The ends are distinctly rounded. Motility is variable; many strains are non-motile except during the earlier hours of growth. Young cultures on gelatin are said to exhibit motility when older growths even in the same medium are motionless except for Brownian movement. Very commonly only a very few organisms in a microscopic field exhibit motion, the remainder being without movement. Four to eight peritrichic flagella are commonly attached to each bacillus; less frequently as many as twelve may be demonstrated. The flagella are somewhat shorter than those of the typhoid bacillus and they are more difficult to stain. Bacillus coli forms no spores nor capsules. It stains readily with the ordinary anilin dyes, and it is uniformly Gram-negative.

Isolation and Culture.—The colon bacillus grows readily on the ordinary media; the superficial colonies on agar plates are clear and colorless and attain a diameter of from 2 to 5 mm. after eighteen hours' incubation at 37° C. If the surface of the medium is moist the edges of the colonies are somewhat irregular in outline; on dry surfaces the colonies are round and slightly convex in section. Viewed by trans-

[1] Die Darmbakterien des Säuglings, Stuttgart, 1886, 63.
[2] Loc. cit., 73, 74.
[3] Deutsch. med. Wchnschr., 1885, Nos. 34 and 35.

mitted light the growths are yellowish-brown; by reflected light they are colorless. Colonies on gelatin develop more slowly and become somewhat brownish in color. The medium is not liquefied. Rapid development occurs in plain and sugar broths. A heavy, brownish spreading growth occurs on the surface of slanted potato.

Bacillus coli is an aërobic, facultatively anaërobic organism which grows best at 37° C. Growth ceases below 8° to 10° C., and above 43° to 45° C. An exposure of fifteen minutes at 75° C. kills them. In general the colon bacillus is somewhat more resistant to physical and chemical agents than the typhoid bacillus.

Products of Growth.—(*a*) *Chemical.*—Bacillus coli produces indol from tryptophan in sugar-free media, and phenolic bodies from

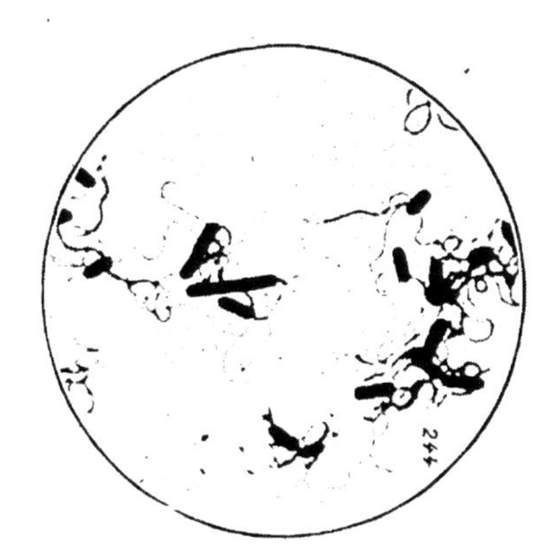

FIG. 48.—Bacillus coli flagella. × 1500. (Kolle and Hetsch.)

tyrosine under the same conditions. Hydrogen sulphide and ammonia, the latter resulting largely from deaminization of proteins and protein derivatives, are also produced in considerable amounts in media containing no utilizable carbohydrates.[1] Similar products may be formed in the intestinal tract under certain conditions. The addition of utilizable carbohydrates to protein media changes the character of the products of metabolism in a noteworthy manner. Under these conditions the protein constituents of the media are practically unchanged; the sugars are fermented with the production of carbon dioxide and hydrogen,[2] lactic acid and smaller amounts of

[1] Kendall, Day and Walker, Jour. Am. Chem. Soc., 1913, xxxv, 1228.

[2] In the proportion $H:CO_2 = \frac{2}{1}$. Theobald Smith, The Fermentation Tube. The Wilder Quarter Century Book, 1893, p. 202. Very exact determinations of the gaseous products of fermentation of B. coli have been made by Harden and Walpole, Proc. Roy. Soc., 1906, 77, 399.

acetic acid and formic acid. Dextrose, lactose and mannite are thus fermented; saccharose is not decomposed by the strains of the colon bacillus commonly found in the intestinal tract. Occasionally a saccharose-fermenting strain is encountered in the feces.[1]

The reactions of the colon bacillus in milk are variable; typical strains produce enough acid from the fermentation of the lactose to cause an acid coagulation in one to three days at 37° C. Neutralization of the acid by alkali redissolves the coagulum and the medium resumes its normal appearance. Occasional strains do not cause coagulation even after boiling the milk.[2] Gas is not produced in appreciable amounts in milk by B. coli, and the organism leaves the milk proteins practically intact even after prolonged incubation—

Fig. 49.—Bacillus coli, broth culture.

the carbohydrate constituents alone are acted upon.[3] Coagulation does not as a general rule occur in litmus milk, but boiling the medium usually causes rapid clotting. The ordinary litmus of commerce contains considerable amounts of calcium carbonate. This may neutralize some of the acid products of fermentation, reducing the acidity below the coagulation point. This explanation does not account for the same phenomenon in milk colored with pure litmus or azolitmin. Gelatin is not liquefied by B. coli. Nitrates are reduced to nitrites.

(*b*) *Enzymes.*—Soluble proteolytic and lipolytic enzymes have not been detected in cultures of Bacillus coli. Buxton[4] has demonstrated

[1] Theobald Smith, Am. Jour. Med. Sc., September, 1895.
[2] Ibid., Fermentation Tube, p. 201.
[3] Kendall, Day and Walker, Jour. Am. Chem. Soc., 1914, xxxvii, 1945.
[4] Am. Med., 1903, vi, 137.

both a maltase and a lactase in maltose and lactose cultures of the organism respectively. The investigations of Franzen and Stuppuhn[1] would suggest that the liberation of gas in sugar broth cultures of B. coli and other aërogenic bacteria depends upon the production of formic acid from the carbohydrate and its subsequent decomposition into carbon dioxide and hydrogen by the action of an enzyme, *formiase*, in accordance with the equation $H.COOH = CO_2 + H_2$.

(*c*) *Toxins*.—Bacillus coli does not produce a soluble toxin. The injection of killed cultures into laboratory animals frequently causes death; if large amounts are introduced intravenously into rabbits there is usually a lowering of the body temperature, diarrhea, collapse and death even within three hours.[2] If the animals survive for a longer time a purulent peritonitis may develop. Living cultures of colon bacilli derived from inflammatory processes in man are generally virulent for guinea-pigs. Old stock cultures are less virulent as a rule. The symptoms of toxemia which are exhibited by laboratory animals following the injection of colon bacilli are probably caused by the liberation of endotoxins from the bacilli.

Pathogenesis.—The colon bacillus is a normal inhabitant of the intestinal tracts of man and the higher animals. Ordinarily it is a harmless parasite, but it may become invasive if conditions arise which weaken the intestinal mucosa. In peritonitis, purulent perforative appendicitis, angiocholitis, and even in occasional cases of pancreatitis the organism is frequently isolated, either in pure culture or in association with other bacteria, as streptococci, typhoid bacilli, or staphylococci. It is difficult to determine with precision the part played by Bacillus coli in these conditions. Occasional cases of enteritis are encountered which appear to be caused by this organism, other bacteria having been ruled out. The careful studies of Coleman and Hastings[3] are of great importance in this connection. They isolated colon bacilli from the blood stream in a small series of cases which presented symptoms indistinguishable from those of typhoid fever. No typhoid bacilli were ever found in these patients, and no specific agglutinins for the typhoid bacillus were demonstrable. Specific agglutinins for the *homologous* strains of B. coli persisted until recovery. Cystitis and pyelonephritis, particularly the former, are frequently found to be a pure colon infection. B. coli is occa-

[1] Ztschr. f. physiol. Chem., 1912, lxxvii, 129.
[2] Escherich, Fort. d. Med., 1885, 521.
[3] Med. and Surg. Report of Bellevue and Allied Hospitals of the City of New York, 1909–1910, iv, 56.

sionally isolated from the centre of gall-stones; it is surmised that the organism, or clusters of them, act as nuclei around which the cholesterin is gradually deposited. Colon bacilli have been isolated in rare instances from purulent cerebrospinal fluids, and they may cause bronchopneumonia: Perirectal abscesses also may contain pure cultures of colon bacilli.

Immunity and Immunization.—The constant occurrence of B. coli in large numbers in the normal intestinal tract is an index of the relative immunity of man to infection with this organism. Occasionally very small numbers of bacilli may gain entrance to the tissues, particularly in young children. The blood serum usually contains agglutinins in small amounts for the organism. In practice no attempt is made to increase the immunity to colon bacilli, except in cases of cystitis or other local infection. Vaccines of the homologous strain of B. coli are occasionally administered in such instances. The results have been variously interpreted.

Bacteriological Diagnosis.—The methods of isolation, identification and significance of B. coli in water supplies will be discussed in the chapter on water. Isolation of colon bacilli from the intestinal contents or feces is readily accomplished by plating methods. The organisms far outnumber any others normally present, and even in severe diarrheal disorders colon bacilli do not entirely disappear. Prolonged starvation does not eliminate B. coli from the intestinal canal.[1] The morphology and staining reactions are not distinctive. Plating methods—principle involved: lactose agar, containing litmus or decolorized fuchsin (Endo medium) as an indicator is infected with material suspected to contain B. coli. The organism ferments the lactose with the production of acid; the acid changes the color of the indicator immediately surrounding the colon bacilli, red if litmus is used, pink if fuchsin is employed. The red colonies are inoculated into broth and incubated to obtain sufficient organisms for their identification by cultural methods.

Cultural Identification.—A Gram-negative bacillus which produces gas in dextrose, lactose and mannite (optionally in saccharose), coagulates but does not peptonize milk, does not liquefy gelatin, and is without action upon starches is Bacillus coli.

[1] At the end of thirty-one days' abstinence from all food, typical colon bacilli were present in the lower part of the large intestine. Kendall, Observations upon the Bacterial Intestinal Flora of a Starving Man, Publication No. 203 of the Carnegie Institute of Washington, 1915, p. 232. This experiment emphasizes the fallacy of "starving out" intestinal bacteria by withdrawing food.

BACILLUS CLOACÆ.

Bacillus cloacæ was isolated from sewage and polluted water by Jordan.[1] The organism appears to be relatively abundant some years and comparatively uncommon other years. When it is abundant in sewage it is found occasionally in the intestinal tract of man.

Morphology.—The bacillus is of moderate size, measuring from 0.6 to 0.8 micron in diameter and from 1 to 2 microns in length. It occurs singly or in pairs, uncommonly in short chains. Young cultures exhibit motility, and the organisms possess peritrichic flagella. No spores or capsules have been demonstrated. Ordinary anilin dyes color the bacilli readily, and they are Gram-negative.

Isolation and Culture.—The colonies on agar plates after eighteen hours' incubation are round, clear and colorless, and measure from 1 to 3 mm. in diameter. There is nothing distinctive in the appearance of the growths.

Products of Growth.—(*a*) *Chemical.*—Indol, phenol, hydrogen sulphide and ammonia are produced in sugar-free broth. The ammonia production is greater than that characteristic of B. coli and less than that ordinarily produced by B. proteus.[2] Acid and gas are produced in dextrose, lactose, saccharose and mannite broths. The gas ratio is somewhat variable, but distinctive; the proportion of carbon dioxide to hydrogen is greater than that produced by other closely-related bacteria.[3] The action of the organism upon lactose is slow, and less gas is produced from this sugar. The amount of gas produced from dextrose and saccharose is greater than that produced by other aërogenic members of the paratyphoid-proteus group. B. cloacæ forms but little acid from the fermentation of sugars, and after one to three days the reaction, even in sugar broth, becomes alkaline, due to the exhaustion of the sugar and the subsequent decomposition of the protein constituents of the broth.[4] Indol and other products of putrefaction are formed as soon as the sugar is exhausted.

Milk is coagulated and slowly peptonized. Freshly isolated cultures usually liquefy gelatin, but this property is lost after prolonged artificial cultivation.

The organism is ordinarily non-pathogenic for man.

[1] Annual Report of Massachusettes State Board of Health, 1890, p. 836.
[2] Kendall, Day and Walker, Jour. Am. Chem. Soc., 1913, xxxv, 1230.
[3] $H:CO_2 = \frac{1}{2} - \frac{1}{3}$. Theobald Smith, Fermentation Tube, 1893, p. 215.
[4] Kendall, Day and Walker, loc. cit.

BACILLUS PROTEUS GROUP.

Synonyms.—Proteus vulgaris, Proteus mirabilis, Proteus Zenkeri, Proteus Zopfii, Proteus fluorescens.

Historical.—The proteus group comprises several closely-related bacilli found commonly in soil, in water rich in organic matter, as sewage, in human feces, and associated with the decay of organic matter. The important members of the group were first isolated in pure culture and described by Hauser.[1]

Morphology.—The proteus bacilli are rod-shaped organisms of variable length which occur singly and in pairs as a rule; less commonly they remain adherent in short chains. The size of individual cells varies considerably, even in the same culture. The limits of variation are comprised within the following dimensions: diameter from 0.6 to 0.8 micron, length from 1.0 to 3.5 microns. Proteus bacilli are actively motile and possess a large number of peritrichic flagella[2] which are frequently seen as a tangled filamentous mass surrounding each individual cell.[3] Special staining methods are required for the demonstration of these flagella. The organisms produce no spores and form no capsules. They stain with ordinary anilin dyes, but somewhat faintly, and they are Gram-negative.

Isolation and Culture.—The members of the proteus group develop rapidly on gelatin at room temperature; the organisms typically liquefy the medium with great rapidity. Some strains liquefy gelatin but slightly or even not at all. The colonies of rapidly liquefying strains in 5 per cent. gelatin are frequently very characteristic; the organisms tend to remain adherent, forming masses of bacilli which slowly move around in an area of liquefied gelatin. Hauser[4] recognized four types of proteus bacilli classified according to their ability to liquefy gelatin: Proteus vulgaris liquefies gelatin rapidly; Proteus mirabilis liquefies gelatin slowly; Proteus zenkeri and Proteus zopfii do not liquefy this medium. The latter, Proteus zopfii, exhibits negative geotropism on slanted solid media. It is now recognized that cultures of B. proteus may gradually lose their gelatin-liquefying power after prolonged cultivation, so that a cultural transition from B. proteus to B. zenkeri may be observed in the laboratory. A dis-

[1] Ueber Fäulnisbakterien und deren Beziehungen zur Septikämie, Leipzig, 1885.

[2] Zettnow, Centralbl. f. Bakt., 1891, x, 689.

[3] Massea (Centralbl. f. Bakt., 1891, ix, 106) states that young bacilli may possess from 60 to 100 flagella.

[4] Loc. cit.

tinction between the three types is no longer made. It is not determined whether B. zopfii is a separate variety of B. proteus.

The organisms grow vigorously in milk, causing slight acidification and peptonization. The development in broth is equally vigorous; acid and gas are produced in dextrose and saccharose broths.[1] Neither acid nor gas is formed in lactose broth.[2]

Proteus bacilli grow slowly at 0° C.[3] and at temperatures not exceeding 43° to 45° C. The optimum temperature is about 25° C., but development is rapid at 37° C. Strains obtained from putrefying organic matter are tolerant of considerable degrees of alkalinity[4] and acidity;[5] those from the human body are somewhat less tolerant. The growth of B. proteus at low temperatures is of considerable prae-

Fig. 50.—Bacillus proteus, flagella stain. × 1500. (Gunther.)

tical importance; several cases of ptomain poisoning have been attributed to foods decomposed by this organism at the temperature of the ice-box. The resistance of the organisms to heat is not great. According to Meyerhof,[6] an exposure of twenty-five to thirty-five minutes at 54° C., five to ten minutes at 56° C., and of one-half a minute at 60° C., kills them. Their resistance to disinfectants is similar to that of B. coli.

Products of Growth.—(*a*) *Chemical.*—Proteus bacilli decompose proteins and protein derivatives energetically. The following substances have been detected among the cleavage products: trimethylamine, betain, phenol, hydrogen sulphide;[7] from the decomposition of

[1] Theobald Smith, Fermentation Tube, Wilder Quarter Century Book, 1893, p. 213.

[2] The bacilli may gradually lose their ability to ferment saccharose; strains which do not ferment this sugar may be mistaken for paratyphoid bacilli, particularly if the gelatin-liquefying power disappears simultaneously. The very considerable production of ammonia in sugar-free broth readily distinguishes the proteus bacilli. Kendall, Day, and Walker, Jour. Am. Chem. Soc., 1913, xxxv, 1231.

[3] Levy, Arch. f. öffentl. Gesundhpf. in Els. Lothr., 1895, xvi, Heft 3.

[4] Deelman, Arb. a. d. kais. Gesamte, 1897, xiii, 374.

[5] Fermi, Centralbl. f. Bakt., 1898, xxiii, 208.

[6] Centralbl. f. Bakt., 1898, xxiv, 20.

[7] Emmerling, Ber. chem. Gesell., 1896, 2711.

casein, deuteroalbumose, peptone, mono- and diamino-acids (histidin and lysin), tyrosin, indol, and skatol.[1] An extensive liberation of ammonia takes place in protein media free from sugars.[2] Ammonia is also formed from the proteins of milk, but more slowly, and in smaller amounts.[3] Carbon dioxide and hydrogen ($H:CO_2 = \frac{2}{1}$) are formed in dextrose and saccharose broths, together with lactic acid and small amounts of formic acid. Lactose is unfermented.[4] Urea is actively decomposed, ammonia and carbon dioxide being liberated.[5] The addition of dextrose prevents the liberation of ammonia and carbon dioxide.[6]

(*b*) *Enzymes.*—B. proteus produces a soluble proteolytic enzyme in protein media containing no utilizable sugars, which liquefies egg albumen, fibrin, blood serum, and gelatin. This enzyme is not produced when utilizable sugars are present in the medium. No other enzymes are known.

(*c*) *Toxins.*—A soluble toxin has not been demonstrated in cultures of B. proteus. At one time "sepsin" (see page 75) was supposed to be an important factor in "ptomain poisoning." This substance is produced in but minute amounts by proteus bacilli, however, and no importance is attached to it. The nature of the poisonous substance produced by B. proteus is unknown.

Pathogenesis.—Several types of disease have been attributed to members of the proteus group. Meat poisoning and ptomain poisoning epidemics caused by eating meats decomposed by the organisms have been reported by Levy,[7] Wesenberg,[8] Silberschmidt,[9] and Pfuhl.[10] Dieudonné[11] has described an epidemic which originated in a potato salad from which proteus bacilli were isolated. B. proteus is one of the very few bacteria which will cause cystitis when it is injected into the urinary bladder. Cystitis in man is frequently caused by B. proteus.[12] Pyelonephritis, frequently of a very purulent type, and abscesses are occasionally caused by members of the group. The organisms do not as a rule grow in normal tissues, but they grow

[1] Taylor, Ztschr. f. physiol. Chem., 1902, xxxvi.
[2] Kendall, Day and Walker, Jour. Am. Chem. Soc., 1913, xxxv, 1232.
[3] Ibid., 1914, xxxvi, 1945.
[4] Theobald Smith, Fermentation Tube, Wilder Quarter Century Book, 1893, p. 213.
[5] Schnitzler, Centralbl. f. Bakt., 1893, xiv, 219.
[6] Brodmeier, Centralbl. f. Bakt., 1895, xviii, 380.
[7] Arch. f. exp. Path. u. Pharm., 1895, xxxiv, 342.
[8] Ztschr. f. Hyg., 1898, xxviii, 484.
[9] Ibid., 1899, xxx, 328.
[10] Ibid., 1900, xxxv, 265.
[11] München. med. Wchnschr., 1903, 2282.
[12] See Meyerhof, Centralbl. f. Bakt., 1898, xxiv, 18, 55, 148.

readily in necrotic tissues, forming much pus which has a fœtid odor. Middle ear infections, characterized by very foul-smelling pus, have been reported.

Bacillus proteus fluorescens, an organism exhibiting many characteristics of the proteus group, has been isolated from several cases of Weil's disease (infectious jaundice) by Jaeger,[1] Conradi and Vogt,[2] and Bruning.[3] Bar and Renon[4] isolated a similar bacillus from a case of jaundice in the newborn. Booker[5] has isolated B. proteus from the feces of a large number of cases of acute summer diarrhea in children. It would appear from his studies that the organisms played a prominent part in the causation of certain types of this illness, particularly those characterized by choleraic symptoms.

Bacillus proteus is not very pathogenic for laboratory animals. The injection of large doses usually causes death.

Bacteriological Diagnosis.—Bacillus proteus is readily isolated upon gelatin plates: the bacilli grow rapidly at room temperature and liquefy the medium around each individual colony. Subcultures in sugar media, gelatin and milk produce the changes outlined above. B. proteus may be confused with B. cloacæ, because the latter organism ferments lactose more slowly than other sugars.[6] B. cloacæ, however, is distinctly less proteolytic than B. proteus,[7] and it produces less acid and more gas from dextrose.

[1] Zeit. f. Hyg., 1892, xii.
[2] Ibid., 1901, xxxvii, 283.
[3] Deut. med. Woch., 1904, 1269.
[4] Sem. méd., 1895, 234.
[5] Johns Hopkins Hospital Reports, vi.
[6] Theobald Smith, Fermentation Tube, 1893, 215.
[7] Kendall, Day and Walker, loc. cit., 1230.

CHAPTER XVIII.

THE MUCOSUS CAPSULATUS GROUP.

THE first member of the bacteria commonly known as the pneumo-Bacillus Group or the Mucosus Capsulatus Group was isolated by Friedländer[1] from pneumonic lungs. At that time he believed his "pneumonia micrococcus" was the causative agent of lobar pneumonia, and it was so regarded until Fränkel[2] and Weichselbaum[3] pointed out its comparative infrequency in lobar pneumonia, and differentiated it clearly from the pneumococcus, the true etiological organism of this disease. Weichselbaum also correctly interpreted its morphology and conferred upon it the name Bacillus pneumoniæ. Subsequent investigations by many observers have added several closely-related bacteria to the group which at the present time comprises the following somewhat imperfectly-differentiated types: Bacillus mucosus capsulatus (Friedländer's pneumobacillus), Bacillus rhinoscleromatis,[4] Bacillus ozænæ,[5] Bacillus lactis aërogenes,[6] and Bacillus acidi lactici.[7]

Morphology.—The members of the Mucosus Capsulatus Group are bacilli which vary in size and shape in the same culture from oval almost coccoid elements to distinctly elongated rods. The limits of size are comprised practically within the following dimensions: diameter, 0.5 to 1.5 microns, length, 0.6 to 3.5 microns. They occur typically singly or in pairs, less commonly united in short chains. Motility is not observed in cultures of any members of the group and they appear to be devoid of flagella. Spores have not been detected. A well-defined capsule, readily demonstrable by capsule stains, surrounds each organism if it is examined in tissues or secretions of the animal body, or in albuminous media. It tends to disappear during

[1] Virchows Arch., 1882, lxxxvii, 319; Fort. d. Med., 1883, i, 719.
[2] Ztschr. f. klin. Med., 1886, x, 401.
[3] Wien. med. Jahrb., 1886.
[4] V. Frisch, Wien. med. Wchnschr., 1882, No. 32.
[5] Abel, Ztschr. f. Hyg., 1896, xxi, 89; Centralbl. f. Bakt., 1893, xiii, 161.
[6] Escherich, Darmbakterien des Säuglings, Stuttgart, 1886, p. 57.
[7] Hueppe, Deutsch. med. Wchnschr., 1884, p. 778.

prolonged cultivation in the usual artificial laboratory media. Ordinary anilin dyes color the organisms readily, and they are Gram-negative.

Isolation and Culture.—The members of the Mucosus Capsulatus Group grow readily on artificial media. The colonies on agar are white or gray, from 1.5 to 3 mm. in diameter, very viscid, and raised; they tend to become confluent. When touched with a platinum needle the growth may be drawn away as a tenacious, sticky filament. In gelatin, a non-characteristic filamentous growth occurs along the line of inoculation and the surface becomes covered with a white, glistening raised colony. The gelatin is not liquefied. Milk is acidified, and

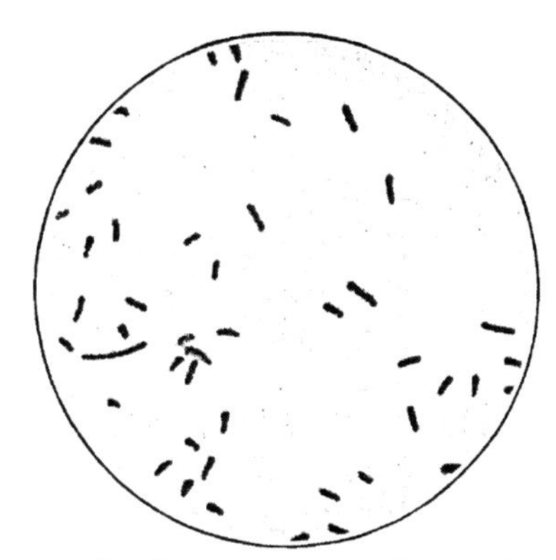

FIG. 51.—Bacillus mucosus capsulatus. × 1000.

frequently the accumulation of acid leads to coagulation. A light-pink color is imparted to litmus milk and coagulation is irregular in this medium. Broth is clouded, and a slimy, viscid sediment collects at the bottom of the tube. A majority of strains produce gas bubbles on potato.

The organisms are aërobic, facultatively anaërobic. Growth takes place at 8° to 10° C., but 37° C. is the optimum temperature. Little or no growth occurs above 43° C.

Products of Growth.—The majority of strains do not form indol, but occasional cultures give a marked reaction for this substance.[1] Practically all strains form a mucinous substance on artificial media.

The reactions of fermentation have been used as a basis for separation into types by Perkins,[2] who groups the organisms in the following manner:

[1] Kendall, Day and Walker, Jour. Am. Chem. Soc., 1913, xxxv, 1237.
[2] Jour. Infec. Dis., 1904, i, 241.

Type I.—All carbohydrates and starch fermented with the production of gas ($H:CO_2 = \frac{2}{1}$) and acid; Bacillus lactis aërogenes.

Type II.—All carbohydrates except saccharose fermented; starch fermented—Bacillus mucosus capsulatus, Bacillus rhinoscleromatis, Bacillus ozænæ.

Type III.—All carbohydrates except saccharose and starch fermented; Bacillus acidi lactici.

Enzymes and toxins have not been demonstrated in cultures of any members of the group.

Pathogenicity.—*Human.*—Bacillus mucosus capsulatus has been isolated in a considerable proportion of cases of lobular pneumonia, but it practically never is the sole incitant of lobar pneumonia. It is occasionally detected in purulent inflammations of the respiratory tract not pneumonic in character, in the purulent secretions of the nasal and frontal sinuses, in occasional cases of pericarditis and pleurisy, stomatitis and otitis media. The normal sputum occasionally contains the organism.

Animal.—Subcutaneous inoculations into mice, rabbits or guinea-pigs frequently lead to abscess formation characterized by thick, viscid pus. Occasionally a generalized infection which results fatally takes place.

Bacillus Rhinoscleromatis.—Rhinoscleroma, characterized by indurated granulomatous nodules of the mucous membrane of the nose, is ascribed to Bacillus rhinoscleromatis by v. Frisch,[1] Paltauf and v. Eiselsberg,[2] and others. A satisfactory demonstration of the etiology of this infection is wanting, but organisms culturally like Bacillus rhinoscleromatis have been isolated from the cells of Miculicz, large, swollen cells with crescentric nuclei characteristically present in rhinoscleroma and demonstrated within them on section.

Bacillus Ozænæ.—Ozena, a disease of the nose characterized by a fetid catarrhal inflammation, is very frequently associated with the presence of large numbers of a member of the Mucosus Capsulatus Group to which Abel[3] gave the name Bacillus ozænæ. The organism has not been sharply separated from Bacillus rhinoscleromatis and Bacillus mucosus capsulatus, and its etiological relationship to ozena is still *sub judice.* Autogenous vaccines of the organism have been used with varying success in the treatment of the disease.

[1] Loc. cit.
[2] Fort. d. Med., 1886, Nos. 19 and 20.
[3] Loc. cit.

Bacillus Lactis Aërogenes.—This organism is an almost constant inhabitant of the upper part of the intestinal tract of nurslings; it is common in the intestinal contents of bottle-fed infants, and it frequently persists in small numbers in the adult intestinal tract. A closely related organism, Bacillus acidi lactici, is found fairly widely distributed in milk, water, and sewage. A sharp differentiation between the two organisms is difficult to establish. There is evidence that the organism, ordinarily a harmless intestinal parasite, may become temporarily pathogenic and incite intestinal disturbance varying in intensity from slight diarrhea to severe enteritis.[1] Occasional cases of cystitis in infants are also associated with the presence of Bacillus lactis aërogenes in pure culture.

It is obvious that the interrelations of the Mucosus Capsulatus Group are at present in an unsatisfactory state—attempts to separate the organisms on the basis of serological reactions have been unsuccessful, partly because of the difficulty of removing the capsules which appear to be somewhat impervious to antibodies. A final arrangement of the group and an ultimate differentiation of the various organisms comprising it awaits future elucidation.

[1] Kendall and Day, Boston Med. and Surg. Jour., 1913, clxix, 753; Kendall, ibid., May 20, 1915.

CHAPTER XIX.

GLANDERS, ANTHRAX, PYOCYANEUS, INFECTIOUS ABORTION: ACIDURIC BACTERIA.

BACILLUS MALLEI.

Historical.—Glanders is a disease primarily of animals having an undivided hoof: horses, asses, and mules. It may be acute or chronic, and two clinical types are recognized: glanders, an initial infection of the nasal mucosa and regional lymphatic glands, later an involvement of the internal organs, more commonly the lungs; and farcy, a cutaneous glanders, in which the cutaneous lymphatics are involved with the formation of nodules (farcy buds) which frequently ulcerate and discharge a cohesive sticky secretion. Man is occasionally infected, the disease being one of the most fatal known. The causative organism, Bacillus mallei, was described by Löffler and Schütz in 1882.[1]

Morphology.—Bacillus mallei is a small bacillus with rounded or somewhat attenuated ends, measuring from 0.5 to 0.75 micron in diameter, and from 2 to 5 microns in length. The organisms occur singly and in pairs in culture media, although long filamentous forms are not uncommon on potato. In pus and from tissues the bacilli occur in groups or clusters. The bacilli frequently appear as short, almost coccoid elements, both in culture and *in vivo*. Older cultures frequently contain many branched forms. The glanders bacillus is non-motile, and possesses no flagella. Capsules and spores have not been observed. The organism stains faintly with ordinary anilin dyes, better with those having an alkaline reaction. It is Gram-negative. Stained with Löffler's alkaline methylene blue, the organism exhibits irregularity of colorable material; the bacilli may even resemble groups of cocci with faintly stainable substance conneeting the deeply stained, round granules. Zeit[2] has called attention to the resemblance of B. mallei in pus and tissue to staphylococci when stained with methylene blue, and the possibility of error in diag-

[1] Deutsch. med. Wchnschr., 1882, No. 52.
[2] Jour. Am. Med. Assn., 1909, lii, 181.

nosis upon morphological examination alone. The Gram stain will distinguish between the two, however.

Isolation and Culture.—Bacillus mallei grows well upon ordinary laboratory media, better if glycerin is added, and upon blood serum and potato. The first growth outside the animal body may be difficult to obtain. Colonies on glycerin agar are small, yellowish and round. At first the growths are translucent, later they become nearly opaque and more deeply colored. Growth in gelatin is slow and not distinctive; no liquefaction takes place. A uniform turbidity appears in broth after twenty-four hours' incubation at 37° C., which gradually settles out as a tenacious, slimy sediment. If the culture is undisturbed, a pellicle gradually forms on the surface of the medium. Litmus milk is slowly acidified, and coagulation may occur after seven to fourteen days' incubation. Growth on old alkaline potato is distinctive; after twenty-four to forty-eight hours' incubation a light brown, translucent layer appears, which has been likened in color and general appearance to a layer of honey. Later the growth becomes darker, even brownish-red in color, and the underlying potato becomes greenish or even brown. Potato that is acid does not exhibit the typical honey yellow growth.

The glanders bacillus is aërobie, facultatively anaërobic; the optimum temperature of development is 37° C., growth ceases above 43° C., and is extremely slow below 25° C. The resistance of the organism to chemical agents is not great, but it remains viable for several weeks when dried in pus or blood and maintained in a cool, dark place. An exposure of naked bacilli to 55° C. for five to seven minutes kills the organisms.

Products of Growth.—*Chemical.*—Bacillus mallei is culturally inert in purely protein media: indol, skatol and other products of degradation of amino acids are not produced. Acid, but no gas, is formed in dextrose broth, and acid is produced in milk.

Enzymes.—No enzymes have been demonstrated in cultures of B. mallei.

Toxins.—Soluble toxins have not been isolated from growths of the glanders bacillus; the poison of the organism belongs to the group of the endotoxins. A substance analogous to tuberculin has been prepared from four to five weeks' glycerin broth cultures of Bacillus mallei, mallein or morvin. The preparation of mallein is essentially the same as for tuberculin. The injection of mallein in moderate doses into normal animals may lead to transient fever and a slight local swelling which quickly subsides. In horses infected with B. mallei a swelling appears within a few hours which is painful and inflamed; it gradually enlarges for twenty-four hours or more, and the lymphatics of the area usually become prominent. The swelling may persist for several days, but gradually diminishes and usually disappears within ten days. The temperature rises with the local swelling and reaches a point 1° to 2° or even 3° above the normal within twenty-four hours. The animal usually exhibits all the signs of a generalized reaction; it becomes listless, the coat roughens, and there is greater or lesser generalized weakness. The temperature usually persists for forty-eight hours or more. The reaction is specific but requires experience for its interpretation. Variations in temperature are caused by strangles, bronchitis and other inflammatory infections, hence the temperature should be observed for some hours before the injection of the mallein. A positive reaction is of more diagnostic value than a negative reaction. It should be borne in mind that mallein interferes with serologic tests, hence the latter should be made before the injection of mallein.

Pathogenesis.—*Animal.*—Cattle appear to be immune to glanders; swine are but slightly susceptible; cats, sheep, goats, field mice and guinea-pigs are susceptible, but white mice are refractory.

Acute glanders in horses and asses begins after an incubation period of from three to six days with an abrupt rise of temperature and a viscid, purulent nasal discharge. The nasal mucosa, at first deeply congested, becomes ulcerated; the regional lymph glands enlarge and may suppurate. The lungs become involved and death usually occurs within six to fourteen days; occasionally the animal lives several weeks. The onset of the chronic form is somewhat more insidious, and the symptoms are less violent. There is usually a nasal discharge which may be blood-streaked, and the superficial glands of the neck are palpable. The cutaneous lymph glands and usually the lymph channels as well become generally enlarged, and they may break down and suppurate. The disease may run a very mild course, hardly noticeable, and frequently terminates in a cure after months or years.

The injection of material from ulcers, nasal secretion, or lymph glands into male guinea-pigs leads, usually within two or three days, to a characteristic lesion, unless the material is grossly contaminated with other organisms, namely, a purulent orchitis; the testicle enlarges until it can not be retracted, and the inflammation spreads from the tunica vaginalis to the epididymis. The peritoneum is inflamed, and if the organism is not very virulent there is joint involvement and gradual emaciation and death. This is known as the Straus reaction.

Human.—The essential lesion in man is similar to that in the horse—a granulomatous nodule made up chiefly of epithelioid cells and many lymphoid cells. The bacilli occur in these nodules in large numbers as a rule. The nodules occur chiefly in the nasal mucosa, or in cutaneous infections under the skin; they break down readily, causing ulceration or abscess formation. A crop of papules, which soon break down, appears on the face, around joints, and frequently upon the arms. The disease terminates fatally in about 65 to 70 per cent. of all cases.

Immunity and Immunization.—Recovery from an attack of glanders does not appear to confer immunity to subsequent infection, and attempts to induce immunity in susceptible animals by vaccines, by the use of mallein, or by sera have been unsuccessful. Specific agglutinins and precipitins are present in the blood serum of infected animals and a diagnosis can be made by the method of complement fixation. The latter procedure, important in horses and other domestic animals, has not been tried very extensively in man, partly because of the comparative rarity of cases.

Bacteriological Diagnosis.—1. *Microscopical Examination.*—Material from the purulent discharges of the nose or scrapings from cutaneous nodules are stained by Gram's method and by Löffler's alkaline methylene blue. The organism is Gram-negative, and frequently exhibits a beaded appearance not unlike the diphtheria bacillus. A diagnosis based upon purely morphological characters is not reliable.

2. *Cultural.*—Scrapings from unopened granulomata or from the organs postmortem should be inoculated upon potato having an alkaline reaction. The characteristic appearance of the growth upon this medium is suggestive, but not conclusive. Bacillus pyocyaneus grows very similarly. Pus must be plated upon glycerin agar or blood serum, because the discharges from ulcers and abscesses are almost invariably contaminated with other organisms. Pure cultures are examined microscopically and injected into male guinea-pigs intraperitoneally.

3. *Animal Injection.*—The intraperitoneal injection of suspected material into the peritoneal cavity of male guinea-pigs leads, in the absence of organisms capable of causing a violent peritonitis, to the

localization of the bacilli in the testes, which become inflamed and swollen—the Straus reaction. The animal usually dies within a week. Potato cultures and microscopical examination of the purulent material in the testes usually suffices to establish the diagnosis. In case the material for examination is contaminated with other bacteria, it is advisable to inoculate it into the subcutaneous tissues of one guinea-pig, and to inoculate a second male pig with material from an enlarged lymph gland of the first pig. A negative examination is inconclusive.

Serological Diagnosis.—(*a*) *Mallein.*—Discussed above.

(*b*) *Ophthalmo Reaction.*—The instillation of a few drops of mallein into the conjunctival sac of a glanderous horse leads to a reaction very similar to the ophthalmo-tuberculin reaction in man, except that in positive cases a purulent discharge as well as a red inflamed conjunctiva results.

(*c*) *Agglutination Test.*—Specific agglutinins for Bacillus mallei appear in the blood of infected animals usually within four to seven days in acute glanders, and there is a rough parallelism between the severity of the disease and the development of the immune bodies. The agglutinins as a rule diminish considerably if the disease becomes chronic, and may become reduced to such a degree that the reaction becomes unreliable. The sera of normal horses frequently contain non-specific agglutinins which may clump glanders bacilli in dilutions of 1 to 100 to 1 to 300. Injections of mallein appear to influence antibodies specific for the glanders bacillus adversely, consequently serological examinations should be made before mallein is injected.

Serum for agglutination tests should be withdrawn in a sterile syringe from the jugular vein in the horse, and from the median basilic vein in man. The serum, separated from the clot, is diluted with a suspension of glanders bacilli to the following degrees: 1 to 500, 1 to 1000, 1 to 2500, 1 to 5000, 1 to 7500. Glanders bacilli, virulent for guinea-pigs (obtained by passing glanders bacilli through a series of animals until the organism kills the animal within five days—intraperitoneal injection), from glycerin agar slants are emulsified in physiological salt solution containing 0.5 per cent. carbolic acid, thoroughly shaken and filtered through a thin layer of absorbent cotton to remove clumps. Salt-phenol solution is added to the suspension until a moderately turbid suspension is obtained. Decreasing amounts of serum from the suspected animal are added to obtain the dilutions mentioned above. A normal serum and a known positive serum are diluted in the same manner to serve as controls. Incubation is continued at 37° C. for seventy-two hours, because the reaction is usually slow in developing. Sterility must be maintained throughout. Strongly positive sera may give a definite clumping in twenty-four hours or less; the supernatant fluid becomes clear, and the organisms collect as a diffuse sediment at the bottom of the tube. A negative reaction is indicated by a turbid supernatant fluid. The reaction may be made microscopically or macroscopically, the latter being preferable.

Attempts have been made to shorten the reaction time by aiding sedimentation with the centrifuge. The various dilutions are incubated for a full hour at 37° C., allowing fifteen minutes for the tubes to reach 37° C. in the incubator; then they are whirled for fifteen minutes at a speed with a twenty-four inch radius not exceeding 1500 revolutions, placed in the ice-box and examined after three hours. The slowly developing reactions may not be definitely positive for twenty-four hours.

A reaction in a dilution of 1 to 500 (horse, ass or mule) is the lowest limit to which a definite reaction may be attributed, and the result should be controlled with a mallein test. Dilutions of 1 to 750 or higher are usually safely regarded as diagnostic. In human cases a positive reaction in a dilution of 1 to 100 is diagnostic.

The method of complement fixation (see page 164 for details) is rapidly becoming a general method for the diagnosis of glanders.

Dissemination and Prophylaxis.—Glanders is transmitted by direct contact, by infection through cutaneous abrasions and cuts, and by feeding paraphernalia, watering troughs and buckets. In man cutaneous infection is more common.

BACILLUS ANTHRACIS.

Bacillus anthracis was first seen by Davaine[1] in 1863, in the blood of animals infected with anthrax. Koch[2] confirmed Davaine's observation, obtained the organism in pure culture, and reproduced the disease with these cultures in other animals, thus establishing the etiology of anthrax. He also demonstrated spore formation by B. anthracis upon artificial media.

Morphology.—Bacillus anthracis is a rod-shaped organism measuring from 1 to 1.50 microns in diameter and from 2 to 4 microns in

[1] *Compt. rend. Acad. Sci.*, 1863, lvii.
[2] *Cohn's Beitr. z. Biol. der Pflanzen*, 1876, ii, 277.

length. Occasionally filaments 20 to 25 microns in length are observed, which exhibit no demonstrable septation; these long rods may be single cells or chains of cells in which septation is imperfect. The ends of the bacilli are square cut and often appear to be concave, particularly when the organisms are examined in a strained preparation made directly from the blood of an infected animal. Occasionally the ends are somewhat thickened, giving the bacillus an appearance which suggests a segment of bamboo. Bacillus anthracis produces short chains of three to eight elements in the bloodvessels of infected animals, and in artificial media it produces long, coiled chains of bacilli which give a characteristic filamentous appearance to the colonies upon solid media. The organism is non-motile, and possesses no flagella. A capsule[1] is formed around the

FIG. 52.—Bacillus anthracis, spore formation. × 1000. (Günther.)

bacilli in the animal body and also in cultures containing albuminous substances, as uncoagulated blood serum.[2] Spores are produced in media freely exposed to the air between the temperatures of 15° C. and 40° C. The lower limit of spore formation has a practical bearing upon the presence of anthrax spores in soil. In the temperate zones a temperature exceeding 15° C. in midsummer is not found at depths greater than five feet, hence anthrax carcasses buried deeply are not likely to cause infection of the soil. It has been stated that earthworms may carry infected material from the deeper layers of the soil to the

[1] The capsule was first seen by Serafini (Progresso Medico, 1888), but Johne (Deutsch. Ztschr. f. Tiermed. u. vergl. Path., 1893, xix, 244; 1894, xx, 426) first called attention to the diagnostic importance of the capsule in the diagnosis of anthrax of the domestic animals.

[2] Haase, Deutsch. Ztschr. f. Tiermed. u. vergl. Path., 1894, xx, 429; Johne, ibid., 1894, xxi, 142.

surface, where sporulation may occur. If these temperatures are exceeded in either direction, spore formation does not occur. The spores, which are oval, are situated at or near the centre of the cell and measure about 0.8 micron in diameter and from 1.2 to 1.4 microns in length. Occasional asporous strains[1] are met with, and spore formation may be suppressed by cultivating the bacteria at 42° C. for several hours or in fluid media containing potassium bichromate in dilutions from 1 to 5000 to 1 to 2000, or small amounts of phenol.[2] Lehmann[3] states that long-continued transfer of cultures from gelatin to gelatin frequently leads to a suppression of spore formation. Some strains become asporeless much more readily than others.[4] Spores

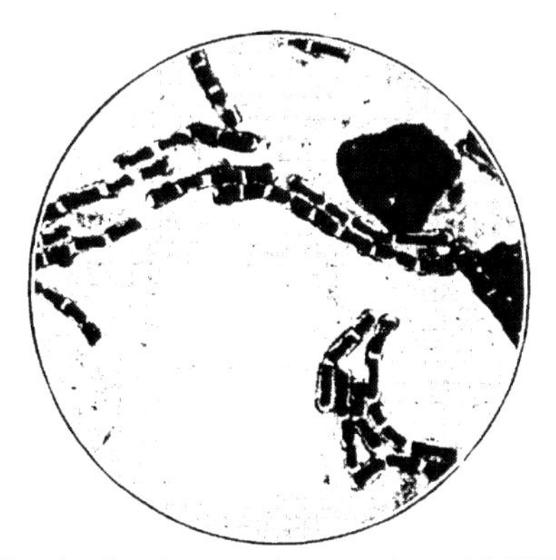

Fig. 53.—Bacillus anthracis, showing capsule formation. × 1000. (Kolle and Hetsch.)

are not formed in the intact animal body. Mature vegetative bacilli emerge from the spores in the presence of oxygen, if the temperature is maintained between 15° and 40° C. The spore membrane merges imperceptibly into the newly formed vegetative cell; no visible rupturing of the spore membrane is detectable.

Bacillus anthracis stains well with ordinary anilin dyes and young cultures are Gram-positive. Older cultures may gradually lose their ability to retain Gram's stain. Spores may be stained with the Ziehl-Neelsen stain. (See Staining of Spores.)

Isolation and Culture.—Bacillus anthracis grows readily upon any artificial media. Material is best obtained from the spleen or liver

[1] Asporous cultures do not necessarily become avirulent (Chamberland and Roux, Compt. rend. Acad. des Sci., 1883, xcvi, 1090).

[2] Roux, Ann. Inst. Past., 1890, 25.

[3] München. med. Wchnschr., 1887, No. 26.

[4] Surmont and Arnould, Ann. Inst. Past., 1894, p. 832.

of dead animals, or from the blood of an infected animal. Gelatin is rapidly liquefied; colonies appear in gelatin plates within eighteen hours after inoculation, which are from 1 to 2 mm. in diameter. They are gray, opaque, and somewhat irregular in size. The organisms develop rapidly, and liquefaction commences within thirty hours as a rule. At this stage of development the edges of the colonies are composed of tangled, radiating chains of bacilli which extend into the surrounding medium, and the colony itself is composed of a mass of twisted filaments which has been likened to a Medusa head. Few, if any, pathogenic bacteria present such an appearance. The growth in stab cultures in gelatin is also characteristic; the organisms grow

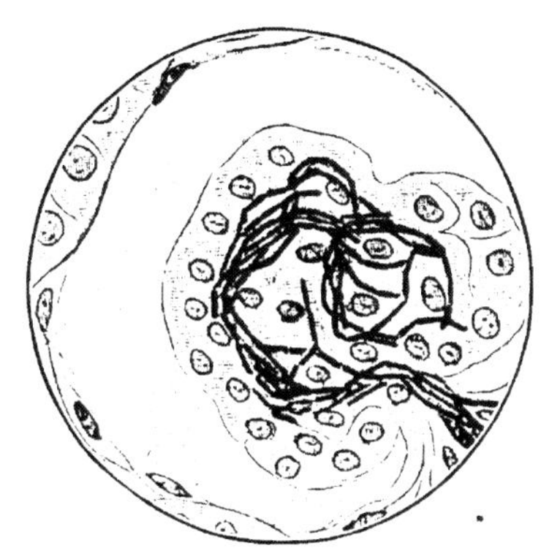

Fig. 54.—Bacillus anthracis, section from kidney, semi-diagrammatic. × 500. (Kolle and Hetsch.)

away from the line of inoculation into the medium as spikelets which resemble an "inverted pine tree." Liquefaction soon takes place. Milk is rendered acid, and the casein precipitated and slowly liquefied. A pellicle forms upon the surface of broth which readily becomes detached from the sides of the tube and settles to the bottom. No turbidity is produced in fluid media.

Bacillus anthracis is a strongly aërobic bacillus, but growth will take place under anaërobic conditions. Growth is very slow at 18° C., and ceases below 15° C. The optimum is about 37° C., and development does not take place at 45° C.

The vegetative (asporeless) organisms are not resistant to heat or drying. The spores are very resistant. Dried spores have remained viable and virulent for eighteen years.[1] Fresh blood containing anthrax

[1] V. Székely, Ztschr. f. Hyg., 1903, xliv, 363.

bacilli may remain viable for two months if relatively thick layers are prepared. Dry heat at 160° C. kills anthrax spores within one and a half hours; live steam (100° C.) kills them within ten minutes. Carbolic acid is not very effective as a germicide, but 1 to 1000 bichloride of mercury kills the spores within half an hour. Direct sunlight kills them within six hours.[1]

Products of Growth.—*Chemical.*—Martin[2] found protoalbumose, deuteroalbumose, a trace of peptone, an alkaloidal substance, and small amounts of leucin and tyrosin in a serum culture of B. anthracis. No acid or gas is produced in any sugar media. The albumoses and peptone caused a febrile reaction in animals, and the alkaloidal substance (anthrax-alkaloid) caused edema and congestion. These results have never been repeated.[3]

Enzymes.—Bacillus anthracis produces a proteolytic enzyme which liquefies gelatin, blood serum and casein. No other enzymes are known.

Toxins.—Soluble toxins have not been demonstrated in cultures of anthrax bacilli, and the nature of the endotoxin is unknown—the cellular substance of the organism is not as toxic as that of many other pathogenic bacteria, and the nature of the action of the bacillus is not clearly determined.

Pathogenesis.—*Animal.*—Anthrax is a disease of cattle, sheep[4] and horses. Swine are less susceptible. Guinea-pigs, rabbits and white mice are very susceptible to inoculation. Rats and dogs succumb to large doses. Birds and cold-blooded animals are naturally immune, although, as Pasteur showed, the immunity may be overcome by reducing the body temperature of birds and by raising the body temperature of cold-blooded animals.

The artificially-induced disease in small laboratory animals is usually a rapidly fatal septicemia; the organisms swarm in the blood-vessels and appear upon section to almost occlude the capillaries. The spleen is greatly enlarged and there is congestion of the other glandular organs. Cattle and sheep readily succumb to infection with pure cultures of the organism. The natural infection in cattle and sheep appears to be chiefly through the intestinal tract. In horses

[1] Momont, Ann. Inst. Past., 1892, 23.

[2] Proc. Royal Soc., London, May 22, 1890. Brit. Med. Jour., March 26, April 2, 9, 1892. Animal Report Local Government Board, Supplement, 1890–91, xx, 255–266.

[3] It is probable that these substances were produced from the serum by the action of the organism; they cannot be regarded as specific toxic products.

[4] Algerian sheep are said to be more resistant to infection than ordinary sheep.

infection may take place through the skin as well. Less commonly cutaneous infection may occur through wounds in cattle and sheep. A localized severe inflammation results which may heal spontaneously or lead to a generalized infection. It is stated that flies, particularly the horse flies (Tabanidæ) may transmit the virus to animals. The disease may also be transmitted experimentally by the inhalation of spores; this method of infection is probably not common in animals.

Human.—Anthrax bacilli or their spores may cause disease in man either by gaining entrance to the body through abrasions of the skin, by inhalation, or by ingestion. Inoculation through the skin may give rise to malignant pustule, characterized by a small papule at the site of infection, which soon becomes vesicular. The process may stop spontaneously with the formation of a scab and the gradual drying up of the vesicle, or the inflammation may spread, producing a wide area of induration in which vesicles appear, often in considerable numbers. The involved area becomes edematous, and the regional glands become enlarged. Death may ensue within five to seven days, or the inflamed area slowly returns to normal. Less commonly edema is the prominent symptom, pustule formation being absent or not conspicuous. The edematous area spreads rapidly and it may be extensive enough to interfere with the nutrition of the part and lead to gangrene. The head, the arms, or the hands are more frequently involved than the lower extremities.

Intestinal anthrax and pneumonic anthrax or woolsorters' disease are usually caused by the ingestion or inhalation of anthrax bacilli or their spores. Intestinal anthrax is uncommon; it is supposed to be an infection through the gastro-intestinal tract resulting from the ingestion of meat or milk of diseased animals. The symptoms are essentially those of meat-poisoning: chill, vomiting, and nausea, diarrhea, and some fever. Woolsorters' disease prevails where hides and wool, particularly from South America, Morocco and Russia, are handled. The symptoms are: a sudden chill, immediate great prostration, intense pain, bronchial irritation, and occasionally death within twenty-four hours. Cerebral symptoms frequently are prominent in those cases which are more protracted. There are no distinctive postmortem changes; the lungs may be edematous and there are scattered patches of lobular pneumonia with inflammation of the regional bronchi.

Immunity and Immunization.—The vulnerability of human tissues to anthrax infection is varied; the skin appears to be relatively resistant, but the lungs are very susceptible. The disease resulting from infection of the lungs by anthrax bacilli is one of the most rapid and fatal known to man. Practically no attempt has been made to immunize man to anthrax, but Sobernheim has prepared a serum obtained by injecting animals immunized by Pasteur's method with virulent anthrax bacilli, which is said to be of some value as a curative agent in malignant pustule.

Animal Immunization.—Pasteur protected animals against anthrax infection by vaccination with attenuated anthrax bacilli. Two vaccines were used; they were prepared in the following manner: Vaccine A was obtained by growing anthrax bacilli at 42.5° C. for six weeks. The organisms are asporeless after this treatment, but they grow luxuriantly. They are avirulent for rabbits and guinea-pigs, but kill mice. Vaccine B was obtained by growing anthrax bacilli at 42.5° C. for two weeks. The organisms kill mice and guinea-pigs, but do not kill rabbits. Vaccine A is injected, and after two weeks Vaccine B is injected, both subcutaneously. The animals are immune two weeks after the last injection to cutaneous infection with anthrax bacilli, but are somewhat less resistant to infection by way of the alimentary tract. The immunity is of about one year's duration, and it must be renewed at the end of that time. Sobernheim[1] has attempted to increase the immunity to ingestion anthrax by injecting his serum (5 to 15 c.c.) and Vaccine B of Pasteur simultaneously. He states that this combined immunizing process brings the resistance of the animal to such a level that ingestion infection rarely or never occurs.

Bacteriological Diagnosis.—The diagnosis of anthrax in man depends wholly upon the identification of the anthrax bacillus.

(*a*) *Morphological Diagnosis.*—Smears from the blood or tissues of animals stained by Gram's method show large, square-ended, Gram-positive bacilli, which occur singly, in pairs, or short chains. The organisms are encapsulated but require special capsule stains for their demonstration. In man similar examination is made from the serous fluid expressed from the malignant pustule, the blood (best obtained from the ear), fluid from edematous areas, sputum from woolsorters' disease, and feces from intestinal cases.

(*b*) *Cultural.*—The material collected aseptically is inoculated into ordinary media. It is well to examine the media after two to three days' incubation for spores if the culture is impure; if spores are

[1] Ztschr. f. Hyg., 1899, xxxi, 89.

found heating the culture to 80° C. for fifteen minutes will destroy all vegetative forms leaving the anthrax spores in excess and frequently in pure culture. The growth on gelatin is fairly distinctive.

(*c*) Inoculate a guinea-pig or a mouse with a small amount of blood or fluid from a suspected lesion; if bacilli are not numerous, incubate the material in broth for twenty-four hours, then inject the enriched culture. The occurrence of typical large Gram-positive bacilli in the blood stream postmortem is sufficient to establish the diagnosis in the light of the clinical history. The principal organisms likely to cause confusion are: B. subtilis and members of the mesentericus group, which do not produce acute death in guinea-pigs by generalized septicemia, and B. edematis maligni and B. aërogenes capsulatus, both of which are obligate anaërobes.

Dissemination.—The spores of anthrax bacilli are extremely resistant to dessication, and they remain alive for years in the soil. Once a pasture or other enclosure is infected with the organisms it is unsafe to permit cattle, sheep or other domestic animals to graze there. The washings from such infected lands may convey infection to other lands.

Prophylaxis in man consists essentially in preventing contact infection with diseased animals or infected material, and particular care in preventing the inhalation of dust from hides or wool of cattle or sheep from countries where the disease is prevalent; this applies particularly to South American, Moroccan and Russian hides and wool.

BACILLUS PYOCYANEUS.

Historical.—Surgeons for many years have noticed that occasional suppurating wounds discharge pus which stains bandages a green or green-blue color. Gessard[1] demonstrated the specific organism, Bacillus pyocyaneus, in pure culture and described it in considerable detail. Somewhat later Charrin[2] studied the pathogenesis of the organism for rabbits (maladie pyocyanique), setting forth clearly the importance of the bacillus as a disease-producing microörganism.

Morphology.—Bacillus pyocyaneus is a moderate-sized organism with rounded ends, usually occurring singly or in pairs, less commonly in short chains. The dimensions vary considerably even in the same culture; the diameter averages about 0.6 micron, although some

[1] Thèse de Paris, 1882.
[2] La maladie pyocyanique, Paris, 1889.

bacilli measure but 0.3 micron and others as much as 1 micron. The length varies between 1.5 and 4 microns, the average being about 2 microns. The organism is actively motile, and possesses a terminal polar flagellum (monotrichic flagellation). Capsules and spores have not been observed. Ordinary anilin dyes color the bacillus with moderate intensity, and it is Gram-negative, although the gentian violet is somewhat less readily removed by alcohol than from a majority of Gram-negative bacilli, as Bacillus coli for example.

Isolation and Culture.—The organism grows readily and rapidly upon ordinary artificial media, producing the characteristic pigments in the presence of oxygen. The colonies on agar are round and measure from 1 to 3 mm. in diameter after eighteen to twenty-four hours' incubation at 37° C. The growth spreads rapidly, and the pigment which becomes visible within eighteen hours dissolves in the medium imparting a blue-green color to it. Gelatin colonies are not characteristic in outline, but rapidly liquefy the medium, which becomes green. A turbidity is visible within eight hours in broth and a pellicle usually forms on the surface. A viscous, gray-brown sediment collects at the bottom of the tube, and an ammoniacal odor is noticeable even within twenty-four hours. The medium, particularly the upper layers in contact with oxygen, becomes blue-green. Milk is coagulated, the coagulum being slimy, and eventually partly or even completely dissolved; the medium, at first yellowish, becomes green, then blue, particularly in the upper layers.

Bacillus pyocyaneus is aërobic, facultatively anaërobic. The optimum temperature is 37° to 38° C.; development is sluggish below 18° C. and practically ceases at 43° to 44° C.

Products of Growth.—*Chemical.*—The organism produces a relatively large amount of ammonia from proteins and protein derivatives,[1] and in milk.[2]

Pigments.—Two pigments are produced by Bacillus pyocyaneus: a water-soluble, green, fluorescent pigment similar in physical properties to that found in cultures of other fluorescent bacteria; and a specific pigment, pyocyanin, which is insoluble in water but soluble in chloroform. Pyocyanin, to which the empirical formula $C_{14}H_{14}NO_2$ has been ascribed by Ledderhose,[3] crystallizes from chloroform solu-

[1] Armaud and Charrin, Compt. rend. Ac. sc., 1891, cxii, 755, 1157; Kendall, Day, and Walker, Jour. Am. Chem. Soc., 1913, xxxv, 1243.
[2] Kendall, Day and Walker, Jour. Am. Chem. Soc., 1914, xxxvi, 1948, 1963.
[3] Deutsch. Ztschr. f. Chir., 1888, xxviii, 201.

tion as blue needles. It forms salts with acids, and exists as a leuco-base in cultures from which oxygen is excluded. The color changes to a brownish-red in old cultures.

Enzymes.—One of the noteworthy products of Bacillus pyocyaneus is a soluble proteolytic enzyme, a protease, which dissolves gelatin, casein, coagulated blood serum and fibrin.[1] Breymann[2] showed that the bodies of the bacteria, freed from culture media, contained the same or a similar enzyme. Emmerich and Löw[3] isolated a proteolytic enzyme, called by them pyocyanase, which possessed the remarkable property of dissolving alien bacteria. This enzyme has been used therapeutically with some success. Whether pyocyanase is identical with the protease mentioned above has never been clearly determined.

No diastatic enzymes have been detected in cultures of Bacillus pyocyaneus.[4]

Toxins.—Wassermann[5] found that filtered cultures of Bacillus pyocyaneus or cultures killed with toluol would kill guinea-pigs when injected intraperitoneally in amounts of 0.2 to 0.5 c.c. The organisms themselves were decidedly less toxic. The toxicity is not attributable to the specific pigment, pyocyanin, but to substances of unknown composition.

Pathogenesis.—*Animal.*—Bacillus pyocyaneus is pathogenic for small laboratory animals, guinea-pigs being the most susceptible. A cubic centimeter or less of an actively growing broth culture introduced into the peritoneal cavity causes death within twenty-four hours as a rule. There is edema, leukocytosis, and the peritoneal fluid increased in amount swarms with the bacilli. Rabbits are less susceptible; rats and mice are relatively refractory. The subcutaneous injection of cultures of the organism, especially if the virulence is not great, leads to a chronic, wasting infection which usually terminates fatally. The subcutaneous tissue becomes edematous and necrotic, and ulceration frequently occurs.

Human.—Besides the focal lesions, abscesses, ulcers, otitis media, less commonly liver abscesses, and bronchopneumonia, Bacillus pyocyaneus occasionally produces severe gastro-intestinal infection, especially in young children, generalized sepsis, and inflammation of serous surfaces, the pleura, pericardium, and peritoneum.

[1] Jakowski, Ztschr. f. Hyg., 1893, xv, 474; Fermi, Centralbl. f. Bakt., 1891, x, 401; Kendall, Day and Walker, Jour. Am. Chem. Soc., 1914, xxxvi, 1966, and others.

[2] Centralbl. f. Bakt., Orig., 1902, xxxi, 481.

[3] Ztschr. f. Hyg., 1899, xxxi, 1.

[4] Fermi, loc. cit.

[5] Ztschr. f. Hyg., 1896, xxii, 263.

Immunity and Immunization.—It is possible to immunize animals both by the cautious injection of the bacilli which stimulate the formation of specific bacteriolysins, and by filtrates of broth cultures of the organisms, which incite the formation not only of bacteriolytic substances, but antitoxic substances as well. No practical use is made of these antibodies in human infections, however.

Bacteriological Diagnosis.—Wounds infected by Bacillus pyocyaneus are usually diagnosed by the blue-green color of the dressings. The bacilli are readily isolated upon gelatin plates, where the development of the blue-green color is very characteristic.

BACILLUS ABORTUS.

Historical.—Infectious abortion is a disease which has for many years been recognized as an important economic one in the cattle industry. Later it was found that the same disease also exists among horses, goats and sheep. The organism was first isolated by Bang.[1]

Morphology.—B. abortus is a small pleiomorphic bacillus, measuring 0.4 to 0.6 micron in diameter, by 0.6 to 2.5 microns in length. It occurs singly and in pairs; rarely short chains of three to six elements are found. The shape varies: some organisms are almost spherical, others are distinctly rod-shaped, the latter being more frequently found in broth cultures and *in vivo*. According to Priesz,[2] branched forms may be found in older cultures. It is non-motile and possesses no flagella, although Brownian movement may be fairly active. It possesses no capsules, and no spores have been demonstrated. It stains readily with ordinary anilin dyes, but somewhat irregularly, some areas staining more intensely than others. Occasionally with the methylene blue stain the organisms may present a bipolar appearance. The organism is Gram-negative.

Isolation and Culture.—Initial growths on artificial media outside the animal body are somewhat difficult to obtain. The organism appears to grow best in a somewhat rarified atmosphere. This has been obtained by Fabyan[3] by growing the organism on an agar slant which is connected by a narrow tube with an agar slant on which B. subtilis is growing. B. subtilis appears to so change the percentage composition of the air in the two tubes that B. abortus grows fairly readily. He also found that a pressure of three to five atmospheres

[1] Ztschr. f. Thiermedizin, 1897, i, 241–278.
[2] Centralbl. f. Bakt., Orig., 1903, xxxiii, 190.
[3] Jour. Med. Research, 1912, xxvi, 441.

would facilitate the growth of the organism. On dextrose agar the colonies are round, normally colorless and transparent, and have a very glistening, pearly sheen. The colonies attain a diameter of from 0.5 to 2.5 mm. The organism grows well on blood serum. On gelatin the growth is usually very slow, probably because of the lowered temperature of incubation. No liquefaction takes place. In milk there is a moderate growth; no acid is formed, and no coagulation or peptonization takes place.

Conditions of Growth.—The organism is killed by an exposure of 59° C. for ten minutes.[1]

Products of Growth.—The organism produces no known ferments and it produces no acid in dextrose or other sugars; on the contrary, the reaction on artificial media in which the organism is growing becomes slightly alkaline. The organism forms no extracellular toxins.

Pathogenesis.—Infectious abortion appears to be an infection of the fetus *in utero* and its membranes, which results in the death of the fetus and its expulsion, or less commonly its expulsion in a living and enfeebled state. The time of expulsion is not definite; it may occur early during the period of gestation, or it may not take place until the normal completion of pregnancy. Ordinarily there is no direct evidence of disease in the mother.

The lesions in experimental guinea-pigs, which have been described very carefully by Fabyan,[2] resemble both macroscopically and histologically those of tuberculosis. As a rule, the muscles are free from lesions and there is a tendency for the organism to localize itself in the perivascular or subcapsular regions of various abdominal organs. The organism may persist in experimental animals for very considerable periods of time without producing manifest symptoms. Fabyan has shown[3] that the organism may remain alive but latent in guinea-pigs for over a year.

Immunity.—Cows which have aborted once or twice appear to acquire an immunity which is supposed to be due to the formation of antibodies in the blood. Although no extracellular toxins have been demonstrated as yet, it is probable that the infected animal is sensitized by endotoxins of the abortion bacillus, for such animals injected with "Abortin" (an extract of the abortus bacillus) usually give a definite reaction.

Of extreme importance is the frequent occurrence of the organism

[1] Fabyan, loc. cit., p. 481. [2] Loc. cit.
[3] Jour. Med. Research, 1913, xxviii, 81.

in milk. Melvin[1] has found B. abortus in eight out of seventy-seven samples of market milk and in the milk of six dairies out of a total of thirty-one examined. As early as 1894 Theobald Smith[2] called attention to peculiar tubercle-like lesions induced in guinea-pigs following the injection of cow's milk. He recognized that the disease was not tuberculosis; later Schroeder[3] made similar observations. In the same year Smith and Fabyan[4] showed that the tubercle-like lesions were caused by B. abortus, and in 1913 Fabyan[5] demonstrated conclusively the extremely important fact that B. abortus is very frequently found in the milk of cows that have aborted. He also showed that pasteurization of milk, if carried out in the proper manner, will certainly destroy the bacillus. Whether certain cases of abortion observed in man are due to the organism is not yet proven.[6]

Bacteriological Diagnosis.—The bacteriological diagnosis is best made by injecting guinea-pigs with suspected milk or material from a diseased animal and observing the development of the characteristic tubercle-like lesions. If the animal does not die within a reasonable time it should be killed and autopsied.

Serological Diagnosis.—The blood serum of infected cattle usually agglutinates B. abortus in dilutions greater than 1 to 50. The value of the agglutination reaction as a method of diagnosis is as yet debatable. The extensive statistics of MacFadyen and Stockmann[7] upon this phase of the subject are representative. An agglutination with B. abortus in a dilution of 1 to 50 was obtained with the sera of 526 out of a total of 535 apparently healthy cows; in the remainder (9) agglutination took place in dilutions greater than 1 to 50. Of 127 cattle, either infected or suspects, an agglutination was not obtained in a dilution of 1 to 50; in 11 agglutination was positive, 1 to 50; in 19 a positive reaction was obtained in a dilution of 1 to 100; and in 20 a reaction in a dilution of 1 to 200. Holth[8] tested the sera of 7 normal cattle with negative results. The sera of 38 animals out of a total of 39, which were plainly infected with B. abortus, gave positive agglutination with the specific organism in a dilution of 1 to 100.

[1] Vet. Jour., 1912, lxviii, 526.
[2] Bureau of Animal Industry, 1894, Bull. 7, 80.
[3] Bur. Animal Industry, Circ. 198, November 2, 1912.
[4] Centralbl. f. Bakt., Orig., 1912, lxi, 549.
[5] Jour. Med. Research, 1913, xxviii, 85.
[6] Recently Larsen and Sedgwick (Am. Jour. Dis. of Child., 1913, vi, 326) have examined blood serum from 425 children by the method of complement-fixation; 73 were positive, 325 were negative.
[7] Jour. Compt. Path. and Therap., 1912, xxv, 22.
[8] Berl. tierärztl. Wchnschr., 1909, 686.

The method of complement fixation, precipitin test, ophthalmo reaction and intracutaneous reaction with various preparations of B. abortus have been tested for their diagnostic value, but the results are not clear cut and definite.[1]

Prophylaxis and Dissemination.—The infection of market milk with B. abortus focuses attention sharply upon the transmissibility of the organism to man. Definite details are lacking, but pasteurization of milk should remove all practical danger from this source.

ACIDURIC BACTERIA.[2]

There is a somewhat poorly defined group of bacilli, chiefly found in the intestinal contents of man and animals,[3] which possesses the unusual property of growing in fermentation media of a degree of acidity incompatible with the development of all other known bacteria. The aciduric bacteria are of two kinds: the true aciduric bacilli, of which Bacillus acidophilus is the best known, and facultatively aciduric bacteria,[4] which are occasionally detected in the intestinal contents of man and animals fed for some time upon carbohydrate. The facultative organisms rapidly lose their acid tolerance upon cultivation in ordinary media, and they are probably to be regarded as examples of bacterial adaptation.

Rahe[5] distinguishes three types of aciduric bacilli, depending upon their action upon carbohydrates. Acid, but no gas is formed, as follows:

Type I. Bacillus bulgarieus (not an intestinal organism) coagulates milk, but does not ferment mannite.

Type II. Coagulates milk and ferments mannite.

Type III. Does not coagulate milk, but ferments mannite.

Bacillus Acidophilus.—Bacillus acidophilus, described by Moro[6] and independently by Finkelstein[7] is a somewhat pleiomorphic bacillus of varying length, which occurs singly or in pairs as a rule. Chain formation is not uncommonly observed in cultures on artificial media. The organism forms no spores or capsules and it is typically Gram-

[1] See Klimmer, Ergebnisse der Immunitätsforsch. u. experimentelle Therap., 1914, i, 143–188, for details.

[2] Kendall, Jour. Med. Research, 1910, xxii, 153—for resumé and literature to 1910. See also Rahe, Jour. Inf. Dis., 1914, xv, 141.

[3] Mereschkowsky, Centralbl. f. Bakt., Orig., 1905, xxxix, 380, 584, 696; 1906, xl, 118.

[4] Kendall, loc. cit., p. 165.

[5] Loc. cit.

[6] Wien. klin. Wchnschr., 1900, v, 114.

[7] Deutsch. med. Wchnschr., 1900, xxii, 263.

positive, although in old cultures a majority of the bacilli are frequently Gram-negative.

Isolation and Culture.—The organism may be isolated directly from suspected material in 2 per cent. dextrose broth containing 0.25 per cent. acetic acid. After two or three days growth becomes apparent and a few loopfuls of the well-shaken culture are transferred to a second and then a third tube of the same medium. Usually the third transfer contains either a pure culture or it is greatly enriched with the specific organism. Pure colonies are obtained by plating upon 2 per cent. dextrose agar unadjusted for reaction, or better, upon dextrose agar containing 0.2 per cent. sodium oleate according to the procedure of Salge.[1]

The colonies are of two types—a round, smooth-edged compact colony, and a thin, semi-translucent colony with delicate filamentous edges.

Products of Growth.—Bacillus acidophilus is carbohydrophilic in its activities; it does not grow well in media containing proteins and protein derivatives only. Indol, phenols and similar products of protein degradation are not found in cultures of this organism. Gelatin is not liquefied and growth is feeble in this medium.

Frequently cultures on sodium oleate agar slants exhibit a clouding of the medium;[2] the cause of the clouding is not known.

Pathogenesis.—Escherich[3] and Salge[4] have described acute diarrheas in young children, characterized bacteriologically by large numbers of Gram-positive bacilli in the feces which are strongly acid in reàction; Escherich applied the name "Blaue Bazillose" to this type of intestinal disturbance, because of the great preponderance of Gram-positive bacilli in Gram-stained preparations prepared directly from the feces. Subsequent investigations have shown that the "blue bacilli" were in all probability Bacillus acidophilus, and it has been shown that a condition apparently identical with that described by Salge may develop in young children fed with too much maltose or malz suppe.[5]

Bacillus Acidophil-aërogenes.—Torrey and Rahe[6] have described a member of the aciduric group of bacteria which produces acid and gas

[1] Jahrb. f. Kinderheilk., 1904, lix, 399.
[2] Kendall, loc. cit., p. 156; Rahe, loc. cit., p. 9.
[3] Jahrb. f. Kinderheilk., 1900, lii, 1.
[4] Kie akute Dünndarmkatarrh des Säuglings, Leipzig, 1906.
[5] Kendall, Boston Med. and Surg. Jour., 1910, clxiii, 322.
[6] Jour. of Inf. Dis., 1915, xvii, 437.

in dextrose, lactose, saccharose and maltose; mannite was not fermented. The morphology of the organism, the types of colonies produced on dextrose agar, and its staining reactions resemble those of Bacillus acidophilus. The production of gas in the sugar mentioned and the relatively feeble growth in milk are its distinguishing cultural characteristics.

Sera of animals immunized to Bacillus acidophil-aërogenes failed to agglutinate Bacillus acidophilus, and *vice versa,* indicating that the two organisms are quite distinct entities.

CHAPTER XX.

THE DIPHTHERIA BACILLUS GROUP.

THE DIPHTHERIA BACILLUS.

Synonyms.—Corynebacterium diphtheriæ, Klebs-Löffler bacillus.

Historical.—A small group of bacteria excrete soluble extracellular toxins which produce specific disease. The first member of the group to be isolated and studied was the diphtheria bacillus. Klebs[1] called attention to the very general occurrence of a bacillus of unusual and characteristic appearance in the gray membranes usually present in the throats of severe and fatal cases of diphtheria, and a year later Löffler[2] isolated the organism in pure culture from several cases of the disease. Löffler also obtained the diphtheria bacillus from the throat of an apparently normal child, which led him to be very guarded in attributing a specific relationship of the organism to the disease. Subsequent studies by innumerable investigators have corroborated these observations in every essential detail, and have demonstrated conclusively that the diphtheria bacillus is the specific etiological organism of diphtheria. Roux and Yersin[3] discovered the soluble toxin of the diphtheria bacillus and reproduced the essential systemic phenomena of the disease in experimental animals by injecting the toxin freed from bacteria by filtration through porcelain. V. Behring and Kitasato[4] made the very important discovery that the blood serum of animals injected with gradually increasing amounts of diphtheria toxin contained a specific antitoxin which would neutralize the toxin; diphtheria antitoxin is one of the very few specific sera possessing curative properties.

Morphology.—The diphtheria bacillus is one of the very few bacteria which possess a characteristic morphology. The organisms are

[1] Verhandl. Kong. Inn. Med., Wiesbaden, II. Abt., 1883, 143.
[2] Mitt. a. d. kais. Gesamte, 1884, ii, 451.
[3] Ann. Inst. Past., 1888, 642.
[4] Deutsch. med. Wchnschr., 1890, xvi, 1113.

highly pleiomorphic bacilli, usually slender straight or slightly curved rods with rounded and frequently swollen ends. The size and shape of the individual organisms vary greatly even in the same culture; they are not uniformly cylindrical as a rule, but have club-like thickenings at one or both ends, or they are swollen in the middle and more or less pointed at the ends. Occasionally one end only is thickened, giving rise to a long, somewhat wedge-shaped rod. The distinctive morphology is best seen in eighteen- to twenty-four-hour growths on Löffler's coagulated blood serum; organisms from growths on agar are more uniform in appearance. Diphtheria bacilli observed directly in diphtheritic membranes are also less pleiomorphic than those from blood serum cultures. The organisms occur singly or in pairs, very uncommonly in short chains. The size is very variable, ranging from 0.3 to 0.8 micron in diameter and from 1.5 to 6 microns in length. The organism as ordinarily seen in diphtheritic membranes is about 0.6 micron in diameter and about 4 microns in length. Branched forms are occasionally seen, particularly in the membrane which forms on old plain broth cultures.

The stainable substance of the organisms is not uniformly distributed, but occurs in somewhat irregular concentration, giving rise to three rather distinct types of bacilli: the granular, the barred, and the solid.[1] Metachromatic granules (Ernst-Babes granules) are also present, and, according to Williams, the diphtheria bacillus reproduces by fission at one of these granules. It was originally supposed that the metachromatic granules were only found in virulent strains and that the non-virulent strains had no granules. Neisser[2] invented a stain which brings out these granules very sharply.[3] It is now known that the granules are not necessarily related to virulence, consequently the Neisser stain is rarely used. Diphtheria bacilli stain well with the ordinary anilin dyes and very characteristically with

[1] Wesbrook, Wilson, and McDaniel, Jour. Boston Soc. Med. Sc., 1900, iv, 75; Trans. Assn. Am. Phys., 1900.

[2] Ztschr. f. Hyg., 1897, xxiv, 443.

[3] The stain is prepared in the following manner:

A.—Methylene-blue (Grübler's)	. . . .	1 gram
Alcohol, 96 per cent.	. . . .	20 c.c.
Glacial acetic acid	. . . .	50 c.c.
Distilled water	. . . .	950 c.c.
B.—Bismarck brown	. . . .	1 gram
Distilled water	. . . .	500 c.c.

The smear, fixed in the flame in the usual manner, is covered with solution A for three to five seconds, washed in water, then covered with B for three to five seconds. After thorough washing in water, the preparation is ready for microscopic examination. The granules are stained blue, the bodies of the bacilli brown.

Löffler's methylene blue. With methylene blue the granules above mentioned are brought out very sharply, and it is observed that these granules exhibit the phenomenon known as metachromatism, that is, they stain mahogany red while the rest of the organism stains blue. Diphtheria bacilli are Gram-positive, but prolonged washing with alcohol removes the Gram-positive stain. Cultures prepared directly from diphtheritic membranes stain more uniformly than organisms obtained from cultures on Löffler's blood serum.

Diphtheria bacilli are non-motile, possess no capsules and form no spores. Very frequently the organisms are arranged in a definite and characteristic manner, occurring very commonly in pairs, each pair of organisms forming a configuration very similar to a capital "L," and a series of these angulated pairs are arranged in parallel,

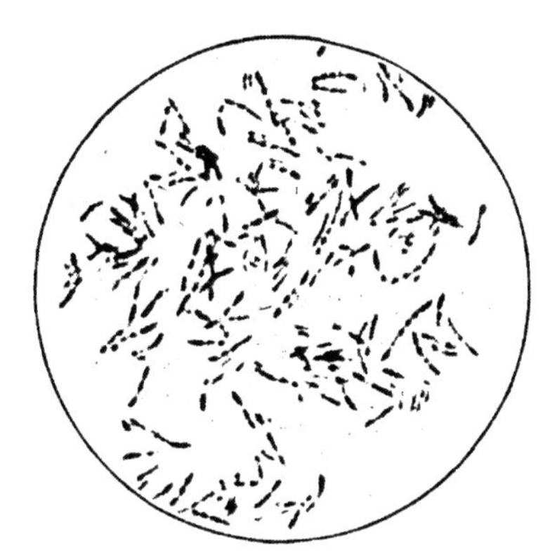

Fig. 55.—Bacillus diphtheriæ, methylene-blue stain. (X 1000.)

very much like chevrons. This angular arrangement of the organisms is due to their method of reproduction.

Isolation and Culture.—The diphtheria bacillus grows best on Löffler's alkaline blood serum and this medium is almost specific for the organism, which during the first nine to eighteen hours' incubation outgrows all other organisms with which it is usually associated in characteristic lesions, except staphylococci. Colonies of diphtheria bacilli on this medium after eighteen hours' incubation at 37° C. are gray-white, round, rather dull, with darker centres, and may attain a diameter of 1 to 1.5 mm. Diphtheria bacilli grow somewhat more slowly on plain agar, forming small, non-characteristic colonies. The organisms produce a well-marked zone of hemolysis around the individual colonies on blood agar, but the hemolytic area is smaller than that characteristic of the streptococcus. Pseudodiphtheria bacilli do not produce hemolysis on this medium[1] The growth of diphtheria bacilli in gelatin is slow, and the organisms do not produce liquefaction of the medium. In plain broth the organism grows rather slowly; repeated transfers are usually followed by the development of a pelliele which floats on the surface.[2] This pellicle may sink, but a new one usually takes its place. The growth in dextrose broth is more luxuriant than in plain broth, but no pellicle forms. There is, however, a well-marked turbidity. The diphtheria bacillus grows well in milk, producing an initial acid reaction during the first two or three days

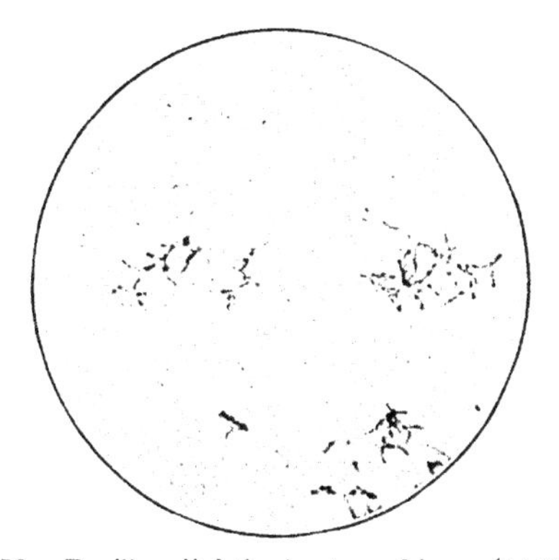

Fig. 56.—Bacillus diphtheriæ, branching. (× 800.)

of incubation, followed by the gradual development of an alkaline reaction.[3] No gross changes, however, are produced in the milk, even with prolonged cultivation. The growth on potato is very slight provided the reaction of the potato is alkaline; no growth at all takes place on acid potato.

The diphtheria bacillus is an aërobic, facultatively anaërobic organism. Its limits of growth are 17° C. as a minimum, 43° C. as a maximum, with the optimum at 37° C. Ten minutes exposure to 60° C., five minutes at 70° C., or one minute at 100° C. readily kills diphtheria bacilli. The organisms are occasionally transmissible through milk, and in this connection it should be remembered that the ordinary method of heating milk in an open vessel will not certainly kill diph-

[1] Mandelbaum and Heinemann, Centralbl. f. Bakt., Orig., 1910, liii, 356; Rankin, Jour. Hyg., 1911, xi, 271.

[2] It is essential for the production of toxin that the organisms be cultivated until they produce a pellicle, leaving the underlying medium perfectly clear and free from bacilli.

[3] Kendall, Day and Walker, Jour. Am. Chem. Assn., 1914, xxxvi, 195.

theria bacilli, for, as Theobald Smith[1] pointed out many years ago, a scum forms on the free surface of the milk, consisting of casein and lime salts, which is a non-conductor of heat. Within this membrane the diphtheria bacilli may resist a long period of heating. Diphtheria bacilli exposed to heat enclosed in a false membrane, as for example, those taken from the throat, may resist an exposure of 95° to 100° C. even for an hour. Organisms dried in this membrane may remain viable at low temperatures, protected from sunlight, from three to five months. Naked germs are readily killed by antiseptics in the ordinary concentrations, but those exposed to the action of antiseptics protected in membranes may resist for some time. Hydrogen peroxide is said to be particularly germicidal for the diphtheria bacillus.

Products of Growth.—*Chemical.*—Bacillus diphtheriæ produces acids, chiefly lactic, together with smaller amounts of formic acid from the fermentation of dextrose and maltose. Lactose, saccharose and mannite are not fermented. Neither indol nor phenols are formed in sugar-free broths,[2] but small amounts of ammonia are produced in this medium, the amount increasing with the age of the culture.[3]

Enzymes.—No enzymes acting upon proteins, carbohydrates or fats have been detected in cultures of the organism.

Toxin.—The most important and characteristic of the products formed by the diphtheria bacillus is a potent, soluble (extracellular) toxin. The potency of the toxin varies somewhat with the culture used, some strains producing more than others. An occasional strain, typical in other respects, fails to form toxin. Prolonged cultivation of the organism in artificial media may lead to a diminution in toxin-producing capacity, but this is by no means a general rule. Williams[4] isolated a diphtheria bacillus from a mild case of tonsillar diphtheria (No. 8) which has retained its toxin-producing power unimpaired up to the present time. This culture is widely used throughout the world in the commercial preparation of diphtheria antitoxin.

Conditions Favoring the Production of Diphtheria Toxin.—1. **Composition of the Medium.**—Park and Williams[5] found that 2 per cent. of peptone added to meat infusion broth increased the yield of toxin very materially, and Theobald Smith[6] made the very important

[1] Theobald Smith, Jour. Exp. Med., 1899, iv, 233.
[2] Ibid., 1897, ii, 543.
[3] Kendall, Day and Walker, Jour. Am. Chem. Soc., 1913, xxxv, 1210.
[4] Jour. Med. Research, June, 1902.
[5] Jour. Exp. Med., 1896, i, 164.
[6] Tr. Assn. Am. Phys., 1896; Jour. Exp. Med., 1899, iv, 373.

observation that the presence of muscle-sugar (dextrose), commonly found in small amounts in meat-juice, prevented the formation of diphtheria toxin; he demonstrated conclusively that small amounts of dextrose (less than 0.2 per cent.) delay the appearance of toxin; in sugar-free broth toxin production increases with the growth of the organisms. Diphtheria toxin is formed from the protein constituents of the medium; when utilizable carbohydrate (dextrose) is present in the medium, the bacilli ferment it instead of attacking the protein.[1] It is customary to add 0.1 per cent. of dextrose to broth for the production of diphtheria toxin; the initial development of the bacilli is greater, and this amount of dextrose is rapidly used up, leaving greater numbers of organisms to form toxin from the protein constituents. The culture must be grown at 37° C. to insure a potent toxin.

2. **Oxygen.**—Free oxygen is an essential factor in the production of toxin. It is customary to distribute the broth in shallow layers with a relatively large surface exposed to the air.

3. **Pellicle.**—Cultures of diphtheria bacilli which grow habitually on the surface of fluid media must be used for the preparation of toxin. Diphtheria bacilli can be "trained" to develop on the surface by repeated transfers in broth.[2] Surface development insures a maximal exposure of the bacilli to the air.

4. **Incubation.**—It requires from seven to ten days' incubation at 37° C. for the maximum accumulation of toxin. Deterioration of the toxin after this time sets in, and the formation of new toxin fails to keep pace with the recession in potency of the toxin already formed.[3]

Storage of Toxin.—At the end of the period of incubation carbolic acid or other preservative is added to the broth to kill the bacilli; they rapidly settle out, leaving a clear supernatant fluid free from bacteria, containing the toxin, which is either decanted off from the bacilli or filtered through unglazed porcelain to remove the bacteria. It is then stored in amber bottles which are completely filled and kept in cold storage. Under these conditions the toxin deteriorates comparatively slowly.

Testing Toxin.—Toxin produced by the diphtheria bacillus kills the ordinary laboratory animals, guinea-pigs, rabbits, dogs, and birds; but it is practically without effect upon rats and mice, unless the toxin is injected directly into the nervous system. The general method

[1] Kendall, Boston Med. and Surg. Jour., 1913, clxviii, 825.
[2] Theobald Smith, Jour. Exp. Med., 1899, iv, 392.
[3] Ibid.

of testing the potency of the toxin is to inject successively smaller graduated doses of it subcutaneously into guinea-pigs of two hundred and fifty grams weight and observe the results. The smallest amount of a toxin which will kill a guinea-pig weighing two hundred and fifty grams in four days is designated the minimal lethal dose (M. L. D.). The minimal lethal dose varies considerably with different strains of bacilli; in general it varies from 0.25 c.c. to 0.001 c.c. The injection of a M. L. D. of toxin leads to an edematous swelling at the site of inoculation and the animal soon exhibits generalized symptoms as well; the temperature rises, the respirations are hurried, and death ensues from the results of the toxemia. The more acute the death, the less striking the symptoms and lesions. Guinea-pigs which have died on the fourth day exhibit a marked congestion of the abdominal and thoracic viscera and of the colon. A hemorrhagic infiltration and enlargement of the suprarenals is almost pathognomonic. Frequently the stomach wall is markedly injected with blood and small ulcerations are demonstrable in the mucosa.[1] The lesions present the same general appearance when both toxin and bacilli are injected, but a false membrane, composed of bacteria and a fibrinopurulent exudate, forms at the site of inoculation. The bacilli do not spread to other parts of the body, however, but remain strictly localized. The changes in the visceral organs are attributable to the absorption of the toxin. A sub-lethal dose of toxin or an attenuated culture of diphtheria bacilli does not cause death; an ulcer forms at the site of inoculation which eventually sloughs away and is completely replaced by scar tissue.

Constitution of Diphtheria Toxin.—The composition of diphtheria toxin is unknown, although many investigations have been made upon it. Attempts to demonstrate that the toxin is non-protein in nature by growing the organisms in protein-free media have not been convincing. Small amounts of toxin have been detected in these cultures, but the well-recognized synthetic powers of bacteria make this line of evidence inconclusive. There are two current theories which receive serious consideration. One theory maintains that diphtheria toxin is enzymic in nature, the other theory assumes that the toxin is related to the proteins, particularly the globulins. The toxin is readily destroyed by exposure to light, heat, protoplasmic poisons and to peptic digestion, consequently moderate amounts of it may be swallowed without apparent harm. Acids destroy the toxin slowly, and oxidizing agents, as hydrogen peroxide, iodin and iodin trichloride, reduce the

[1] Rosenau and Anderson, Jour. Inf. Dis., 1907, iv, 1.

toxicity very materially. An exposure to 60° C. for ten hours, or at 70° C. for two hours, attenuates the toxin, and it is rapidly inactivated or destroyed at 100° C.

Protein precipitants, as ammonium sulphate and alcohol, precipitate the toxin from the broth in an insoluble state with but little reduction in potency. The precipitate, after dialysis to remove the salts, is soluble in water. A further reduction in volume and partial purification can be attained by evaporating the broth to one-tenth its original volume *in vacuo* at a temperature not exceeding 25° C. (in the dark), precipitating with alcohol, filtering and dissolving in water, and again precipitating, then drying the precipitate *in vacuo.*

Physiological Action.—The chemical constitution of the toxin molecule is unknown, and toxin can not be detected or assayed chemically. It provokes, however, a definite physiological response in susceptible animals, as guinea-pigs, and its presence is detected and its strength determined by injecting graduated doses into them, as mentioned above. From its physiological action the toxin molecule appears to consist of three components in varying amounts: (*a*) Toxin, which causes the acute symptoms of intoxication, parenchymatous degeneration and death when injected into susceptible animals. This fraction of the toxin, according to Ehrlich, has a special affinity for the antitoxin. (*b*) Toxone: the toxone causes edema at the site of inoculation and the postdiphtheritic paralyses. It combines with antitoxin more slowly than the toxin. (*c*) Toxoid: diphtheria toxin rather readily loses its toxic properties on standing, retaining its power of combining with antitoxin unimpaired, however. Toxin which is devoid of toxic power but which combines with antitoxin is called "toxoid."

Antitoxin.—*Preparation.*—The injection of the soluble toxin of the diphtheria bacillus in sublethal doses into experimental animals stimulates the formation of specific antitoxin which has both curative and prophylactic value. Antitoxin is obtained from horses because they are less susceptible to the action of the toxin than smaller animals. The serum of horses, at least in single doses, is innocuous for man, and horses furnish large amounts of blood (containing antitoxin) without injury to the animal. Young animals free from glanders, tuberculosis and other diseases are used for the purpose. Several methods are available for immunization, but the one commonly selected is carried out in the following manner: an initial injection of diphtheria toxin, either mixed with an excess of antitoxin or attenuated by iodin trichloride, is made and about a week later a second injection containing an increased amount of toxin follows. At regular intervals the injections are repeated, each time increasing the amount of toxin in regular progression until after three to four months as much as 250 to 300 c.c. of unaltered toxin is introduced at one time. After about two weeks following the last injection the animal is bled and the potency of the serum tested. If it contains one hundred and fifty units or more of antitoxin to the cubic centimeter, from two to five liters of blood are removed from the jugular vein with sterile precautions into sterile receptacles, and the animal is again treated with toxin to induce further immunization. As a rule, about two-thirds of the volume of blood taken is regained in antitoxin-containing serum. It is customary in large establishments to immunize several horses at the same time and mix this serum, for experience has shown that the serum of certain animals contains substances which cause erythematous rashes in man which are disagreeable and irritating although not necessarily harmful. Pooling the blood reduces this possibility. The serum is stored in sterile containers in a dark cold place and retains its antitoxic properties well. It deteriorates less rapidly than toxin.

Concentration.—Atkinson[1] noticed that the globulins of the horse serum increased and the albumins diminished as the antitoxin content of the blood increased, and he effected a partial purification of the antitoxin fraction by removing the albumin with protein precipitants. Gibson[2] carried the process further and obtained a serum which was about three times as rich in antitoxin per unit volume as the original horse serum. Banzhaf[3] has reduced the proportion of non-specific protein as far as is practical by purely physical agents. This reduction of non-specific proteins is important for two reasons: first, because it reduces the danger of anaphylaxis due to sensitization of the patient; and, secondly, because the rashes and joint swellings are notably reduced when the concentrated antitoxin is used instead of the whole horse-serum. It is possible to obtain the same therapeutic effect by the injection of about one-third the amount of solution when concentrated antitoxin is administered.

Properties.—Diphtheria antitoxin specifically neutralizes diphtheria toxin both *in vitro* and *in vivo*. It has little neutralizing value for the toxone, however; consequently in severe cases when it is used late,

[1] Jour. Exp. Med., 1899, iii, 649.
[2] Jour. Biol. Chem., 1906, i, Nos. 2 and 3.
[3] Collected studies from the Research Laboratory, New York City Board of Health, vols. v and vi.

it will not prevent the development of postdiphtheritic paralyses. It has both prophylactic and curative properties. It is not bacteriolytic and exhibits no agglutinins for diphtheria bacilli. Nothing is definitely known of the nature of diphtheria antitoxin. If the diphtheria toxin is a ferment, the antitoxin would appear to be an antiferment. The fact that it is precipitated with the globulin fraction of the blood serum would suggest that it may be either closely related to the proteins or a true protein itself.

Diphtheria toxin varies considerably in its potency due to the fact that it deteriorates; the antitoxin, on the contrary, is more stable. Consequently for purposes of comparison and standardization a standard antitoxin is used. Two such standard antitoxins are recognized officially: one prepared by Ehrlich in Germany; the other prepared by the United States Public Health Laboratory in Washington, D. C. Both of these antitoxins were prepared on a very large scale and preserved in a cold, dark, dry place in packages of convenient size. When the supply of one or the other of the standards is nearly exhausted a new lot of antitoxin will be prepared and carefully compared with the old. Small amounts of the standard antitoxin containing a definite number of antitoxin units are sent out regularly by the central laboratories to interested laboratories for testing purposes.

Standardization of Antitoxin.—The antitoxin unit may be defined as "that amount of antitoxin which just suffices to protect a guinea-pig of 250 grams weight against 100 times the minimal fatal dose of diphtheria toxin." The process of standardization of antitoxin of unknown potency is carried out in the following manner: diphtheria toxin, prepared as described above, is mixed in gradually diminishing amounts with a definite amount of the standard antitoxin (containing a known number of antitoxic units) and allowed to stand for twenty to thirty minutes to permit union of the toxin-antitoxin to take place. The mixtures are then injected subcutaneously into guinea-pigs of 250 grams weight. The greatest dilution of toxin which kills a guinea-pig in four days is said to be the L+ dose—that amount of toxin which will neutralize (say) 100 antitoxin units and leave an excess of toxin just sufficient to kill the animal. Having found the L+ dose of toxin (which standardizes its toxicity in terms of standard antitoxin), the same process is repeated, using this L+ dose of toxin mixed with gradually diminishing amounts of the antitoxin to be standardized. That dilution of antitoxin of unknown potency which will neutralize all except sufficient toxin to kill a 250 gram guinea-pig in four days contains 100 antitoxin units in the example cited. Knowing the dilution of the antitoxin, it is a simple problem to determine the number of units in 1 c.c. A good antitoxic serum should contain from 200 to 700 units per cubic centimeter of the unconcentrated product.

Curative Value of Diphtheria Antitoxin.—Diphtheria antitoxin should be used as early as possible in order to obtain the maximum curative effect. This is clearly set forth in the following table.[1]

Day of illness.	Treated.	Cured.	Died.	Cures. Per cent.
1	7	7	0	100
2	71	69	2	97
3	30	26	4	87
4	39	30	9	77
5	25	15	10	60
6	17	9	8	47
7-14	41	21	20	51
Indefinite	3	2	1	
Totals	233	179	54	77

According to Dönitz and others, the initial dose of antitoxin should be large; it is believed that with large doses of antitoxin even some of the toxin attached to the tissue cells can be neutralized. For this purpose 4000 units is a minimal initial dose, and severe or desperate cases are given 10,000 to 100,000 units. The antitoxin should be repeated the following day if necessary. It is better to administer too much than too little antitoxin. Antitoxin given subcutaneously is least dangerous so far as danger from anaphylaxis is concerned but the absorption is slow. Intramuscular injections, particularly in the gluteal region, are said to be more efficient curatively. In desperate cases intravenous injections of antitoxin (without carbolic acid as a preservative if possible) are indicated.

Active Immunization with Toxin-Antitoxin Mixtures.—Following a suggestion of Theobald Smith,[2] Von Behring[3] has attempted to create active immunity to diphtheria in man by subcutaneous injections of toxin-antitoxin mixtures which are neutral or but slightly toxic for guinea-pigs. The few observations which are available are on the whole encouraging, but do not justify a formal opinion of the practical value of this procedure.

[1] Quoted from Citron, Immunity.
[2] Jour. Med. Research, 1907, xvi, 359.
[3] Deutsch. med. Wchnschr., 1913, p. 873.

The Schick Reaction.—Available evidence indicates that immunity to infection with the diphtheria bacillus depends largely upon the antitoxin content of the blood, and systematic studies of the antitoxin content of the blood of infants, children and adults by Schick,[1] Park, Zingher and Serota,[2] Park and Zingher,[3] Kolmer and Moshage,[4] Bundesen,[5] and Moody[6] indicate that a large percentage—nearly 80 per cent. of young infants, 50 per cent. of children, and nearly 90 per cent. of adults—exhibit sufficient antitoxin to protect them against the disease. The demonstration of antitoxin in the blood has been simplified greatly by Schick, and modified somewhat by Park.[7] It is made in the following manner: an amount of diphtheria toxin equivalent to one-fiftieth the minimal fatal dose for a guinea-pig is made up to a volume of 0.2 c.c. in sterile salt solution and is injected subcutaneously, or preferably intracutaneously, in the flexor surface of the forearm. Immediately the skin is raised somewhat as the fluid enters the tissues. The reaction elicited depends upon the antitoxin content of the blood, a positive reaction indicating that antitoxin is absent, or present in minimal amounts appears within twenty-four hours as a circumscribed area of redness and a more diffuse area of induration measuring from one-half an inch to more than an inch in diameter. The maximum reaction appears within forty-eight hours and disappears within a week. The blood of a patient reacting in this manner contains less than one-thirtieth of a unit of antitoxin per cubic centimeter. A fainter reaction is frequently exhibited, which is interpreted to mean that the antitoxin content of the blood lies approximately between one-fortieth and one twenty-fifth of an antitoxin unit per cubic centimeter. If the antitoxin content is at least one-twentieth of a unit per cubic centimeter, the reaction is negative; only a slight reaction results due to the wound itself.

Practically, it has been found that individuals giving a negative reaction possess sufficient antitoxin to protect them from infection; nurses, doctors, ward orderlies, and patients who react negatively do not need to be immunized with antitoxin if they have been, or are, exposed to the infection. Persons giving a mild or severe reaction should be immunized with prophylactic doses of antitoxin.

[1] München. med. Wchnschr., 1913, lx, 2608.
[2] Arch. Pediatrics, July, 1914.
[3] Proc. New York Path. Soc., N. S. 1914, xiv, 151.
[4] Am. Jour. Dis. Child., 1915, p. 189.
[5] Jour. Am. Med. Assn., 1915, lxiv, p. 1203.
[6] Ibid., 1915, lxiv, p. 1206.
[7] Loc. cit.

Pathogenesis.—*Experimental Evidence of Pathogenesis.*—Löffler[1] appears to have been the first to attempt to establish the etiological relationship of the diphtheria bacillus to the disease. He succeeded in producing diphtheritic membranes on the mucous surfaces of animals by rubbing cultures on the previously injured surface. Numerous laboratory accidents, where the organisms have been inadvertently swallowed with the subsequent development of typical clinical diphtheria, complete the proof of the etiological relationship of the organism to the disease.

Animal Pathogenesis.—Laboratory animals, excepting mice and rats, are very susceptible to the diphtheria toxin. Guinea-pigs are particularly susceptible, and the subcutaneous injection of fatal or nearly fatal doses of broth cultures is followed after one to three days by the appearance at the site of inoculation of a membrane, edema, and a serosanguineous exudate. A pleuritic and frequently a pericardial exudate are found as well. There is hyperemia of the abdominal organs and a very characteristic swelling and hyperemia of the adrenals. The kidneys are also usually hyperemic. Often there are ecchymoses and even ulcers in the gastric mucosa. No bacilli are found in the internal organs. Intraperitoneal injections are less severe as a rule than subcutaneous inoculations of the same dose. There is usually some peritoneal effusion which frequently contains diphtheria bacilli. Intratracheal inoculation after mechanical injury is commonly followed by the appearance of a false membrane and the animal dies of toxemia;[2] intravaginal injection after injury of the mucosa frequently leads to a necrotic inflammation with membrane formation.[3] Repeated applications of diphtheria toxin to the conjunctiva of rabbits cause a marked conjunctivitis with membrane formation.[4]

Human.—In man diphtheria bacilli are usually localized in the false membranes, chiefly on the tonsils or pharynx, and these membranes may extend to the nose, larynx, and mouth. The organisms occasionally invade the blood stream. They may even extend into the lungs causing a true bronchial pneumonia. Occasionally diphtheria bacilli may cause rhinitis fibrinosa or simple rhinitis.[5] They also are found in occasional cases of vulvitis gangrenosa and noma

[1] Loc. cit.
[2] Fraenkel, Deutsch. med. Wchnschr., 1895, 176.
[3] Roux and Martin, Ann. Inst. Past., 1894, p. 625.
[4] Morax and Elmassian, Ann. Inst. Past., 1898, p. 219.
[5] Neumann, Centralbl. f. Bakt., 1902, xxxi, 33.

faciei.[1] Rarely, false membranes are found on the genitalia or in cutaneous wounds, in the latter case producing a true wound diphtheria.[2] The association with certain other organisms, particularly the streptococcus, the staphylococcus, and B. coli, appears to increase the virulence of the diphtheria bacillus.[3]

Diphtheria is a generalized toxemia with a local infection. The bacilli cause coagulation necrosis of the superficial cells, and an inflammatory membrane consisting of a serofibrinous exudate in which fibrin and leukocytes are prominent, together with epithelial cells, pyogenic cocci, and diphtheria bacilli in the deeper layers adjacent to the denuded epithelium. At times the membrane strips off without serious injury to the underlying epithelium, but in severe cases the membrane tears away, leaving a bleeding raw surface.

There are three principal types of diphtheria: the faucial, laryngeal, and tracheal. The incubation period is from two days to a week. An important sequela is the postdiphtheritic paralysis, which is supposed to be caused by the toxone component of the diphtheria toxin. This is anatomically a toxic neuritis and it occurs in from 10 to 20 per cent. of all cases of diphtheria from 2 to 4 weeks after the attack. There is no apparent relation between the severity of the attack and the paralysis. The pharynx is most commonly affected, next in order the eyes, leading to strabismus (ptosis). In a smaller number of cases the heart is affected. When the heart is affected the patients not infrequently drop dead as the result of cardiac failure. The early use of antitoxin usually prevents or greatly modifies the development of postdiphtheritic paralysis.

Bacteriological Diagnosis.—The principle involved in the bacteriological diagnosis of diphtheria (and the diagnosis can only be definitely established by bacteriological examination) is to make cultures from the suspected lesions on Löffler's alkaline blood serum, to incubate the culture from twelve to eighteen hours at 37° C., to stain the resulting growth with Löffler's methylene blue, and to diagnose the organisms by their characteristic morphology.

The Technic of Inoculation.—Rub a sterile swab on the under surface of the diphtheritic membrane, avoiding extraneous organisms and avoiding touching the tongue or other parts of the mouth. Smear this infected swab gently over the surface of the serum, rotating the

[1] Freymouth, Deutsch. med. Wchnschr., 1898, No. 15.
[2] Schottmüller, Deutsch. med. Wchnschr., 1895, p. 273.
[3] Theobald Smith, Medical Record, May, 1896.

swab while doing so to bring every part of it in contact with the medium. The serum is incubated at 37° C. for twelve to eighteen hours. It is customary in many laboratories to make a preliminary examination of the growth on the serum after five hours' incubation, and also to make a smear from the swab itself after the serum has been inoculated with it. By these preliminary examinations from 30 to 60 per cent. of diagnoses may be correctly anticipated. During the first eighteen hours of incubation diphtheria bacilli outgrow practically all other organisms. After this time the other organisms tend to outgrow the diphtheria bacillus.

Results.—1. *Negative.*—Negative results may be due to several factors: (*a*) the absence of diphtheria bacilli; (*b*) lack of care in taking the culture, either failure to touch the infected membrane, or making preparations immediately after the use of antiseptic gargles; (*c*) improper smears and improper stains; (*d*) poor media; (*e*) improper interpretation.

2. *Positive.*—Positive results do not necessarily prove that the patient has diphtheria for carriers of diphtheria bacilli are fairly numerous and appear to be responsible, in part at least, for the spread of the disease. From 1 to 3 per cent.[1] of healthy people harbor fully virulent bacilli in their mouths, and about 2 per cent. of all school children in large cities have them. Positive results may also be obtained with avirulent strains of diphtheria bacilli. In order to determine the virulence it is necessary to isolate the organism in pure culture and to inject two guinea-pigs respectively with a forty-eight-hour broth culture. The isolation is best made from cultures on Löffler's blood serum which microscopic examination has shown to contain diphtheria bacilli. Such a culture is emulsified in broth and streaked out on an agar plate, or, better, upon blood agar plates. After twenty-four hours' incubation diphtheria colonies are removed to plain (sugar-free) broth and incubated two days. One-half a cubic centimeter of this forty-eight-hour broth culture per 100 grams weight of guinea-pig is injected into Pig A, and a similar amount of the broth culture, mixed prior to inoculation with an excess of antitoxin, allowing half an hour for the antitoxin to unite with the toxin prior to inoculation, injected into guinea-pig B. Guinea-pig A should die in from one to five days, and an autopsy should present a typical

[1] Recent observations by Moss, Guthrie, and Gelien (Tr. XV Congress on Hyg. and Demog., 1912, iv, 156) indicate that the number of carriers of virulent diphtheria bacilli may greatly outnumber the actual cases of the disease. Their observations showed that carriers were about four times as numerous as the cases.

picture of diphtheria poisoning. Guinea-pig B should live because the diphtheria toxin is neutralized by the antitoxin.

The diagnosis of diphtheria by serological methods is not practical.

Dissemination and Prophylaxis.—Diphtheria bacilli are spread chiefly by contact or by carriers. Occasionally milk appears to be a vehicle of transmission. As a prophylactic agent for destroying diphtheria bacilli, antitoxin is one of the greatest blessings which bacteriology has conferred on medicine. Diphtheria antitoxin is used in two ways: (*a*) prophylactically; (*b*) curatively. If diphtheria breaks out in a household or a hospital, those in contact with the patient should receive prophylactic doses of antitoxin: that is, from 500 to 1500 units of antitoxin repeated after fourteen days or until all danger is over, provided the Schick test is faintly or markedly positive. (See Schick test.) Curatively, from 3000 to 15,000 units, or in severe cases 20,000 units, or even more, are used. In severe and desperate cases the antitoxin should be introduced intravenously, preferably using antitoxin prepared without preservatives for this purpose. Antitoxin must be used early. If it is used early the mortality is reduced more than 50 per cent. If the serum is used within the first twenty-four hours, the prognosis is favorable in at least 95 per cent. of the cases. The general death rate prior to the introduction of antitoxin was from 25 to 33 per cent.; since the use of antitoxin it varies from 3 to 14 per cent.

In a certain proportion of cases of diphtheria treated with antitoxin, usually from eight to fourteen days after the administration of antitoxin, rashes and painful joints develop, together with fever, angioneurotic edema, swollen lymph glands, and albuminuria. This is the so-called serum disease, which is usually particularly severe in asthmatics, in whom there occasionally develops a true bronchial spasm with respiratory embarrassment. In a few cases, less than one in ten thousand, sudden death may occur within five to fifteen minutes after the injection. At autopsy there is usually found a persistent thymus. These are cases of status lymphaticus. This sudden death is not due to the antitoxin, but to the proteins in the horse serum in which the antitoxin is contained.[1] If there is reason to suspect that the administration of antitoxin will result seriously, a few drops (not more than a quarter of a cubic centimeter) should be injected, and

[1] It should be remembered in this connection that man is less susceptible than a guinea-pig to serum diseases, and, furthermore, it ordinarily takes about 5 c.c. of horse serum to bring about the anaphylactic reaction in sensitized guinea-pigs. Proportionately, it would take 200 c.c. to induce the same symptoms in man.

the remainder after half an hour. The first small injection indicates the susceptibility of the patient; if no symptoms appear the full dose may be given with impunity; even if symptoms do appear the anaphylactic shock is aborted by the first injection and the remainder may be given at the end of an hour.

BACILLI SIMILAR TO THE DIPHTHERIA BACILLUS.

There is a group of bacteria closely related to Bacillus diphtheriæ, but differing from it either in virulence, morphology, or both. Certain of these organisms exhibit the characteristic morphology, staining and cultural reactions of the diphtheria bacillus, but do not form toxin; these strains, which are occasionally found in healthy and diseased throats, may be tentatively regarded as non-toxin-producing variants of the type organism.

In addition to the non-virulent but morphologically typical diphtheria bacilli, other bacteria have been described which resemble Bacillus diphtheriæ superficially, but differ from it in certain important details. Two principal types have been recognized: Bacillus hofmanni and Bacillus xerosis.

Bacillus Hofmanni.—Bacillus hofmanni appears to have been first observed by Löffler;[1] somewhat later Hofmann[2] studied it in considerable detail.

Morphologically the Hofmann bacillus is somewhat shorter and relatively thicker than Bacillus diphtheriæ, and more uniform in size and shape. Stained with Löffler's methylene blue, but a single unstained area is observed typically, the organism being somewhat diplococcoid in form under these conditions.

Growth is relatively more luxuriant in artificial media than that of the diphtheria bacillus, and no toxin is produced in sugar-free broth. The organism ferments no sugars, not even dextrose.

Bacillus hofmanni is found not infrequently in normal and diseased throats, and occasionally in the nasal secretion.

Bacillus Xerosis.—Bacillus xerosis, first observed by Bezold,[3] was obtained in pure culture from several cases of a chronic type of conjunctivitis known as xerosis by Kirschbert and Neisser.[4] Recently the organism has been isolated repeatedly from the healthy conjunc-

[1] Centralbl. f. Bakt., 1887, ii, 106.
[2] Wien. klin. Wchnschr., 1888, Nos. 3 and 4.
[3] Berl. klin. Wchnschr., 1874, p. 408.
[4] Breslauer ärztl. Ztschr., 1883, No. 4.

tiva and the nasal secretion. The morphological similarity between Bacillus hofmanni and Bacillus xerosis has doubtless led to confusion in the past. Knapp[1] has studied the fermentation reactions of the group and has shown that within the diphtheria group three cultural types are recognizable, as follows:

	Dextrose.	Saccharose.
Bacillus diphtheriæ	acid	alkaline
Bacillus hofmanni	alkaline	alkaline
Bacillus xerosis	acid	acid

Bacillus Hodgkini.—Hodgkin's disease. a malignant granulomatous lymphatic infection long regarded as a special type of infection with the tubercle bacillus, is now generally regarded as an infectious entity

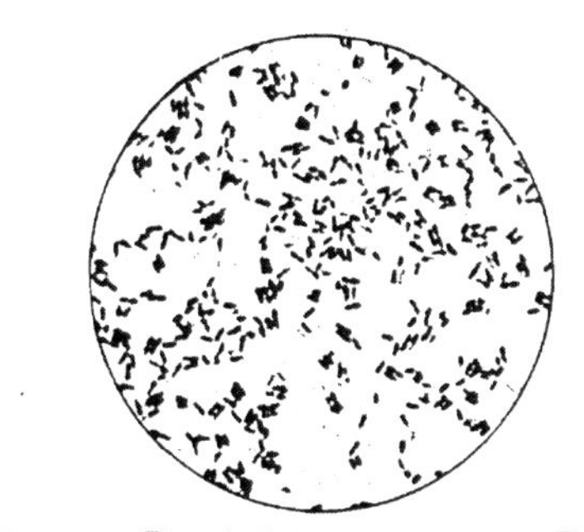

FIG. 57.—*Pseudodiphtheria bacilli.* (*Park.*)

quite apart from tuberculosis. The etiology remained obscure until Negri and Mieremet[2] published a description of a pleiomorphic, diphtheroid bacillus obtained from two undoubted cases. The organism, which was found to be Gram-positive, received the name Corynebacterium granulomatis maligni. Bunting and Yates[3] have recovered a similar pleiomorphic bacillus from several cases of Hodgkin's disease. Initial cultures were obtained upon Dorset's egg medium. Subsequent development upon ordinary media gave the following cultural reactions: gelatin not liquefied; little or no change in litmus milk; an adherent growth in broth tubes with the gradual accumulation of a slimy sediment. The colonies upon serum and agar are not characteristic.

[1] Jour. Med. Research, November, 1904, vol. xii, 475.
[2] Centralbl. f. Bakt., Orig., 1913, lxviii, 292.
[3] Arch. Int. Med., 1913, xii, 236. See also Bull. Johns Hopkins Hosp., 1915, xxvi, 376, for relation of pseudodiphtheria bacilli to leukemia, pseudoleukemia, and Banti's dlsease.

Morphologically the organism is variable in shape. Bacillary forms predominate in young cultures, but the bacilli exhibit a marked tendency toward coccoid elements after prolonged cultivation.

The etiological relationship of the organism,[1] which received the name Corynebacterium hodgkini, to Hodgkin's disease is as yet undetermined, but vaccines injected into several typical cases caused a definite recession in the size of the enlarged glands. The permanence of this recession must await final reports.

Bunting and Yates[2] injected their organism into monkeys, and a chronic lymphadenitis with an increased mononuclear and eosinophile count resulted. A polymorphonuclear leukocytosis was not observed. They conclude that the anatomical lesions were very similar to those observed in the early stages of Hodgkin's disease in man.

[1] See excellent résumé by Bloomfield, Arch. Int. Med., August, 1915, p. 197.
[2] Jour. Am. Med. Assn., 1913, lxi, 1803; ibid., 1914, lxii, 516.

CHAPTER XXI.

THE HEMORRHAGIC SEPTICEMIA GROUP.

The Hemorrhagic Septicemia or Pasteurella Goup of bacilli comprises a number of organisms which possess in common peculiarities of morphology, similarity of cultural characters and great pathogenicity for animals.

Morphologically they are short, ovoid bacilli of relatively large diameter, measuring about 0.5 to 0.8 micron in the widest part, and varying in length from 0.8 to 1.5 microns. The organisms usually exhibit marked pleiomorphism in old lesions and in old cultures. They are non-motile, uniformly Gram-negative, and exhibit a marked tendency to bipolar staining; the stainable substance is collected at the ends of the bacillus, separated by a central, faintly stainable area.

The hemorrhagic septicemia bacilli grow well upon ordinary cultural media, and they are chemically relatively inert. Indol is produced by certain types, but not by all. Gelatin is not liquefied. Acid, but no gas, is formed in dextrose, lactose and many hexoses. The fermentation of other sugars and starches has not been thoroughly studied.

The type of infection induced is usually an acute generalized septicemia, which, because of punctate hemorrhages on serous surfaces, and in the internal organs, is called hemorrhagic septicemia. Inflammation of the intestinal tract and frequently the respiratory tract is usually an important feature of the infection.

The most important animal diseases are chicken cholera, swine plague, rabbit septicemia, and a similar disease of cattle and wild herbivora. Plague is the disease of man which most closely approaches hemorrhagic septicemia of the lower animals.

The lesions caused by Bacillus pestis in experimental animals and in the naturally occurring disease in rodents present many similarities to the hemorrhagic septicemias, and occasionally a distinction must be made between the plague bacillus and other members of the group. The Indian Plague Commission state that Bacillus pestis may be differentiated from the other members of the group by its ability to develop and produce acid (but no gas) in dextrose and mannite media containing bile salts, particularly sodium taurocholate; the other organisms will not grow in this medium.

Bacillus Pestis.—Plague, the most dreaded of the acute epidemic diseases, has at somewhat irregular intervals swept over parts of the Orient, and during the earlier centuries of the Christian era even invaded Europe. The great epidemics of the third and the fourteenth centuries caused widespread death; literally millions perished, and the effect upon the residual population was most distressing. The disease has recently become endemic on the western coast of the United States, the reservoir of infection being certain rodents.

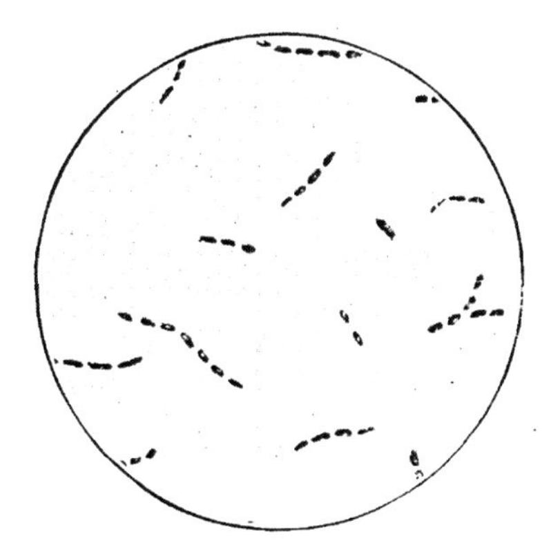

Fig. 58.—Plague bacillus, bouillon culture, methylene-blue stain showing bipolar staining. × 1000. (Kolle and Hetsch.)

The causative organism, Bacillus pestis, was isolated and described almost simultaneously by Kitasato[1] and Yersin[2] from the purulent contents of buboes, the lymph glands, the blood and the cerebrospinal fluid. Later the specificity of the organism was established by laboratory accidents and by the very comprehensive studies of the British Indian Plague Commission.

Morphology.—Bacillus pestis is a small thick bacillus with rounded ends, which occurs singly or in pairs as a rule, although short chains of three to six elements are occasionally seen. The organism is not characteristically rod-shaped, rather it approaches in outline a somewhat ovoid cell. The size varies within the comparatively narrow limits of 0.5 to 0.7 micron in diameter at the widest part and 1.5 to 1.8 microns in length. The bacilli are very pleiomorphic and exhibit great variation in size and shape according to the medium and age

[1] Lancet, 1894. [2] Ann. Inst. Past., 1894, p. 662.

of the culture. In young cultures and fresh lesions the typical ovoid shape predominates, but in older cultures and lesions considerable variation in size and outline is very common. The addition of 2 to 3 per cent. of salt to artificial media greatly increases the proportion of involution forms. Bacillus pestis is non-motile and possesses no flagella. Spores are not produced. Zettnow[1] and Albrecht and Ghon[2] state that the organism forms a capsule. The organism stains readily with anilin dyes, and it is Gram-negative. Dilute methylene blue colors the bacilli in a characteristic manner. This is best observed when the bacilli are fixed with absolute alcohol for thirty minutes in place of heating.[3] The centre of the cell is practically uncolored and the stainable substance is seen as a deeply colored granule at each

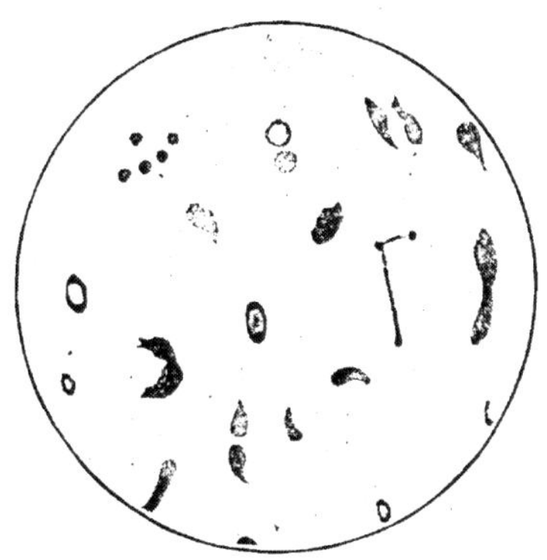

Fig. 59.—Plague bacillus. Involution forms from culture on 3 per cent. salt agar. × 1000. (Kolle and Hetsch.)

end of the rod—bipolar staining. Pleiomorphic forms are usually stained faintly or scarcely at all by this method.

Isolation and Culture.—The plague bacillus grows readily on ordinary media and pure cultures are usually readily obtained from the aspirated contents of unopened buboes or other lesions, and frequently from the blood stream in septicemic cases. Colonies on agar after twenty-four hours' incubation are small, somewhat irregular in outline, translucent, and not distinctive. Similar growths appear upon gelatin after two to three days' incubation. The medium is not liquefied. The bacilli develop with considerable luxuriance in broth forming a granular sediment in the bottom of the tube and frequently adhering to the sides. The addition of a drop of neutral oil—as cocoa-

[1] Ztschr. f. Hyg., 1896, xxi, 165. [2] Centralbl. f. Bakt., 1899, xxvi, 362.
[3] Kossel and Overbeck, Arb. a. d. kais. Gesamte, 1901, xviii, 117.

nut oil—provided the culture is maintained free from all vibration, causes a characteristic "stalactite" growth; the organisms grow down from the oil droplets as filaments (which have been likened to stalactites) until they even reach the bottom of the tube. Chains of bacilli are most characteristically developed in this medium. The addition of 2 to 3 per cent. common salt to broth or agar stimulates the formation of very irregular involution forms. Milk is not coagulated, but a slight permanent acidity gradually develops. Growth on coagulated blood serum or ascitic agar, although somewhat more luxuriant than on ordinary laboratory media, is not characteristic.

Bacillus pestis is an aërobic organism; it fails to develop with its customary vigor in the absence of oxygen. Unlike a majority of pathogenic bacteria, the optimum temperature of growth is about 30° C.; growth ceases below 10° C. and above 40° C. The viability of the organism in cadavers is considerable; they may remain alive for several weeks. In pus and sputum viable cultures may be obtained after one or even two weeks. Exposure to sunlight kills the bacilli within a few hours, and naked germs (unprotected by mucus or other protein envelope) are rapidly killed by drying. An exposure to 58° C. for an hour, or 100° C. for a few minutes is fatal: 5 per cent. carbolic acid and 1 to 1000 bichloride of mercury kill the organisms within fifteen minutes. The virulence of the bacilli diminishes rather rapidly in artificial media as a rule.

Products of Growth.—Indol is not produced in sugar-free broth. Acids, but no gas, are produced in dextrose, lactose, galactose, mannite and maltose, but not in saccharose, sorbite, dulcite, and inulin.

No *enzymes* have been demonstrated in cultures of plague bacilli.

Toxins.—Filtered cultures of plague bacilli possess little or no toxicity, although old broth cultures, freed from bacteria by filtration through unglazed porcelain, may exhibit slight toxic action. It is probable that this toxicity is referable to some endotoxin which has been liberated in the medium during the gradual autolysis of the organisms. The symptoms of plague are attributed to the action of endotoxins which are liberated within the host as the organisms disintegrate. The virulence of plague cultures is variable. Freshly isolated strains may occasionally exhibit almost no virulence for experimental animals, although as a rule they are very virulent. Prolonged cultivation upon artificial media usually results in a decided lowering of virulence, although here again exceptions are met with.[1]

[1] McCoy and Chapin, Pub. Health Bull., January, 1912, No. 53, p. 1.

Pathogenesis.—*Animal.*—"Plague is primarily a disease of rodents, and secondarily and accidentally[1] a disease of man."[2] The reservoir of plague appears to be certain rodents; the disease exists in chronic form in the marmot (Arctomys bobac) of India, and has recently been discovered in the western United States as a chronic disease in the ground squirrel (Citellus beechyi) by Wherry,[3] whose observations have been confirmed by McCoy. McCoy[4] and Chapin[5] have found that a smaller member of the squirrel family, Ammospermophilus lecurus) is also susceptible to infection with Bacillus pestis. The various members of the genus Mus—Mus norwegicus, Mus rattus, Mus alexandrinus, and probably Mus musculus, "are the producers of acute outbreaks, the conduit for the carriage of the virus from its perpetuating reservoir to the body of man." Guinea-pigs are somewhat more susceptible to infection with the plague bacillus than rodents; the disease appears to have a seasonal distribution among rodents[6] in India, and these epidemic periods coincide in time with epidemics in man. Immediately following epidemic periods considerable numbers of rodents appear to be relatively non-susceptible to infection. Rabbits[7] and monkeys[8] are also susceptible. Dogs and cats are more refractory; herbivora appear to be practically immune, at least to natural infection.

The lesions observed in rats are striking and important because plague epidemics usually appear about two weeks earlier among these animals than in man. Infection may take place through infected fleas from other rats, from ingestion of dead plague-infected animals, or by inhalation. The lesions are those of an hemorrhagic septicemia; upon laying open the animal,[9] the inguinal and axillary glands are usually enlarged (buboes), markedly injected and frequently hemorrhagic. The contents may be firm, or, less commonly, purulent. The peritoneal and pleural surfaces are red and injected, and there is an excess of fluid in both cavities. The spleen is enlarged, engorged and moderately soft, and the liver usually presents a mottled appearance, due to punctiform hemorrhages and areas of necrosis which

[1] This statement may possibly require modification in connection with the pneumonic type of plague in man (see human pathogenesis).
[2] Rucker, Public Health Reports, July 19, 1912, p. 1130.
[3] Public Health Reports, 1908, xxiii, 1289; Jour. Inf. Dis., 1908, v, 485.
[4] Public Health Bull., April, 1911, No. 43.
[5] Ibid., January, 1912, No. 53, p. 15.
[6] See Jour. Hyg., 1908, viii, 266, for details.
[7] McCoy, Public Health Bull., April, 1911, No. 43.
[8] Wyssokowitsch and Zabolotny, Ann. Inst. Past., 1897, xi, 665.
[9] Which should be done after treatment with an insecticide to kill ecto-parasites.

appear yellowish in contrast to the hemorrhagic points. A simple inspection suffices as a rule to establish a correct diagnosis, although cultures and smears should be prepared. Occasionally rats are submitted for diagnosis which are badly decomposed. The rapid overgrowth of adventitious bacteria makes the isolation of Bacillus pestis difficult by ordinary methods. Albrecht and Ghon[1] have shown that the plague bacillus, even in the presence of large numbers of contaminating bacteria, may be obtained in pure culture by rubbing the suspected material upon the freshly shaved abdomen of a guinea-pig. The plague bacillus readily penetrates the skin and causes a rapidly fatal generalized infection with characteristic lesions. It may be obtained in pure culture from the internal organs. Fritsche[2] has found that other bacteria, even of the hemorrhagic septicemia group, fail to penetrate the skin and infect the animal. This cutaneous test is of great diagnostic importance.

McCoy and Chapin[3] have described a disease superficially resembling plague in its pathological anatomy caused by Bacillus tularense. The disease is readily transmitted to guinea-pigs, rabbits and mice, less readily to rats. Wherry and Lamb[4] have recently isolated the organism from an epizoötic among wild rabbits and from a human case presenting corneal ulcerations and lymphadenitis.

Man.—The atria of infection are chiefly the skin and the respiratory tract, giving rise to two general types of the disease, glandular and pneumonic plague. Rarely a localized cutaneous lesion, plague carbuncle, is met with where the focus of localization of the organisms appears to be very circumscribed. Cases of pneumonic plague which develop sporadically during epidemics of the bubonic type do not as a rule appear to spread rapidly; on the contrary, during epidemics in which the pneumonic type predominates the infectivity from man to man is very great. No theory has been presented in explanation of this very unusual phenomenon, and the origin of the pneumonic type of the disease is not definitely known.[5] Typical pneumonic plague resembles lobar pneumonia in its symptomatology, and the fatalities

[1] Denkschrift der math-Naturw. Klasse der kaiserl. Akad. d. Wissensch., Wien., 1898, lxvi.

[2] Arb. a. d. kais. Gesamte, 1902, vol. xviii.

[3] Jour. Inf. Dis., 1912, x, 61; Public Health Bull., April, 1911, No. 43; ibid., January, 1912, 53.

[4] Jour. Inf. Dis., 1913, xv, 331; Jour. Am. Med. Assn., 1914, lxiii, 2041.

[5] Animal experiments suggest that attenuated cultures of Bacillus pestis which fail to kill guinea-pigs by subcutaneous inoculation may give rise to a fatal infection when inoculated into the respiratory tract, and that the virulence of cultures may be recovered by this process. The high mortality observed during epidemics of pneumonic plague in man may be a similar phenomenon.

are very great. Death usually intervenes in less than a week. The marked cardiac depression which is a feature of this type of plague is of some differential diagnostic value. One or more lobes are infected, the inflammation being catarrhal in nature. Enormous numbers of plague bacilli are coughed up with the sanguinoserous exudate, which are readily transmitted to doctors, nurses and attendants by droplet infection. Very frequently a generalized invasion of the blood stream occurs.

Bubonic plague, the most common type of the disease in man, is essentially a localization of plague bacilli which have gained entrance to the tissues through the skin in the regional lymph glands. The inguinal glands are more commonly invaded; next in order of frequency are the axillary, then the cervical glands. The glands are violently inflamed, and not infrequently soften and caseate. A generalized blood infection—septicemic plague—may occur either secondarily following the development of a bubo or of pneumonic plague, or, less commonly, as an initial generalized invasion following very shortly after infection and before the bubo becomes conspicuous. In such cases the organisms may be obtained in pure culture from the blood stream, and in about 20 per cent. of cases may be actually demonstrated in stained preparations of the blood by microscopical examination.

Immunity and Immunization.—Recovery from one attack of plague in man almost always confers lasting immunity. Immunity has been induced in animals—monkeys, rats and guinea-pigs—by inoculation with living avirulent cultures. Usually a moderate reaction is noticed, and a bubo may even form, but the animal recovers, and even after months successfully resists several times the fatal dose of virulent organisms. This method is far too dangerous for human practice. Bacillus pseudotuberculosis rodentium, an organism that occasionally is found in diseased rats producing lesions superficially not unlike plague, will immunize rats against Bacillus pestis and *vice versa*. This bacillus must be sharply differentiated from the plague bacillus in the microscopic diagnosis of plague in rodents. It fails to infect guinea-pigs by the cutaneous method, however, and is less virulent for laboratory animals. Cultures of plague bacilli heated to 50° C., or killed with chemicals, as phenol or alcohol, have also been employed successfully, but the degree of resistance to subsequent infection is less.[1] Specific bacteriolysins and agglutinins develop during the immunizing process.

[1] Kolle and Otto, Ztschr. f. Hyg., 1903, xlv, 507.

Active immunization of man against plague has been accomplished by Haffkine, using broth cultures of plague bacilli grown in shallow layers of broth containing droplets of cocoanut or other neutral oil on the surface to increase the development of the organisms by stalactite formation; after about six weeks' incubation, during which time several crops of bacilli develop and sink to the bottom, the culture is heated to 60° to 65° C. for an hour, and 0.5 per cent. phenol is added. 2 to 3.5 c.c. of the killed culture are injected subcutaneously into adults, proportionately smaller amounts into children. Usually a second injection is given, somewhat larger in amount, after ten days. The German Plague Commission[1] used forty-eight-hour agar cultures of virulent plague bacilli emulsified in salt solution and sterilized at 65° C.; 0.5 per cent. phenol was added as a preservative. The amount for injection into an adult was the equivalent of one agar culture of the organism. Available evidence indicates that prophylactic inoculation against plague reduces materially both the morbidity and mortality of the disease. The protection, as the statistics show, is by no means absolute, and it appears that the duration of resistance to infection is indeterminate, probably on the average several months. A serum obtained from horses immunized against plague bacilli has been prepared, but its use in man has on the whole been irregular and disappointing; the chief practical use appears to be in those cases where exposure to infection is reasonably certain, as for example, in those attending plague patients. The excessive cost of the serum is prohibitive for general use.

Transmission and Plague.—The Interim Report of the Advisory Committee for Plague Prevention in India[2] contains a very excellent summary of the mechanism of plague transmission by the flea. In bubonic plague, the most common type seen in man, the plague bacilli are locked up in the body, as it were, and can not of themselves escape to other hosts. The rat is usually the source of infection in plague epidemics, and rat fleas, Xenopsyllus cheopis, transmit the disease from rat to rat and from rat to man. When the host dies (rat or man) its ectoparasites escape if possible to living hosts. It was shown by the Indian Commission[3] that fleas from plague-infected rats frequently contained large numbers of plague bacilli in their intestinal tracts,

[1] Gaffky, Pfeiffer, Sticker, and Dieudonné, Bericht u. d. Thätigkeit der zur Erforschung der Pest im Jahre 1897, etc., Berlin, 1899.

[2] Jour. Hyg., 1910, x, No. 3.

[3] See Jour. Hyg., 1906, 1907, 1908, 1910, for a most complete discussion of the relation of fleas to the transmission of plague.

and that the bacilli were present at least three weeks after the last feeding. In the absence of fleas no infection takes place, at least in man. The bite of the infected flea may result in infection, or, since the feces of the flea are usually deposited during feeding, laden with plague bacilli, the irritant flea bite may lead to scratching of the area, resulting in the "rubbing in" of the bacilli deposited with the rat feces. Epidemics of pneumonic plague are spread by droplet infection.

Preventive measures include the appropriate care of the patient and measures to reduce the rat population. This is accomplished by careful disposal of all garbage, rat proofing all houses and granaries, and an active campaign against rats by poison, destruction of nests and runways, and the creation of rodent-free zones of considerable magnitude around settlements.

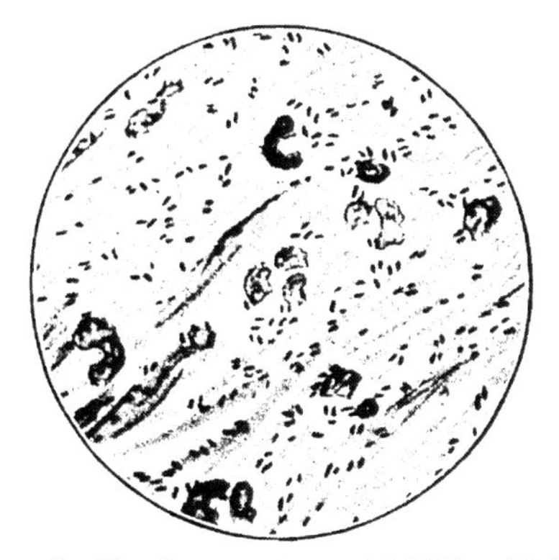

Fig. 60.—Influenza bacillus from sputum. × 1200. (Kolle and Hetsch.)

Bacteriological Diagnosis.—*Human.*—The juice of buboes,[1] of lymph glands, of petechiæ, the blood, the sputum from pneumonic cases, and occasionally the urine contain plague bacilli in large numbers. They may be obtained readily from the spleen, liver, lungs and kidneys of the cadaver.

Animal.—The postmortem appearance of plague-infected rats is very characteristic.

Microscopical Diagnosis.—The presence of Gram-negative ovoid short bacilli in considerable numbers in films prepared from material outlined above is very suggestive, but not conclusive evidence of infection with Bacillus pestis. In man the evidence is stronger than

[1] Buboes which have suppurated frequently do not contain plague bacilli, or plague bacilli in association with extraneous organisms. Even if buboes have not developed, the lymphatic glands usually contain the bacilli.

in rats, for in the latter Bacillus pseudotuberculosis rodentium, Bacillus tularense, and other organisms may be present, which produce lesions superficially not unlike those of plague.

Cultural Diagnosis.—Prepare agar plates from the contents of enlarged glands or other material incubate at 30° C. (or 37° C.), and isolate colonies in pure culture. Blood, collected aseptically, should be plated out on agar. From the pure colonies inoculate 3 per cent. salt agar and examine after twenty-four hours for involution forms; make the "stalactite" test in broth containing a few drops of neutral oil. (The culture must be kept in an absolutely quiet environment to obtain stalactite growth.) This reaction is not absolutely distinctive for other members of the Hemorrhagic Septicemia Group may also develop in this manner.

Animal Inoculation.—A small amount of culture inoculated at the root of the tail of a rat subcutaneously or intranasally in a guinea-pig will cause death within three to five days with characteristic lesions. The organism may be recovered from the internal organs. If the material for inoculation be mixed with adventitious bacteria, the cutaneous method of inoculation[1] of guinea-pigs will give positive results and the organisms may be recovered in pure culture from the internal organs postmortem.

[1] Albrecht and Ghon, loc. cit.

CHAPTER XXII.

HEMOGLOBINOPHILIC BACILLI: KOCH-WEEKS, MORAX-AXENFELD AND DUCREY BACILLI.

BACILLUS INFLUENZÆ.

Bacillus influenzæ was isolated in pure culture and described by Pfeiffer.[1]

Morphology.—It is an extremely small bacillus, one of the smallest known, measuring from 0.2 to 0.3 micron in diameter and from 0.5 to 1 micron in length. The ends are rounded and it occurs singly or in pairs, rarely in short chains. The organism is non-motile, and no flagella have been demonstrated. Spores and capsules are not produced. Ordinary anilin dyes do not color the organism readily, but Pfeiffer[2] has shown that dilute carbol-fuchsin[3] stains it readily. Stained with methylene blue or dilute carbol-fuchsin, the ends of the bacilli are colored somewhat more deeply than the centre, suggesting a bipolar distribution of the cytoplasm similar to that exhibited by the bacteria of the hemorrhagic septicemia group. The organism is Gram-negative.

Isolation and Culture.—Bacillus influenzæ is an obligately hemoglobinophilic organism; it does not grow outside the body in the absence of hemoglobin, although the amount of this substance required to encourage development may be so small in amount that it is invisible to the eye.[4] The organism may be isolated from bronchial mucus by Pfeiffer's method. The mucus is washed several times with sterile water to remove extraneous bacteria, then spread upon blood agar plates. Human, pigeon or rabbit blood added to neutral plain agar creates a favorable medium for the bacillus. The colonies which appear after twenty-four to forty-eight hours' incubation at 37° C. are very minute, clear and colorless. They may require a lens for their recognition. The hemoglobin is not visibly changed in appearance and no hemolysis occurs. Massive cultures of influenza bacilli

[1] Deutsch. med. Wchnschr., 1892, No. 2.
[2] Ztschr. f. Hyg., 1893, xiii, 357.
[3] One part carbol-fuchsin, 9 parts water.
[4] Ghon and Preyss, Centralbl. f. Bakt., 1902, xxxii, 90; 1904, xxxv, 531.

may be obtained in blood bouillon. One c.c. of sterile defibrinated pigeon or rabbit blood is added to 50 c.c. of neutral nutrient broth. After incubation to demonstrate its sterility, the medium is ready for inoculation.[1] Attempts to grow the bacilli on hemoglobin-free media have been uniformly negative; no development occurs in ordinary media.

Bacillus influenzæ is an aërobic bacillus. It has not been grown in the absence of oxygen. Growth does not take place below 25° C. nor above 42° to 43° C. The optimum temperature is 37° C. It is possible to maintain cultures by transplanting them upon fresh hemoglobin media at intervals not greater than five days. Drying is rapidly fatal to influenza bacilli; dried in mucus the organisms are not viable after one to three days. They may remain alive in moist mucus for nearly two weeks, however. Ten minutes' exposure at 57° C. kills them and ordinary chemical disinfectants, bichloride of mercury 1 to 1000, and 5 per cent. carbolic acid, destroy them in a few minutes.

Products of Growth.—The nature of the products of metabolism of the influenza bacillus are unknown. Enzymes have not been detected in cultures of the organism and soluble toxins have not been demonstrated. There is evidence that the cell substance of the bacilli is toxic; it is probable that this toxic substance is endotoxic in character.

Pathogenesis.—The direct evidence of the etiological relationship of B. influenzæ to the disease influenza rests upon a single laboratory infection with a pure culture of the organism. The hands were contaminated, and within twenty-four hours a typical attack of influenza developed. The organisms persisted in the sputum for two months.[2]

Animal.—Influenza is essentially a disease of man. Pfeiffer[3] has shown that mice, rats, guinea-pigs, swine, dogs, and cats are refractory to infection with the living organisms. The introduction of hemoglobin broth cultures of B. influenzæ through the chest wall of monkeys frequently causes a transient febrile reaction, and a catarrhal bronchitis which, however, is not clinically comparable to influenza of man. The animals recovered rapidly. Rabbits are susceptible to the endotoxin of the influenza bacillus.[4] The injection of large numbers of living or killed organisms causes dyspnea and a paralysis of the leg muscles. Frequently the animals die.

[1] Delius and Kolle, Ztschr. f. Hyg., 1897, xxiv, 327.
[2] Tedesco, Centralbl. f. Bakt., Orig., 1907, xliii, 323.
[3] Ztschr. f. Hyg., 1893, xiii, 357.
[4] Pfeiffer, loc. cit.; Cantani, Ztschr. f. Hyg., 1896, xxiii, 265.

Man.—Influenza occurs pandemically at infrequent intervals: during interpandemic intervals the organism causes somewhat localized epidemics of "grippe," and not infrequently appears to be the causative factor in "grippe colds." The bacilli persist in the respiratory tract as "opportunists" and are frequently detected in the lungs of consumptives. Invasion takes place through the respiratory tract, usually by droplet infection, and frequently spreads by continuity to the lungs, where a purulent broncho- or lobular pneumonia develops in typical eases. Pleurisy is a frequent complication, usually caused by a secondary infection with pneumococci or streptococci. The influenza bacillus rarely causes pleurisy. Enormous numbers of bacilli are coughed up in the sputum. The incubation period is brief, from one to three days as a rule. Influenzal meningitis,[1] pharyngitis and laryngitis[2] and conjunctivitis[3] are not uncommon. The occurrence of influenza bacilli in the blood has been a matter of controversy. Canon,[4] Bruschettini[5] and Ghedini[6] have isolated bacilli from the blood of patients at the height of the disease which they believe to be B. influenzæ. Slawyk[7] has reported a case of generalized infection with the influenza bacillus which would appear to confirm these observations. Other investigators have questioned the accuracy of this work and lay stress upon the incomplete diagnosis of the organisms obtained from the patients. The question can not be regarded as definitely settled at the present time.

Immunity.—Attempts to induce immunity in experimental animals have been unsuccessful. Relapses are common in man, and there is no evidence of immunity as the result of recovery from the disease.

Bacteriological Diagnosis.—1. Sputum raised from the deeper air passages is spread upon slides, air dried, fixed, and stained with dilute carbol-fuchsin. Large numbers of minute organisms colored pink with a tendency toward bipolar staining are suggestive of the influenza bacillus. There is no tendency toward a definite arrangement of the bacilli. They are frequently found in leukocytes.

2. *Cultural.*—Blood agar plates are made by depositing a generous drop of human, rabbit or pigeon's blood in the centre of an agar plate.

[1] *Pfuhl*, Ztschr. f. Hyg., 1897, xxvi, 112; Fränkel, Ztschr. f. Hyg., 1898, xxvii, 329; Jundell, Jahrb. f. Kinderheilk., 1904, lix, 777.

[2] Treitel, Arch. f. Laryngol., 1902, xiii, 147.

[3] *Pretori*, Arch. f. Augenheilk., 1907, lvii, 97; Possek, Wien. klin. Wchnschr., 1909, No. 10.

[4] Deutsch. med. Wchnschr., 1892, No. 3; Arch. f. Anat. u. Phys., 1893, cxxxi, 401.

[5] Riforma Med., 1893, viii, 783.

[6] Centralbl. f. Bakt., Orig., 1907, xliii, 407.

[7] Ztschr. f. Hyg., 1899, xxxii, 443.

Mucus raised from the deeper air passages is thoroughly washed in sterile water and emulsified in broth or water, selecting for the purpose purulent masses by preference, and streaked out radially from the drop of blood. After twenty-four to forty-eight hours' incubation the plate is examined with a lens for very minute, clear, homogeneous colonies which should be removed to blood agar slants. When growth occurs transfer some of it to plain agar. No further development occurs unless some blood has been removed with the organisms. The failure of the bacteria to develop on media free from hemoglobin is distinctive.

3. *Serological.*—The serum diagnosis of influenza has been unsuccessful.

Dissemination and Prophylaxis.—Influenza bacilli are distributed chiefly by droplet infection. Carriers are said to be common. Prophylaxis is the same as for any respiratory disease.

BACILLUS PERTUSSIS.

The etiology of pertussis (whooping-cough) has been a subject of controversy for several years. The problem is complicated by the rather general occurrence of influenza-like bacilli in the sputum and bronchial exudate from cases of whooping-cough. A clean-cut differentiation between these influenzoid bacilli and Bacillus pertussis described by Bordet and Gengou[1] has been difficult and has doubtless led to confusion in the past. It is now generally conceded that the Bordet-Gengou bacillus is worthy of serious consideration as the etiological factor of whooping-cough.

Morphology.—B. pertussis is somewhat larger than B. influenzæ, measuring 0.3 micron in diameter and varying in length from 0.5 to 1.5 microns, the average length being about 1 micron. It occurs singly and in groups, less commonly in pairs. The organism has rounded ends; frequently it is almost ovoid in shape. The organism is nonmotile and possesses no flagella. Neither capsules nor spores have been demonstrated. It stains poorly with ordinary anilin dyes and is Gram-negative. Carbol methylene blue, carbol toluidine blue and dilute carbol-fuchsin stain it readily. Methylene blue is also a satisfactory stain. The organisms stain irregularly, particularly when grown in artificial media. In young cultures and in sputum they appear frequently with the ends stained more deeply than the centre, resembling in this respect the influenza bacillus.

[1] Bull. Acad de Méd. Belgique, July, 1906; Ann. Inst. Past., 1906, xx, 731.

Isolation and Culture.—Unlike the influenza bacillus, B. pertussis can be made to grow in media which do not contain hemoglobin. For initial growths outside of the human body, however, Bordet and Gengou have recommended a potato-glycerin-blood agar medium which is claimed to be far more efficient than blood agar.[1] The Bordet-Gengou bacillus is more readily isolated from the bronchial secretion during the first paroxysms[2] than later in the disease. Cultures are obtained from bronchial mucus which has been washed several times in sterile water, then spread on the surface of the potato medium and incubated at 37° C. After twenty-four to forty-eight hours' incubation colonies appear as very minute, transparent growths which resemble dew drops; colonies of B. influenzæ frequently develop at the same time, but the colonies of the latter are somewhat larger than those of B. pertussis. Secondary transplantations of B. pertussis upon fresh potato-glycerin-blood agar grow more luxuriantly than of B. influenzæ, however, and after repeated transfers the Bordet-Gengou bacillus will grow upon ascitic agar. The influenza bacillus will not grow in media free from hemoglobin. Ordinary media, unless ascitic fluid or blood serum is added, are wholly unsuited for the growth of B. pertussis.

B. pertussis, like the influenza bacillus, is an aërobic organism. Anaërobic development has not been obtained. The optimum temperature of growth is 37° C.; Wollstein[3] states that slight development takes place even at 5° to 10° C. An exposure of thirty minutes at 57° to 60° C. prevents further development in artificial media. The organisms may remain viable upon the potato-glycerin-agar medium for two months.

Products of Growth.—According to Wollstein,[4] no acid is produced in dextrose, lactose, saccharose or mannite serum broth. No enzymes have been demonstrated in cultures of the organism, and it produces no visible changes in hemoglobin. Extracellular toxins have never been demonstrated, but autolyzed cultures introduced intravenously

[1] It is prepared as follows: 100 grams finely chopped potatoes are boiled in 200 c.c. of 4 per cent. glycerin for a short time, then cooled. To every 100 c.c. of the potato glycerin extract there is added 300 c.c. of 7.5 per cent. agar containing 0.5 per cent. NaCl. The glycerin potato extract replaces the usual peptone-meat-juice nutrients of nutrient agar. The mixture is heated to boiling, filtered, and sterilized in test-tubes about 3 c.c. of medium per tube. (Old potatoes which are slightly alkaline in reaction are much better than new potatoes which are usually acid in reaction, in preparing the glycerin extract.) To each tube of the sterilized glycerin potato agar medium is added an equal volume of sterile, defibrinated human or rabbit's blood, while the medium is still warm, 45° to 50° *C*. Then the mixture is cooled in an inclined position.

[2] Wollstein, Jour. Exp. Med., 1909, xi, 41.

[3] Loc. cit.

[4] Loc. cit.

into rabbits frequently kill them within twenty-four to forty-eight hours. Subcutaneous injections of autolysates may cause local necrosis, but generalized symptoms fail to appear. Similar results have been obtained with endotoxin obtained by grinding the bacilli to an impalpable powder and injecting a saline suspension of it.[1]

Pathogenesis.—*Animal.*—Klimenko[2] and Fränkel[3] produced a catarrh of the respiratory mucosa of monkeys and young dogs by intratracheal injections of B. pertussis suspended in salt solution. A febrile reaction appeared after three to four days and several of the animals died within two to three weeks. The bacilli were recovered from the bronchial mucus, bronchi, and from the areas of bronchopneumonia which developed in the lungs. No characteristic paroxysms were induced, although Klimenko stated that sneezing and coughing were noticed. Wollstein[4] has pointed out a possible source of error in the dog experiments: she finds that those dogs which die after injections of B. pertussis succumb to canine distemper; the lesions of the respiratory tract are readily accounted for on this basis, and the blood of the animals fails to react specifically with the Bordet-Gengou bacillus.

Human.—There are no postmortem lesions characteristic of whooping-cough. Bronchopneumonia is the most common complication seen at autopsy. Mallory and Hornor,[5] and Mallory, Hornor and Henderson[6] have advanced an interesting explanation for the paroxysms of whooping-cough. They find that the ciliated epithelium of the respiratory tract is denuded in places and the cilia plastered down to such an extent as to interfere with the free removal of mucus by the mechanical action of the bacteria. When mucus accumulates in sufficient amount, it is forcibly expelled by a prolonged violent paroxysm of coughing. These experiments were made upon animals; the frequent occurrence of B. bronchosepticus or a closely-related bacillus in the respiratory tracts of laboratory animals, which produces similar lesions to those seen in canine distemper, should be borne in mind in interpreting these results.

Immunity.—Whooping-cough is more commonly a disease of children, and recovery from one attack appears to confer lifelong immunity as a rule.

[1] Bordet and Gengou, Centralbl. f. Bakt., Ref., 1909, xliii, 273.
[2] Centralbl. f. Bakt., Orig., 1909, xlviii, 64.
[3] Münch. med. Wchnschr., 1908, p. 1683.
[4] Loc. cit.
[5] Jour. Med. Research, 1912, xxvii, 115.
[6] Ibid., 1913, xxvii, No. 4.

Bacteriological Diagnosis.—1. *Morphological.*—The diagnosis of whooping-cough by a microscopical examination of bronchial discharges is not satisfactory. Influenza bacilli are frequently present in the mucus and sputum from cases of pertussis, and no method is available at the present time which will distinguish with certainty between the two bacilli.

2. *Cultural.*—The isolation of B. pertussis from bronchial mucus upon potato-glycerin-blood agar and its ability to grow upon ascitic media free from hemoglobin separates the Bordet-Gengou organism from B. influenzæ.

3. *Serological.*—The sera of animals highly immunized to B. pertussis agglutinate the organism in high dilution, but fail to agglutinate B. influenzæ and *vice versa.* The serum of patients during and after recovery from whooping-cough, however, agglutinates B. pertussis irregularly, and the method has no general diagnostic importance. The method of complement fixation similarly has not been successful as applied to the diagnosis of the disease in man, although the reaction is clear-cut when applied to the sera of immunized animals.[1]

The etiology of whooping-cough has not been definitely established; the Bordet-Gengou bacillus, however, is found in the majority of cases of pertussis. Up to the present time it has not been isolated from healthy subjects.

THE KOCH-WEEKS BACILLUS.

Acute contagious conjunctivitis or, as it is popularly known, pink-eye is generally considered to be an infection of the conjunctiva by a small bacillus first described by Koch.[2] Somewhat later Weeks[3] described the organism anew and succeeded in growing it in artificial media, probably in association with other organisms. Kartulis[4] isolated it in pure culture on blood serum, and Kamen[5] published a more complete study of the cultural characters of the organism.

Morphology.—The Koch-Weeks bacillus is a small rod-shaped organism resembling the influenza bacillus. It is of about the same diameter as the influenza bacillus, 0.25 micron, but somewhat longer, measuring from 1 to 2 microns in length. It occurs singly and in

[1] Wollstein, loc. cit.
[2] Wien. klin. Wchnschr., 1883, 1550; Arb. a. d. kais. Gesamte., 1887, iii, 19.
[3] Arch. of Ophthalmology, 1886, xv, No. 4.
[4] Centralbl. f. Bakt., 1899, i, 449.
[5] Ibid., 1889, xxv, 401, 449.

pairs, but short chains of bacilli are not uncommonly seen in growths on artificial media. Involution forms, which are atypical in form and size, are also found in cultures outside of the body. The organism is non-motile and it has no flagella. Capsules and spores have not been demonstrated. The Koch-Weeks bacillus stains with ordinary anilin dyes, but not intensely. It is Gram-negative.

Isolation and Culture.—The organism grows best in a medium of semi-liquid consistency. 5 per cent. agar containing blood or ascitic fluid appears to be the best for this purpose. Material for inoculation is conveniently obtained first by flushing the conjunctiva thoroughly with sterile salt solution then removing some of the secretion which soon accumulates with a sterile swab which is immediately rubbed upon the surface of the blood agar. After twenty-four to forty-eight hours' incubation at 37° C. colonies usually appear which are very minute and colorless. They die rapidly.

Resistance.—The Koch-Weeks bacillus is very susceptible to drying and to heat; chemical disinfectants very rapidly destroy the organism outside the human body.

Nothing is known of the products of growth.

Pathogenesis.—Attempts to produce conjunctivitis in animals with the organism have been uniformly negative but inoculations upon the healthy conjunctiva of man usually reproduce the disease.

The disease is very contagious; it is spread chiefly by contact.

MORAX-AXENFELD BACILLUS.

In 1896 Morax[1] described a diplobacillus which he observed repeatedly during an epidemic of subacute conjunctivitis. The year following, Axenfeld[2] published an excellent description of the same organism, which is commonly referred to as the Morax-Axenfeld bacillus or the diplobacillus of subacute conjunctivitis.

Morphology.—The organisms, as the name implies, occur typically in pairs; less frequently they may remain adherent to form short chains. The individual bacilli are of average size, measuring from 1 to 2 microns in length, and about 1 micron in diameter as an average. The ends of the bacilli are rather square cut. Cultures on artificial media are somewhat variable in size and shape; chain formation is not uncommon and involution forms are frequent. The organisms are non-motile and possess no flagella. Neither spores nor capsules have

[1] Ann. Inst. Past., June, 1896.

[2] Centralbl. f. Bakt., 1897, xxi, 1.

been demonstrated. Ordinary anilin dyes color the bacilli readily, and the Gram stain is negative.

Isolation and Culture.—Growth occurs only in media containing blood serum or ascitic fluid; Löffler's alkaline blood serum is a favorable substrate. The colonies on blood serum after twenty-four to forty-eight hours' development at 37° C. are slightly sunken, due to the liquefaction of the medium. After several days the serum is almost completely liquefied. Colonies grown on ascitic agar are small, colorless and transparent even after several days' incubation. Oxygen is essential for the growth of the bacilli; no growth occurs when oxygen is excluded from the media.

Prognosis of Growth.—A proteolytic enzyme which liquefies coagulated blood serum is the only enzyme which has been described. No other products of growth are known.

Pathogenesis.—Attempts to reproduce the characteristic subacute conjunctivitis in experimental animals have utterly failed. In man the typical disease is a subacute catarrhal conjunctivitis with comparatively little pus formation, differing in this respect sharply from the acute conjunctivitis which is produced by the Koch-Weeks haeillus. The angles of the eye are inflamed, particularly the caruncles. The organisms are best detected in the secretion which collects during the night. They occur both in pus cells and free, frequently in considerable numbers. Morax appears to have reproduced the essential lesions by inoculating a drop of culture upon a healthy conjunctiva.

BACILLUS OF DUCREY.

The soft chancre, chancroid, or soft sore must be sharply differentiated from the hard chancre with which it has nothing in common. The soft chancre is a non-specific, ulcerating sore common to both sexes, particularly among the unclean. It begins as a small red spot which rapidly develops into a pustule. This pustule soon breaks down, leaving a spreading ulcer in which necrosis is a prominent feature. The ulcer spreads with considerable rapidity and is difficult to control. The adjacent and regional lymph glands usually become involved and they soon soften and ulcerate.

Ducrey[1] first called attention to a bacillus (which bears his name), which he found regularly in chancroids. In 1900 Besançon, Griffon and Le Sourd[2] succeeded in growing the organism in pure culture upon

[1] Monatsch. f. prakt. Dermat., 1889, ix, No. 9.
[2] Gaz. des Hôp., 1900, No. 14; Ann. de Dermat. et Syph., 1901, ii, 1.

blood agar. The organism is often referred to as the Streptobacillus of Ducrey.

Morphology.—The bacillus of Ducrey is a small bacillus, measuring about 0.5 micron in diameter and from 1 to 2 microns in length. It occurs characteristically both in chancroids and in culture in chains of considerable length. Frequently these streptobacilli are found in dense masses. The organism stains with ordinary anilin dyes; usually the stain is more intense at the ends of the rod, the centre being nearly devoid of color. This gives the organism a diplococcoid appearance. It is Gram-negative.

Isolation and Culture.—The Ducrey bacillus does not grow upon ordinary media, but cultures may be obtained by the method of Davis,[1] which consists essentially in sterilizing the skin over an unbroken bubo and aspirating the contents with a hypodermic needle which has been maintained at body temperature.[2] The material is introduced directly upon the surface of blood agar.[3] If the ulcers or buboes have opened, they may be cleaned with sterile gauze, dried, then painted with tincture of iodin and covered with sterile gauze. Inoculations upon blood agar are made from the pus which collects under the gauze within twenty-four hours.

The colonies are usually visible after twenty-four hours' incubation at 37° C. They appear as raised, shining, grayish droplets with a pearly lustre. They die out rapidly at room temperature, but may be kept alive at body temperature for some days. The colonies are removed from the medium with some difficulty, for they tend to slip away from the platinum needle. Subcultures tend to increase somewhat in luxuriance of growth, and by frequent transfer the organism may be kept alive for weeks provided the growths are maintained at 37° C.

The bacillus of Ducrey is an aërobic organism which is not resistant to drying. The pus becomes non-infective after one or two days' desiccation. Weak antiseptics quickly destroy it.

The products of growth are unknown.

Pathogenesis.—It is non-pathogenic for ordinary laboratory animals. Tomasczewski[4] claims to have reproduced a chancroid in a monkey

[1] Jour. Med. Research, 1893, ix, 401.

[2] Both the syringe and the blood agar should be warmed to body temperature before use, because the organism rapidly loses its viability at room temperature.

[3] Blood agar for this purpose is prepared by adding defibrinated human or rabbit blood to agar; the medium is heated to 56°-60° *C.* for thirty minutes to destroy natural bactericidal substances, and incubated for twenty-four hours to insure sterility.

[4] Deutsch. med. Wchnschr., 1903, No. 26.

(Macacus) by the injection of a blood agar culture obtained from a bubo. This same culture also produced a chancroid when inoculated into a man. Several successful inoculations in a man are recorded which appear to establish satisfactorily the etiological relationship of the bacillus of Ducrey to the lesion.

Bacteriological Diagnosis.—1. *Microscopical.*—If material be removed carefully from the base of an ulcer and spread gently upon a glass slide to prevent the breaking up of the characteristic arrangement of the bacilli in long intertwined chains, a definite diagnosis may frequently be made by direct observation of the Gram-stained preparation under the microscope.

2. *Cultural.*—Material preferably obtained from an unopened bubo should be spread upon the surface of blood agar, employing the technic outlined above. As much material as possible should be inoculated to insure growth of the bacilli.

3. *Inoculation of Patient.*—The forearm of the patient is thoroughly cleaned, then scarified with a platinum needle infected with material from the ulcer or from a pure culture. The lesion appears within twenty-four hours and it is typically developed in from three to five days. It is obvious that little or no immunity is produced, because autoinoculation results in infection. The possibility of syphilis must be borne in mind in inoculation experiments, particularly in transferring material from one subject into another. Syphilis and chancroid may exist in the same patient.

CHAPTER XXIII.

THE TUBERCLE BACILLUS GROUP: HUMAN, BOVINE, AND AVIAN.

THE ACID-FAST GROUP.

THERE is a well-defined group of bacteria characterized by the relatively large amounts of lipoidal substances contained within their bodies. These lipoidal or waxy substances confer upon the members of the group distinctive staining reactions; ordinary dyes do not stain them at all, or at best slowly. The more intense stains containing a mordant, as carbol-fuchsin, penetrate the waxy envelope, especially when heat is applied; once stained the bacteria retain the dye tenaciously even after treatment with mineral acids. This resistance to decolorization with acids has led to the name—the Acid-fast group.

Included within the group of acid-fast bacteria are saprophytic types found rather commonly in hay and manure; parasitic organisms found upon the surface of the body, as the smegma bacillus and the nasal secretion bacillus of Karlinski; and exquisitely pathogenic bacteria, Bacillus tuberculosis and Bacillus lepræ. The basis of classification therefore is chemical rather than morphological, and in this sense the definition of the acid-fast organisms does not follow a strictly natural system.

Types of Tubercle Bacilli.—Four types of tubercle bacilli are recognized which are pathogenic respectively for man, cattle, birds, and for fishes and reptiles; the human, bovine, avian, and ichthic varieties. Considerable discussion has arisen concerning the identity of the human, bovine, and avian varieties, some authorities claiming that they are identical, although modified somewhat by their continuous sojourn in a series of animals of the same kind. The evidence for this view is arrayed around the observation that tubercle bacilli of undoubted bovine type occasionally are isolated from tuberculous lesions in man (chiefly in children, infrequently in adults). On the other hand human bacilli are less commonly found in progressive tuberculous lesions of cattle. In spite of many attempts to change

PLATE II

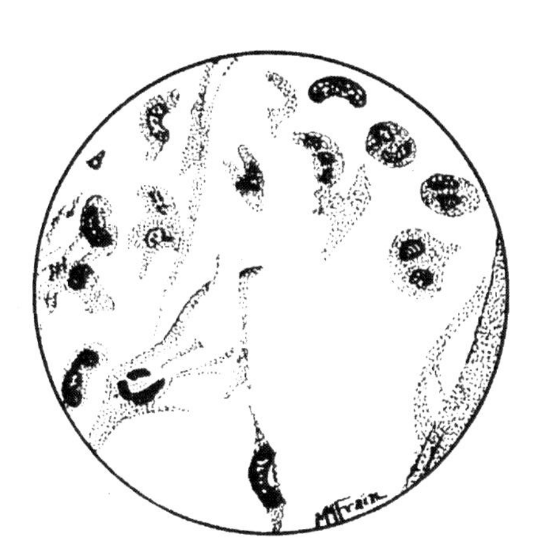

Tubercle Bacilli; Ziehl-Neelsen Stain.

one type into the other, no experiments have been reported up to the present time which are sufficiently conclusive and extensive to warrant the assumption that one variety has been permanently changed into the other. Loss or increase of pathogenic properties of one strain does not suffice to bridge the gap between it and another strain habitually pathogenic for another animal. It is very probable that the human, bovine, and avian strains had a common ancestor and that acclimatization in different animals has led to the perpetuation of three culturally and pathogenically stable varieties. The ichthic type is much more closely related to the non-pathogenic grass and dung bacilli than to the true tubercle bacilli.

TUBERCLE BACILLUS.

Historical.—One of the greatest chapters in the history of medicine was inaugurated by the isolation of the tubercle bacillus in pure culture, and the demonstration of its etiological relationship to tuberculosis. The credit for this work, which in every detail marks an important epoch in bacteriology, belongs to Robert Koch.[1]

Morphology.—The tubercle bacillus is a slender, straight or slightly curved rod measuring from 0.2 to 0.6 micron in diameter and from 1.5 to 6 microns in length. The size and appearance of the organism varies somewhat according to the source. In sputum it frequently occurs in small clumps, often with the long axes of the bacilli parallel. Occasionally a pair of bacilli are arranged at an angle like the letter "V." The bacilli are typically isodiametric, but irregularities of outline are not uncommon; these irregularities are due to nodules which cause the organism to swell or bulge wherever they occur. These nodules frequently stain deeply, and between them are areas which stain lightly or not at all, thus giving the rod a beaded or vacuolated appearance which may be so marked that the organism resembles a chain of cocci. These "beaded" forms are frequently observed in the sputum of consumptives and occasionally in old growths on artificial media.

True branching is also occasionally exhibited by tubercle bacilli derived both from the sputum and from culture.[2] Some observers have classed the tubercle bacillus with the group of Actinomyces on the basis of this branching.[3]

[1] Berl. klin. Woch., 1882, No. 15; Mitt. a. d. Kais. Ges.-Amte, 1884, ii, 1.
[2] Klein, Centralbl. f. Bakt., 1890, vii, 794.
[3] Babes and Levaditi, Arch. de méd. expér. et d'anat. path., 1897, ix, 1041.

The tubercle bacillus is non-motile and possesses no flagella. It forms no capsule but possesses a waxy envelope which confers upon the organism unusual resistance to desiccation and to the action of chemicals. No spores have been definitely demonstrated, but Koch[1] believed that the deeply staining granules found in the bacillus might be true endospores. The generally accepted view is opposed to this supposition.[2]

Staining.—Tubercle bacilli and closely related organisms possess in common a relatively large amount of waxy substance[3] which is relatively impervious even to the more intense stains, as carbol-fuchsin. Ordinary anilin dyes do not stain members of the tubercle bacillus

Fig. 61.—Tubercle bacilli, beaded forms.

group. They are Gram-positive, but it requires several hours for the anilin-oil gentian violet to color the organisms. When a stain has penetrated the substance of the tubercle bacillus it is retained with great tenacity; alcohol and even mineral acids in moderate concentration fail to remove it except after long exposure. The members of the tubercle bacillus group vary somewhat in this resistance to decolorization; the true tubercle bacilli are both "alcohol-" and "acid-fast;" other organisms in the group may be either "alcohol-" or "acid-fast." Young tubercle bacilli are frequently non-acid-fast.[4]

[1] Mitt. a. d. Kais. Gesamte, 1884, ii, 22.

[2] See Wherry, Centralbl. f. Bakt., Orig., 1913, lxx, Heft 3–4. Conditions which favor the formation of "spores" in certain acid-fast bacteria.

[3] For chemical composition of fatty substance of the tubercle bacillus see: de Schweinitz and Dorset, Jour. Am. Chem. Soc., 1898, xx, No. 8, p. 618; 20th Annual Report, Bur. Animal Ind., 1903. Levene, Med. Record, December 17, 1898, 873; Jour. Med. Research, 1901, vi, 120; 1904, xii, 251. Kresling, Centralbl. f. Bakt., 1901, xxx, 897.

[4] Wolbach and Ernst, Jour. Med. Research, 1903, x, No. 3.

The best and most universally applicable stain for the tubercle bacillus is the Ziehl-Neelsen stain.[1] It is used as follows:

1. A thin smear of the material to be examined for tubercle bacilli is prepared and fixed in the usual manner, then flooded with carbol-fuchsin and steamed gently (not boiled) for five minutes. The preparation must be flooded continuously with the stain.
2. Wash thoroughly with water to remove the excess of stain.
3. Decolorize with 90 per cent. alcohol containing 3 per cent. hydrochloric acid until the pink color has practically disappeared.
4. Wash with water.
5. Counterstain lightly with Löffler's alkaline methylene blue.
6. Wash, dry, examine.

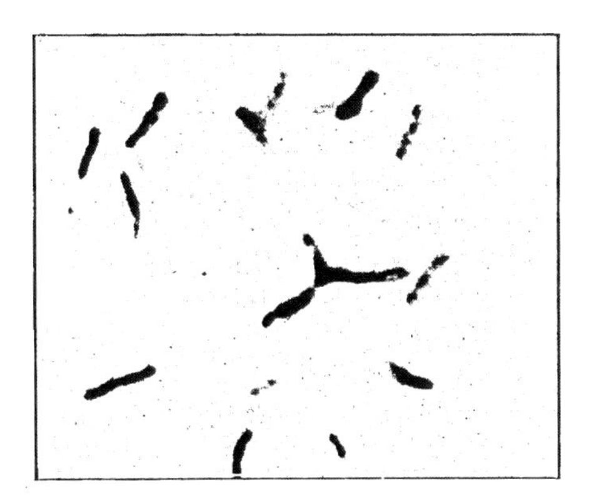

Fig. 62.—Tubercle bacillus showing branching. × 1800. (Wolbach and Ernst.)

Tubercle bacilli are colored red; non-acid-fast bacteria are colored blue. It should be remembered that spores may also be stained red by this method, but they are not likely to be confused with tubercle bacilli; they are round or oval; tubercle bacilli are much longer.

The decolorization and counterstaining may be accomplished by one operation, according to the Fränkel-Gabbett method.[2] The preparation of the smear and staining with carbol-fuchsin is carried out as above (Steps 1 and 2). Decolorization and counterstaining are accomplished by flooding the preparation with the Fränkel-Gabbett solution (100 c.c. water, 25 c.c. sulphuric acid, 50 c.c. saturated alcoholic solution of methylene blue) for three to five minutes, then wash

[1] Ziehl, Deutsch. med. Wchnschr., 1882, 451; Neelsen, Fortschr. d. Med., 1885, 200.
[2] Fränkel, Berl. klin, Woch., 1884; Gabbett, Lancet, 1887.

with water, dry and examine. Acid-fast bacteria are stained red, all other organisms are blue.

Much Granules.—Certain granules are found in old caseous foci and occasionally in the pus of cold abscesses which do not contain tubercle bacilli demonstrable by the acid-fast stain. Material containing these granules introduced into guinea-pigs causes a rapidly fatal tuberculosis. Much[1] states that these granules are living fragments of tubercle bacilli which develop into the typical bacillus when environmental conditions are optimum. They are Gram-positive and non-acid-fast, but may regain their acid-fastness. When they are non-acid-fast they do not multiply. The exact significance of these granules (Much granules) is as yet undetermined; whether they are identical with the "splinters" described by Spengler[2] is problematical. The "splinters" are usually colored red with fuchsin, and they frequently appear in tubercle bacilli that do not stain uniformly, appearing as rows of red, acid-fast granules. According to Spengler they may be found in sputum or other tuberculous material as heaps of small granules, even if the bacilli themselves cannot be demonstrated.

Isolation and Culture.—It is difficult to cultivate the tubercle bacillus directly from lesions upon artificial media and it is even more difficult to obtain pure cultures directly from sputum, feces, or lung cavities where tubercle bacilli are growing in the presence of other organisms which develop much more rapidly on artificial media. The initial growth on artificial media is the most difficult to obtain. Either coagulated dog's serum[3] or the Dorset egg medium[4] is best for this purpose. Tissue containing tubercles is removed from the animal with aseptic precautions to sterile Petri dishes. The tissue is minced somewhat and then distributed over the slanted surface of either the serum or the egg medium. At the end of a week or ten days the bits of tissue are moved around to fresh surface areas; at the end of two to four weeks the tubercle bacilli appear as minute gray nodules which gradually spread, forming eventually a wrinkled dull gray-yellow growth covering the greater part of the medium. Subcultures from the original culture grow better on artificial media than the original culture, although even subcultures grow very slowly.

[1] Beitr. z. Klinik d. Tuberkulose, 1907, viii, 85, 357, 368; 1908, xi, 67; 1913, Supp. Bd. vi.

[2] Deutsch. med. Wchnschr., 1907, p. 337.

[3] Coagulated at 75° C.; Theobald Smith, Jour. Exp. Med., 1898, iii. 647; Trans. Assn. Am. Phys., 1898, xiii, 417.

[4] Bureau of Animal Industry Annual Report, 1902, p. 574.

The coagulated serum or the egg medium may be used for subcultures; glycerin agar or glycerin potato is also suitable for this purpose. It is essential to protect the cultures from evaporation and to incubate them in a slanting position. This is best accomplished by sealing the slant cultures after they are made, either with paraffin or with corks which have been charred to kill off moulds or other organisms, then covered with lead foil. Tubercle bacilli grow fairly readily on the surface of glycerin broth after they have become accustomed to artificial media. A fresh thin film from egg medium floated on the surface of the broth is the best method of obtaining the growth in this medium. The organisms must be floated on the surface of the broth, otherwise growth does not take place. If the growth sinks to the bottom all development ceases. Tubercle bacilli do not grow readily in gelatin or other artificial media not containing glycerin or proteins derived from blood serum or egg.

Cultures of tubercle bacilli which have been grown on artificial media for some time may be gradually accustomed to develop in media of simple composition. Proskauer and Beck[1] grew the organism upon the Uschinsky medium to which glycerin was added: Wherry[2] and Löwenstein[3] have employed media in which ammonium salts were the only source of nitrogen. Kendall, Day and Walker[4] have corroborated these results. Tuberculin appears to be produced even in these simple media. The tubercle bacillus grows in milk, producing a gradual solution of the casein.[5]

The tubercle bacillus is aërobic, although it will develop slowly anaërobically. Its temperature range is rather limited, the organisms growing between 30° C. and 42° C., with an optimum temperature of 37° C. Growth below 35° C. is slow. Tubercle bacilli are fairly resistant to drying, naked germs being killed by dry heat at 100° C. only after forty-five minutes. With moist heat an exposure to 60° C. kills them in thirty minutes, 65° C. in fifteen minutes, 70° C. in five minutes, 80° C. in one minute, and 100° C. in half a minute. The organisms enclosed in mucus are much more resistant, dry heat (100° C.) killing them only after an exposure of from two to three

[1] Ztschr. f. Hyg., 1894, xviii, 128.
[2] Jour. Inf. Dis., 1913, xiii, 144; Centralbl. f. Bact., Orig., 1913, lxx, 115.
[3] Centralbl. f. Bakt., Orig., 1913, lxviii, 591.
[4] Jour. Inf. Dis., 1914, xv, 428.
[5] Klein, Centralbl. f. Bakt., Orig., 1900. xxviii, 111. Monvoisin, Compt. rend. Acad. Sci., October, 1909, xxvi; Rev. de Méd. vétерin., 1910, lxxxvii, 16. Mossu and Monvoisin, Compt. rend. Soc. Biol., 1907, lxii, No. 26. Kendall, Day and Walker, Jour. Am. Chem. Assn., 1914, xxvi, 1959.

hours, 70° C. after seven hours, and 60° C. after ten hours. In sterile water the organisms may remain alive for over two months. They are quite resistant also to putrefaction. Instances are on record where tuberculous lungs have been buried for six months and yet contained virulent organisms. Schottelius claims that a tuberculous lung buried two years contained virulent tubercle bacilli at the end of that time.

The thermal death point in milk is 60° C. for thirty minutes. There is a source of error in determining the thermal death point of the tubercle bacillus or of any other organism in milk. If the experiment is carried out in milk which is not enclosed in such a manner as to prevent surface evaporation the results are inaccurate; the scum which forms on the surface of the milk as the result of evaporation contains casein and salts; they are non-conductors of heat and protect the organisms so that they apparently resist a much higher temperature than would otherwise be the case.[1]

Tubercle bacilli in sputum are killed in twenty-four hours by mixing the sputum with an equal volume of 5 per cent. carbolic acid. Mercuric chloride is not suitable for this purpose because it precipitates mucus, forming a compound with it which renders its germicidal action *nil*. Rooms containing tubercle bacilli may be disinfected either by burning four pounds of sulphur to 1000 cubic feet in a moist atmosphere, or by evaporating 500 c.c. of formaldehyde to every 1000 cubic feet under the same conditions. The room should not be opened up until after eight hours have elapsed.

Direct sunlight kills tubercle bacilli even when they are enclosed in sputum, but the rapidity with which they are killed depends somewhat upon the season; a longer exposure is required in winter than in summer. Sputum exposed out of doors in indirect light may remain infectious for some time. In order to determine that tubercle bacilli are killed it is necessary to inoculate the material containing them into guinea-pigs, the guinea-pig being far more sensitive than artificial media for this purpose. Theobald Smith[2] has shown that it takes at least 1500 times as many tubercle bacilli to infect artificial media as it does to infect a guinea-pig. It must be remembered that even killed tubercle bacilli, as Prudden and Hodenpyl[3] have shown, produce tubercles in guinea-pigs, but that these tubercles are not transmissible

[1] Theobald Smith, Jour. Exp. Med,. 1899, iv, 233.
[2] Jour. Med. Research, 1913, xxviii, 91.
[3] New York Med. Jour., Jun e6, 1891, p. 20.

to other guinea-pigs; consequently it is necessary to inoculate a second set of guinea-pigs from the tubercles developing in the first set of pigs in order to be certain that the bacilli are killed.

Products of Growth.—*Enzymes.*—Tubercle bacilli do not produce soluble proteolytic enzymes. No carbohydrate-splitting enzymes have been observed.

Carriere,[1] Wells and Corper[2] have shown that the bodies of tubercle bacilli contain a lipase of moderate activity. Kendall Walker and Day[3] have demonstrated that the filtrates of cultures of human and bovine tubercle bacilli contain a soluble esterase; the action of the enzyme upon fats is relatively slight. This esterase is produced in an active form in media of very simple composition.[4]

Winternitz and Meloy[5] have shown that the lipase (esterase ?) activity of the blood is decreased in tuberculosis. Bauer states it is increased in the early stages of the disease.

Hemolysis.—Raybaud and Hawthorn[6] state that cultures of tubercle bacilli will not hemolyze the erythrocytes of normal guinea-pigs; the erythrocytes of tuberculous pigs are hemolyzed.

Tubercle bacilli do not form indol or the ordinary products of bacterial decomposition in ordinary media. They do not liquefy gelatin nor do they coagulate milk. Theobald Smith[7] has called attention to a very constant differential character between the human and bovine types of the tubercle bacillus. In glycerin broth the human tubercle bacillus causes a permanent acid reaction, while the bovine bacillus under the same conditions causes the medium to become alkaline if the growth conditions are suitable. Tuberculin prepared from human cultures consequently is acid in reaction, while that prepared from bovine cultures is alkaline. The organism liberates a moderate amount of ammonia incidental to its metabolism of proteins or amino acids.[8] Old cultures of tubercle bacilli occasionally are very gelatinous.[9] Vaughan,[10] White and Avery[11] and White,[12] using

[1] Compt. rend. Soc. Biol., 1901, liii, 320.
[2] Jour. Inf. Dis., 1912, xi, 388.
[3] Ibid., 1914, xv, 443.
[4] Kendall, Walker and Day, Jour. Inf. Dis., 1914, xv, 455.
[5] Jour. Med. Research, 1910, xxii, 107.
[6] Compt. rend. Soc. Biol., 1903, No. 55.
[7] Trans. Am. Phys., 1903, xviii, 108; Am. Jour. Med. Sc., 1904, cxxviii, 216; Jour. Med. Research, 1905, xiii, 253, 405.
[8] Kendall, Day and Walker, Jour. Inf. Dis., 1914, xv, 417, 423, 428, 433.
[9] Weleminsky, Berl. klin. Wchnschr., 1912, xlix, 1320. Götzl, Wien. klin. Wchnschr. 1913, 1614. Kendall, Day and Walker, Jour. Inf. Dis., 1914, xv, 428.
[10] Protein Split Products, 1913.
[11] Jour. Med. Research, 1912, xxvi, 317.
[12] Trans. 9th Ann. Meet. Nat'l. Assn. Study and Preven. Tuberculosis.

the method of Vaughan, have isolated a non-specific poisonous substance from fat-free tubercle bacilli which kills guinea-pigs with symptoms typical of anaphylaxis. The mineral constituents of tubercle bacilli have been determined by de Schweinitz and Dorset.[1]

Toxins.—The tubercle bacillus appears to elaborate both an endotoxin and an extracellular toxin.[2] The endotoxin causes necrosis, caseous degeneration and general cachexia and stimulates tubercle formation. The extracellular toxin causes fever and the acute inflammatory reaction observed around tubercles and tuberculous tissue in tuberculous animals. Little or no effect is produced in healthy animals except emaciation. The toxins liberated by the tubercle bacillus are apparently on the whole rather mild, because they produce as a rule only local lesions. This would indicate that the diffusion of toxin is somewhat limited. Furthermore, the kidneys do not ordinarily exhibit anatomical changes which could be definitely ascribed to the elimination of a tuberculous toxin through them. Whether the cachexia, which is a prominent feature of advanced cases of tuberculosis, is to be regarded as a purely toxic phenomenon is not clear. Holmes[3] has suggested that the fatty acids of the tubercle bacillus cause a lymphocytosis.

Pathogenesis.—*Human.*—According to Naegeli,[4] rather more than 90 per cent. of adults who come to autopsy show scar tissue at the apices of the lungs, which he believed were healed tubercles. Later observations have not fully confirmed these figures, but it appears that fully 50 per cent. of adults have healed tubercles at this site.[5] Frequently virulent tubercle bacilli have been isolated from the centre of this scar tissue, but it should be remembered that occasionally virulent tubercle bacilli have been isolated from bronchial lymph nodes which appear to be normal.

Modes of Infection.—*Hereditary Transmission.*—Transmission of the tubercle bacillus through the sperm has never been established; transmission through the ovum is also not definitely established. The maternal blood, on the contrary, appears to be a vehicle through which tubercle bacilli may pass, or grow through the placental barrier and thus reach and infect the fetus.[6]

[1] Jour. Am. Chem. Assn., 1898, xx, 618.
[2] Armand-Delille, Monographies Cliniques en Médicine, etc., 1911, No. 66, Paris.
[3] Guy's Hospital Reports, 1909, lix, 155.
[4] Virchow's Arch., 1900, clx, 426.
[5] Lubarsch, Virchow's Arch., 1913, ccxiii, 417.
[6] See Gärtner, Ztschr. f. Hyg., 1893, xiii, 126–139, for summary and discussion of early literature. Also, Schmorl and Geifel, München. med. Wchnschr., 1904, 1676.

Latency Theory.—Baumgarten[1] believes that tubercle bacilli may lie dormant in the body for months or years and become active when the "resistance" of the body is lowered. The evidence is on the whole opposed to this view, partly because congenital tuberculosis is uncommon, chiefly because the organs of fetuses of tuberculous mothers do not cause infection in guinea-pigs. More recently v. Behring[2] has advanced the theory that infection takes place in childhood, probably by ingestion of milk containing tubercle bacilli, and that the manifestations of infection become apparent later in life.

Inoculation Theory.—Direct inoculation through the skin is rare.

Inhalation Theory.—Droplet infection and infection by dust containing viable tubercle bacilli appear to be the most common methods of transmission of the organism.

Ingestion of milk or meat containing tubercle bacilli must be considered as important methods of transmission of the organism.

Trauma, establishing a locus minoris resistentiæ to which tubercle bacilli lying dormant in the body may be transported and set up infection, is probably uncommon.

Conditions Favoring Infection.—Overcrowding with its attendant evils of dark, damp rooms, poor food and general unhygienic conditions appears to be a most potent factor in the spread of tuberculosis. No age is exempt, although the disease is somewhat less frequent between the ages of five and ten years, greatest between sixteen and thirty-five years. The sexes are about equally infected. Negroes are especially prone to the disease, possibly because of their surroundings and manner of living rather than any inherent lack of resistance. Tuberculosis is relatively uncommon among the aboriginal negro races in Africa. Jews appear to be relatively immune to the disease. Those occupations in which dust is generated in large amounts exhibit a higher incidence of the disease than occupations in which dust is not a feature. Catarrhal infections of the respiratory tract appear to predispose to pulmonary tuberculosis as do measles, whooping-cough and influenza.

Atria of Invasion.—The respiratory and digestive tracts (including the tonsils) and the skin are the three portals through which tubercle bacilli enter the tissues of the body. Of these the respiratory tract is more frequently involved. Droplet infection is by far the most common method of transmission of tubercle bacilli; dust-borne infection is probably relatively uncommon.

[1] Deutsch. med. Wchnschr., 1882, No. 22.
[2] Ibid., 1903, 692; 1904, 194.

Tubercle bacilli also enter the intestinal tract and they may pass through the intestinal mucosa without leaving any trace of their passage, particularly if they be suspended in fatty menstrua, as butter or cream.[1] The bovine type of the tubercle bacillus may enter through the tonsils, or the digestive tract occasionally. Rarely, tubercle bacilli enter through the skin, usually causing somewhat localized epidermal proliferations containing tubercle bacilli in small numbers, which are sometimes called butcher's warts, postmortem warts or verruca necrogenica. Usually they remain localized.

COMBINED TABULATION, CASES REPORTED AND OWN SERIES OF CASES. (PARK AND KRUMWIEDE.)

Diagnosis.	Adults 16 years and over		Children 5 to 16 years.		Children under 5 years	
	Human	Bovine.	Human.	Bovine.	Human.	Bovine.
Pulmonary tuberculosis	568	1?	11	..	12	
Tuberculous adenitis, axillary or inguinal	2	..	4	..	2	
Tuberculous adenitis, cervical	22	1	33	20	15	20
Abdominal tuberculosis	15	3	7	7	6	13
Generalized tuberculosis, alimentary origin	6	1	2	3	13	10
Generalized tuberculosis	28	..	4	1	28	5
Generalized tuberculosis, including meninges, alimentary origin	..	..	1	..	3	8
Generalized tuberculosis, including meninges	4	..	7	..	45	1
Tubercular meningitis	..	..	2	..	14	2
Tuberculosis of bones and joints	18	1	26	1	21	
Genito-urinary tuberculosis	11	1	1			
Tuberculosis of skin	1	..	1	..	1	
Miscellaneous cases:						
Tuberculosis of tonsils	..	..	..	1		
Tuberculosis of mouth and cervical nodes	..	1				
Tuberculous sinus or abscesses	2					
Sepsis, latent bacilli	..	..	..	..	1	
Totals	677	9	99	33	161	59

Mixed or double infections, 4 cases.
Total cases, 1042.

For some years much discussion has centred upon the incidence of bovine tubercle bacillus infection in man. Koch was inclined to the view that infection with this organism was so rare as to be practically negligible. Later he modified his opinion. Weber[2] studied 628 cases

[1] Nicolas and Descos, Jour. Physiol. et Path. gen., 1902, iv, 910; Ravenel, Jour. Med. Research, 1903, x, 460.
[2] Tuberkulose, Arbelten a. d. kais. Gesamte, 1910, Heft x.

(284 children, 335 adults, 9—age unstated), all of whom had drunk the milk of cows having tuberculosis of the udder, or had consumed uncooked products made from the milk. Only two patients, both very young children, were definitely shown to be infected with bovine tubercle bacilli. Both had enlarged caseous cervical glands from which the organism was isolated and identified. Six children and one adult had glandular swelling in the neck, but the evidence was not conclusive that bovine infection had taken place. The general conclusion was that there was relatively little danger from drinking milk containing viable bovine tubercle bacilli.

Much more convincing are the studies of Park and Krumwiede.[1] The accompanying table (see preceding page), which is their summary of their own extensive investigation and a recapitulation of authentic observations of others, shows very definitely that infection with bovine bacilli is relatively common in children and young adults up to sixteen years of age, but relatively uncommon in adults.

Bovine bacilli are found not only in unpasteurized market milk,[2] but also in the glandular organs of a considerable proportion of cattle and swine. The muscles are usually not invaded. No meat from tuberculous animals can be offered for sale in the public markets, however.

Lipschutz[3] has reported a case of cutaneous infection by the avian tubercle bacillus in man which resembled leprosy anatomically. The diagnosis was arrived at only after an exhaustive study of the organism. Infection of man with the avian tubercle bacillus is uncommon.

The mechanism of infection with the tubercle bacillus has been the subject of much controversy. It is apparent that the acid-fastness of the organism *per se* does not confer pathogenic properties upon the organism because other non-pathogenic acid-fast bacteria are unable to induce progressive disease from man to man or from animal to animal. Acid-fastness, however, may be an initial factor in pathogenism, an opening wedge as it were, for it appears to be well established that acid-fast bacteria are relatively insoluble in body fluids and remain unchanged for considerable periods of time when they are introduced into the animal body. Theobald Smith[4] has advanced a tentative hypothesis which explains satisfactorily many of the

[1] Trans. 6th Ann. Meet. Nat'l. Assn. Study and Preven. Tuberculosis.
[2] See Kober, Trans. Am. Phys. for literature to 1903. Hess, Jour. Am. Med. Assn., 1909, lii, No. 13, 1011. Moore, Jour. Med. Research, 1911, xxiv, 517.
[3] Arch. f. Dermat. u. Syph., June, 1914, cxx.
[4] Jour. Am. Med. Assn., April 28, 1906.

phenomena. As tubercle bacilli reach the body (and as they escape from the body) they are surrounded by a protective envelope which causes the organism to behave somewhat as an inert foreign body until it finally settles down in some structure where it can grow. The envelope is then slowly removed or modified by the action of normal tissue fluids and growth commences. In this connection it is interesting to note that young tubercle bacilli are frequently non-acid-fast,[1] and that the tissues usually invaded by the bacilli—lymphoid tissue and the lungs—contain active lipase.[2] If this supposition is correct, the tubercle bacillus may remain latent in the body until the fatty capsule is removed or modified, perhaps by a fat-splitting enzyme (lipase); then development takes place. It should be remarked parenthetically that polymorphonuclear leukocytes which occasionally engulf tubercle bacilli do not contain lipase;[3] these leukocytes may transport the organisms to lymphoid tissue or other tissue where eventually the bacilli escape, thus establishing new foci. Mononuclear leukocytes appear to contain lipase, as do certain fixed phagocytic cells in the alveoli of the lungs.

Lenk and Pollak[4] and Wiener[5] appear to have found active proteolytic ferments in tuberculous exudates. Opie and Barker[6] have shown that the mononuclear epithelioid cells contain an enzyme which digests protein in a slightly acid medium; it is practically inert in an alkaline medium. Jobling and Petersen[7] have found that the inhibition of enzyme action in caseous tubercle foci is apparently due to unsaturated fatty acids. Saturation of these acids with iodin causes an acceleration of the activity of the ferments.

The primary lesions usually tend to progress slowly. Secondary invasion by tubercle bacilli through the lymph and bloodvessels frequently occurs, causing tuberculous foci in various ducts and glands of the body, as the bronchi, alveoli of the lungs, spleen, liver, tubules of the kidney, and in the genito-urinary system, particularly the epididymis and testicle of the male and the Fallopian tubes in the female. The glandular organs are those most commonly infected, and of these the lungs and lymph nodes are most frequently involved;

[1] Wolbach and Ernst, Jour. Med. Research, 1903, x, 313.
[2] Bradley, Jour. Biol. Chem., 1913, xiii, No. 4. Briscoe, Jour. Path. and Bact., 1907, xii. Bartel and Neumann, Centralbl. f. Bakt., Orig., 1909, xlviii, 657. Zinsser and Carey, Jour. Am. Med. Assn., 1912, lviii, 692.
[3] Fiessinger and Marie, Compt. rend. Soc. Biol., 1909, lxvii, 177. Bergell, München med. Wchnschr., 1909, lvi, 64.
[4] Deutsch. Arch. klin. Med., 1910, cix, 350.
[5] Biochem. Ztschr., 1912, xli, 149.
[6] Jour. Exp. Med., 1908, x, 645; 1909, xi, 686.
[7] Ibid., 1914, xix, 383.

also the spleen, kidneys, liver, meninges both of the cord and brain, the pleural and pericardial cavities, the genito-urinary apparatus, and, less frequently, joints and bones. The muscles are only very rarely invaded. Various clinical names have been applied to tuberculosis of different tissues: tuberculosis of the lungs is commonly designated consumption; of the spine, Pott's Disease; of the cervical lymph glands, scrofula; and of the skin, lupus. The characteristic initial lesion is a small nodule or tubercle which may undergo secondary changes, as caseation, calcification, ulceration, or various types of sclerosis. In the lungs the first organisms that reach the alveoli may leave no trace. They are dissolved there apparently, but may produce no progressive lesion. A second invasion in the same area frequently causes a local inflammation which usually results in infection, apparently because the body has been sensitized by the first bacilli that entered, and in some way is rendered locally susceptible to the organism.

The irritation caused by the extracellular toxin excreted by the tubercle bacillus brings about a response on the part of the tissues which is protective, as is manifested by a walling off of the bacilli. First there is a proliferation of the connective tissue which forms a spherical mass of epithelioid cells around the focus of infection. Outside of the epithelioid cells there is usually an infiltration of lymphocytes. The tissue is avascular and the young tubercles contain little or no fats.[1] The central part of the tubercle soon begins to undergo coagulation necrosis, probably due to the action of the intracellular toxin, and it is gradually converted into a homogeneous, cheesy mass. In many tubercles giant cells are found, which are formed either by the coalescence of several epithelioid cells, or by atypical cell division, the nucleus dividing faster than the cytoplasm. The nuclei of the giant cell are arranged peripherally as a rule, either completely around the cell, or in the shape of a horseshoe. The centre of the giant cell likewise may undergo caseous degeneration, and tubercle bacilli are not infrequently found in the middle of these cells.[2] According to Zeit, giant cells are essentially blind blood capillaries which have extended into the tuberculous area, but have not become true vessels because the toxins of the organisms have prevented the final development of functional blood channels. Besides these small

[1] Joest, Virchow's Arch., 1911, cciii, 451.

[2] See Evans, Bowman, and Winternitz, Jour. Exp. Med., 1914, xix, 283, for a critical experimental study of the histogenesis of the miliary tubercle in vitally stained rabbits for the finer details of the process.

miliary tubercles, larger areas of caseation may develop; epithelioid cells, lymphocytes, and giant cells are usually found closely packed around these areas.

The destruction of the capillaries and the resulting avascular tissue helps in the necrosis of the tubercle by cutting off the blood supply.

What is generally known as consumption or destruction of the lung tissue is probably not due to the action of the tubercle bacillus alone, but to secondary infection and liquefaction of tissue by other organisms, as the streptococcus, staphylococcus, pneumococcus, or Micrococcus tetragenus. If a caseous necrotic tubercle located near a bronchus ruptures into this bronchus, a large amount of tuberculous material is suddenly swept into the regional areas of the lung, overwhelming it and setting up a rapidly fatal infection which is known as galloping consumption or phthisis florida. If a caseous tubercle ruptures into a lymph or bloodvessel, the material may be carried very widely through the body, causing generalized miliary tuberculosis, which resembles typhoid fever clinically. Hemorrhage not infrequently takes place from the lung, due to the erosion and subsequent bursting of a bloodvessel which may have been included in the caseous area. In the human lung it is practically always possible to find old lesions at the apices when the infection is due to the human type of the tubercle bacillus. Uncommonly no old healed tubercles can be found, and the lungs are filled with miliary tubercles, in which case the infection is usually caused by the bovine type of the tubercle bacillus. Tubercle bacilli ingested with milk or other foods may cause tubercle formation in the mesenteric glands with lesions in other parts of the body. Metastatic nodules are found occasionally in the brain, meninges, and epiphyses of bones.

Pathogenesis for Lower Animals.—Generally speaking, the human type of the tubercle bacillus is less virulent for lower animals than the bovine type. Monkeys in captivity, however, are susceptible to both types, and even infection with the avian type has been found in them. The course of the disease, which is spontaneous, runs similar to that of human consumption, with, however, a greater tendency toward generalized invasion. Goats, sheep, and horses are not as a rule infected with the human tubercle bacillus. Cattle are very rarely infected with the human type. Dogs and cats are said to be infected occasionally.

Rabbits.—Rabbits are not as susceptible to the human tubercle bacillus as the guinea-pig. Subcutaneous injections of the human tubercle bacillus usually causes only local lesions, which in the vast majority of instances are not fatal and clear up after some weeks. Massive doses, however, usually produce lesions. Intravenous injection, unless massive doses are given, also fails to kill rabbits as a rule. Occasionally, however, a generalized tuberculosis with fatal termination results. Intraperitoneal inoculations only occasionally bring about a generalized fatal tuberculosis. Usually slight lesions are produced which clear up spontaneously. Ingestion of human organisms rarely leads to infection.[1]

Guinea-pigs.—Guinea-pigs are very susceptible to infection with either the human or bovine tubercle bacilli, although the disease rarely appears spontaneously. Theobald Smith[2] has shown that it requires but one fifteen-hundredth as much tuberculous material to infect a guinea-pig as is required to infect artificial media. This susceptibility of the guinea-pig to inoculation with the tubercle bacillus explains the well-attested fact that inoculation of suspected tuberculous material into these animals is a far more delicate test for their presence than attempts to grow the organisms from the same material on artificial media. It must be remembered in this connection that even dead tubercle bacilli stimulate tubercle formation in guinea-pigs,[3] hence for an absolutely safe diagnosis whatever tubercles are produced in the first guinea-pig must be ground up and injected into a second guinea-pig. If viable tubercle bacilli are present a successful infection will take place, otherwise the experiment is negative.

Subcutaneous Inoculation.—After ten to fourteen days a small hard nodule appears at the site of inoculation, and very soon afterward the regional lymph glands begin to enlarge and the animal begins to lose weight. The animal usually dies in from two to four months. Postmortem, the spleen is enlarged, yellowish-brown in color, and studded with tubercles, some minute and gray, others larger, yellowish and frequently caseous. The regional lymph nodes also have usually undergone caseation, particularly the inguinal glands, less commonly the axillary glands. The liver usually has a few rather large caseous or fibrinous tubercles, particularly on the free border. The kidneys also may have a few tubercles. If the animal has lived two or three months the thoracic cavity is also invaded, and scattered miliary tubercles may be seen on the lungs. The mesenteric, bronchial, sternal and cervical glands are invaded.

[1] Theobald Smith, Jour. Med. Research, 1905, xiii, 253.
[2] Jour. Med. Research, 1913, xxviii, 91.
[3] Prudden and Hodenpyl, New York Med. Jour., 1901, and others.

Intraperitoneal Inoculation.—The disease runs a more rapid course, death usually taking place in from three to eight weeks. The peritoneum is chiefly involved, particularly when death takes place early. The omentum is thickly studded with tubercles which tend to become confluent and to caseate. Certain mesenteric glands also enlarge and become caseous. As in the subcutaneous inoculation, the inguinal and axillary glands may be involved, but the lesions do not progress so far.

Ingestion.—The lesions usually resemble those of intraperitoneal infection, with, as a rule, more marked lung involvement.

Inhalation and Pulmonary Inoculation.—The lungs contain confluent tubercles, many of which are caseated. Not infrequently one or more entire lobes may be involved. Cavity formation, however, is uncommon. The abdominal viscera, particularly the spleen, are involved, as well as the regional lymph glands.

Products of Clinical Importance Derived from the Tubercle Bacillus.—Old Tuberculin (O. T. Koch).—Four to six weeks' pure culture of the tubercle bacillus grown in 5 per cent. glycerin broth is killed by heating to 110° C. for half an hour, and then evaporated to one-tenth its original volume on the steam bath. It is then filtered through sterile, unglazed porcelain filters. The resulting fluid, which is dark brown in color, syrupy in consistency, and which keeps *in the undiluted condition* in the cold and away from sunlight for months apparently unchanged, is known as old tuberculin. Old tuberculin contains the water and glycerin-soluble products of metabolism of the tubercle bacillus and products of autolysis of tubercle bacilli which are not precipitated by heat, as well as unchanged concentrated constituents of the broth and about 50 per cent. of glycerin: 0.25 to 0.50 per cent. tricresol is added as a preservative. The nature of the reactive substance or substances in tuberculin is unknown. The composition of tuberculin even when prepared by a uniform technic appears to be variable.[1] Tuberculin prepared from the human type of the tubercle bacillus is acid in reaction; that from the bovine type is alkaline.[2]

THE NATURE OF TUBERCULIN.—Tuberculin[3] appears to be a true product of the metabolism of the tubercle bacillus. It is thermostabile, dialyzable, insoluble in alcohol, gives no biuret reaction, and is pre-

[1] White and Hollander, The Chemical Composition of Commercial Tuberculins, Trans. 9th Ann. Meet. Natl. Assn. Study and Preven. Tuberculosis.

[2] Theobald Smith, Jour. Med. Research, 1905, xiii, 405.

[3] The word "tuberculin" appears to have been used first by Pohl-Pincus, Deutsch. med. Wchnschr., 1884, 108.

cipitated by certain alkaloidal precipitants as tannic acid, potassium mercuric iodide and mercuric chloride in acid solution. It is decomposed by pepsin HCl and by trypsin in alkaline media. Proskauer and Beck,[1] and Löwenstein and Pick,[2] and others have shown that tuberculin is produced by the tubercle bacillus when this organism is grown in a protein-free medium. They suggest that it is probably a polypeptid.

Whether tuberculin contains a true toxin or an endotoxin, or a mixture of both toxin and endotoxin is not clearly settled. Pick believes it contains a true toxin secreted by the tubercle bacillus.

Variants of Old Tuberculin.—A number of observers, impressed with the possibility that the reactive substances of tuberculin might be changed by heat, have attempted to produce tuberculin which has been unheated.

(*a*) *Bouillon Filtrate Denys* (*B. F.*).—This is unheated, unconcentrated old tuberculin prepared as above and sterilized by passage through sterile porcelain filters.

(*b*) *Vacuum Tuberculin.*—The six weeks glycerin broth culture of the tubercle bacillus is concentrated to one-tenth its volume *in vacuo* and filtered. By so doing the advantages of concentration are obtained without the disadvantages of heating.

The action of old tuberculin and its variants would suggest that it does not contain all the necessary elements for the establishment of true immunity, and this has led to the production of a series of new products, new tuberculins, which attempt to retain the more specific products of the tubercle bacillus. The principle involved is to grind up dried tubercle bacilli in ball mills to an impalpable powder, and to suspend or partially dissolve this powder in salt solution with or without the addition of glycerin.

New Tuberculin (T. R. Koch).—Young virulent tubercle bacilli are dried first between sheets of sterile filter paper, then *in vacuo* over sulphuric acid and phosphorus pentoxide until thoroughly anhydrous, then ground in a mortar until a dry powder is obtained. This powder is suspended in water, thoroughly mixed, and then centrifugalized. The first supernatant fluid obtained (T. O.) is rejected. This preliminary grinding with water is intended to wash out the water-soluble substance. The residue is then repeatedly ground and centrifugalized, saving the supernatant liquid each time until all of it has passed into

[1] Ztschr. f. Hyg., 1894, xviii, 128.
[2] Biochem. Ztschr., 1911, xiii, 142.

solution and suspension. This constitutes new tuberculin. The new tuberculin is finally prepared of such a strength that 1 c.c. of the dried residue will contain 0.002 gram of solid material. It is customary to add glycerin and a small amount of formaldehyde to the preparation before it is finally made up to strength.

At times new tuberculin has been found to contain living, virulent tubercle bacilli, although they are killed by prolonged exposure to formalin. New tuberculin was originally intended for curative purposes, as it was found to be relatively free from the toxic substances which are found in old tuberculin. The theoretical inherent dangers of this preparation, however, have tended in the past to limit its use.

Bacillus Emulsion (**B. E.**).—This is an emulsion of untreated tubercle bacilli which are washed and dried as for new tuberculin. The organisms are ground thoroughly and then suspended by continual grinding in physiological salt solution containing about 20 per cent. of glycerin. From 0.25 to 0.5 per cent. carbolic acid is added to kill whatever tubercle bacilli or other organisms may have been included in the preparation. For use it is standardized so that 1 c.c. of the solution contains the equivalent of 0.001 gram of dried tubercle bacilli. It will be seen that the bacillus emulsion contains both new tuberculin and a certain amount of the water-soluble products of the tubercle bacillus or old tuberculin.

Alkaline Tuberculin (**T. A.**).—Virulent tubercle bacilli freed from culture media are extracted with 10 per cent. caustic soda for three to four days at 20° C. The bacilli and their fragments are then removed by filtration through filter paper. The filtrate is neutralized by the careful addition of hydrochloric acid and again filtered. The clear fluid (T. O.) gives similar but somewhat more severe reactions than the regular tuberculin. It often leads to sterile abscess formation.[1]

DIAGNOSIS OF TUBERCULOSIS.

A. Clinical, by tuberculin reaction.
  1. Action of tuberculin on healthy animals and man.
  2. Action of tuberculin on tuberculous animals and man.
  3. Principle of tuberculin reaction.

[1] For a general survey of the nature and composition of various tuberculins see Kuthy and Wolff-Eisner, Die Prognosenstellung bei der Lungentuberkulose, Berlin and Vienna, 1914, pp. 438–446.

4. Technic of tuberculin reaction.
    (*a*) Subcutaneous test (Koch).
    (*b*) Cutaneous test (von Pirquet).
    (*c*) Percutaneous test (Moro).
    (*d*) Detre test (human and bovine tuberculin to detect the type of infection).
    (*e*) Ophthalmo reaction (Calmette and Wolff-Eisner).
5. Specificity of the tuberculin reaction.

B. Serological.
1. Opsonic index.
2. Agglutination.
3. Complement fixation.

C. Bacteriological.
1. Principle involved.
    (*a*) Microscopical.
    (*b*) Cultural.
    (*c*) Animal inoculation.
2. Technic.

A. **Clinical Diagnosis.**—1. **Action of Tuberculin on Healthy Animals and Man.**—In healthy laboratory animals, as guinea-pigs and rabbits, as much as 1 c.c. of old tuberculin may be injected with no apparent harm other than a somewhat transient rise in temperature. In normal man even as small an amount as 0.01 c.c. of old tuberculin may cause violent symptoms: chill, temperature, vomiting, malaise, and even diarrhea. These effects, however, are usually transient. Man appears to be far more sensitive to tuberculin than the guinea-pig.

2. **Action of Tuberculin on Tuberculous Animals and Man.**—In tuberculous animals very small amounts of tuberculin injected subcutaneously may cause marked symptoms; 0.2 to 0.5 c.c. will almost invariably kill a guinea-pig which has been injected with tubercle bacilli from four to five weeks before by the subcutaneous route. Intracerebral inoculation of tuberculin will kill tuberculous guinea-pigs in much smaller amounts. Postmortem there is intense congestion around the tuberculous foci and ecchymotic hemorrhages in the viscera. Twenty-five hundredths (0.25) of a cubic centimeter of old tuberculin would be extremely dangerous to inject into a tuberculous man. It would probably result fatally.[1]

3. **The Principle Involved.**—The reaction obtained in tuberculous man or animals by the injection of tuberculin or other products of

[1] Deist, Beitr. z. Klinik d. Tuberkulose, 1912, xxii, 547, has observed albumoses in the urine of tuberculous patients following the injection of tuberculin.

the tubercle bacillus depends upon the fact that the tuberculous subject is sensitized to the proteins of the tubercle bacillus as the result of infection with this organism.[1] The presence of the proteins of the tubercle bacillus and perhaps other products of growth of the tubercle bacillus stimulate certain body cells of the host to produce specific proteolytic ferments which dissolve these proteins within the body.[2] When tuberculin is introduced into the tuberculous host these specific proteolytic ferments liberate from the tuberculin a poisonous cleavage product which has three specific effects: a focal effect, characterized by intense irritation and inflammation around the tuberculous foci, a local effect, and a general effect which is characterized by a rise of temperature and other general systemic reactions. The tuberculin reaction is an anaphylactic reaction according to Vaughan.[3]

4. **Technic of the Tuberculin Reaction.**[4]—The tuberculin reaction elicited in man by the introduction of tuberculin is of two types, depending upon the method of injection employed. If introduced subcutaneously so that the tuberculin enters the lymph or blood streams, even in minute amounts, the reactions consist of three rather distinct phases: a general, a local, and a focal reaction respectively. If, on the contrary, the injection is purely superficial in the epidermis or on the conjunctiva the reaction is almost exclusively local.

(*a*) *The Subcutaneous Reaction* (*Koch*).—The characteristic response following the subcutaneous injection of appropriate amounts of old tuberculin is a generalized reaction consisting of a rise in temperature of *at least* one-half a degree above the highest temperature exhibited before the inoculation. This rise in temperature usually begins twelve to eighteen hours after the injection; it may be delayed to twenty-four or even forty-eight hours. The temperature should be taken at half-hourly intervals. There is frequently an initial chill following the introduction of tuberculin and in addition malaise, headache and restlessness; even nausea or vomiting may be observed. The focal reaction consists essentially of hyperemia and a distinct inflammatory reaction around active foci. In superficial foci, as in lupus, this inflammatory reaction may be distinctly seen, and in deeper foci

[1] For discussion of theories of the tuberculin reaction, see Kuthy and Wolff-Eisner, Die Prognosenstellung bei den Lungentuberkulose, Berlin and Vienna, 1914.

[2] White, Jour. Med. Research, 1914, xxx, 393, has shown that lipoids of the tubercle bacillus neither sensitize nor induce anaphylaxis in experimental animals.

[3] Protein Split Products, Philadelphia, 1913. See Baldwin, Yale Med. Jour., February, 1909, for excellent summary of present status of subject.

[4] The diagnostic and particularly the therapeutic use of tuberculin requires much skill and experience. For details, see Baldwin and Brown, Osler's Modern Medicine, iii, 137, 361.

it can be frequently demonstrated or at least inferred by an increase of local clinical signs. The local reaction at the site of inoculation consists essentially of a reddened, swollen, circumscribed area of inflammation. The specificity of the reaction is dependent upon the size of the dose of tuberculin. Too large a dose may cause a reaction even in a non-tuberculous subject. It is obvious that patients already exhibiting a febrile reaction due to intercurrent disease or otherwise are unfit subjects for injection, for the rise in temperature is the chief diagnostic symptom relied upon in establishing a diagnosis.

The subcutaneous injection of old tuberculin is used only for adults and for children over five years of age in the very early stages of the disease. This reaction is claimed by some observers to be more delicate than any other tuberculin test. In practice tuberculin is introduced subcutaneously either in the breast or preferably in the back, and a control inoculation, using dilute glycerin containing 0.5 per cent. carbolic acid, is made in another area. If a nodule and congestion appear at the site of inoculation of the tuberculin and the control area remains practically unchanged, the reaction is considered positive if the temperature chart taken at half-hour intervals shows at least half a degree rise in temperature above that exhibited previously for several days. The size of the dose to be administered depends upon the age and condition of the patient and upon the potency of the old tuberculin.

(b) *The Cutaneous Test* (*von Pirquet*).—The patient's forearm is sterilized and two drops of undiluted old tuberculin are placed upon the skin about 8 to 10 c.m apart. A light scarification is made, preferably with the von Pirquet scarifier, through each drop of tuberculin. A control scarification is made midway between the drops, but no tuberculin is applied here. A small pledget of cotton is placed over each drop of tuberculin and allowed to remain ten minutes to prevent the tuberculin from spreading beyond the site of scarification. The amount of cotton used should be small enough to prevent any considerable absorption of the tuberculin. No dressing is required. During the first few hours of vaccination the control and vaccinated areas appear the same, a slight area of inflammation due to trauma surrounding each. The specific reaction appears first as a slightly elevated red area around each drop of tuberculin, which increases in size and somewhat in elevation until it reaches a diameter of from 1 to 3 or even 4 cm., the former being the more common. The maximum intensity is usually reached within forty-eight hours, the first signs appearing in from four to twenty-four hours. In scrofulous children small raised follicular swellings commonly appear around the central specific area, the so-called scrofulous reaction. It should be remembered that a second injection of tuberculin in the same area on the same arm following an initial negative reaction may be positive. This is not an indication of infection, however; it is rather a manifestation of local sensitization.[1]

(c) *The Percutaneous Test* (*Moro*).—Moro has modified the von Pirquet test in such a manner as to exclude the traumatism incidental to scarification. This is accomplished by rubbing into the skin of the abdomen or the chest an ointment made of 5 c.c. of old tuberculin mixed intimately in 5 grams of anhydrous lanolin. In practice a bit of the ointment one-half a centimeter in diameter is rubbed over an area of about four square inches for about half a minute. The ointment is left on the skin to absorb gradually. A positive reaction consists in the development, usually within twenty-four to forty-eight hours, of a number of small red papules within the area of inunction. Ordinarily but a few papules are formed; less commonly a considerable crop appear. Rarely the skin in the immediate area is reddened and there may be slight itching. The papules are few in number in a mild reaction, usually from two to eight or ten; they are red and from 1 to 2.5 mm. in diameter. A moderate reaction is characterized by many red papules, from 10 to 100, which are rather closely crowded together and usually from 0.25 to 2.5 mm. in diameter. The interpapular areas of skin may or may not be reddened. If the skin is reddened it frequently itches somewhat. In a severe reaction papules appear within a few hours after inunction, many in number with a markedly hyperemic background. Itching is a very disagreeable feature of a severe reaction.

(d) *The Detre Test.*—Human tuberculin is rubbed into one arm and bovine tuberculin is rubbed into the other, using preferably the cutaneous reaction of von Pirquet. It was supposed by Detre that the tuberculin reaction elicited will indicate the type of infection, whether it be of the bovine or the human bacillus. This test is of doubtful value for this purpose.

(e) *The Ophthalmo Reaction of Calmette.*—One drop of a 1 per cent. dilution of old tuberculin purified by precipitation with alcohol is instilled in the conjunctival sac. In tuberculous subjects this instil-

[1] The von Pirquet reaction is usually negative in tuberculous children during the acute stage of measles, and occasionally in whooping-cough.

lation is followed by a reddening of the caruncle and usually the conjunctiva as well. The reaction varies in intensity from a very mild local reddening of the caruncle to a conjunctivitis. The reaction becomes visible usually in from four to eight hours. The maximum intensity is reached about the twelfth hour and it usually disappears within twenty-four to forty-eight hours. If the first test is negative and it is desirable to repeat it with a 2 to 4 per cent. concentration of tuberculin, the second instillation must be made in the unused eye. The eye first used is sensitized by the first instillation and will react in a severe manner even if the patient has no tuberculosis.[1] This reaction, however, can only be elicited after ten to fourteen days following the first instillation.

If tuberculin treatment is to be instituted, the ophthalmo reaction is not indicated, for a reaction is almost certain to take place when tuberculin treatment is established. The ophthalmo reaction should not be used in old people or in individuals having other than perfectly normal eyes.

5. **Specificity of the Tuberculin Reaction.**—The tuberculin reaction is not absolutely specific as an index of the occurrence within the host of an active tuberculous focus. From 30 to 60 per cent. of adults known to contain no clinically active foci of tuberculosis, who come to autopsy, show evidences of healed tubercles. In these individuals the tubercle bacillus has been dissolved and its products have sensitized them. The tuberculin reaction, consequently, will be positive in the majority of these individuals because they are sensitized to the proteins of the tubercle bacillus. Young children are much less likely to possess these healed or latent tuberculous foci, and provided they are not too young, or that they are not born of tuberculous mothers, a positive tuberculin reaction is much more conclusive in them. Individuals having advanced tuberculous lesions occasionally do not give a tuberculin reaction.

Müller[2] has studied the von Pirquet reaction in young children. His results follow:

0– 3 months	8.1 per cent. cases with positive reaction
3– 6 months	7.0 per cent. cases with positive reaction
6–12 months	11.7 per cent. cases with positive reaction
12–24 months	24.4 per cent. cases with positive reaction

In 8 cases with a positive reaction which came to autopsy all were found to be tuberculous; 46 negative cases which came to autopsy

[1] Rosenau and Anderson, Jour. Am. Med. Assn., 1908, l, 961.
[2] Arch. f. Kinderheilk., l, 18.

were all free from tuberculosis, except one which had subsequently developed miliary tuberculosis. His general conclusion is that the unreliability of this reaction increases with age.

Von Ruck[1] has collected a large series of cases from the literature with the following results:

Subcutaneous reaction, 8108 cases in all.	
4803 tuberculous patients (clinical)	4318 positive reaction
485 suspected cases (clinical)	318 positive reaction
2820 clinically non-tuberculous	1444 positive reaction
Cutaneous reaction, 6571 cases.	
2192 tuberculous patients (clinical)	1851 positive reaction
865 suspected cases (clinical)	563 positive reaction
3514 clinically non-tuberculous	1047 positive reaction
Conjunctival reaction, 6788 cases.	
2834 tuberculous patients (clinical)	2370 positive reaction
1188 suspected cases (clinical)	685 positive reaction
2766 clinically non-tuberculous	407 positive reaction

SUMMARY.

	Tuberculosis	Suspicious	Non-tuberculosis
Subcutaneous reaction . .	89.2 % +	65.5 % +	51.2 % +
Conjunctival reaction . . .	83.6 % +	56.6 % +	14.7 % +
Cutaneous reaction . . .	84.4 % +	65.0 % +	29.8 % +
Cutaneous reaction (excluding children)	..	..	33.9 % +

The tuberculin test, therefore, when positive gives no absolutely definite distinction between healed, latent, or active foci of infection with the tubercle bacillus. Furthermore, no quantitative evidence of the nature of the reaction or extent of the lesions is elicited by a positive reaction. A negative reaction properly performed, however, in non-cachectic subjects or those who have not had progressive treatment with tuberculin is fairly conclusive. The subcutaneous test of Koch is moderately reliable in adults provided the dosage is correctly selected—a matter requiring unusual skill and experience. The von Pirquet reaction is quite unreliable as an index of an active focus in adults, because from 30 to 60 per cent. of all persons over five years of age react positively in varying degrees. It is a fairly conclusive test in children under five years of age. Similar fallacies are to be expected in the ophthalmo and Moro reactions.[2]

B. **Serological Diagnosis.**—1. **Opsonic Index.**—Wright and his pupils have followed the opsonic index in tuberculous patients and they

[1] Beitr. z. Klinik d. Tuberkulose, 1909, xiii, Heft 1.

[2] For a detailed summary of the Koch, Wassermann, von Pirquet, and Wolff-Eisner theories of the tuberculin reaction and its importance for therapy, see Kuthy and Wolff-Eisner, Die Prognosenstellung bei den Lungentuberkulose, Berlin and Vienna, 1914, 393–399. Also, Wolff-Eisner, Die Tuberkulinbehandlung, 1913.

believe that the changes in opsonic index furnish a reliable index for treatment with tuberculin or other products of the tubercle bacillus. The inherent and unavoidable errors of the opsonic index determination make it unreliable for general use.[1]

2. **Agglutination.**—Agglutinins occur in tuberculous patients,[2] but the agglutinating reaction is unreliable, partly because of the difficulty in obtaining a proper suspension of tubercle bacilli. It is practically never used in practice.[3]

3. **Complement Fixation.**—Practically never used at the present time.[4]

C. **Bacteriological Diagnosis.**—1. **Principle Involved.**—(*a*) *Microscopical Examination.*—Fluids, tissues or exudates suspected to contain tubercle bacilli are stained preferably by the Ziehl-Neelsen method for the demonstration of acid-fast bacilli having the morphology of the tubercle bacillus. Other acid-fast bacilli may be confused with the tubercle bacillus and their presence must be borne in mind when microscopical examinations are made. These will be discussed under their appropriate headings.

It is essential to use absolutely new slides for examination of the tubercle bacillus; old slides which have been used for this purpose not infrequently retain tubercle bacilli. It is also advisable not to make a positive diagnosis unless ten typical tubercle bacilli can be demonstrated in the preparation. Among thousands of smears which have been examined by Arms, formerly of the Boston State Board of Health Laboratory, only one has failed to show ten tubercle bacilli, if *any* were present, after careful search. This one case was shown by many subsequent examinations to be negative for tubercle bacilli. The microscopical examination is the most rapid method of diagnosing tuberculosis bacteriologically.

(*b*) *Cultural.*—Cultures of typical bacilli may be obtained either directly from lesions, fluids, tissues, or exudates, or from animals following inoculation with suspected material. In order to obtain cultures of tubercle bacilli from material containing other organisms as well, it is necessary to treat the material first with antiformin[5] from two

[1] Trudeau, Am. Jour. Med. Sc., 1907, cxxxiii, 813; Baldwin, New York Med. Jour., June 27, 1908.

[2] Romberg, Deutsch. med. Wchnschr., 1901, 275, 295.

[3] Eisenberg and Keller, Centralbl. f. Bakt., 1903, xxxiii, 549, for literature.

[4] Stimpson, Bull. 101, Hygienic Laboratory, 1915, for literature.

[5] Paterson, Antiformin for the Detection of Tubercle Bacilli in Sputum. Studies from the Saranac Laboratory for the Study of Tuberculosis, 1904–1910. Sodium carbonate, 600 grams; fresh chlorinated lime, 400 grams; distilled water, 4000 grams. Dissolve the sodium carbonate in 1000 c.c. distilled water; triturate the chlorinated lime in 3000 c.c. distilled water; filter, and mix with the sodium carbonate solution. Filter again.

to three hours so as to kill off the contaminating bacteria. Antiformin does not as a rule kill tubercle bacilli during this time. It is necessary to remove the antiformin by repeated washing and centrifugalization of the tubercle bacilli before the latter are inoculated into artificial media, preferably Dorset's egg medium. At best the cultural procedure is an unsatisfactory one.

(c) *Animal Inoculation.*—Animal inoculation is the most delicate test for demonstrating the presence of tubercle bacilli. Guinea-pigs are the animals selected, and the method of inoculation depends upon the nature of the material. If the material is suspected to contain tubercle bacilli only, it is introduced directly under the skin, or, better, intraperitoneally. If other organisms are associated with the tubercle bacillus, the material mav either be mixed with antiformin and shaken for two hours to kill off or weaken the other organisms, then washed to remove the antiformin and the sediment injected, or the material may be introduced, contaminating organisms and all, subcutaneouslv into a guinea-pig in the following manner. A subcutaneous pocket is made on the flank of the guinea-pig and the suspected material is introduced beneath the skin and pushed forward. The cut is left open and whatever pus-producing organisms are present cause suppuration; the pus drains away and the initial inflammation is recovered from before the tubercle bacilli kill the animal. Tubercle bacilli then may be recovered from the regional lymph nodes and the internal organs. When the inguinal glands are well enlarged the animal is chloroformed. The skin is sterilized with bichloride of mercury and sterile instruments are used in performing the autopsy. Bits of tissue from the lymph glands, spleen, or other organs are removed aseptically and dropped on the surface of specially-prepared media, preferably the Dorset egg medium. A microscopical examination is also made in the usual manner.

Sputum.—It should be remembered that apparently normal individuals may infrequently have acid-fast bacilli in their sputum and, rarely, tubercle bacilli, without producing apparent symptoms. Sputum from suspected tuberculous patients may be examined directly by stained smears, in which case the early morning sputum coming from the depths of the lungs is to be employed. The caseous or purulent masses which are characteristic of tubercular sputum are removed, spread upon slides and examined after staining with carbol-fuchsin and decolorizing in the usual manner. If the result is negative, the sputum may be mixed with caustic soda or antiformin, shaken, and the sediment examined by staining with carbol-fuchsin. If this method proves negative, cultures from the sputum may be made after treating it with antiformin for two to three hours. Some of the sediment, after washing to remove the antiformin, should be injected into guinea-pigs. It is usually difficult to find tubercle bacilli in the sputum during hemoptysis, but they are found very frequently in the blood-streaked sputum following hemoptysis.

Blood.—Tubercle bacilli are usually not found in the peripheral blood. If they do occur they are there almost invariably in very small numbers. Occasionally positive results have been reported in examinations of stained preparations made directly from the blood when every precaution has been taken to preclude the inclusion of extraneous acid-fast organisms. It is far more satisfactory, however, to inoculate the blood subcutaneously into guinea-pigs.

The Nasal Cavity.—Tubercle bacilli have been found occasionally in the nasal passages of healthy individuals, particularly those who have been closely in association with tuberculous patients. It must be remembered that the acid-fast organism described by Karlinsky as the "nasal secretion bacillus," is found occasionally both in the noses of tuberculous patients and in healthy individuals. This organism grows readily on artificial media and should not be confused with the true tubercle bacillus.

Pus and Exudates.—It is frequently difficult or impossible to detect tubercle bacilli in the pus of cold abscesses. Material derived from this source is very frequently caseous, and although tubercle bacilli can not be demonstrated by staining methods, the so-called Much granules are found occasionally, which are Gram-positive but not acid-fast. In order to demonstrate the infectiousness of this material it is inoculated into guinea-pigs. Tubercle bacilli may be found quite frequently in exudates, but in order to make the diagnosis reliable the material should be inoculated into guinea-pigs.

Urine.—As a rule it is difficult to find tubercle bacilli in urine, and the probability of contamination with the smegma bacillus or the Lustgarten bacillus must constantly be borne in mind. Sedimenting large volumes of urine and injecting the sediment into guinea-pigs is the most practical method of detecting tubercle bacilli in this excretion, for it rules out both the Lustgarten and smegma bacilli, which do not produce lesions in guinea-pigs. The slight differences in acid- and alcohol-fastness between these organisms and the tubercle bacillus make diagnosis by the direct smear method of doubtful value.

Feces.—Tubercle bacilli may appear in the feces either because they have been swallowed with the sputum or because of the existence of tuberculous ulcers in the intestinal tract.[1] Acid-fast bacilli which are not true tubercle bacilli are quite common in the feces, and for this reason animal inoculation after treatment of the feces with antiformin is the best method for demonstrating the organism.

Milk.—Tubercle bacilli are very infrequent in human milk, although they are said to be relatively common in unpasteurized cow's milk as it is sold in large cities. The organisms get into the cow's milk far more frequently through the contamination with feces than from direct infection through the udder. Microscopical examination of the sediment of milk or cream is usually valueless, as is the examination of the cream layer itself. Acid-fast bacilli which are not tubercle bacilli very frequently cause confusion. Among these organisms are those described by Petri and Rabinovitch, which are called butter bacilli. Inoculation of the sediment and of the cream of milk into guinea-pigs is the only safe test.

Immunity and Immunization.[2]—The disproportion between the incidence of "healed tubercles" in cadavers which do not exhibit symptoms of tuberculosis *ante mortem* and the actual number of clinical cases suggests that the average individual possesses a certain degree of refractoriness to progressive invasion by the tubercle bacillus, that is to say, the clinical cases of the disease are considerably outnumbered by those in whom the organism has gained entrance, but failed to develop sufficiently to cause symptoms. Early, uncomplicated cases of tuberculosis frequently react favorably when placed in a favorable environment. Spontaneous recovery from tuberculosis complicated by secondary infections with other bacteria is more tedious and the prognosis is generally less favorable.

Active immunization of man with various products of the tubercle bacillus has been one of the greatest problems of medicine. Up to the present time the solution of this problem has not been realized. At least a generation must elapse before final judgment can be passed upon any system of human immunization, for the disease tuberculosis progresses slowly and results to be trustworthy must be numerous and of long duration.

[1] Laird, Kite, and Stewart, Jour. Med. Research, 1913, xxix, 31, for summary and literature.

[2] For an excellent summary of Immunity in Tuberculosis, see Baldwin, Am. Jour. Med. Sc., 1915, cxlix, 822.

Active immunization has been attempted with the following types of preparation:

1. Tuberculins.
2. Vaccines.
    (*a*) Killed cultures.
    (*b*) Soluble vaccines.
3. Living tubercle bacilli.
    (*a*) Virulent organisms, Webb method.
    (*b*) Attenuated viruses.
    (*c*) Alien acid-fast bacilli.
4. Sera.

It has long been recognized that various tuberculins do not confer immunity upon experimental animals, although in skilled hands they possess undoubted curative value.[1] Killed cultures of tubercle bacilli have not been satisfactory, and their use has been greatly restricted by the non-solubility of the organisms which produce local indurations of refractory nature. Soluble vaccines including "bacillus emulsions" and various proteins of the bacillus do not appear to confer definite immunity upon susceptible animals. The superiority of living viruses over the various preparations of killed organisms and their products for protective inoculation is conceded by the great majority of investigators. Webb and Williams[2] have attempted to induce artificial active immunity in experimental animals by injecting virulent tubercle bacilli, beginning with one or two organisms and gradually increasing the number. Their results, while few in number, appear to be worthy of serious consideration. The use of attenuated cultures and of alien acid-fast bacilli have not been successful up to the present time. Sera have been equally unsatisfactory.

Bovine Tubercle Bacillus.—Cattle and swine are susceptible to infection with the bovine tubercle bacillus and the disease is widespread among dairy herds. Statistics indicate that in certain parts of the United States the incidence of tuberculosis in swine increases almost proportionately to the spread of the disease among cattle. This condition is brought about partly through the practice of feeding slaughter-house offal to swine, chiefly through dairies and cheese factories where the skimmed milk or whey forms a not inconsiderable part of the rations of swine. Scrofula or tuberculosis of swine thus is a true ingestion disease. It is possible to infect swine by feeding human

[1] Trudeau, Osler's Modern Medicine, 1907, iii, 434.
[2] Jour. Med. Research, 1909, xx, 1; 1911, xxiv, 1.

tubercle bacilli; the spontaneous disease, however, is almost invariably an infection with the bovine type of the tubercle bacillus.[1]

The most common initial lesion in cattle is an involvement of the retropharyngeal glands; the lungs are frequently infected, and occasionally the liver and serous membranes are invaded rather early in the disease. A peculiar and characteristic type of infection of cattle, known as Perlsucht or pearly disease, which progresses slowly and is recognizable only in the later stages, is distinguished by the occurrence upon serous surfaces of thick fibrous tumors containing much connective tissue. The lesions in infections of the peritoneal surfaces consist of large numbers of solitary or clustered tubercles varying in size from 1 to more than 10 mm. in diameter. They may be attached to the surface by tough, fibrous pedicles or they may rest directly upon the membrane itself. These tumors may become calcified or caseated and they are larger than tubercles found in human tissues. Morphologically their structure is fundamentally not unlike human tubercles.

Theobald Smith[2] was the first to clearly point out the differences between the human and bovine tubercle bacilli. His evidence was based upon morphological, cultural and pathological characters. He showed that the human tubercle bacillus, grown on serum, was longer and slenderer than the bovine type and frequently curved. The growth is more luxuriant, forming a thick, wrinkled membrane upon glycerin broth. The bovine type commonly develops feebly in this medium and produces a thin, delicate pellicle. The reaction curves of the two types on glycerin broth are distinctive and characteristic. The bovine type gradually creates an alkaline reaction; the human type leaves the reaction acid and tuberculin made with the bovine type consequently is alkaline in reaction; that of the human type is acid. The pathogenic action of the two types is distinctive; the bovine type is highly pathogenic for rabbits and calves; the human type is only slightly pathogenic for these animals. One milligram of a human culture fails to kill rabbits, but 0.1 milligram of a freshly isolated bovine culture results fatally.

The important differential characters are summarized in the following table:

Human Types.	Bovine Types.
MORPHOLOGY.	
On serum or glycerin bouillon, long, slender, slightly curved rods which usually stain uniformly; occur in clusters usually lying parallel.	On serum or glycerin bouillon, relatively short thick rods irregularly arranged; frequently exhibit slight irregularity in staining.

[1] Theobald Smith, Boston Medical and Surgical Journal, 1909, clix. 707.
[2] Jour. Exp. Med., 1898, iii, 451.

CULTURAL CHARACTERS.

Glycerin bouillon, after two to four weeks' growth, dense wrinkled membrane.	Glycerin bouillon, delicate membrane exhibiting occasional wrinkling of the surface.
Reaction remains permanently acid.	Reaction gradually becomes alkaline.
Tuberculin has acid reaction.	Tuberculin has alkaline reaction.
Growth on blood serum relatively luxuriant and develops with comparative rapidity.	Growth on blood serum relatively meagre—develops slowly.

ANIMAL PATHOGENESIS.

Guinea-pigs very susceptible. Young cattle, rabbits and swine resistant to infection.	Guinea-pigs, young cattle, rabbits and swine very susceptible to infection.

Bovine infections in man are much more common in children than in adults[1] Milk is a frequent vehicle for the transmission of the virus to man; the origin of the bacilli in milk has been summarized by Moore[2] as follows:

"1· Cows with tuberculous udders eliminate tubercle bacilli with the milk. In such cases these organisms are usually present in large numbers.

2. Cows with glandular or pulmonary tuberculosis, in which the lesions are discharging into the bronchi, eliminate tubercle bacilli with the feces and with the droolings. In cases of intestinal tuberculous ulcers the organisms are excreted with the feces.

3. Milk is usually infected with tubercle bacilli when it is taken from cows with tuberculous udders. It may, through contamination with feces or uterine discharges, be infected when drawn from cows with open lesions in the respiratory and digestive tracts or organs of reproduction.

4. Tubercle bacilli are not, as a rule, present in milk of cows that react to tuberculin and which, on careful physical examination, exhibit no evidence of disease."

The identification of tubercle bacilli in milk presents no insurmountable difficulties, but certain precautions must be observed. Prudden and Hodenpyl,[3] Straus and Gamaleia[4] and others have shown that the injection of killed tubercle bacilli into guinea-pigs by the intraperitoneal route will induce tubercle formation even if the organisms have been heated in the autoclave. These tubercles, however, if crushed and injected into fresh animals, do not reproduce tubercles. It is obvious that the injection of pasteurized milk containing dead tubercle

[1] Statistics by Park and Krumwiede, p. 438.
[2] Jour. Med. Research, 1911, xxiv, 517.
[3] New York Med. Jour., 1891, liii, 637, 697.
[4] Arch. méd. expér., 1891, iii, 705.

bacilli may lead to false conclusions unless this possibility be borne in mind. The bacillus of infectious abortion induces lesions in guinea-pigs closely simulating tuberculosis.[1] This organism is rather widely distributed in unheated milk.[2]

Method.—Ten to 20 c.c. of milk are centrifuged for half an hour and from 5 to 10 c.c. of the sediment and lower portion of the sample are injected subcutaneously into guinea-pigs. The cream layer is also injected subcutaneously into a second pig. After three to six weeks, if the animal shows signs of emaciation, 0.5 cm. of undiluted bovine tuberculin is injected. This injection usually results fatally. In any event the animal is killed and the bovine tubercle bacilli are identified in the usual manner.

Immunity.—Cattle and swine exhibit little or no natural immunity to infection with the bovine tubercle bacillus. The most satisfactory prophylaxis consists in isolating infected animals from the herd, disinfection of the stables and testing all apparently sound animals with tuberculin.

Tuberculin Test.—The preparation of bovine tuberculin is precisely similar to that of human tuberculin, except that the bovine organism is used. The test is carried out in the following manner:

The temperature of the animal is taken at frequent intervals for twenty-four hours, then tuberculin is injected subcutaneously, preferably over the fore-shoulder, about 10.00 P.M. Temperatures are taken from 6.00 A.M. of the following morning at two-hour intervals until 10 P.M. An elevation of temperature of from one to three degrees occurs within a few hours in positive cases and a hot swollen area of induration appears around the site of inoculation. Both the febrile reaction and the indurated area slowly become normal. The International Commission on the Control of Bovine Tuberculosis[3] states:

"1. That tuberculin, properly used, is an accurate and reliable diagnostic agent for the detection of active tuberculosis.

2. That tuberculin may not produce a reaction under the following conditions:

(*a*) When the disease is in the period of incubation.

(*b*) When the progress of the disease is arrested.

(*c*) When the disease is extensively generalized.

[1] Theobald Smith, Bureau of Animal Industry, 1894, Bull. 7, 80 Smith and Fabyan, Centralbl. f. Bakt., Orig., 1912, lxi, 549.

[2] Melvin, Vet. Jour., 1912, lxviii, 528. Fabyan, Jour. Med. Research, 1913, xxviii, 85.

[3] Forty-seventh Annual Report of the American Veterinary Medical Association, September, 1910.

The last condition is relatively rare and may usually be detected by physical examination.

3. On account of the period of incubation and the fact that arrested cases may sooner or later become active, all exposed animals should be retested at intervals of six months to one year.

4. That the tuberculin test should not be applied to any animal having a temperature higher than normal.

5. That an animal having given one distinct reaction to tuberculin should thereafter be regarded as tuberculous.

6. That the subcutaneous injection of tuberculin is the only method of using tuberculin for the detection of tuberculosis in cattle, which can be recommended at the present time.

7. That tuberculin has no injurious effect upon healthy cattle."

Avian Tubercle Bacillus.—Hens, pheasants and other birds are subject to a spontaneous disease which is anatomically very much like tuberculosis of other warm-blooded animals. Koch[1] believed that the organism of avian tuberculosis was identical with the bovine tubercle bacillus, but later work has not confirmed this assertion. It is believed at the present time that the bovine and avian tubercle bacilli are distinct entities.

The morphology of the avian tubercle bacillus and its staining reactions are quite similar to those of the bovine organism, except that pleiomorphism is more marked in the former, particularly when it is grown at 40° to 42° C. It forms no spores and no capsules, is non-motile and has no flagella. It grows more readily than either the human or the bovine strains; the addition of glycerin to media, while not essential, increases the luxuriance of the growth. On coagulated blood serum or agar after six to ten days the organisms appear as small white colonies with a waxy luster. A second transfer to artificial media results in a more luxuriant growth which spreads and increases in luxuriance, eventually covering the whole medium. The growth is moist and may become slimy, differing markedly in this respect from the human and bovine types. The pellicle formed on broth cultures is less friable and more tenacious than that characteristic of the mammalian strains. The range of growth is from 35 to 45 degrees; 40° is the optimum temperature, but development is luxuriant at 37° C. An exposure of two hours at 65° C. usually fails to kill avian tubercle bacilli—but fifteen minutes at 70° to 72° C. is fatal. The organisms are very resistant to drying, remaining alive

[1] Mitt. a. d. Kais. Ges. Amte, 1884, ii, 4.

for several months both in cultures and in uncontaminated material from infected birds. Unlike the human or the bovine disease, avian tuberculosis is transmitted ordinarily as a congenital infection. Birds are comparatively readily infected artificially, however, by injection of the organisms. Edwards[1] has infected hens with the excrement of infected birds. The liver and the spleen are the organs more commonly involved.[2]

Among laboratory animals rabbits appear to be more susceptible than guinea-pigs, although Edwards[3] appears to have successfully infected several guinea-pigs with pure cultures of the organism. Moore[4] was unsuccessful in causing the bacilli to multiply in guinea-pigs but observed that the animals frequently died of marasmus, apparently from the absorption of toxins from the bacilli; Mohler[5] has induced infection in swine by feeding them the carcasses of tuberculous hens. Himmelberger[6] has made the important observation that calves may be susceptible to infection with the avian tubercle bacillus.

[1] Ont. Agricult. College Bull., No. 193.
[2] De Jong, Ann. Inst. Past., 1910, xxiv.
[3] Loc. cit.
[4] Jour. Med. Research, xi, 521.
[5] Twenty-fourth Annual Report, Bureau of Animal Industry.
[6] Centralbl. f. Bakt., Orig., 1914, lxxiii, 1.

CHAPTER XXIV.

LEPROSY AND ACID-FAST BACTERIA OTHER THAN THE TUBERCLE GROUP.

THE BACILLUS LEPRÆ.

THE first definite observation of the organism now called Bacillus lepræ was that of Hansen,[1] who described rod-shaped organisms in leprous tissue. Somewhat later Neisser[2] succeeded in staining them. Sticker[3] made the important discovery that the nasal mucosa and nasal secretion of a large percentage of lepers (140 out of a total of 153 cases examined) contain large numbers of leprosy bacilli. The organism described by Hansen is generally accepted as the causative agent of leprosy.

Morphology.—Morphologically Bacillus lepræ resembles Bacillus tuberculosis. It is a slender, rod-shaped organism, measuring from 4 to 6 microns in length. Usually it is somewhat shorter than the tubercle bacillus, curved forms are less common, and the ends of the bacilli are not infrequently somewhat enlarged. They occur characteristically as clusters of bacilli, grouped together in bundles like cigars, lying within large cells—the so-called lepra cells. The organisms from young lepra nodules are more acid-fast than the tubercle bacillus; in old, degenerating nodules they tend to lose their acid-fastness and other staining properties, and the cytoplasm becomes vacuolated, giving the bacilli a beaded appearance. Bacillus lepræ is Gram-positive as well as acid-fast. Available evidence indicates that the organism is non-motile, possesses no flagella, and forms no capsules.

[1] Virchow's Arch., 1880, lxxix, 32.
[2] Ibid., 1881, lxxxiv, 514.
[3] Berl. klin. Wchnschr., 1897, 518; Arb. a. d. kais. Gesamtc, 1899, xvi, 357.

Bacillus lepræ differs somewhat from Bacillus tuberculosis in its staining reactions and occurrence in lesions.

Bacillus Lepræ.	Bacillus Tuberculosis.
1. Stains with aqueous solutions of basic anilin dyes.	Does not stain with these dyes.
2. Stains readily by Gram's method.	Stains with difficulty.
3. Resists decolorization by the Ziehl-Neelsen stain.	Somewhat more readily decolorized.
4. Large numbers of acid-fast bacilli occur within swollen cells—lepra cells.	Relatively fewer acid-fast bacilli found together as a rule. No cells resembling lepra cells.

These differences are quantitative rather than qualitative, however, and can not be individually relied upon to establish an absolute differentiation between the two organisms. The most distinctive difference is the "lepra cell" with its large number of acid-fast bacilli in groups with their long axes arranged in parallel.

Isolation and Culture.—Clegg[1] cultivated an acid-fast organism from lepers by growing the organism upon agar symbiotically with amebæ, then killing the amebæ by an exposure to 60° C. for thirty minutes. The bacilli, once acclimatized to artificial media, grew readily. Duval[2] has also cultivated acid-fast bacilli directly from lepers. Animal experimentation and serological studies have been inconclusive thus far. Kedrowski[3] described a pleiomorphic streptothrix-like organism which grew in artificial media as a non-acid-fast streptothrix, but tended to change to a pleiomorphic, diphtheroid, acid-fast bacillus. (Plate III.) When this diphtheroid bacillus is injected into rats or mice it becomes acid-fast and resembles Bacillus lepræ in detail. Bayon[4] states that Kedrowski's organism is the true leprosy bacillus, and that the diphtheroid form is one stage of the typical acid-fast type seen in leprous nodules. The entire subject of the cultivation of Bacillus lepræ on artificial media must be regarded as *sub judice.*

Products of Growth.—Deycke and Reschad[5] isolated a streptothrix (Streptothrix leproides) from a leper; from cultures of this organism a fatty acid-glycerin ester was prepared, to which the name nastin was applied. Rost[6] isolated an acid-fast organism from lepers, upon salt-free media. A substance, leprolin, was prepared from broth cul-

[1] Philippine Jour. Sci., 1908, iv, 403.

[2] Jour. Exp. Med., 1910, xii, 649; 1911, xiii, 365; Jour. Am. Med. Assn., 1912, lviii, 1427.

[3] Ztschr. f. Hyg., 1901, xxxvii, 52.

[4] Centralbl. f. Bakt., Ref., 1913, lvi, 592.

[5] Deutsch. med. Wchnschr., 1907, 89. Deycke, Lepra, 1907, vii, 174.

[6] British Med. Jour., 1905, p. 2302.

PLATE III

Lepra Bacilli; Ziehl-Neelsen Stain.

tures of this organism precisely as tuberculin is prepared. Neither nastin nor leprolin have been successful clinically judging from available information.

Pathogenesis.—*Human.*—McCoy and Goodhue[1] have summarized observations of the infectiousness of lepers in the leper settlement in the Hawaiian Islands as follows: of 119 men, practically all Hawaiians living in the same house with lepers, 5 (4.40 per cent.) developed leprosy; of 106 women, practically all Hawaiians, living in the same house with lepers, 5 (4.71 per cent.) developed leprosy; of 12 women, all Caucasians, nurses and members of religious orders, living among lepers, none contracted the disease; but of 23 Caucasian males, three contracted leprosy. The shortest period in which the disease appeared

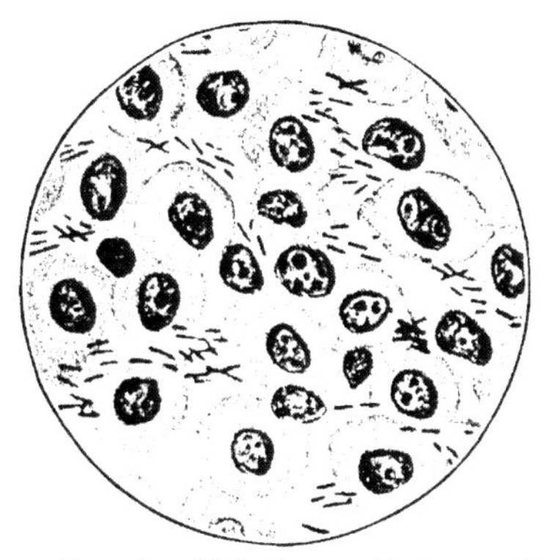

FIG. 63.—Lepra bacilli in liver. (Kolle and Hetsch.)

after exposure was three years (2 cases); the longest seventeen years. Arning[2] inoculated a condemned criminal with leprosy bacilli derived from a leper and the criminal developed leprosy. Several other investigators have made similar experiments, a few of which have resulted positively. The majority of such attempts have been failures, and the consensus of opinion is that the few positive results are to be explained by the existence of leprosy in the early stages.

The earliest lesion appears to be an ulcer at the junction of the bony and cartilaginous septum of the nose. When the bacilli are carried to any tissue in the body they excite the usual inflammatory reaction which, however, is continued to excessive tissue proliferation. Granular tissue containing bloodvessels is formed, providing a good vascular supply, so that proliferation continues until a fair-sized

[1] Public Health Bull., 1913, No. 61.

[2] Centralbl. f. Bakt., 1889, v, 672.

nodule or tumor is produced. If the seat of infection is a nerve, a spindle-shaped thickening is produced about the nerve, which causes pressure, irritation, inflammation, degeneration, and, finally, atrophy, which leads to anesthesia of the area of distribution of the nerve. The bacilli are found both in the neuroglia and in nerve cells, particularly the ganglion cells. When the cell is thus infected it undergoes degeneration, sometimes accompanied by swelling and formation of vacuoles. If the bacilli become attached to arteries a proliferating inflammation results which causes the walls of the vessel to become greatly thickened and the lumen narrowed.

The disease often displays a selective action for the skin, especially in the face, the extensor surfaces of the nose and elbows, and on the backs of the hands and feet. These areas are very apt to undergo ulceration. The organism makes its appearance in the skin by producing red spots which either disappear, leaving pigmented areas, or become elevated in nodules of a brown-red color. In the region of these nodules the subcutaneous tissues contain large numbers of bacilli. These eruptions are probably to be regarded as general inflammatory reactions in response to the irritation produced by the bacillus. The nodules may remain small and hard, or they may become enlarged, in which case large protuberances appear which destroy the symmetry of the face and give the victim a lion-like appearance, hence the name facies leontina.

Animal.—It must be remembered that tubercle bacilli are very frequently found in lepers; in fact, a not inconsiderable number of lepers die of tuberculosis rather than leprosy. Consequently it is not surprising to find that many experiments on animals have resulted in the production of lesions from which acid-fast bacilli have been obtained. These experiments must be interpreted with a great deal of caution for this reason. Nicolle[1] claims to have successfully infected a monkey (Macacus) with leprous material. Lesions appeared which were nodular in character, but they disappeared spontaneously within six months. The result is questionable and successful inoculations of leprosy bacilli from lepra nodules into the lower animals are not definitely proven as yet.

Portal of Entry.—Leprous lesions appear early and are fairly constant in the nasal passage, where an ulcer appears at the junction of the bony and cartilaginous septum. This has given rise to the belief that the nasal passages are the chief portals of entry of the organism

[1] Semaine Médicale, 1905, 110.

and that infection takes place through dust or droplets. Recently a case has been described which shows extensive involvement of the liver and spleen, in which the intestinal tract was considered the atrium of infection. Cases also are on record in which the primary lesions appear to have occurred on the feet, indicating that abrasions of the skin may also be portals of entry.

Diagnosis.—Scrapings from the nasal ulcer may give an early diagnosis. It should be remembered, however, that an acid-fast organism described by Karlinski (nasal secretion bacillus) is fairly common, not only in the nasal passages of lepers and tuberculous individuals, but also in normal individuals as well. Karlinski's organism grows readily on artificial media and is in no way related

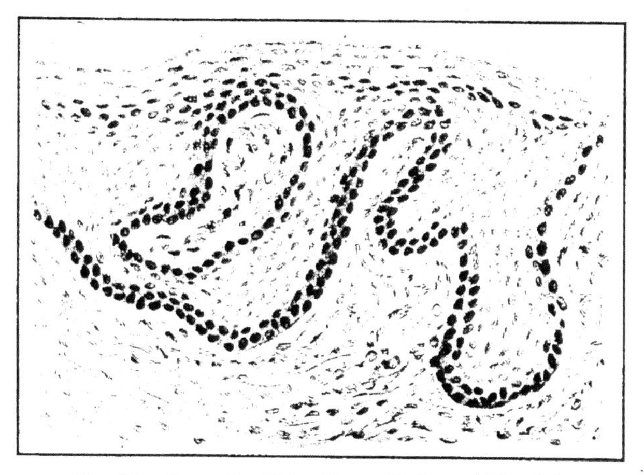

FIG. 64.—Lepra bacilli in skin. (Kolle and Hetsch.)

to the leprosy bacillus. As a matter of practice, a clinical diagnosis of leprosy is more important than a microscopical diagnosis; the latter merely confirms the former. From ulcerated lepromata or from intact tuberculoid nodules material may be gathered and stained in the usual manner for the presence of acid-fast organisms. Inasmuch as the leprosy bacilli occur in large numbers in these tubercles it is best to excise a small portion of one, cut sections and stain it for leprosy bacilli. This will give the characteristic arrangement of the organisms and make the diagnosis very much more certain. Leprosy bacilli can be definitely distinguished from tubercle bacilli; when injected into guinea-pigs they do not produce lesions.

Prophylaxis.—*Heredity.*—Whether leprosy is a germinal infection or not is not known, although the bacilli have been found both in the ova and in the testicles. It is suspected that children born of leprous parents are probably infected immediately before or shortly after birth.

Method of Dissemination.—The exact method of transmission of leprosy is still unknown. It is generally believed that the bacillus might be transmitted either by droplet infection or by direct contact. The organism, however, does not appear to be very infectious, in adults at least, for experience has shown that prolonged and intimate association with a leper does not as a rule result in infection.

Leprosy of Rats.—Stefansky[1] has reported a disease of rats which resembles human leprosy in a striking manner; like the disease in man the lesions, which consist essentially of glandular enlargement, subcutaneous infiltration and induration, alopecia, and frequently deep-seated cutaneous ulcerations, contain large numbers of acid-fast bacilli which resemble Bacillus lepræ both morphologically and in their collection in large numbers in the localized swellings. Dean[2] has corroborated the observation and also found acid-fast bacilli in the nasal secretion of the rats. He also isolated a diphtheroid bacillus similar to that of Kedrowski. The disease is wide-spread among rats, being reported in Russia,[3] Berlin,[4] Australia,[5] and the United States.[6] It is found in areas free from human cases, as human leprosy is found in locations where the disease is not found in rodents. Whether the human and rat leprosy bacilli are identical or not is not finally decided; Mezinescu[7] and Schmitt[8] have shown by the method of complement fixation that the sera of human and rat lepers mutually exhibit complete reactions.

ACID-FAST BACILLI OTHER THAN BACILLUS TUBERCULOSIS AND BACILLUS LEPRÆ.

Following the discovery of the tubercle and leprosy bacilli, which exhibit the striking phenomenon of "acid-fastness," a number of bacteria presenting the same general staining reactions have been described. They are somewhat widely distributed in nature, being

[1] Centralbl. f. Bakt., Orig., 1903, xxxiii, 481.
[2] Ibid., xxxiv, 222.
[3] Stefansky, loc. cit.
[4] Rabinovitsch, Centralbl. f. Bakt., Orig., 1903, xxxiii, 577.
[5] Bull. Jour. of Australasia, 1907, 263.
[6] Wherry, Jour. Am. Med. Assn., 1908, l, 1903. McCoy, Public Health Rep., July 10, 1908.
[7] Compt. rend. Soc. biol.. 1908, lxiv, 514; 1909, lxvi, 56.
[8] University of California Pub. in Path., 1911, ii, 29.

occasionally isolated from water, grass or manure, and they are also found in association with man and the higher domestic animals, frequently occurring as parasites upon the skin, less commonly in the nasal secretion or sputum.

Morphologically the acid-fast bacteria of the non-pathogenic type are somewhat shorter and relatively thicker than the tubercle bacillus (human type), and in culture they not infrequently grow in filaments and exhibit branching. Upon artificial media, furthermore, development is relatively rapid; growth usually appears within forty-eight hours, even at 25° to 27° C. The usual type of growth is an irregular, wrinkled layer, waxy in appearance and of variable color from gray to yellow, orange or even brown.[1] The injection of considerable amounts of the bacteria into guinea-pigs may lead to the formation of granulation tissue nodules, for the organisms are very insoluble in the body juices; superficially these nodules may resemble tubercles, but they differ in two important particulars—they do not develop progressively but are limited to the site of inoculation, and they tend to soften gradually and eventually to suppurate and heal spontaneously with scar tissue formation.

The best-known members of the group are: Bacillus phlei, including the various bacilli isolated from grasses and manure; the smegma bacillus, which grows on the genitalia and the cerumen; and the nasal secretion type found occasionally on the skin, in the nasal secretion, the sputum, tonsillar exudates and rarely in gangrene of the lungs. It is very probable that the tubercle bacilli of cold-blooded animals—(ichthic tubercle bacilli)—fish, turtles, snakes, and the "Blindschleiche" bacillus belong to this group.

The Smegma Bacillus.—Alvarez and Tarbel[2] found an organism on the external genitalia and around the anus which is very similar morphologically and in staining reaction to the tubercle bacillus. Moeller[3] and others have confirmed this observation. The organism was called the smegma bacillus. It has been regarded by many as identical with a bacillus described in 1884 by Lustgarten as the causative organism of syphilis.

The cultivation of both of these organisms in artificial media is difficult, and it is not definitely proven that it has been accomplished.

The practical importance of these organisms lies in the fact that

[1] Tuberculin is not produced in cultures in artificial media.
[2] Arch. d. phys. norm. et path., 1885, No. 7.
[3] Centralbl. f. Bakt., Orig., 1902, xxxi, 278.

they may be confused with the tubercle bacillus in the examination of urine or feces for the latter. The organisms are not pathogenic for guinea-pigs and a distinction between the smegma bacillus and the tubercle bacillus may be effected in this way.

The Nasal Secretion Bacillus.—Karlinski[1] isolated an organism from the nasal secretion of a man which possessed morphological and staining peculiarities very similar to those of the tubercle bacillus. Similar or identical organisms have been isolated from tonsillar exudates, from a few cases of pulmonary gangrene and from sputum.

The organism grows readily on ordinary media. It presents no definite peculiarities of staining which would distinguish it from the tubercle bacillus, and its occasional occurrence in the nasal and oral secretions necessitates great care in distinguishing it from that organism.

The organism is non-pathogenic for guinea-pigs and in suspicious cases a differentiation between the nasal secretion bacillus and the tubercle bacillus can be made through this animal.

Bacillus Phlei.—**Synonyms.**—*Grass bacillus, Timothy grass bacillus, Mist bacillus.*

Historical.—The most important investigations of the saprophytic acid-fast bacilli are those of Moeller.[2] The members of this group, designated as Grass bacillus I and II, from hay infusions, and the Mist bacillus from manure, are very similar in their general staining and cultural reactions—so similar that the slight differences noticed are of insufficient magnitude to warrant their separation into distinct types. For the present they are best regarded as variants of the same organism.

Morphology.—Bacillus phlei resembles the tubercle bacillus (human type) in its morphological characters, except that it is somewhat shorter and relatively thicker. Occasionally isolated organisms exhibit swollen, club-shaped ends, and branching is frequently observed in cultures in artificial media. They stain with difficulty and resist the combined decolorizing action of mineral acids and alcohol.

Isolation and Culture.—The organisms grow readily and rapidly on ordinary media, and after three or four days' incubation, the colonies are round, somewhat waxy in appearance, and vary in diameter from 2 to 5 mm. Typically colonies are yellowish to a dark orange in color. Subcultures are obtained very readily.

[1] Centralbl. f. Bakt., 1901, xxxix, 525.

[2] Deutsch. med. Wchnschr., 1898, No. 24; Centralbl. f. Bakt., 1899, xxv, 369; 1901, xxx, 513.

Pathogenesis.—Bacillus phlei is not pathogenic for man, so far as is known, but the introduction of large numbers of the organisms into the peritoneal cavity of guinea-pigs leads to the formation of localized nodules which eventually soften and contain a purulent, somewhat caseous mass. Typical tubercles with giant-cell and epithelioid-cell formation are not observed. Moderate doses do not cause death, but very large doses frequently lead to fatal results. The inoculated animals fail to give any reaction whatsoever with tuberculin derived from human or bovine cultures.

The organisms are of practical importance because they may be confused with the tubercle bacillus. A simple microscopic examination may in rare instances lead to error, but the correct differentiation between these organisms and the tubercle bacillus may be safely arrived at by their injection into guinea-pigs and the subsequent negative reaction with a fairly large dose of tuberculin.

The Butter Bacillus.—This organism was first described by Rabinovitsch,[1] and subsequently her observations were confirmed and extended by Petri.[2]

Morphologically the organisms are very similar to tubercle bacilli, but they are relatively less acid-fast. Differentiation between the butter bacillus and the tubercle bacillus, however, can not be made upon this basis. The organisms grow in culture media very like the grass bacilli. In broth the medium remains clear and the organisms form a thick, wrinkled pellicle on the surface. Very frequently there is a distinct ammoniacal odor to the broth, and it is said that they form small amounts of indol.

So far as is known the butter bacilli are non-pathogenic for man and the lesions they induce in guinea-pigs are very similar to those produced by the grass bacilli. They are chiefly confusing when they are found in milk and butter because of their resemblance to the bovine tubercle bacillus. A distinction between the butter bacillus and the bovine tubercle bacillus can be definitely made by injection into guinea-pigs. The lesions are not tubercular in nature, and the animals fail to react to tuberculin.

[1] Ztschr. f. Hyg., 1897, xxvi, 90.

[2] Hyg. Rund., August 15, 1897.

CHAPTER XXV.

ANAËROBIC BACTERIA.

BACILLUS TETANI.

THE infectious nature of tetanus was first clearly demonstrated by inoculating rabbits subcutaneously with pus from a human case of the disease. This experiment, which reproduced the essential clinical features of the disease and killed the animals, was performed by Carle and Rattoni[1] in 1884. The same year Nicolaier[2] saw the tetanus bacillus in laboratory animals which were inoculated subcutaneously with garden soil, at the site of injection. It remained for Kitasato,[3] however, to grow the tetanus bacillus in pure culture and to definitely transmit the disease to laboratory animals through pure cultures of the organism.

Morphology.—Bacillus tetani is a long, slender bacillus with rounded ends, measuring from 0.3 to 0.8 micron in diameter and from 2 to 5 microns in length, which commonly occurs singly and in pairs in young cultures; in older cultures the organisms tend to form long chains. It tends to degenerate in older cultures, leaving free spores and involution forms. The bacillus is slightly motile in recently inoculated cultures and possesses from sixty to eighty peritrichic flagella.[4] Capsules are not produced by Bacillus tetani. It stains readily with ordinary dyes and is Gram-positive. Spores are readily formed under anaërobic conditions, which are so characteristic in appearance and constant in occurrence that they are of diagnostic importance. The spores are spherical, greater in diameter than the bacillus (measuring 1 to 1.5 microns in diameter) and occur at one end of the rod, giving it the appearance of a drumstick or plectridium. The rate of spore formation in artificial media appears to be greatly influenced by the temperature of incubation: at 20° C. spores appear in from seven to eight days; at 37° C. they are usually found in large numbers after one to two days; at 43° C. the organisms grow slowly and form but few spores; but little toxin is produced at this temperature.

[1] Giornale della R. accad. di med. di Torino.
[2] Deutsch. med. Wchnschr., 1884, No. 52; Inaug. Diss., Göttingen, 1885.
[3] Deutsch. med. Wchnschr., 1889, No. 31; Ztschr. f. Hyg., 1889, vii, 225.
[4] Schwarz, Lo sperimentale, 1891, p. 373. Grandi, Centralbl. f. Bakt., Orig., 1903, xxxiv, 97.

Isolation and Culture.—Pure cultures of tetanus bacilli are difficult to obtain from the soil or from other sources, where it exists in association with other bacteria. Kitasato[1] succeeded in isolating pure cultures by alternately incubating alkaline broth containing tetanus and other bacteria anaërobically at 37° C. for forty-eight hours, then heating the culture to 80° C. for thirty minutes to destroy non-spore-forming organisms. Theobald Smith[2] has devised a method for obtaining pure cultures of spore-forming anaërobes, including tetanus bacilli, which is far more successful in practice than the Kitasato method. Fermentation tubes containing sugar-free broth slightly alkaline in reaction and bits of sterile tissue (kidney or liver from rabbits or

Fig. 65.—Bacillus tetani, spore formation. × 1000. (Günther.)

guinea-pigs) are inoculated with the suspected material and incubated at 37° C. for forty-eight hours. The growth of anaërobic organisms is much more luxuriant in this tissue medium than in similar media without the tissue.[3] Tetanus spores are formed abundantly around the bit of tissue, and after forty-eight hours of incubation the culture is again heated to 80° C. for thirty minutes to kill non-spore-forming organisms. The spore-containing medium which collects at the lowest part of the fermentation tube around the bit of tissue is reinoculated into a fresh fermentation tube of the same medium, and the process is repeated in detail. It is advisable to use a pipette to remove the material for inoculation in order to insure an abundance of spores. The success of the procedure is readily controlled by stained prepara-

[1] Loc. cit.

[2] Jour. Boston Soc. Med. Sci., 1899, 340; Jour. Med. Research, 1905, xiv, 193.

[3] This method has been rediscovered by Tizzoni (Centralbl. f. Bakt., Orig., 1905, xxxiv, 619), and others

tions made from the material inoculated each time, and the process is repeated until microscopical examination reveals a sufficient number of bacilli of characteristic appearance. Finally, the enriched culture, after a final heating to 80° C. for thirty minutes, is plated out anaërobically upon blood agar plates.

Tetanus bacilli characteristically produce a wide zone of hemolysis around the colonies on blood agar, and the colonies themselves tend to spread rapidly.[1]

Growth in Media.—The tetanus bacillus is typically an obligate anaërobe, although various successful attempts to induce aërobic development have been recorded.[2] The characteristic reactions and products of the organism, however, are detected only in anaërobic cultures. On anaërobic agar plates the colonies are filamentous; under the lower powers of the microscope they resemble densely matted strands of cotton fiber. Gelatin colonies are quite similar, except that in sugar-free gelatin liquefaction takes place after three to five days' incubation. The growth in deep stab cultures is distinctive; the organisms grow away from the line of inoculation at right angles, producing an appearance which has been likened to an inverted pine tree. The growth fails to reach the surface of the medium, however, indicating the anaërobic nature of the bacteria. Milk appears to be a favorable medium for their development. A slight acidity is produced, but no coagulation or peptonization. Slightly alkaline, sugar-free broth overlaid with a layer of paraffin or paraffin oil[3] is a favorable medium; the organisms produce a well-defined turbidity after twenty-four to forty-eight hours' incubation at 37° C. which increases in intensity for about fourteen days, at the end of which time the growth begins to settle to the bottom of the flask. Cultures of tetanus bacilli usually possess a very disagreeable odor.

Conditions of Growth.—Bacillus tetani is an obligate anaërobe, but strains may be gradually accustomed to oxygen so that eventually they will grow slowly even in the presence of air. They lose their toxin-producing powers, however, under these conditions.[4] The

[1] Boulton and Fisch, Trans. Am. Phys., 1902, 463. It is occasionally possible to isolate tetanus bacilli directly from mixtures by inoculating dextrose agar with dilute suspensions of the suspected material, which has previously been heated to 80° *C.* for thirty minutes. The agar is drawn up into long, sterile glass tubes of approximately 5 mm. bore, and the ends sealed by heating. After forty-eight hours' incubation at 37° *C.* characteristic colonies are visible through the glass. The outside of the tube is carefully sterilized and cut with a file close to the desired organisms, which may be removed by a sterile capillary pipette.

[2] Ferran, Centralbl. f. Bakt., 1898, xxiv, 28.

[3] Park, Centralb. f. Bakt., 1901, xxix, 445.

[4] Ferran, loc. cit.

organisms grow well in a vacuum or in an atmosphere of hydrogen or nitrogen; they grow poorly or not at all in an atmosphere of carbon dioxide. Growth does not take place below 14° C. or above 45° C.; the optimum is 37° C. Growth is slow at 20° C. and spore formation proceeds sluggishly. At 37° C. growth and spore formation are optimum. Growth is fairly rapid at 43° C., but spore formation is greatly interfered with, and above 45° C. growth ceases.

The spores are very resistant to drying; when kept in the dark and cool they may survive for years. Henrijean[1] has found that tetanus spores may remain viable and virulent for nearly eleven years. The resistance of spores to heat is a subject on which there is great difference of opinion, Theobald Smith[2] has studied the resistance of tetanus spores under varying conditions, and his results are the most trustworthy available. In gelatin sporulation is relatively feeble and spores formed in this medium do not appear to be very resistant. He states that a majority of tetanus spores survive an exposure to flowing steam for forty minutes, occasionally for sixty minutes; and in one experiment a seventy-minute exposure did not destroy all spores. Morax and Marie[3] have found that dried spores are killed by an exposure to dry heat at 125° C. for twenty minutes. A 5 per cent. solution of carbolic acid kills tetanus spores in about ten hours; mercuric chloride in a dilution of 1 to 1000 kills them in three hours; the addition of 0.5 per cent. hydrochloric acid increases the germicidal action of both carbolic acid and mercuric chloride. A 1 per cent. solution of silver nitrate kills tetanus spores in one minute, and a 0.1 per cent. solution in five minutes. Iodoform is said to be particularly efficient.

Products of Growth.—Among the products of metabolism of the tetanus bacillus in sugar-free media are indol, hydrogen sulphide, and mercaptan, which impart an extremely disagreeable odor to cultures of the organism. Bacillus tetani ferments dextrose and maltose, producing acid, partly lactic, as well as considerable amounts of carbon dioxide and hydrogen. Bioses other than maltose, and polysaccharides are not fermented.

The most characteristic and striking metabolic product, however, is an extremely potent, soluble, extracellular toxin. This toxin, as Ehrlich has shown,[4] contains at least two distinct components in varying

[1] Ann. de la Soc. méd.-chir. de Liège, 1891, 367.
[2] Jour. Am. Med. Assn., 1908, l, 929.
[3] Ann. Inst. Past., 1902, 421.
[4] Berl. klin. Wchnschr., 1898, No. 12.

proportions, which may be recognized by their respective physiological actions: tetanospasmin, a neurotoxin, which is relatively thermostable and produces the characteristic tonic contractions or spasms which characterize the disease tetanus; and tetanolysin, a relatively thermolabile hemotoxin which dissolves red blood cells. It is doubtful if the tetanolysin is ordinarily of clinical importance.

Tetanus toxin appears to be produced only in sugar-free media under anaërobic conditions. Buchner[1] seems to have detected small amounts of true tetanus toxin in cultures of tetanus bacilli grown in a modified Uschinsky medium containing asparagin and certain inorganic salts. Brieger, on the contrary,[2] maintains that the toxin is produced only when the organisms are grown in albuminous media.

Tetanus toxin is best prepared in slightly alkaline peptone-meat infusion broth containing 0.1 per cent. of dextrose. The dextrose is added to insure a large initial development of bacteria which as soon as the sugar is exhausted (within twenty-four hours) attacks the protein constituents of the medium, forming from them the tetanus toxin.[3] It is essential to heat the medium to the boiling point and cool it rapidly immediately before inoculation to drive out all traces of oxygen. Anaërobic conditions are most easily obtained and maintained by overlaying the broth with pure paraffin oil, according to the method of Park.[4] Incubation should be maintained at 37° C. for seven to ten days. The toxin appears to lose somewhat in potency if incubated for a longer period. The potency of the toxin prepared in this manner varies considerably, being influenced by the composition and reaction of the medium and the degree of anaërobiosis. Tetanus bacilli retain their ability to produce toxin with great tenacity and regularity, even after prolonged artificial cultivation. At the end of the period of incubation the broth is rapidly filtered through sterile unglazed porcelain filters into dark-colored bottles, which are completely filled to exclude oxygen after the addition of 0.5 per cent. carbolic acid. It should be kept in a cool, dark place under anaërobic conditions. A small amount of the broth containing toxin thus obtained, freed from bacteria, will liquefy gelatin, thus showing that a peptonizing ferment is present in the filtrate, either inherent in the toxin or in association with it. According to Fermi and Pernossi,[5]

[1] München. med. Wchnschr., 1893, No. 24, 450.
[2] Ztschr. f. Hyg., 1895, xix, 102.
[3] Kendall, Boston Med. and Surg. Jour., 1913, clxviii, 825.
[4] Loc. cit.
[5] Centralbl. f. Bakt., 1894, xv, 303.

the gelatin-liquefying ferment (peptonizing ferment) has nothing to do with the toxin; it is quite distinct from it.

Properties of Tetanus Toxin.—Tetanus toxin is unstable. Exposure of broth filtrates containing tetanus toxin to 55° C. for an hour and a half, twenty minutes at 60° C., or five minutes at 65° C., reduces the potency to a very considerable degree.[1] For the complete destruction of the toxin, however, considerable heating is necessary. The toxin is particularly susceptible to light. According to Fermi and Pernossi,[2] fifteen to eighteen hours' exposure to daylight destroys it.[3] Tetanus toxin is destroyed by gastric and by tryptic digestion.[4] Dried tetanus toxin is more stable to physical agents and to heat than toxin in solution. Morax and Marie[5] have shown that dried tetanus toxin is not destroyed by an exposure to dry heat of 120° C. for fifteen minutes.

Purification of Toxin.—Tetanus toxin may be obtained in a partially purified state by precipitating the broth in which it is contained with saturated ammonium sulphate, dialyzing the salts from the precipitate and drying the salt-free residue in vacuo.[6] The dried toxin, if kept in a cool, dark place, remains potent for many months.

Tetanus toxin is one of the most potent known: as little as 0.0001 c.c. of the toxic broth frequently kills a 15-gram mouse. Purified toxin, prepared by precipitation with ammonium sulphate, will kill a mouse of the same weight if but 0.00005 gram is injected. Man and the horse are very susceptible to the tetanus toxin. Knorr[7] estimated that a gram of horse was twelve times as susceptible to the tetanus toxin as a gram of mouse, and three hundred times as susceptible as a gram of hen. The reptilia are practically non-susceptible: toxin injected into these animals circulates in the blood stream without causing symptoms and it is finally eliminated.

Action of Tetanus Toxin.—Even when massive doses of toxin are injected into susceptible animals, a latent period exists between the time of inoculation and the appearance of symptoms, which can not be reduced below eight hours.[8] The incubation period increases when

[1] Kitasato, Ztschr. f. Hyg., 1891, x, 267. [2] Ztschr. f. Hyg., 1894, xvi, 385.

[3] For a full discussion of the physical properties of tetanus toxin see Fermi and Pernossi, Centralbl. f. Bakt., 1894, xv, 303.

[4] Baldwin and Levene, Jour. Med. Research, 1901, vi, 120.

[5] Ann. Inst. Past., 1902, 419–420. [6] Brieger and Cohen, Ztschr. f. Hyg., 1893, xv, 8.

[7] München. med. Wchnschr., 1898, 321.

[8] Courmont and Doyen, Arch. de Phys., 1893; Goldschneider and Flatau (Kong. f. inn. Med., Berlin, June 11, 1897; Deutsch. med. Wchnschr., 1897, Vereinsbl. No. 18, 129; Fort. d. Med., 1897, 609) have noticed however, that changes in the anterior horn ganglion cells of the spinal cords of rabbits are demonstrable two hours after injection of tetanus toxin.

smaller amounts of toxin are used; symptoms may not appear until two or three days, or even a week after inoculation. Subfatal doses of tetanus toxin administered to experimental animals give rise to local symptoms which are frequently the only signs observed. The incubation period of the natural infection in man is usually about fourteen to sixteen days. It may be stated as a general rule that the shorter the incubation period, the higher the mortality. The site of inoculation of the tetanus toxin influences the character of the symptoms and the incubation period quite materially. Subcutaneous injections are usually followed by symptoms (spasms) which affect the muscles nearest the site of inoculation as a rule. Intravenous injections usually cause a generalized spasm.[1] When toxin is introduced directly into the central nervous system smaller doses cause death and the symptoms develop much more rapidly. There is great restlessness in these cases before the characteristic spasms occur, and the spasms are epileptiform in character. The toxin is supposed to exert a harmful effect on the central nervous system, which it reaches by way of the nerve trunks. Dönitz,[2] and Wassermann and Takaki[3] have shown that mixtures of brain tissue (especially the gray substance) and tetanus toxin are practically without effect when they are injected into susceptible animals, indicating that a firm union has taken place between the tissue and the toxin. This union will take place *in vitro*. The spleen, liver, kidney and other non-nerve-containing tissue have little or no neutralizing power for tetanus toxin. Metchnikoff[4] and Blumenthal[5] have determined experimentally that the brain tissue of pigeons and hens, which are almost refractory to tetanus toxin, possess but little neutralizing power for it.[6] Asakawa[7] has corroborated these results and has also shown that the toxin may circulate for some

[1] Ransom, Deutsch. med. Wchnschr., 1893. Marie and Morax, Ann. Inst. Past. 1902, xvi, 818.

[2] Deutsch. med. Wchnschr., 1897, 248.

[3] Berl. klin. Wchnschr., 1898, xxxv, 5.

[4] Ann. Inst. Past., 1898, 81.

[5] Deutsch. med. Wchnschr., 1898.

[6] There appears to be some combining power of the brain tissue of non-susceptible animals, as hens and pigeons, for tetanus toxin, however. A possible explanation for this phenomenon is furnished by Landsteiner and Von Eisler (Centralbl. f. Bakt., Orig., 1903, xxxiv, 567; 1905, xxxix, 315). They found that lipoids would combine with tetanus toxin at least to a limited degree. Levene (Biochem. Ztschr., 1911, xxxiii, 225; xxxiv, 495) has shown that tetanus toxin will unite not only with lipoids but with fats and similar substances. Marie and Tiffeneau (Ann. Inst. Past., 1908, xxii, 289, 644) have discovered that although a small amount of tetanus toxin may be bound by lipoidal substances in the brain in susceptible animals, the greater part of it is bound by albuminous substances. They believe that the essential albuminous substances necessary for this union are absent or inactive in non-susceptible animals.

[7] Centralbl. f. Bakt., 1898, xxiv, 166, 234.

time in the blood of these animals before it is excreted. Dönitz[1] and Knorr[2] have shown that tetanus toxin disappears rather rapidly from the blood stream of susceptible animals, on the contrary, and almost coincidently with its disappearance the symptoms become manifest. Wolff[3] states that the injection of tetanus toxin into experimental animals in small doses produces a lymphocytosis.

How Tetanus Toxin is Absorbed.—The brilliant researches of Meyer and Ransom[4] have shown that tetanus toxin is absorbed by the peripheral nerve end-organs and travels along the axis cylinders of the nerves to the central nervous system. The spasms, which are characteristic of tetanus, are supposed to be of central origin, and the experiments of Gumprecht[5] would suggest that this is the case. He cut the motor nerves to a limb and thus prevented the tonic contractions in that part. Zupnik[6] believes that the spasms may be either of peripheral or central origin, the symptoms elicited depending largely upon the reflex irritability of the medulla or cord. This view has not been substantiated.

Tetanus Antitoxin.—The injection of tetanus toxin in very small, sub-fatal doses, which are gradually increased, or of toxin weakened by chemicals, as iodine trichloride, induces immunity in horses or other susceptible animals, which is manifested by the gradual appearance of a specific antitoxin in the blood. This antitoxin will neutralize tetanus toxin both *in vitro* and *in vivo;* it will prevent the development of tetanus in experimental animals, provided it is given before or immediately following the injection of toxin. Dönitz[7] has shown that as many as twelve fatal doses of toxin may be neutralized by 1 c.c. of a 0.001 to 0.002 dilution of antitoxin, provided the toxin and antitoxin are mixed before injection. Four minutes after injection of 1 c.c. of toxin, 1 c.c. of 1 to 600 dilution of antitoxin is required for neutralization; eight minutes after injection of the same amount of toxin, 1 c.c. of 1 to 200 dilution of antitoxin is required to protect the animal, and fifteen minutes after the injection of 1 c.c. of toxin, 1 c.c. of 1 to 100 dilution of antitoxin is required. These experiments illustrate clearly the necessity of administering tetanus antitoxin at the earliest possible moment to obtain favorable results.

Inasmuch as the toxin appears to reach the central nervous system

[1] Deutsch. med. Wchnschr., 1897, No. 27.
[2] München. med. Wchnschr., 1898, Nos. 11 and 12.
[3] Berl. klin. Wchnschr., 1904, xli, 1273.
[4] Arch. f. exp. Pharm. u. Path., 1903, xlix, 369.
[5] Pflüger's Archiv, 1895.
[6] Deutsch. med. Wchnschr., 1900, 837.
[7] Ritchie, Jour. of Hyg., ii.

by way of the nerves, while the antitoxin circulates in the blood stream, it is not surprising, as Welch[1] has pointed out, that tetanus antitoxin has been disappointing as a curative agent. Used prophylactically it is very much more satisfactory. Flooding the nerves near the site of inoculation with antitoxin, or the intracerebral injection of antitoxin in desperate cases is sometimes successful.[2] Tetanus antitoxin has also been administered intraneurally and subdurally in desperate cases. Subcutaneous injection is comparatively inefficient. The subcutaneous injection of two hundred or more units at the site of infection, or, better, after exposure of the regional nerves, is said to be very efficient in preventing the development of tetanus. Calmette has used dried tetanus antitoxin to dust the navel of the newborn in the tropics and the deaths from tetanus neonatorum have been very greatly reduced by this procedure.[3] Bockenheimer has made a dressing composed of an ointment mixed with tetanus antitoxin, which is also said to be very efficient not only for the treatment of the umbilicus of the newborn, but for other wounds as well.

Tetanus antitoxin is less efficient than the diphtheria antitoxin for several reasons. First, the diphtheria antitoxin has a greater affinity for its toxin *in vitro* than the tetanus antitoxin has for tetanus toxin. Second, diphtheria toxin appears to infect principally the parenchymatous and lymphatic organs. The cells comprising these organs are less susceptible to toxin than are nerve cells, which are energetically attacked by tetanus toxin. The diphtheria toxin has less affinity for parenchymatous cells than it has for its antitoxin, and the diphtheria toxin, furthermore, circulates in the blood stream where the antitoxin also circulates when it is injected. Treatment, therefore, with diphtheria toxin is successful even after symptoms develop. Fourth, tetanus toxin has a considerably greater affinity for nerve cells than it has for its own antitoxin. The tetanus antitoxin is "picked up" by the end-organs of the nerves and reaches the central nervous system by the axis cylinders, while the antitoxin circulates in the blood and is not carried to the central nervous system by way of the nerves. Treatment with tetanus antitoxin, consequently, is rarely successful after symptoms appear and practically never successful after the symptoms have been developed for twenty-four hours.

[1] Bull. Johns Hopkins Hosp., July, 1895.

[2] Roux and Borrel, Ann. Inst. Past., 1898, No. 4. Chauffard and Quenu, La Presse Méd., 1898, No. 5.

[3] It must be remembered that the albuminous substances contained in the antitoxin, mixed with serum from the wound, make a favorable culture medium for many bacteria; the dressings must be sterile and watched carefully to safeguard the patient.

The Tetanus Antitoxin Unit.—The tetanus antitoxin unit of the United States may be defined as "ten times the minimal quantity of tetanus antitoxin necessary to protect a 350-gram guinea-pig against a standard dose of tetanus toxin obtained from the United States Public Health and Marine Hospital Laboratory." It has theoretically the power to neutralize one thousand minimal lethal doses of tetanus toxin, and it has, consequently, ten times the theoretical strength of the diphtheria antitoxin unit.

Distribution of Tetanus Bacilli in Nature.—Under ordinary conditions the tetanus bacillus appears to be a saprophyte, and man is not necessary for its continued existence. The organisms are found very commonly in the excrement of the herbivora, notably horses and cattle.[1] Sormani[2] has even claimed that the virulence of the tetanus bacillus is maintained by frequent passages of the organism through the intestines of the herbivora. Pizzini[3] has found tetanus bacilli in the feces of peasants who tended horses. Not all observers, however, subscribe to the intestinal theory. Hoffmann,[4] for example, found the organism only once out of twenty-two samples of feces from twenty-two different horses.

Tetanus spores are found widely distributed in nature, particularly in the upper layers of the soil; in temperate climates their distribution is somewhat irregular, but in the tropics they appear to be very widely disseminated. Tetanus spores also occur in gelatin occasionally, and they have even been detected in cat gut. Levy and Bruns,[5] and Anderson[6] have all found tetanus spores in commercial gelatin. The potential dangers attending the use of gelatin as a hemostatic are apparent.[7] Tetanus spores have also been found in vaccine virus in the past,[8] and Carini[9] has found spores in vaccine virus; and at least two outbreaks of tetanus, one in this country and one in Europe, have resulted from the infection of diphtheria antitoxin with tetanus spores. Rabinovitch[10] has also found tetanus spores in washings from strawberries sold in Berlin.

[1] Sanchez, Toledo, and Baillon, La Semaine Méd., 1890, No. 45; Centralbl. f. Bakt. 1890, ix, 18.
[2] Beband. der 10th Intern. med. Kong., Berlin, 1890, v, 152.
[3] Riv. d'igiene e san. publ., 1898, x, 170.
[4] Hyg. Rund., 1905, xv, 1233.
[5] Grenzgeb. der Med. u. Chir., 1902, x, 235; Deutsch. med. Wchnschr., 1902, 130.
[6] Mar. Hosp. Lab. Bull., 1902, ix.
[7] Zibell, München. med. Wchnschr., 1901, 1643, for literature.
[8] McFarland, Lancet, September, 1902.
[9] Centralbl. f. Bakt., Orig., 1904, xxxvii, 48.
[10] Arch. f. Hyg., 1907, lxi, 103.

Pathogenesis.—Tetanus occurs spontaneously in man, horses, cattle and sheep, rarely in dogs and goats. Birds and reptilia are highly refractory to experimental inoculation. The disease tetanus both in man and animals is purely toxic in character; notwithstanding the wide distribution of tetanus spores, it is relatively uncommon. It may follow traumatism, particularly deep, narrow wounds and contused wounds to which tetanus spores, together with other organisms gain entrance. In the tropics an infection of the umbilicus of the newborn (tetanus neonatorum) is very common.[1] Postpartum infections, particularly of the uterus (tetanus puerperalis), were also at one time very common.[2]

The lesions observed in tetanus are very slight and postmortem there may be no marked changes other than a slight congestion of the internal organs. Bacilli may occasionally be found at the site of inoculation, but they do not as a rule penetrate deeply into the body, although Hochsinger[3] and Creite[4] have found the organisms at autopsy in a very few instances in the spleen and heart blood.

Tarozzi[5] and Canfora[6] have studied the fate of tetanus spores after subcutaneous inoculation into guinea-pigs and rabbits very carefully. They find the spores may be transmitted rather rapidly to the parenchymatous organs, liver, spleen, and kidneys principally, where they may remain alive but latent for seven to eight weeks. If trauma or injury resulting in inflammation occurs during this time, acute or chronic tetanus may result. These observations suggest a possible explanation for the so-called cryptogenetic, ideopathic, or rheumatic tetanus; the intestinal tract is supposed to be an occasional portal of entry, thus explaining another source of cryptogenetic tetanus.

Experimental Pathogenesis in Animals.—The disease tetanus may be produced in susceptible animals by injecting soil or active cultures of tetanus bacilli, spores mixed with tetanus toxin, or tetanus toxin alone. If, however, tetanus spores carefully freed from toxin are injected alone, tetanus frequently fails to develop. Vaillard and Vincent[7] and Vaillard and Rouget[8] have furnished an interesting

[1] Anders and Morgan, Jour. Am. Med. Assn., 1906, xlvii, 2083.
[2] Stern, Deutsch. med. Wchnschr., 1892, No. 12. Heyse, Berl. klin. Wchnschr., 1893, No. 24.
[3] Centralbl. f. Bakt., 1887, ii, 145. Hohlbeck, Deutsch. med. Wchnschr., 1903, 172.
[4] Centralbl. f. Bakt., Orig., 1904, xxxvii, 312.
[5] Ibid., 1905, xxxviii, 619.
[6] Ibid., 1908, xlv, 495.
[7] Ann. Inst. Past., 1891, 24.
[8] Ann. Inst. Past., 1892, 428; Centralbl. f. Bakt., xvi, 208.

explanation for this possibility. They find that phagocytosis plays an important part in the removal of tetanus spores which are injected without tetanus toxin or other irritating substances. Polymorphonuclear leukocytes engulf free tetanus spores. If, however, the spores are introduced into the body in collodion capsules, thus protecting the organisms from the leukocytes, the tetanus spores develop into bacilli there, form toxin, and produce tetanus. If tetanus spores are mixed with lactic acid, with tetanus toxin, or with other irritants, or even injected with saprophytic bacteria, the spores develop into tetanus bacilli, produce toxin and kill the animal.

Bacteriological Diagnosis.—1. *Microscopical.*—Smears made from the pus of wounds in suspected cases of tetanus may show the characteristic spores of the tetanus bacilli. The organisms, however, are usually present in very small numbers and several smears should be made. Negative results do not prove the absence of the tetanus bacillus.

2. *Cultural.*—Pus from wounds scraped out with sterile curettes, or suspected material is placed in fermentation tubes containing bits of sterile tissue, according to Theobald Smith's method mentioned above, incubated for forty-eight hours and examined microscopically for typical spores. If these are found the material is heated to 80° C. for thirty minutes to kill vegetative forms and then reinoculated to obtain growths of the organism.

3. *Toxin.*—Inoculation of material containing tetanus bacilli and other organisms into slightly alkaline broth (sugar-free) grown anaërobically for six or eight days will lead to toxin formation even if other bacteria are present. Inoculation of this toxic broth into mice will frequently give positive results. Broth obtained according to the Theobald Smith method in Step 2 also should be inoculated into mice if the preliminary microscopic examination shows tetanus spores.

4. At times tetanus toxin occurs even in the blood of the patient, provided no antitoxin has been administered; 1 c.c. of this blood inoculated into a mouse may occasionally produce characteristic tetanic phenomena.

Prophylaxis.—Any wound likely to be a suitable portal of entry for the tetanus bacillus should be regarded as potentially dangerous and tetanus antitoxin should be administered *promptly* as a prophylactic measure. Fifteen hundred units of tetanus antitoxin is the ordinary prophylactic dose in such cases. For curative doses 3000 to 20,000 units have been injected locally, intraneurally or subdurally, depending upon the condition of the patient and the time which has elapsed since infection took place. The results are usually unsatisfactory if symptoms of tetanus have developed, but the treatment should be carried out energetically.

BOTULISM OR ALLANTIASIS.

A rather definite train of symptoms consisting of gastro-intestinal irritation, nervous disturbances, bulbar paralysis, dysplagia and protrusion of the eyeballs with, however, no fever, has occasionally followed the consumption of uncooked or imperfectly cooked meats or fish. Uncooked products, particularly ham and sausages, are more commonly the source of these intoxications. The mortality is fairly high in such cases, amounting to as much as 25 per cent. in various epidemics. Patients retain consciousness to the end as a rule.

The best-studied epidemic of this type was one which occurred in Ellezelles, Belgium. Von Ermengem[1] investigated this epidemic very thoroughly and found that all the cases had partaken of an imperfectly cured ham, from which he isolated an organism which he called B. botulinus. He established the relationship of the organism to the disease which resulted from the ingestion of the toxins of this bacillus by animal experimentation.

Morphology.—Bacillus botulinus is a rather large bacillus, measuring from 0.9 to 1.2 microns in diameter by 4 to 6 microns in length, with rounded ends; it occurs singly or in pairs, less commonly in short chains of three to six elements. Old cultures of this organism and those incubated above 36° C. show involution forms which are usually long, intertwined filaments. The organism is sluggishly motile and has from 4 to 8 peritrichic flagella. It forms oval spores, slightly greater in diameter than the rod and situated near one end of it. The organism stains readily with anilin dyes and is Gram-positive.

Isolation and Culture.—Bacillus botulinus grows most characteristically in slightly alkaline dextrose gelatin incubated at 25° C. under strictly anaërobic conditions. The colonies, which grow with moderate rapidity, are light yellow in color, nearly transparent, and are composed of coarse granules. These granules after a few hours' growth exhibit a slow but constant motion in a zone of liquefied gelatin. As

[1] Centralbl. f. Bakt., 1896, xix, 442; Ztschr. f. Hyg., 1897, xxvi, 1.

they reach their maximum development the colonies become brown and opaque, and only those granules at the periphery of the colony remain motile.

Growth in Artificial Media—The organism grows well in the ordinary nutrient media, better when dextrose is added, but only under anaërobic conditions. A strong odor of butyric acid is characteristic of growths of the organism in artificial media. It is essential to transfer large amounts of material to insure growth of the organism. Gelatin is liquefied. The growth on agar is very similar to that in gelatin, except that no liquefaction takes place and no motile granules appear in the colonies. A slight turbidity is developed in plain broth after twenty-four hours' incubation, a heavy turbidity in dextrose broth. The organism grows well in milk, producing a slightly acid reaction but neither coagulation nor peptonization.

The organism is an obligate anaërobe, whose optimum temperature of growth is 22° to 25° C. It grows but slowly at 25° C. Incubation at the latter temperature leads to the development of involution forms and an inhibition of spore formation and toxin production. The spores are not particularly resistant to heat or disinfectants and cultures die out in three to four weeks unless transferred to fresh media within that time. The spores are killed by an exposure at 80° C. for sixty minutes. Five per cent. carbolic acid kills them in twenty-four hours and pickling in 10 per cent. salt solution kills them within a week. If the spores are protected from oxygen and sunlight they retain their vitality for several months, either in a moist condition or dried.

Products of Growth—The organism produces an active soluble gelatinase in plain broth cultures and in gelatin, particularly the latter. It forms acid and gas in dextrose broth; bioses and polysaccharides are not fermented. The acid formed is partly butyric, and the gas consists principally of carbon dioxide and hydrogen.

The most important product of B. botulinus, however, is a potent extracellular toxin which is readily prepared by growing the organisms anaërobically in sugar-free broth at 25° C. for two weeks. The broth is filtered through sterile porcelain filters, preferably in an atmosphere of hydrogen, and the toxin is found in the filtrate, from which it can be precipitated by the addition of a 3 per cent. aqueous solution of zinc chloride in the proportions of two parts of zinc chloride to one of broth.[1] The toxin deteriorates rather rapidly if it is exposed

[1] Brieger and Kempner, Deutsch. med. Wchnschr., 1897, xxxiii, 521.

to sunlight or oxygen. If it is kept in the dark in sealed, full bottles and kept cool it retains its potency for some months. It keeps still better dried in the absence of light and moisture. Heat promptly inactivates it. An exposure at 58° C. for three hours, or at 80° C. for thirty minutes utterly destroys its potency. It is not, however, destroyed by putrefaction or by gastric digestion, a point of great importance clinically, for poisoning with the toxin of B. botulinus almost always results from its absorption from the intestinal tract. The toxin has also been isolated from hams in which the organisms have grown. The hams are macerated with water in a cool, dark place, filtered through porcelain, and the filtrate is found to contain the toxin. The toxin also is produced when the organisms grow under proper conditions in vegetables.[1] The toxin causes death when injected subcutaneously or fed to experimental animals. There is a latent period which elapses between the time of administration of the toxin and the appearance of symptoms. This latent period when large doses are administered is from twelve to twenty hours; with moderate doses it is about thirty-six hours. One-thousandth c.c. of broth containing toxin injected subcutaneously into guinea-pigs usually kills them in three to four days; 0.1 to 0.5 c.c. of the same toxin absorbed in bread and fed to rabbits results fatally in from four to six days. It is toxic for man, white rats, mice, kittens, guinea-pigs, rats, and even monkeys in relatively small doses. In larger doses it is also pathological for cats and doves. The toxin is bound by the gray matter of the central nervous system. Cholesterin, lecithin, and fats such as butter and oils are believed to bind the toxin as well.

Antitoxin.—Kempner[2] has succeeded in immunizing goats to the toxin of B. botulinus, and has identified in their serum a specific antitoxin which has considerable potency both curatively and prophylactically. Wassermann has been able to immunize horses with the same results. The antitoxin neutralizes the toxin both *in vivo* and *in vitro*.[3] Leuchs has shown that dilute acids will split up the toxin-antitoxin combination into the two components, both of which may be recovered.

Pathogenesis—The lesions produced by the toxin both in man and in animals are very similar, and the symptoms produced are referable to the action of the toxin on the medulla and cord.[4] There is bulbar paralysis, paralysis of the eye muscles, great muscular weakness,

[1] Landmann, Hyg. Rundschau, 1894, 449.
[2] Ztschr. f. Hyg., 1897, xxvi, 482.
[3] Forssman and Lundstrom, Ann. Inst. Past., 1902, 294.
[4] Kempner and Scheplewsky, Ztschr. f. Hyg., 1898, xxvii, 214.

profuse nasal and oral discharge, aphagia, aphonia, and interference with the workings of the cardiac and respiratory centres. Microscopically there are degenerative changes limited chiefly to the cells of the gray matter of the medulla, cord and salivary glands.[1]

The disease produced by ingestion or injection of toxins of B. botulinus in experimental animals reproduces faithfully the symptom-complex seen in the naturally acquired disease in man. The organism itself does not appear to grow in the tissues of warm-blooded animals except just before and after death, hence it is logical to conclude that the ingestion of food containing the toxins of this organism rather than the generation of the toxin in the tissues of the host is the source of intoxication.

Bacteriological Diagnosis.—The bacteriological diagnosis can not be made ordinarily in man. It is necessary in the vast majority of instances to obtain the meat in which the organisms have grown.

(*a*) *Microscopic.*—This is usually not feasible.

(*b*) *Cultural.*—Make anaërobie dextrose gelatin plates from the suspected meat, selecting portions which are removed from contaminated surfaces, as follows: (1) Rapidly make a maceration of some of the meat in sterile salt solution. (2) Heat some of the opalescent fluid to 60° C. for thirty minutes, and make plates. (3) Add some of the opalescent fluid to fermentation tubes according to Theobald Smith's method (see page 473) with bits of sterile animal tissue. (4) Plate some of the opalescent fluid directly without heating into dextrose gelatin plates. (5) Examine the media for characteristic colonies.

(*c*) *Identification of Toxin.*—1. Filter some of the macerated meat rapidly through sterile filter paper and inject 0.5 to 1 c.c. subcutaneously into a rabbit or guinea-pig. The protruding eyeballs and respiratory failure usually suffice to establish the diagnosis, which may be confirmed by staining sections of the central nervous system and identifying the lesions. (2) Add 2 to 5 c.c. of the filtrate to some bread and feed a rabbit with it. Note the symptoms. (3) Filter some of the broth from the fermentation tube in Step 3 of the cultural identification and inject subcutaneously or feed to a rabbit and observe symptoms.

(*d*) *Inspection of Suspected Meat.*—It is difficult usually to detect anything abnormal in meat in which B. botulinus has grown. Occa-

[1] Marinesco, Compt. rend. soc. de biol., 1896. Kempner and Pollak, Deutsch. med. Wchnschr., 1897, xxxiii, 521.

sionally a slight odor of butyric acid is noticed; usually there is no sign recognizable either by smell or taste which will furnish a clue to the unfitness of the meat for food.

Prophylaxis.—The disease is not contagious and patients are not a source of danger to others. The organisms are not as a rule found in man. The toxin is thermolabile; consequently thorough cooking of foods will eliminate all danger. Hams, similar meats and meat products alone cause the disease. If such meats are cured by pickling they should be immersed in the pickle not less than a week and the pickle should contain the equivalent of 10 per cent. salt solution.

BACILLUS AËROGENES CAPSULATUS.

Historical.[1]—This organism was first described by Welch in 1891, and later in detail by Welch and Nuttall.[2] It appears to be identical with Bacillus phlegmonis emphysematosæ,[3] B. perfringens,[4] B. emphysematis vaginæ,[5] and possibly B. enteritidis sporogenes[6] and Granulobacillus saccharo butyricus immobilis liquefaciens.[7] It is commonly referred to as the "gas bacillus." The organism has been described most commonly in the past as the causative agent of the so-called "foamy organs." It was isolated by Welch from such a case in 1891, and it has been isolated many times since from similar lesions.

Morphology.—B. aërogenes capsulatus is a rather large bacillus, measuring from 1 to 1.2 microns in diameter and from 2 to 5 microns in length, with somewhat square-cut ends, occurring usually singly or in pairs; in artificial culture media rarely in short chains. According to Welch, the organism tends to form chains in bloodvessels. The organisms under these conditions may be somewhat shorter than those typically found in artificial media, frequently being but 1.5 to 2 microns in length.

The organism is non-motile and possesses no flagella. It forms capsules in the animal body and occasionally in albuminous media. It also forms spores, first observed by Dunham.[8] The spores are

[1] For an excellent study and critical summary see Simonds, Monograph V, Rockefeller Institute for Medical Research, September 27, 1915.
[2] Johns Hopkins Bull., 1892, iii, 81.
[3] Fränkel, Centralbl. f. Bakt., 1893, xiii, 13.
[4] Veillon and Zuber, Arch. de méd. éxper. et d'anat. path., 1898, x, 517.
[5] Lindenthal, Wien. klin. Wchnschr., 1897, x, 3.
[6] Klein, Centralbl. f. Bakt., 1895, xviii, 737.
[7] Schattenfroh and Grassberger, Centralbl. f. Bakt., ii abt., 1899, v, 209; München. med. Wchnschr., 1900, Nos. 30–31; Wien. klin. Wchnschr., 1900, No. 48.
[8] Johns Hopkins Bull., 1897, viii, 68.

oval, somewhat less in diameter than the vegetative form of the organism, and are usually situated near one end of the rod. But one spore is found in a single organism. Spores are apparently not formed in the tissues of the body. The organism stains readily with ordinary anilin dyes. It is Gram-positive, although old cultures on artificial media exhibit irregularities in staining, probably due to beginning degeneration.

Isolation and Culture.—The organism is an obligate anaërobe. It grows well in all ordinary media containing dextrose or lactose. From tissues it is best obtained on anaërobic agar plates, where the colonies are round, semi-translucent and colorless, and not characteristic. Many strains hemolyze blood and on blood agar the colonies are surrounded by a rather narrow zone of hemolysis. From the intestinal

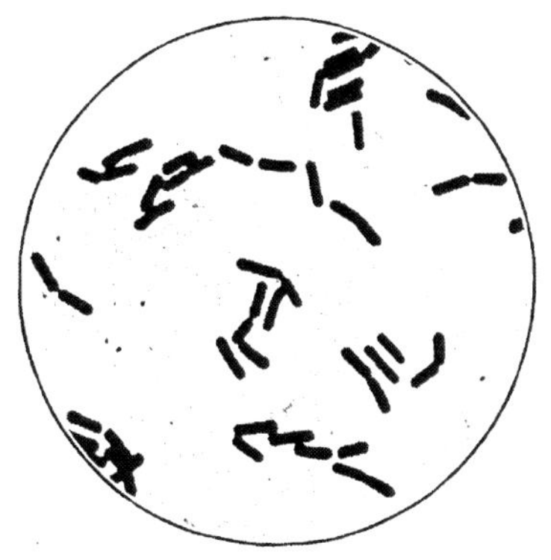

Fig. 66.—Bacillus aërogenes capsulatus from pure milk culture. × 1000.

contents the organism is best isolated in milk. A thin suspension of feces is emulsified in milk (whole milk) after the milk has been boiled and rapidly cooled to remove all oxygen. The milk is heated to 80° C. for twenty minutes to kill vegetative organisms, and then it is incubated at body temperature for eighteen to twenty-four hours. At the end of that time the milk exhibits a characteristic stormy fermentation. The casein is reduced in amount and the residual casein is full of holes and is usually slightly pink in color. The whey is usually colorless, gas bubbles are seen at the top of it, and there is characteristically an odor of rancid butter—butyric acid. The organism may be obtained from the milk culture directly by plating anaërobically on agar, or it may be obtained by injecting some of the whey into the ear vein of a rabbit, killing the animal after five minutes and incubating it for twelve to eighteen hours.[1] The rabbit will be found to be enormously distended with gas. The tissues, particularly the muscles, will be found to be soft, partly liquefied, and the course of the blood-vessels will be marked out by rows of gas bubbles. The organisms are found in greatest abundance in the liver, which is light colored[2] and in typical cases so thoroughly fermented that it appears to be a colleetion of gas bubbles. The gas bubbles found in the blood stream and in the muscles and particularly in the liver are the result of the decomposition of the muscle sugar and glycogen by this organism.

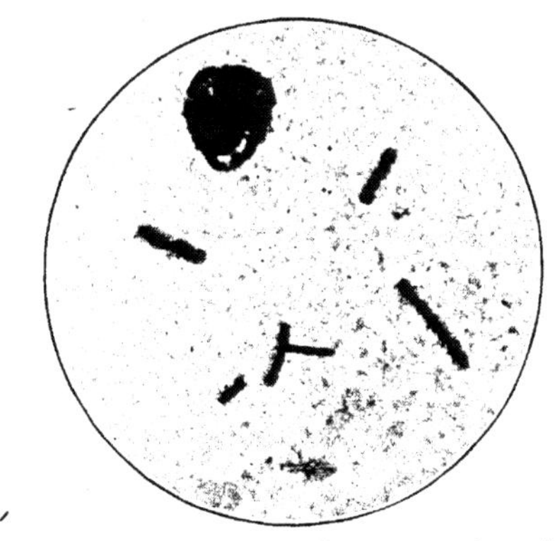

Fig. 67.—Bacillus aërogenes capsulatus, capsule stain. × 1000.

Growth on Artificial Media.—Anaërobic growth on gelatin is variable; some strains do not grow in this medium, others produce a slight liquefaction. In plain broth there is a slight turbidity; in broth containing dextrose or lactose the turbidity is marked. The reaction in milk has been described previously, the characteristic features being a stormy fermentation, a slight pink color to the undissolved casein, and gas bubbles together with a slight odor of butyric acid. If the milk has not been heated sufficiently to remove all oxygen the organism frequently produces coagulation, but no stormy fermentation and no gas.

Conditions of Growth.—The organism is an obligate anaërobe which does not grow below 20° C., or above 45° C. The optimum temperature of growth is 37.5° C. The spores are quite resistant; five minutes' boiling usually fails to kill them. They are extremely resistant to

[1] This procedure is frequently known as the "Welch Nuttall Test."

[2] The absence of darkening of the liver tissue indicates that little or no proteolysis is taking place; otherwise the liver would be discolored, due to the production of sulphide of iron from the liberation of H_2S of protein and its reaction upon the blood.

drying, particularly in the absence of sunlight. Viable spores have been obtained from dust in a vault which had not been opened for fifteen years. Sporulation does not take place, as a rule, in the tissues; spores are frequently found in the intestinal tract. They do not form readily in media containing utilizable carbohydrates, but are found on the surface of slanted blood serum[1] and in protein media. Simonds[2] finds that an acidity greater than 1 per cent. to phenolphthalein inhibits spore formation.

Products of Growth.—B. aërogenes capsulatus forms a gelatinase in the absence of utilizable sugars. In dextrose, lactose and saccharose media it produces an energetic fermentation, the products being

FIG. 68.—Bacillus aërogenes capsulatus, smear from liver of rabbit. × 1000.

butyric and lactic acids, carbon dioxide and hydrogen in the proportions $H:CO_2 = \frac{1}{2}$ approximately;[3] comparatively little acid is formed. Welch and Nuttall state that the organism decomposes protein with the formation of carbon dioxide and hydrogen and nitrogen gas; but it is probable that little or no gas is formed from protein. According to Brown,[4] the organism forms a toxin in sugar-free broth, which is pathogenic for guinea-pigs. The toxin is not formed in broth containing utilizable carbohydrates.

Simonds[5] has distinguished four distinct types or subgroups of Bacillus aërogenes capsulatus, which differ essentially in their fermentation of certain sugars and in their sporulation as follows:

[1] Dunham, loc. cit.
[2] Loc. cit., p. 31.
[3] Smith, Brown, and Walker, Jour. Med. Research, 1905–1906, xiv, 193.
[4] Annual Report, Massachusetts State Board of Health, 1909.
[5] Loc. cit., p. 13.

Sub group.	Fermentation.[1]								Spores.[2]				
	Dextrose.	Lactose.	Saccharose.	Maltose.	Starch.	Glycerin.	Inulin.	Mannite.	Mannite broth.	Inulin broth.	Glycerin broth.	Plain broth.	Plain broth, egg albumen.
I	+	+	+	+	+	+	+	−	+	−	−	+	+
II	+	+	+	+	+	+	−	−	+	+	−	+	+
III	+	+	+	+	+	−	+	−	+	−	+	+	+
IV	+	+	+	+	+	−	−	−	+	+	+	+	+

Pathogenesis.—The pathogenesis of B. aërogenes capsulatus is very variable. The production of emphysematous gangrene in contused wounds and compound fractures is the best known of its pathogenic properties. According to Achalme,[3] the organism has been isolated from the blood stream in cases of acute articular rheumatism. This observation has been made by others also. It has not been proven, however, that the organism causes acute articular rheumatism. In the intestinal tract[4] the organism occasionally produces disease which varies in severity from a mild diarrhea to an extremely acute dysenteric diarrhea. Epidemics of such diarrhea appear to have been traced in a few instances to milk.[5] The organism appears to cause an intense irritation in the intestinal tract, probably due to the production of butyric acid, but there is no evidence that the intestinal infection is a true toxemia.[6] Usually the organism is not invasive, but in a few cases the mucosa of the large intestine has been distinctly involved. The mucosa was enormously swollen and edematous and the organisms were found deep in the submucosa. In one instance at least the organism has been isolated from tonsils in a case of chronic

[1] Gas and acid.

[2] This table is in harmony with the view that the organism does not, as a rule, sporulate in media containing utilizable sugar. It is probable that the acid products of fermentation inhibit sporulation, as Simonds has shown.

[3] *Compt. rend.*, Soc. de biol., 1891, xliii, 651; 1897, xlix, 276; Ann. Inst. Past., November 25, 1897.

[4] Howard (Johns Hopkins Hosp. Rep., 1900, ix, 461) states that the organism may develop in the gastric or intestinal mucosa, especially under the folds of the valvulæ conniventes, and cause disintegration of the tissue.

[5] Klein, Annual Report of the Medical Officer of the Local Government Board, London, 1897–1898, No. 27, p. 210.

[6] Kendall and Smith, Boston Med. and Surg. Jour., 1911, clxiv, 306. Kendall, Day and Bagg, ibid., 1913, clxix, 741. Kendall and Day, ibid., 753. Kendall, ibid., May 20, 1915.

hypertrophic tonsillitis; the organisms were deep in the tissues, where they could be distinctly seen in sections, and they did not form spores; at least none could be demonstrated by the ordinary methods. When the organism was isolated from the tonsillar tissue it appeared to have lost its fermentative powers to a very considerable degree, but rapidly regained them with repeated transfer in artificial media. Similarly, the organism has been isolated from the petrous portion of the temporal bone bilaterally in an infant which died of a severe gas bacillus infection of the intestinal tract.

Distributions.—The organism is found in sewage, in impure water, in dust, very frequently in the intestinal tract of man, and probably of animals. In the past B. aërogenes capsulatus has almost undoubtedly been confused with the bacillus of malignant edema. Thus, Grigorjeff and Ukke[1] found an organism complicating typhoid fever, which produced typical foamy organs; this they identified as the bacillus of malignant edema. It is very probable that this organism was in reality the gas bacillus, as was the organism described by Brieger and Ehrlich[2] in a somewhat similar case.

BACILLUS ŒDEMATIS MALIGNI.

Historical.—The bacillus of malignant edema is the oldest known anaërobic organism of which there is an authentic description. It was described by Pasteur, Joubert and Chamberland,[3] later by Koch.[4] Pasteur and his associates obtained the bacillus from a localized epidemic of acute septicemia in small animals, characterized by a local edema at the site of infection. They reproduced the disease by inoculating the organism into other animals, or by the injection of putrescent animal tissues. The bacillus was called Vibrion septique. Koch studied malignant edema in larger animals and called attention to the localized edema and the absence of generalized sepsis, which are the characteristic features of the disease. Koch called the organism Bacillus œdematis maligni.

Morphology.—B. œdematis maligni is a slender rod, 0.8 to 1 micron in diameter by 2 to 10 microns in length, with rounded ends, frequently occurring in long chains, particularly in the animal body. It is motile under anaërobic conditions and possesses numerous peritrichic flagella, usually about twenty. No capsule has been observed. Sporulation

[1] Centralbl. f. Bakt., 1899, xxv, 253. [2] Berl. klin. Wchnschr., 1882, No. 44.
[3] Bull. de l'acad. de Science, 1878, lxxxvi, 1038.
[4] Mitt. a. d. kais. Gesamte, 1881, i, 52.

takes place readily, and the spores occur typically in the centre of the rod, giving it a slightly swollen appearance. The organism stains readily with ordinary anilin dyes and is usually regarded to be Gram-negative, although some claim it is Gram-positive.[1]

Isolation and Culture.—B. œdematis maligni is a strict anaërobe, and the organisms are best obtained in pure culture from the edematous lesions produced in rabbits or in guinea-pigs by inoculation of them with garden soil. The organism grows readily under anaërobic conditions on dextrose agar, and the colonies produced are very filamentous.

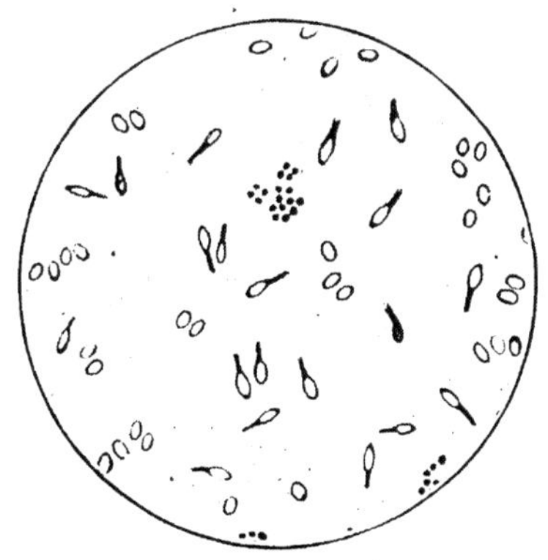

FIG. 69.—Bacillus œdematis maligni, spore formation. × 1000. (Kolle and Hetsch.)

The organism grown anaërobically on gelatin produces similar colonies; the gelatin is liquefied in from three to five days.[2] The colonies are rather small on this medium, exhibit radiating edges, and are surrounded by a liquefied zone.

Milk is both coagulated and peptonized, but no gas is formed in it.

Blood serum is rapidly liquefied and it is an excellent medium for the development of this organism.

Artificial cultures possess an offensive odor. Broth (anaërobic) is clouded by the organism after twelve to twenty-four hours' incubation. but usually clears up after three to six days. The organisms grow much better in albuminous media, particularly those containing blood serum.

Growth does not take place below 15° C. nor above 42° C. The optimum temperature is 37° C. The organism is an obligate anaërobe and sporulation only takes place anaërobically.

[1] Kutscher, Ztschr. f. Hyg., 1894, xviii, 339. Claudius, Ann. Inst. Past., 1897, 335.
[2] Liborius, Ztschr. f. Hyg., 1886, i, 159.

Resistance to Physical Agents.—The vegetative cells are not resistant to heat, three to five minutes' exposure to 60° C. killing them. The spores are very resistant to drying and heating; an exposure to 80° C. for several hours is necessary to kill them, and from thirty to sixty minutes' exposure to 90° C. Sunlight will not kill the organisms even after several days' exposure, and in the dark the spores may remain alive for many years.[1]

Products of Growth.—B. œdematis maligni forms a gelatinase, casease, and apparently a non-specific proteolytic ferment as well. The disagreeable odor noticed in protein media is due to indol, hydrogen sulphide and probably mercaptans. Acid, chiefly butyric and lactic, and gas, probably carbon dioxide and hydrogen, are produced in dextrose broth.

Toxin.—It has been claimed that the bacillus of malignant edema produces a soluble toxin. It is found that these organisms grown anaërobically in plain broth for several days do develop a slight toxicity, which can be demonstrated by filtering the broth through sterile unglazed porcelain filters and injecting several cubic centimeters of the filtrate into guinea-pigs; they die after a longer or shorter time. It is also claimed that the organism produces a leukocidin which destroys leukocytes.

Pathogenesis.—The virulence of cultures of the malignant edema bacillus varies very considerably. Infection rarely or never occurs in man. Brieger and Ehrlich[2] have reported two cases of typhoid fever which terminated fatally after an invasion by an organism morphologically like the malignant edema bacillus, which produced rather extensive edema in the tissues. It is quite possible that this organism was in reality, however, the gas bacillus. In small laboratory animals, as rabbits and guinea-pigs, the organism typically produces a rapidly fatal septicemia with considerable edema at the site of injection. In larger animals, horses, cattle, sheep and swine, the edema is more pronounced, as Koch[3] pointed out, and the organism tends to remain localized at the site of inoculation. As a rule there is no general septicemia. In wound infections with this organism the incubation period is from one to two days. Infection only takes place in deep or contused wounds where oxygen is absent.

Like the tetanus bacillus, the spores of the malignant edema bacilli, freed from adherent culture media or other organisms, do not as a

[1] von Székely, Ztschr. f. Hyg., 1903, xliv, 363.
[2] Berl. klin. Wchnschr., 1882, No. 44.
[3] Loc. cit.

rule lead to infection. The spores are taken up by phagocytes. If the spores are mixed with culture filtrates, with weak acids, or with other organisms, infection usually takes place. The lesions vary with the virulence of the organism. Organisms of moderate virulence produce edema at the site of inoculation which tends to spread. The regional muscles are very hyperemic with bubbles of gas in them, the tissue crepitates, and there is a disagreeable odor. The edema is less marked and death takes place in a few hours when organisms of greater virulence are injected. Animals appear to be immune after one attack.

Distribution.—The organism appears to be very widely distributed in the soil and in dust.

Prophylaxis.—Prophylaxis consists essentially in immediate surgical treatment of wounds to which the organism might gain entrance.

BACILLUS ANTHRACIS SYMPTOMATICI.

Historical.—The disease variously known as black leg, quarter evil, symptomatic anthrax, or Rauschbrand is a disease of cattle chiefly. It is less commonly found in sheep and goats. The organism was first obtained in pure culture by Arloing, Cornevin, and Thomas.[1] The organism is also known as B. chauvei and B. sarcophysematis bovis.

Morphology.—Morphologically, it is a rod-shaped bacillus, 0.6 to 1 micron in diameter and from 2 to 5 microns in length, occurring singly and in pairs. It practically never forms chains, differing in this respect from the bacillus of malignant edema. The organisms are straight and rigid and have square-cut ends. They are motile and possess many peritrichic flagella and form no capsules. Spores occur in the centre of the organism typically, less commonly nearer one end, and the organism is slightly swollen because the spores are slightly greater in diameter than the rod itself. It stains readily with ordinary anilin dyes and is Gram positive.

Isolation and Culture.—The bacillus of symptomatic anthrax is an obligate anaërobe which grows rather poorly in artificial media, particularly in the first transfer from the animal body. Albuminous media, as blood serum, or blood agar, are better adapted for its isolation than ordinary media. It grows particularly well in fermentation tubes containing sterile tissue, according to Theobald Smith's method.

[1] Le Charbon, Symptomatique du Boeuf, Paris, 1887.

Material for inoculation is best obtained from the heart's blood, the local swelling, or the peritoneal exudate of an animal dead of the disease. The material should be sown anaërobically on ascitic or blood agar plates or upon dextrose agar, the latter medium not being as satisfactory. Pure cultures may be obtained readily by inoculating guinea-pigs with morbid material and transferring some of the heart's blood of the animal immediately after death to artificial media.

Growth on Artificial Media.—The organism grows to a limited extent in plain broth if oxygen is excluded. It grows better in dextrose broth. On anaërobic dextrose gelatin and dextrose agar plates the colonies are round, oval, grayish, and possess distinctly filamentous

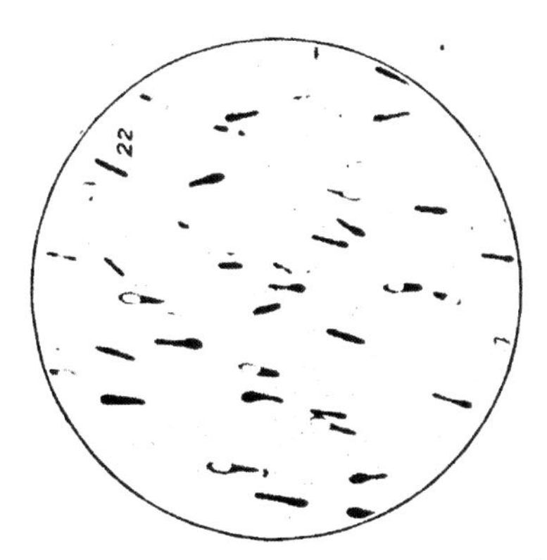

Fig. 70.—Bacillus of symptomatic anthrax spore formation. × 1000.

edges. Gelatin is liquefied in from two to four days. Milk is a good medium; the organism forms a slight amount of acid, but no coagulation or peptonization takes place.

Conditions of Growth.—B. anthracis symptomatici is an obligate anaërobe which does not grow below 14° C. nor above 44° C. The optimum temperature is 37° C. The spores are extremely resistant to heat; half an hour's exposure to 100° C. does not always kill them. The spores appear to be able to remain latent in the animal body. The virulence of vegetative organisms developing from spores is said to be greatly reduced by heating the spores to 100° C. for two to three minutes.

Products of Growth.—The organism forms a gelatinase. In dextrose broth it produces carbon dioxide, hydrogen, and traces of methane, as well as butyric and lactic acids.

Toxin.—According to Leclainche and Vallée,[1] and Grassberger and Schattenfroh,[2] the filtrates of broth cultures of the bacillus are slightly toxic to guinea-pigs in large doses.

Pathogenesis.—The organism is not, so far as is known, pathogenic for man. At the site of inoculation in animals there is a rapidly spreading edema which appears to be very painful. Usually the most prominent naturally occurring lesion is a swelling of the front or hind quarters; the lesion practically never extends below the knee. The edematous area is almost black, due apparently, in part at least, to changed blood pigment, and the area is surrounded by a zone of hyperemia. The hair over the edematous area falls out easily. There is considerable degeneration of the muscular tissue in the edematous zone, and there is in it a sanguineous exudate which contains relatively few leukocytes. The edematous area is crepitant, due to accumulated gas bubbles, and there is a rather strong odor of butyric acid. The incubation period is from one to three days.

Sporulation does not take place in the tissues of the living animal, but it is said to take place in from twenty-four to forty-eight hours after death. If the spores are washed free from toxin and other bacteria they are said not to be infective for experimental animals, according to Leclainche and Vallée.[3]

Vaccine.—One attack of symptomatic anthrax appears to confer immunity to subsequent attacks. Young cattle are usually infected; older ones appear to be more resistant to infection. A vaccine has been prepared which protects the animal from infection. The general process of manufacture is to remove the infected tissues of animals dead of symptomatic anthrax and dry them under aseptic conditions at 37° C.[4] From this dried tissue two vaccines are made up, the first being prepared by mixing the dried powder with sterile water[5] to form a paste, which is heated to 100° C. for six hours. This is the first vaccine, which will not kill experimental animals. It is injected at the tip of the tail. In seven days a second vaccine (prepared from the same powder and heated to 94° C. for four hours) is injected in the same manner. This vaccine will ordinarily kill small experimental animals. These two vaccines or modifications of them are widely used for protecting cattle against blackleg.

[1] Ann. Inst. Past., 1900, 202.
[2] Über das Rauschbrandgift, 1904. [3] Loc. cit.
[4] This temperature does not diminish the virulence of the bacteria; the potency of the dried virus remains unimpaired for eighteen to twenty-four months.
[5] Two parts sterile water to one part of dried powder.

CHAPTER XXVI.

THE CHOLERA GROUP.

CHOLERA VIBRIO.
 Vibrio of Finkler and Prior (Vibrio Proteus).
 Vibrio Metchnikovi.
 Vibrio Massaua.
 Vibrio Tyrogenum (Spirillum Deneke).

MANY vibrios have been described which possess in common with the cholera vibrio a number of cultural characters. They are all comma-shaped organisms, Gram-negative, possess a terminal flagellum, form no spores or capsules, and liquefy gelatin more or less rapidly. They differ among themselves culturally chiefly with respect to the intensity with which these reactions occur. Some produce nitroso indol in sugar-free culture media, others produce indol only. They may be sharply differentiated from the true cholera vibrio by serum reactions. So far as is known, none of these organisms will agglutinate with a specific cholera immune serum in high dilution, 1 to 2000 to 1 to 5000, depending upon the titre. None of these organisms are dissolved by cholera immune serum (Pfeiffer reaction). The true cholera vibrio gives these serum reactions. Most of these organisms have been isolated from water. Even within the group of the true cholera cultures, that is, those which react with a specific cholera immune serum, there appear to be varieties which are distinguishable from the type organism with great difficulty. The principal variants are described below.

CHOLERA VIBRIO.

Synonyms.—Vibrio choleræ asiaticæ, Spirillum choleræ asiaticæ, comma bacillus, cholera vibrio.

Historical.—The cholera vibrio was first isolated in pure culture by Koch in 1883.[1] For some years the organism was not universally accepted as the causative agent in Asiatic cholera, and some weight was attached to the frequent isolation of vibrios very similar in morphological and cultural characters to the true cholera vibrio from the dejecta of normal individuals. These cholera-like vibrios were not

[1] Deutsch. med. Wchnschr., 1883, 615, 743; 1884, 63, 111, 221, 499, 519; 1885, No. 37a; British Med. Jour., 1884, ii, 403, 453.

sharply differentiated from the true cholera vibrio with the imperfect methods available in the early days of bacteriology, when these observations were made. It is now universally held that the cholera vibrio is the causative organism of the disease.

Morphology.—The typical cholera vibrio is a distinctly curved rod, the curvature being in three planes of space. It measures 0.5 to 0.6 micron in diameter by 1 to 3 microns in length, occurring singly or in pairs, less commonly in longer spiral chains of several elements. Pairs of organisms frequently appear as S-shaped spirilla, the curvature being in three planes of space in the living vibrios. Freshly isolated vibrios have slightly but distinctly pointed ends which are

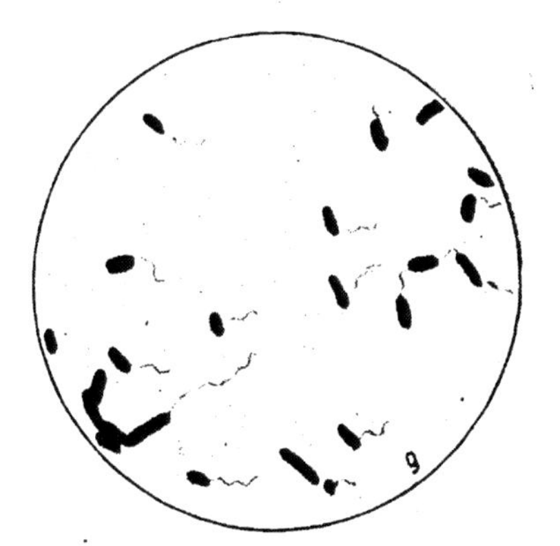

FIG. 71.—Cholera vibrios, showing flagella.

best observed in stained specimens made directly from cholera dejecta. Cultures grown for some time on artificial media lose their original uniformity of size and shape and tend to become less curved, many individuals even appearing as straight rods. The passage of these old cultures through animals is said to restore their original morphology. Cultures in artificial media several days old frequently exhibit involution forms which are irregularly swollen or even coccoid in outline. Bacillary forms and even true spirillum forms also are not uncommonly seen.

Cholera vibrios are actively motile and they possess a single polar flagellum—monotrichic flagellation.[1] No capsule has been demonstrated and no spores are produced, although involution forms which stain somewhat irregularly may suggest spores.

The cholera organism stains with ordinary anilin dyes, although less

[1] Löffler, Centralbl. f. Bakt., 1889, vi, 209.

readily than the majority of pathogenic bacteria. This is particularly the case in freshly isolated cultures. Older cultures are more uniform in this respect. The organism is invariably Gram-negative.

Isolation and Culture.—Cholera vibrios grow rapidly upon all ordinary artificial media, even at 20° C. Their nutritional requirements with respect to nitrogenous substances are less exacting than those of many pathogenic and non-pathogenic bacteria commonly found in the intestinal tract. Also the true cholera organisms are tolerant of a degree of alkalinity which is unsuited for the development of ordinary bacteria. Advantage is taken of these nutritional peculiarities in isolating cholera vibrios from the dejecta of cholera patients. A small portion of fecal mucus is emulsified in slightly alkaline Dunham's solution[1] and incubated for six to eight hours at 37° C. The cholera organisms increase in numbers with great rapidity and they will be found at the surface of the medium in considerable concentration, for they are strongly aërobic. The isolation of them in pure culture by plating is readily accomplished if the material for inoculation is taken from the surface of such a peptone culture.[2]

Growth in Artificial Media.—Colonies of cholera vibrios which appear on agar plates after twelve to eighteen hours' incubation at 37° C. are round, very thin and transparent, and when viewed by transmitted light they are nearly colorless. Colonies of colon and other intestinal bacteria are usually yellowish brown under the same conditions. The colonies of freshly isolated cholera vibrios are even more transparent than colonies of typhoid, paratyphoid, or dysentery bacilli. Older cultures do not exhibit this transparency to such a degree.

Colonies on gelatin plates present a somewhat characteristic appearance. After twenty to twenty-four hours' incubation the organisms have produced a slight liquefaction which gives the surface of the medium a "ground-glass" appearance when the plate is viewed at an acute angle. Liquefaction proceeds rapidly. The cultures which have been grown on artificial media for a long time liquefy gelatin more slowly and eventually may lose this property. In sugar-free gelatin stab cultures an "air bubble," so called, frequently forms just below the surface of the medium. This probably is the result of the evaporation of water from the liquefied medium. No liquefaction takes place in sugar gelatin.

[1] Dunham solution: Peptone 1 gram, NaCl 0.5 gram, potassium nitrate 0.25 gram, sodium carbonate (cryst.) 0.5 gram, water 100 c.c.

[2] See Bacteriological diagnosis for details.

Blood serum is liquefied. Broth is densely clouded and in plain broth or in Dunham's solution a pellicle is usually formed after twelve to twenty-four hours' growth. A pellicle does not ordinarily develop in sugar-containing broth.

Milk is acidified, the degree of acid produced varying greatly with the strain of organism. Some cultures produce enough acid to cause acid coagulation of the milk. No peptonization takes place. Litmus milk is not coagulated.

The production of hemolysis (erythrocytolysis) by cholera vibrios is a subject of controversy. It was formerly maintained that vibrios which agglutinate at high dilution with specific cholera sera of high

Fig. 72.—Cholera vibrios from feces.

potency were non-hemolytic. The consensus of opinion at the present time concedes that a moderate proportion of typical cholera vibrios are hemolytic, although the active hemolysin can not always be obtained in a soluble form. This property is shared by many cholera-like organisms. A group of vibrios, of which two strains, Vibrio Nasik, and Vibrio El Tor, are the best known, are so closely related to the cholera vibrio that they have caused much study and speculation. The former fails to agglutinate with a specific cholera serum, but is strongly hemolytic; the latter also fails to agglutinate at high dilution, although it acts as an antigen with cholera serum in the complement-fixation test. It produces a thermostabile soluble toxin.[1]

The organisms are aërobic, facultatively anaërobic. They were formerly considered to be strongly aërobic; it is doubtful, however, if they are markedly more aërobic than other intestinal bacteria. The

[1] See Kraus and Pribram, Wien. klin. Wchnschr., 1905, No. 39.

limits of growth are 10° C. and 43 to 45° C. respectively, the optimum being 37° C. They are very sensitive to drying; according to Günther,[1] three hours' drying kills them. They remain alive, however, for weeks in culture media. An exposure to 60° C. for thirty minutes usually kills them. Freezing at 10° C. has little effect even if the exposure is prolonged. They will remain viable in impure water for from one to two weeks on the average.

In feces they may remain alive for seven to nine months if air is excluded, according to Zlatogoroff.[2] Under ordinary conditions, however, they remain viable for much shorter periods of time in feces. According to Forster,[3] the organisms are very sensitive to acids and to germicides. According to his observations, a dilution of 1 to 300,000 bichloride of mercury kills them in five minutes, and 1 to 3,000,000 in ten minutes. These results have not been corroborated and it is very likely that they are not markedly more sensitive to disinfectants than the ordinary pathogenic intestinal bacteria, as the typhoid bacillus. Behring[4] has found that 0.5 per cent. carbolic acid will nearly kill cholera organisms after an exposure of an hour. Bichloride of mercury in a dilution of 1 to 1000 kills them in ten minutes, and 5 per cent. carbolic in less than fifteen minutes.

Products of Growth.—Cholera organisms produce in sugar-free protein media an active soluble gelatinase which dissolves gelatin and also blood serum. Some strains elaborate a soluble hemolysin.[5] No other enzymes are known.

One of the striking reactions of the organism is the so-called "cholera-red reaction," or the nitroso indol reaction. The addition of acid, either sulphuric or hydrochloric or nitric, to a forty-eight-hour culture of cholera vibrios grown in sugar-free nutrient broth or in peptone solution, will develop the well-known reddish-brown color indicative of the indol reaction. The organisms appear to form nitrites from the protein constituents of the medium. The reactive substance was regarded by Poehl[6] as a skatol derivative. This view appears to have been accepted by Bujwid[7] and Dunham.[8] Brieger,[9] however, regards it as an indol derivative. It is probable that Brieger's explanation is the correct one. The substance formed is nitroso indol,

[1] Bakteriologie, p. 644.
[2] Centralbl. f. Bakt., 1911, lviii, 14.
[3] Hyg. Rund., 1893, 722.
[4] Ztschr. f. Hyg., 1890, ix, 400.
[5] See Public Health Reports, 1912, xxvii, No. 11, for full details.
[6] Ber. d. deutsch. chem. Gesell., 1886, xix, 1162.
[7] Centralbl. f. Bakt., 1888, iv, 494.
[8] Ztschr. f. Hyg., 1887, ii, 340.
[9] Deutsch. med. Wchnschr., 1887, No. 15.

the indol radical being derived from the decomposition of tryptophan. The same reaction may be obtained from the rice water stools of cholera patients. The cholera-red reaction is not produced in media containing utilizable carbohydrates.[1] The nitroso indol or cholera-red reaction is not specific for the cholera vibrio. Other closely related bacteria also give the same reaction. On the other hand, not all true cholera vibrios form nitroso indol.

Besides nitroso indol, cholera vibrios produce considerable amounts of ammonia and hydrogen sulphide in sugar-free media.[2] All true cholera vibrios produce acid in dextrose and lactose. The production of acid in saccharose and mannite is somewhat less constant. The acids produced are levolactic acid,[3] also acetic and butyric acids.[4]

Toxin.—The nature of the poison or poisons produced by the cholera vibrio is still a subject of controversy, although the disease cholera appears to be a toxemia, for the organisms do not commonly invade the tissues of the body even in fatal cases. Pfeiffer's view[5] was that the toxin is an endotoxin which is liberated by autolysis from the organisms themselves. Behring and Ransom,[6] on the contrary, claim to have separated a soluble toxin from broth cultures of true cholera vibrios which in doses of about 0.5 c.c. will kill guinea-pigs in twenty-four hours. They further claim to have immunized guinea-pigs and goats to the toxin by injecting gradually increasing doses. The antitoxin thus obtained protects non-immune animals against the toxin or from infection with living cholera vibrios. The toxin is unaffected by moderate heat, chloroform, toluol, or carbolic acid.

Metchnikoff Roux and Taurelli-Salimbini[7] enclosed peptone cultures of cholera vibrios in collodion sacs which were placed in the peritoneal cavities of guinea-pigs. As controls, killed cultures of cholera vibrios and sterile uninoculated peptone respectively were placed in other guinea-pigs in collodion sacs. The guinea-pigs which received only sterile peptone solution in capsules failed to show symptoms; those containing killed cultures of cholera vibrios in capsules showed a slight febrile reaction and some emaciation; the guinea-pigs which received the collodion capsules containing living cholera vibrios died after three to five days with symptoms of choleraic

[1] Gorini, Centralbl. f. Bakt., 1893, xiii, 790. Kendall, Boston Med. and Surg. Jour., 1913, clxviii, 825.
[2] Kendall, Day and Walker, Jour. Biol. Chem., 1913, xxxv, 1240.
[3] Kuprianow, Arch. f. Hyg., 1893, xix, 288.
[4] Gosio, Arch. f. Hyg., 1894, xxi, 120; 1894, xxii, 11.
[5] Centralbl. f. Bakt., Ref., 1892, xi, 568.
[6] Ibid., 1895, xviii, 314.
[7] Ibid., 1896, xx, 627.

intoxication. These observers concluded from these experiments that the cholera organism produced a soluble toxin which was diffusible through collodion sacs. The toxicity of these cultures was not destroyed by the boiling temperature, 100° C. They were able to immunize guinea-pigs, rabbits, goats, and horses with this so-called soluble toxin, and found the serum of these animals was antitoxic and protective against several times the fatal dose of toxin or of the living organisms. Antitoxic sera prepared by this method have not been successful in the clinical treatment of cholera in man. It is not unlikely that the soluble toxic substance or substances produced in artificial cultivations of the cholera vibrio play a less important part in the disease than the endotoxins, which appear to be liberated from the organism with unusual readiness.

The extremely brief period which elapses between infection and death, twelve hours in unusual cases, would suggest that the incubation period of the cholera toxin, if such play a part in the disease, is very much less than that of any other known soluble bacterial toxin.

Pathogenesis.—*Animal.*—Different strains of cholera vibrios vary greatly in their virulence for experimental animals; prolonged cultivation on artificial media tends to diminish their pathogenicity as a rule. Virulent cultures injected intraperitoneally in experimental animals, particularly guinea-pigs, frequently cause acute peritonitis; the animal gradually sinks into a state of coma, the temperature falls, and death intervenes with or without convulsions. At autopsy the peritoneum is reddened, the peritoneal surface of the intestines is greatly congested, and there are usually small ecchymoses. There is some increase in the peritoneal fluid, which frequently contains vibrios. They may also be found in the blood stream as well. Subcutaneous injections of like amounts of culture may or may not result fatally. The organisms, however, as Theobald Smith pointed out many years ago, tend to migrate to the intestinal tract, suggesting that some chemotactic influence attracts them there. Intravenous injection, particularly in young rabbits, may lead to lesions in the intestinal tract, suggesting those characteristic of cholera in man, but as a rule far less severe. The organisms may also be found in the intestinal contents and gall-bladder following intravenous injection.

Feeding experiments in the ordinary way are not successful. Koch[1] succeeded in infecting young guinea-pigs with cholera vibrios by first

[1] Deutsch. med. Wchnschr., 1885, No. 37a, 5-6.

administering sodium carbonate by mouth to neutralize the gastric acidity, then introducing by mouth 10 c.c. of a broth culture of the vibrios directly into the stomach with a catheter. The animals died usually in about two days with symptoms, and particularly intestinal lesions which resembled those of cholera in man. There were diarrhea, bloody rice-water stools with abundant organisms in them, collapse and death. Issaeff and Kolle[1] made similar experiments in young rabbits, and Wiener[2] has successfully infected kittens in the same way.

Human.—(*a*) *Experimental Evidence of Disease.*—In man infection takes place usually by ingestion of food or water contaminated with cholera vibrios. The first accidental laboratory infection is probably that mentioned by Koch[3] of a doctor who accidentally swallowed part of a culture and contracted the disease. Hasterlik,[4] Metchnikoff,[5] Reners,[6] Kolle[7] and Voges[8] have also reported laboratory infections of man with cholera vibrios which resulted in typical disease in each instance, thus establishing beyond reasonable doubt the etiological relation of the cholera vibrio to the disease cholera.

(*b*) *Natural Infection.*—The incubation period of the naturally acquired disease cholera may be very short; the patient may be infected and die within twelve hours, so-called cholera sicca.[9] Ordinarily the incubation period is from one to two days.[10]

The important clinical symptoms are extremely painful cramps, great withdrawal of water from the tissues, due to the violent diarrhea, resulting in shriveling of the skin of the extremities and increased viscosity of the blood. The urine after the first day is scanty in amount, the stools are very fluid, "rice-water stools," and there is profound collapse. The most noteworthy lesions postmortem are in the small intestine, particularly the lower half. The mucosa is swollen and congested particularly about Peyer's patches; the contents of the intestinal tract are fluid and contain shreds of mucus. There is parenchymatous degeneration of the liver, kidneys and spleen. The intestinal contents swarm with vibrios. In the markedly chronic cases there may be extensive necrosis and serofibrinous exudation on the surface of the intestinal mucosa.

[1] Ztschr. f. Hyg., 1894, xviii, 17.
[2] Centralbl. f. Bakt., 1896, xix, 205.
[3] Deutsch. med. Wchnschr., 1885, No. 37a, 7.
[4] Wien. klin. Wchnschr., 1893, 167.
[5] Ann. Inst. Past., 1893, No. 7.
[6] Deutsch. med. Wchnschr., 1894, 52.
[7] Ztschr. f. Hyg., 1894, xviii, 17.
[8] Centralbl. f. Bakt., 1895, xviii, 629.
[9] Metchnikoff, Ann. Inst. Past., 1893, 581.
[10] Banti, Lo Sperimentale, 1887. Günther, Deutsch. med. Wchnschr., 1892, 841.

Immunity.—As a rule one attack confers lasting immunity.

Artificial Immunity.—Attempts have been made to induce artificial active immunity:

1. By subcutaneous inoculation of virulent cholera vibrios in man, either directly or after exaltation of their virulence for guinea-pigs or rabbits.[1] (2) By the injection of autolyzed cultures of cholera vibrios, heated at 60° C. for an hour to kill them, then suspended in distilled water at 37° C. for three to four days, and filtered through porcelain.[2] (3) Vaccines. (*a*) Killed cultures (Kolle); (*b*) sensitized cultures (Besredka); (*c*) bacterial extractives.

The only method thus far which has yielded encouraging results is that of Haffkine.[3] This consists in the injection of from 0.25 to 0.5 c.c. of a suspension of an agar culture of cholera vibrios suspended in 5 c.c. of sterile saline solution. This is introduced subcutaneously. The results reported from India are claimed to be favorable.

Bacteriological Diagnosis.—Isolation and identification of the cholera vibrio.

1. *Microscopic.*—The feces may be examined directly for cholera vibrios. Large numbers of slightly curved or S-shaped actively motile vibrios, which when stained with dilute carbolfuchsin exhibit slightly tapered ends, are very suggestive. A bit of mucus (a "grain of rice" from a rice-water stool) is particularly good for microscopical examination. The organisms frequently exhibit a marked parallelism of their long axes, resembling a school of fish in their arrangement if the material is not roughly handled during the preparation of the smear.

2. *Culture.*—(*a*) Schottelius' method. The principle involved: The cholera vibrio grows particularly well in alkaline peptone solution (Dunham solution). Bacillus coli and other intestinal organisms grow less readily.

Technic.—A loopful of feces,[4] or preferably a small piece of mucus is emulsified in a tube of Dunham's peptone solution and incubated at 37° C. for six to eight hours. The cholera organisms are very aërobic and actively motile, and collect in large numbers at the surface of the medium, therefore two or three loopfuls of material from the surface of the Dunham tube are inoculated into a second

[1] Haffkine. [2] Haffkine and Ferran. [3] Bull. Inst. Past., iv, 697, 737.

[4] If the preliminary microscopical examination fails to reveal a preponderance of vibrios of characteristic morphology, a larger amount of fecal material must be taken. Several grams of feces emulsified in 100 to 500 c.c. Dunham solution may give positive results in exceptional cases when smaller samples are negative.

tube and the process repeated in a third tube, when a nearly pure culture of cholera vibrios will frequently be obtained. The organisms may be plated directly from the enriched growth in the first, second, or, best, from the third tube, and the pure cultures agglutinated with a high potency specific anticholera serum in dilutions from 1 to 500 to 1 to 5000.[1] The nitroso indol test should be made on each of the three Dunham tubes after removal of the organisms, for a positive nitroso indol reaction, while not diagnostic, is very suggestive.

(*b*) A small amount of feces or a flake of mucus is emulsified in broth and inoculated on the surface of alkaline agar plates,[2] which have previously been poured and hardened. The very thin transparent colonies which develop within twelve to eighteen hours are either transferred to broth and after twelve hours' incubation agglutinated, or the colony is emulsified directly in a high potency specific serum diluted five hundred times and a macroscopic or microscopic examination made. Controls are made using either normal serum diluted twenty-five times, or normal salt solution. Cholera vibrios will agglutinate rapidly while the controls remain actively motile.

3. *Agglutination of Organism.*—(*a*) A pure culture of cholera vibrios will agglutinate in high dilutions with a high potency specific cholera serum either by the microscopic or macroscopic agglutination method. The macroscopic agglutination test can be made either by preparing successive dilutions of the antiserum in small tubes, 1 to 250 up to 1 to 2500, and adding an equal volume of broth culture of cholera vibrios to each, or by making dilutions of the specific serum 1 to 500 up to 1 to 5000 in small tubes and emulsifying in each tube a small amount of culture from an agar slant. Appropriate controls should be made in either case. A positive diagnosis of cholera vibrios should only be made if agglutination takes place with a specific anticholera serum in a dilution of at least 1 to 500.

(*b*) A flake of mucus containing many vibrios is emulsified directly in specific anticholera serum, diluted at least 1 to 500, and a suitable control is made with normal serum. This is best carried out by the microscopic agglutination method. A positive agglutination under

[1] The anticholera serum is best obtained from rabbits which have been immunized by repeated injections of known cholera vibrios. The titre of the serum should be at least 1 to 4000. A final diagnosis should be made preferably only when the suspected organism agglutinates in a dilution of at least 1 to 2000, although clumping of freshly isolated vibrios at a dilution of 1 to 500 is fairly conclusive. The sera of horses and other large animals are less suitable for agglutination with cholera vibrios; natural antibodies occur which cause clumping in relatively high dilutions.

[2] The necessary degree of alkalinity may be attained by adding 3 c.c. of a 10 per cent. solution of sodium carbonate to each 100 c.c. of neutral (litmus) agar.

these conditions is fairly conclusive. It should be remembered that an occasional strain of the cholera vibrio is met with which does not agglutinate when freshly isolated; prolonged cultivation in artificial media frequently leads to a typical agglutination.

4. *Identification of Cholera Vibrios by the Pfeiffer Phenomenon.*—If cholera vibrios are introduced directly into the peritoneal cavity of an immunized guinea-pig and samples of the peritoneal exudate containing vibrios are removed from the peritoneal cavity with a capillary pipette after ten minutes, sixty minutes and ninety minutes, it will be found that usually after ten minutes, almost invariably within an hour, the vibrios will become very much granulated and will eventually dissolve. A normal guinea-pig similarly infected intraperitoneally with a mixture of cholera vibrios and immune serum will exhibit the same granulation and lysis of the organisms. The reaction does not occur when the vibrios alone are introduced into the peritoneal cavity of a normal pig. It is much simpler to introduce the immune serum and vibrios into test-tubes, incubate them at 36° C. and examine the contents of the tubes for granulated and partly dissolved organisms after intervals up to two hours. The test is carried out as follows: a series of dilutions of fresh immune serum, 1 to 50 to 1 to 500, is prepared in small sterile test-tubes, 0.5 c.c. to each tube. A suspension of cholera vibrios, one loopful of an eighteen-hour agar slant growth to 1 c.c. of sterile salt solution, is also prepared; usually 10 c.c. are sufficient. This is thoroughly shaken and 0.5 c.c. added to each tube of diluted serum. Control tubes of normal serum and bacterial suspension are incubated uuder parallel conditions. The entire set of tubes is incubated at 37° C. and examined at intervals up to four hours. The control tubes swarm with vibrios. The immune serum tubes up to the limits of potency contain vibrios in various stages of solution. Only true cholera vibrios will be thus dissolved. The various cholera-like vibrios are unaffected.

The bacteriological diagnosis of the cholera vibrio is one of the most difficult known to bacteriology. The large number of closely related forms introduces complications in the diagnosis which have frequently led to error. In general it may be stated that a vibrio which agglutinates $\frac{1}{2000}$ with a specific anticholera serum of high potency, and exhibits the Pfeiffer phenomenon in a perfectly typical manner may be safely diagnosed as positive. Departure from this standard should cause the organism to be regarded with suspicion, but should not lead to relaxation of appropriate hygienic measures in relation to the case.

5. *Complement Fixation.*—Besche and Kon,[1] Neufeld and Haendel,[2] and others have been successful in diagnosing cholera and identifying cholera vibrios by means of the complement-fixation test. This method has not been generally used, however.

6. *Agglutination by Serum of Patient.*—The agglutination reaction is not of much value for an early diagnosis of Asiatic cholera. Agglutinins occasionally appear in the blood serum of cholera patients as early as the third or fourth day; usually, however, they are not demonstrable until later. A dilution of at least 1 to 50 should be obtained with the patient's serum to warrant a positive diagnosis. Even in chronic cases and in cholera carriers this reaction is too inconstant to serve practical needs.

Dissemination.—Cholera vibrios are found in the fecal discharges of cholera patients, but practically never in the urine, so far as is known. The disease is spread through contaminated water and sewage, occasionally by uncooked vegetables and by fomites, rarely by milk. Dissemination by flies is probably fairly common, particularly in those countries, as India, where the dejecta are not properly disposed of. Those in contact with the dejecta of cholera patients, particularly doctors, nurses, and especially laundresses, are quite likely to contract the disease. The sacred rivers of India, the Ganges and the Jumna, are regarded by many as the home of the cholera vibrio, and it has been accepted in the past that drinking the water of these rivers by pilgrims who visited them in large numbers yearly has been responsible to a large degree for the spreading of the disease, particularly in India. Hankin[3] has made the astonishing statement that the waters of these rivers kill cholera vibrios in two to four hours, it being surmised that some soluble acid substance is the bactericidal agent. This observation, if corroborated, would discredit the spreading of cholera by pilgrims who bathe in the sacred rivers.

Cholera Carriers.—The observations of Greig, who found cholera vibrios in the gall-bladders of 81 out of 271 cholera cadavers, and of Kulescha,[4] who described pathological changes in the gall-bladder and biliary passages caused by cholera vibrios, have attracted attention to the importance of cholera carriers in the spreading of the disease. Zeidler[5] found cholera organisms in the feces of a patient

[1] Ztschr. f. Hyg., 1909, lxii, 161.
[2] Arb. a. d. kais. Gesundamte., 1907, xxvi.
[3] Ann. Inst. Past., 1896, 175, 511.
[4] Centralbl. f. Bakt., 1909, l, 417.
[5] Med. Klinik., 1907, Nos. 48 and 49.

ninety-three days after recovery, suggesting that these carriers might be of hygienic concern for months after recovery. Zlatorgoroff[1] and others have made similar observations. Even healthy individuals who are in contact with cholera patients may have cholera organisms in their feces without symptoms. It must be remembered in this connection, however, that curved bacilli morphologically like cholera organisms, but not giving specific serum reactions, are not uncommon in the feces of healthy people. Generally speaking, cholera carriers are somewhat less likely to occur than typhoid carriers.

Isolation of Cholera from Water.—The simplest method of isolating cholera vibrios from water is to prepare a sterile stock solution containing 10 per cent. of peptone and 5 per cent. of salt; to every 100 c.c. of water to be examined 10 c.c. of this stock solution are added, which practically converts the suspected water into a culture medium. The isolation then is carried out by the Schottelius method described above. The initial culture being the water itself, successively inoculating Dunham's tubes from the surface growth obtained in the water culture after it has been incubated at 37° C. for forty-eight hours, and finally making agglutination tests with a high potency serum for the final agglutination of the organisms is almost invariably successful.

Vibrio of Finkler and Prior (Vibrio Proteus).—The organism was first isolated and described by Finkler and Prior.[2] It was obtained from the dejecta of a case of acute enteritis and subsequently isolated from the dejecta of patients having cholera nostras.

Synonyms.—*Vibrio Proteus.*—Perhaps identical with Miller's vibrio found in carious teeth in 1884.[3]

Morphology.—Very much like the cholera vibrio except that the organism is somewhat larger, exhibits a greater degree of curvature, and is said to have slightly pointed ends. The organism occurs singly and in pairs, rarely in long spirals. Involution forms, however, are very common. There is a single polar flagellum, and the organism is actively motile. It stains readily with the ordinary anilin dyes and is Gram-negative.

Isolation and Culture.—The organism liquefies gelatin with great rapidity, otherwise there is nothing characteristic about the growth in gelatin.

[1] Centralbl. f. Bakt., 1911, lviii, 14.
[2] Deutsch. med. Wchnschr., 1884, x, 632-657.
[3] Miller, Mikroörganismen d. Mundhöhle.

Growth on Artificial Media.—Gelatin stab cultures are rapidly liquefied. There is not the "air bubble" appearance which is characteristic of stab cultures of the cholera organism ordinarily. On agar there is a rapidly spreading growth which becomes thick, moist and slightly viscid. Broth is clouded and there is a heavy sediment and a pellicle. Blood serum is rapidly liquefied and milk is coagulated. Acid is formed in dextrose.

Products of Growth.—The nitroso indol reaction is given very slightly, frequently not at all. Indol, however, is produced in large amounts. Proteolytic ferments dissolving gelatin, serum and casein are formed by Vibrio proteus. Cultures have a foul odor. According to Kuprianow[1] levorotatory lactic acid is formed from dextrose.

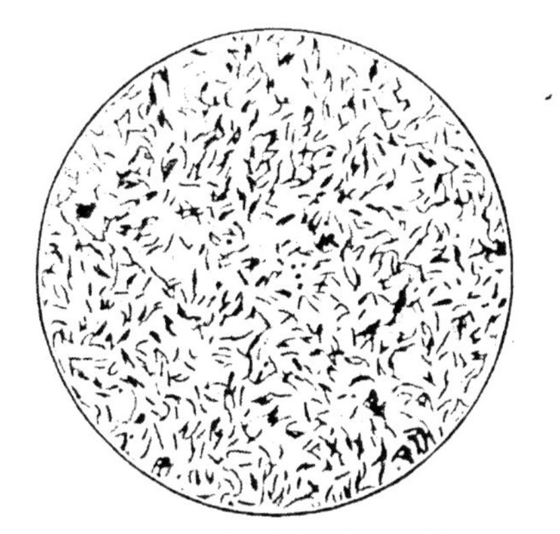

FIG. 73.—Vibrio metchnikovi, bouillon culture. × 1000.

Bacteriological Diagnosis.—Diagnosis depends upon the isolation of curved organisms resembling the cholera vibrio, which do not react with a cholera immune serum.

Pathogenesis.—*Human.*—According to Metchnikoff, an agar culture eaten by man may result in a slight intestinal disturbance. This, however, probably has no significance.

Animal.—The intraperitoneal inoculation of cultures of Vibrio proteus causes a fatal peritonitis. According to Metchnikoff,[2] by feeding cultures to animals previously treated with sodium carbonate and laudanum to reduce the acidity and intestinal peristalsis, irregular results are obtained. Occasionally a profuse diarrhea results, but it is rarely or never fatal. In pigeons inoculation into the pectoral

[1] Arch. f. Hyg., 1893, xix, 288.
[2] Ann. Inst. Past., 1893, 570.

muscles very frequently produces death. The organism is of interest chiefly because it is one of the classical organisms for study. It is rarely confused with the cholera vibrio and has no significance pathogenically.

Vibrio Metchnikovi.—A spirillum found in the feces of fowls suffering from acute enteritis by Gamaleia.[1]

Morphology.—Practically identical with cholera. Staining, culture reactions, products of growth, the same as cholera. It is non-pathogenic for man. If it is ingested by man it is harmless. It does not agglutinate with the cholera immune serum, and is not dissolved by the cholera immune serum. According to Pfeiffer and Nocht,[2] the intrapectoral injection of this organism into pigeons kills them with symptoms of acute septicemia. There is extensive edema at the site of inoculation. If it is fed to young fowls it frequently kills them with symptoms of enteritis.

Vibrio Massaua.—Pasquale isolated this organism at Massaua from a case of clinically doubtful cholera.[3] Pathogenically it is quite similar to Spirillum metchnikovi, and produces septicemia in birds when inoculated intrapectorally. It does not react with cholera immune serum either by agglutinating or by lysis.

Vibrio Tyrogenum (Spirillum Deneke).—Deneke[4] isolated this organism from an old cheese, and it has since been found in butter. Culturally it is very similar to the spirillum of Finkler and Prior, except that the cholera-red reaction is usually negative. Intraperitoneal injection into guinea-pigs and intrapectoral injection into pigeons cause death. According to Metchnikoff, a moderate diarrhea may be induced in man by feeding cultures of this organism.

[1] Ann. Inst. Past., 1888.
[2] Ztschr. f. Hyg., 1889, vii, 259.
[3] Giorn. Med. de r. Eserc. ed. R. Marina, Roma, 1891.
[4] Deutsch. med. Wchnschr., 1885, iii.

CHAPTER XXVII.

TREPONEMATA AND SPIROCHETA.

TREPONEMATA.

Treponema Pallidum.—Synonym.—Spirocheta pallida.

Historical.—The organism which is now universally conceded to be the infective agent in syphilis was first described by Schaudinn and Hoffmann.[1] It was named Spirocheta pallida by these observers, but it presents certain peculiarities of structure which are of sufficient magnitude to separate it from the group of the spirochetes. It has been placed in a newly established group, the Treponemata, of which it is the type organism.

Morphology.—Treponema pallidum is a long, very thin, delicate, closely coiled, flexous spiral organism which measures from 0.25 to 0.4 micron in diameter, and, on the average 7 to 8 microns in length. The length, however, may vary from 3 microns in very young organisms to 15 microns. The spirals, which are very regular in outline, are ordinarily from six to twelve in number per organism; they may be as few as three to five in the shorter forms or as numerous as twenty in the longer forms.

Noguchi[2] has described three morphologically recognizable types of Treponema pallidum: an average or normal type; a type thicker than the average; and a type thinner than the average; each of which induces somewhat different lesions in experimental animals. The three types present no noteworthy cultural differences. Noguchi suggests that these morphological and pathological variations observed

[1] Arb. a. d. kais. Gesamte, 1905, xxii; Deutsch. med. Wchnschr., 1905, Nos. 42–43.
[2] Jour. Exp. Med., 1912, xv, 201.

PLATE IV

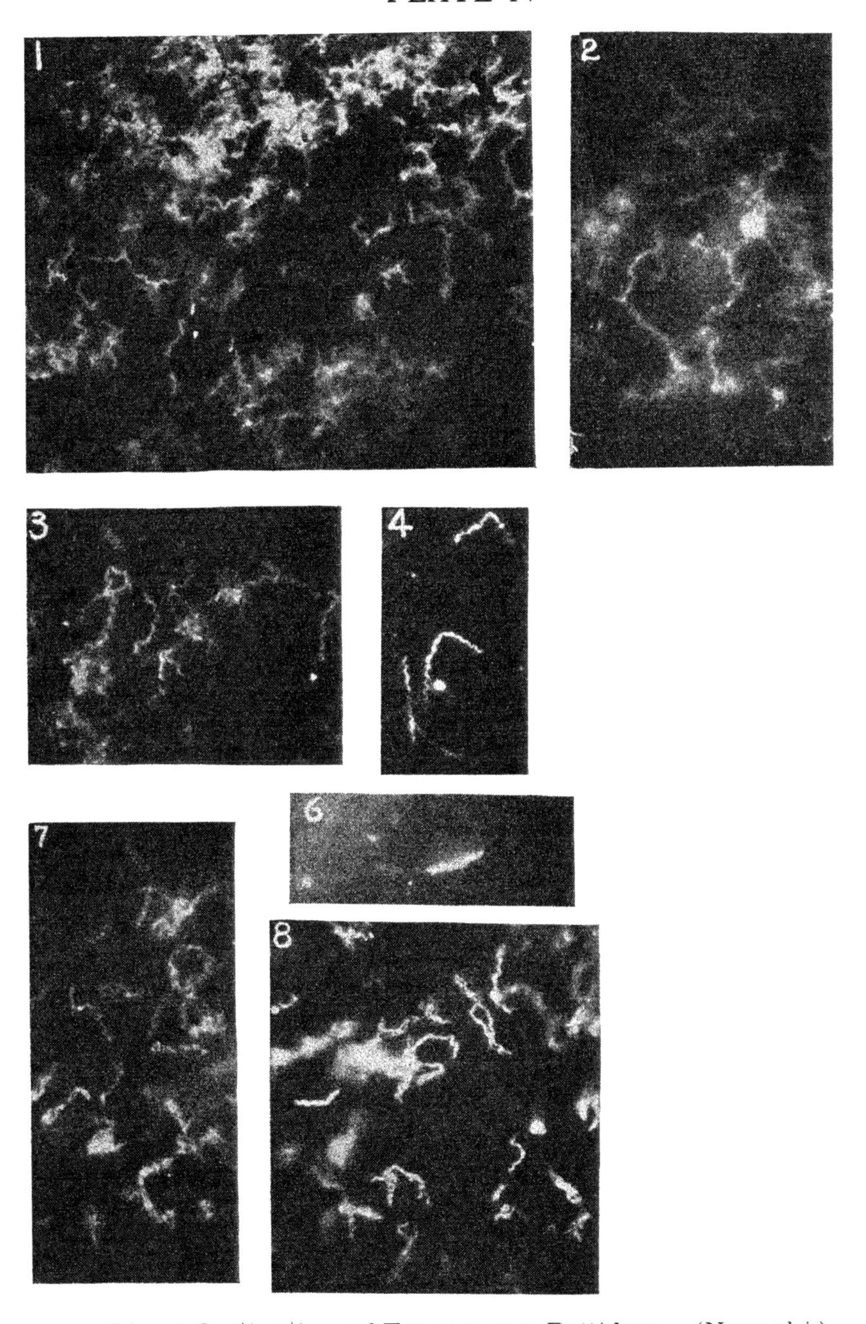

Direct Cultivation of Treponema Pallidum. (Noguchi.)

FIG. 1.—Treponemata pallida from a young pure culture in ascitic agar tissue medium, fo old, at 37° *C.* Dark field. × 1400.

FIGS. 2 and 3.—The same after two weeks.

in cultures of Treponema pallidum may constitute racial difference within the species.[1]

The ends of the organisms are attenuated and merge almost perceptibly into polar flagella, one at each end. The morphology of the Treponemata varies somewhat in artificial media, according to the conditions of growth. According to Noguchi, the typical organisms are only observed in special media where the conditions of culture are strictly anaërobie. The admission of even slight amounts of oxygen produces changes in their appearance. Reproduction, according to Schaudinn[2] and Noguchi,[3] takes place typically by longitudinal fission rather than by transverse fission, as was claimed by Levaditi and others. This would suggest a relationship with the protozoa rather than with the true bacteria.

Treponemata are actively motile in young cultures, particularly in media which are fluid or semi-fluid. In agar of ordinary density the motility is considerably lessened or even absent. The motility is brought about by the activity of the polar flagella mentioned above. The character of the motion is twofold: a rotation about the long axis, and a true progressive motion. The resultant motion is like that of a corkscrew. Undulatory contractions of the organisms have also been observed. No capsules have been discovered and no spores are produced. It has been claimed that an undulatory membrane has been demonstrated on Treponema pallidum, but this observation has not been adequately confirmed. Treponema pallidum does not stain with ordinary anilin dyes, it is non-acid-fast and can not be stained by Gram's method. The organisms may be demonstrated in the living state on suitable material scraped from syphilitic lesions or stained by special methods (*vide infra*).

Isolation and Culture.—Various successful attempts to induce multiplication of Treponemata, both *in vivo* and *in vitro*, are on record. Brucker and Gelasesco[4] and Sowade[5] injected material from syphilitic lesions into the testicles of rabbits and observed considerable prolifera-

[1] Nichols (Jour. Exp. Med., 1914, xvii, 362) has described a Treponema isolated from the spinal fluid of a syphilitic which conforms morphologically to the "thick type" of Noguchi. It produces a rapidly developing lesion in the male rabbit when inoculated into the testicle. The incubation period is about two weeks and one-half, and the organism tends to cause generalized secondary lesions in the eye, and the skin. It is not known whether this tendency toward generalized infection is peculiar to this particular strain or whether the thicker organisms possess in common this property.

[2] Arb. a. d. kais. Gesamte, 1907, xxvi, 11.

[3] Journ. Exp. Med., 1912, xv, 90.

[4] Compt. rend. Soc. de biol., Paris, 1910, lxviii, 648.

[5] Deutsch. med. Wchnschr., 1911, xxxvii, 682.

tion of the organisms there. Schereschewsky[1] grew the organisms in impure culture in anaërobic cultures of gelatinized horse serum, that is, horse serum which has been heated to 60° C. for some hours. To Noguchi, however,[2] belongs the credit of obtaining Treponema pallidum in pure culture, and of demonstrating its etiological relationship to the disease syphilis.

The medium which gave the best results is prepared in the following manner: Two per cent. slightly alkaline agar is melted and quickly cooled to 45° to 50° C. and sterile ascitic or hydrocele fluid is added in the proportion of two parts of agar to one part of fluid. At the same time a small piece of sterile tissue from a rabbit's testis or kidney is introduced. The medium is rapidly cooled to room temperature and

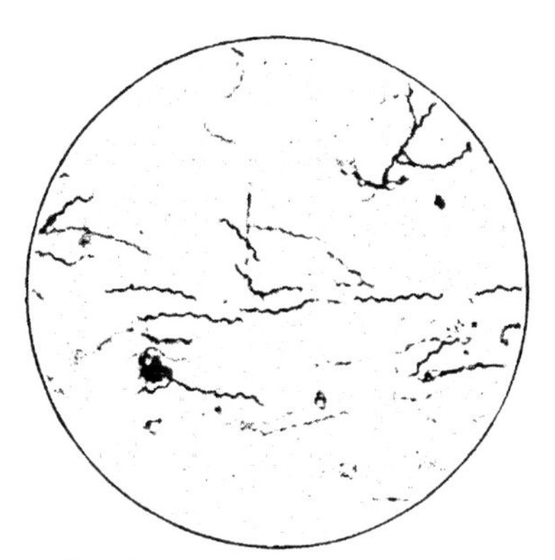

Fig. 74.—Treponema pallidum.

covered with a layer of sterile paraffin oil, 2 to 3 c.c. deep, to keep out the air. The medium is incubated for two days to ensure sterility and is then inoculated with appropriate material, after first being certain that the material contained the organisms. The syphilitic tissue, prior to inoculation, is macerated under sterile conditions with 1 per cent. sodium citrate solution and then introduced deeply into the agar-ascitic fluid-tissue media. Incubation is maintained at 37° C. for two to three weeks. The Treponemata in virtue of their motility move away from the line of inoculation and cause a more or less uniform, faint clouding of the medium. The associated contaminating organisms are for the most part confined chiefly to the line of inoculation. At the end of the period of incubation the tube

[1] Deutsch. med. Wchnschr., 1909, xxxv, 835, 1260.
[2] Jour. Am. Med. Assn., 1911, xvii, 1.02; Jour. Exp. Med., 1912, xv, 90.

is broken at an appropriate level with sterile precautions and some of the turbid medium removed into fresh tubes of the same kind, and the process repeated until pure cultures of the organisms are obtained. If gas-producing bacteria are present the results are unsuccessful as a rule.

Products of Growth.—The products of growth are unknown. Treponema pallidum does not produce a characteristic and disagreeable odor which distinguishes it from cultures of other spirochetes in artificial media.[1]

Pathogenesis.—*Animal.*—In 1903 Metchnikoff and Roux[2] transmitted syphilis to a chimpanzee, and later infected other monkeys

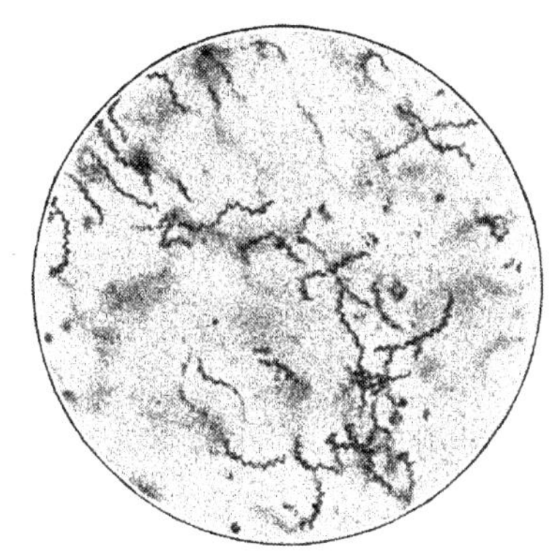

FIG. 75.—Treponema pallidum, congenital syphilitic liver.

with material from primary or secondary lesions in man. These results have been amply confirmed by other investigators. The incubation period averages about three to four weeks. It may be as brief as two weeks or as prolonged as seven weeks. The lesion, histologically indistinguishable from a chancre, appears soon after the end of the incubation period at the site of inoculation; the regional glands become enlarged and indurated. Secondary lesions appear in about 50 per cent. of successful inoculations, usually four to five weeks after a chancre appears. Skin lesions are somewhat indefinite, but the mucous patches are readily recognized. No tertiary lesions have been demonstrated in experimental inoculations into animals with the virus of syphilis up to the present time. Recently Noguchi[3]

[1] Noguchi, Jour. Exp. Med., 1912, xv, 99.
[2] Deutsch. med. Wchnschr., 1903, No. 50.
[3] Loc. cit., p. 96.

has successfully inoculated two monkeys (Macacus rhesus and Sercopithecus callitrichus) with pure cultures of Treponema of human origin, and reproduced in them the initial lesions of the disease. The blood of these monkeys gave a positive Wassermann reaction, thus confirming the relation of the Treponema pallidum to the disease in man. Rabbits have been successfully infected with the virus. Bartarelli[1] produced localized eye lesions by introducing virus from man into the anterior chamber of the eye. A small swelling of the cornea took place about ten days after inoculation and there was a considerable development of Treponemata. Brucker and Gelasesco[2] and Sowade[3] have corroborated these results. Hoffmann has produced a specific orchitis in rabbits. Localized limited growths have been reported in various other experimental animals, guinea-pigs, dogs, sheep and cats. These inoculations have usually been made on the cornea by scarification, and slight nodules have developed.

Human.—Treponema pallidum is present in the hard chancre, in which it can be found in practically every case; also it is found in the enlarged regional glands. The organisms have also been found in the secondary lesions, particularly in the mucous patches and papules. According to Bandi and Simmonelli,[4] the organisms are occasionally found in the blood, and they have also been observed in blister fluid by Levaditi and Petresco.[5]

In the lesions of tertiary syphilis the organisms are present in but small numbers, although usually these lesions are infective for monkeys. Noguchi has found the organisms in the cerebral cortex in many cases of general paresis, and Reuter has demonstrated the organisms in the walls of the larger bloodvessels in an individual infected with syphilitic aortitis. The organisms are present in enormous numbers in the liver, spleen and internal organs of cases of congenital syphilis.

Bacteriological Diagnosis.—*Collection of Material.*—The distribution of the organism in syphilitic tissues is quite irregular, the organisms being very numerous in some cases, in other apparently similar cases so few in number that they may be readily overlooked. In congenital syphilis the organisms are extremely numerous; in the lesions of acquired syphilis the organisms are best observed either in the primary or secondary stages.

[1] Centralbl. f. Bakt., 1906, xli, 320.
[2] Loc. cit.
[3] Deutsch. med. Wchnschr., 1911, xxxvii, 1546.
[4] Centralbl. f. Bakt., 1905, xl, 64.
[5] Presse Médicale, 1905.

Primary Lesion.—Clean the surface of the chancre with brisk rubbing, then make an abrasion in the skin deep enough so that there is an exudation of serum. Films are prepared from this exudate.

Secondary Lesion; Mucous Patches and Papules.—Material is removed from the mucous patch after cleaning the surface, or from the papule by slight curetting. If the material thus obtained is too dense to spread readily it may be macerated in a drop or two of sterile ascitic fluid.

1. *Morphology.*—The organisms may be seen in the living state with the dark-ground illuminating apparatus. The juice of a mucous patch or primary lesion is examined directly and the organisms appear on a black background as light yellowish closely coiled spirals, which are actively motile. The presence of Spirocheta (Treponema) refringens must be borne in mind, this organism being frequently associated with Treponema pallidum. The former is thicker than Treponema pallidum, and the spirals are less numerous and coarser.

India Ink Method.—The juice from a chancre or mucous patch is intimately mixed with india ink[1] and a cover glass placed over the mixture. The organisms appear as white spirals against a black background.

Other Staining Methods.—The material collected as above is spread on slides in thin layers and stained either by Schaudinn and Hoffmann's original method or by the silver impregnation method.

Method of Schaudinn and Hoffmann.—The films are fixed for fifteen to thirty minutes in absolute methyl alcohol or for a few seconds in the vapor of osmic acid, then they are stained from one to three hours in the following solution, which must be freshly prepared each time; Giemsa's solution, 10 drops; 1 per cent. aqueous solution of potassium carbonate, 10 drops; distilled water, 10 c.c. The films, after staining, are washed in distilled water. If overstaining has taken place the film may be left in distilled water for some minutes until a sufficient amount of stain has been removed. The preparation is then dried and examined. The organisms appear as purple or violet spirals on a bluish background.

Silver Impregnation Method.[2] Not generally used.

2. *Cultural Diagnosis.* Not practical for routine.

3. *Serum Diagnosis.* (For Technic see page 161.)

[1] Burri, Wien. klin. Wchnschr., July 1, 1909.

[2] See Levaditi, Compt. rend. de Soc. de Biol., 1905, lix, 326, for details.

For a time the specificity of the Wassermann reaction for syphilis was questioned, because it was found that alcoholic extracts of normal heart could be substituted for extracts of luetic organs as antigens. A careful study of thousands of cases has shown, however, that a vast majority of active syphilitic infections, especially those in the secondary stage, give a positive Wassermann reaction. During the earlier part of the primary stage the reaction is frequently negative. In the tertiary stage the reaction is frequently positive and the spinal fluid frequently gives a positive reaction as well. Occasionally cases of frambesia and of leprosy give a positive Wassermann reaction, but these diseases are rare in temperate climates. For a time it was believed that the serum in cases of scarlet fever gave a Wassermann reaction, but this view has not been fully substantiated.

Statistics indicate that the Wassermann reaction disappears when a cure is effected, but it reappears if the disease again becomes active. It is important to remember that the mercurial treatment tends to diminish the intensity of the reaction, and it may even disappear temporarily. Treatment with salvarsan and neosalvarsan may accentuate the reaction, temporarily at least.

There is no doubt that the Wassermann reaction carefully executed by competent workers is the most delicate and reliable diagnostic method for syphilis known at the present time.

4. *Luetin reaction.*[1]

Preparation of Luetin.—A culture of Treponema pallidum is ground until the organisms are thoroughly disrupted, then heated to 60° C. for an hour and suspended in sterile salt solution to which is added 0.5 per cent. carbolic acid as a preservative. The reaction induced in syphilitics is essentially like the tuberculin reaction. It consists in the development of a vesicle or a pustule at the site of inoculation with temperature and pain. The control inoculation should exhibit but a slight reddening. The reaction is specific, except for old, advanced cases, where the reaction may fail. It is most marked in the later tertiary and congenital cases where the Wassermann reaction is said to be more likely to be negative.

Clinical Methods of Serum Diagnosis.—*Method of Porges and Meyer.*—The principle of this reaction depends on the production of a precipitate when syphilitic serum is mixed with lecithin. Normal serum does not produce a precipitate under these conditions. The technic consists in thoroughly triturating 0.25 grams of lecithin (ovolecithin) in

[1] Noguchi, Jour. Exp. Med., 1911, xiii, 557.

PLATE V

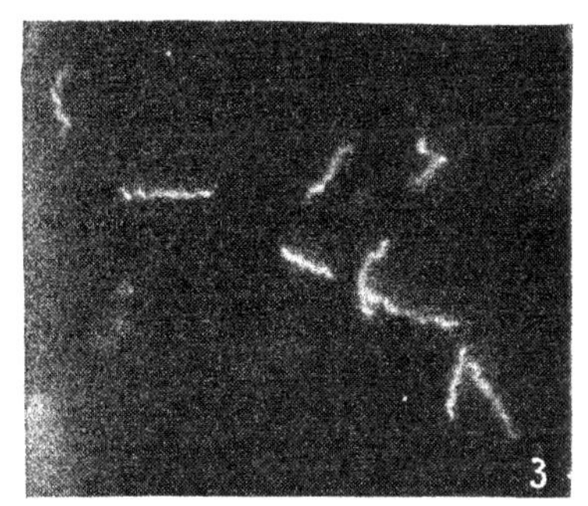

Cultivation of Spirocheta Refringens. (Noguchi.)

ig. 1.—A schematic drawing of Spirocheta refringens from pure cultures (dark field).
ig. 2.—Spirocheta refringens from a three weeks' old pure culture in ascitic tissue agar, a
C. (dark field). × 1100.

100 c.c. of normal saline solution. One c.c. of this lecithin emulsion is added to each of a number of test tubes of 5 mm. diameter. To half of the tubes add 1 c.c. of the suspected syphilitic serum; to the remaining tubes add to some a known syphilitic serum, to others normal serum, and incubate the entire number about four hours at 37° C. The tubes are then removed from the incubator, cooled to room temperature, and those containing syphilitic serum will show a precipitate, which appears to develop first at the surface. It is best observed against a dark background. This method has been modified by Porges by the substitution of a solution of sodium glycocholate for the lecithin. A freshly prepared, 1 per cent. solution of sodium glycocholate is made in distilled water. The test is carried out precisely as in the above method, except that the suspected serum, the known serum, and the normal serum are heated to 55° C. for thirty minutes before being added to the solution. One c.c. each of heated serum and sodium glycocholate are mixed together and kept at room temperature for twenty-four hours. A precipitate forms at the surface of the tubes containing the syphilitic serum, but does not form in the tubes containing normal serum.

Treponema Refringens.—Synonym.—Spirocheta refringens.

Schaudinn and Hoffmann[1] observed Treponema refringens both in syphilitic lesions in association with Treponema pallidum, and in non-syphilitic lesions as well, particularly in superficial lesions of the genitalia. This association of Treponema refringens with Treponema pallidum in syphilitic lesions and its common occurrence in non-specific genital lesions emphasize the necessity of its recognition and differentiation from Treponema pallidum.

Morphology.—Observed under the dark-field microscope, Treponema refringens is noticeably thicker than Treponema pallidum, measuring, according to Noguchi,[2] 0.5 to 0.75 micron in diameter and 6 to 24 microns in length. The ends are somewhat sharply attenuated and they are continued as moderately stiff, delicate spiral flagella. Not infrequently the middle third of the organism is slightly wavy in outline, the end thirds being more closely coiled. Usually the spirochetes are more uniformly curved. Occasionally two or three organisms may be joined end to end. As a rule there are from three to eight complete spirals in each organism.

The organisms are actively motile, and observed with the dark-

[1] Arb. a. d. kais. Gesamte, 1905, xxii, Heft 2.
[2] Jour. Exp. Med., 1912, xv, 467.

field illumination they are golden yellow, contrasting in this respect with the pale yellow appearance of Treponema pallidum. The staining reactions are similar to those of Treponema pallidum.

Isolation and Culture.—Noguchi was the first to grow Treponema refringens in pure culture using the agar ascitic fluid tissue medium with which he isolated Treponema pallidum. The organism is a strict anaërobe, but it may be obtained in the agar ascitic medium without the sterile tissue, although the growth is much more feeble when the tissue is omitted from the medium. The original growths from lesion to artificial media are usually contaminated with other organisms. Purification is accomplished by the same technic as that used for purifying Treponema pallidum. Pure cultures of Treponema refringens produce no odor in growths on artificial media.

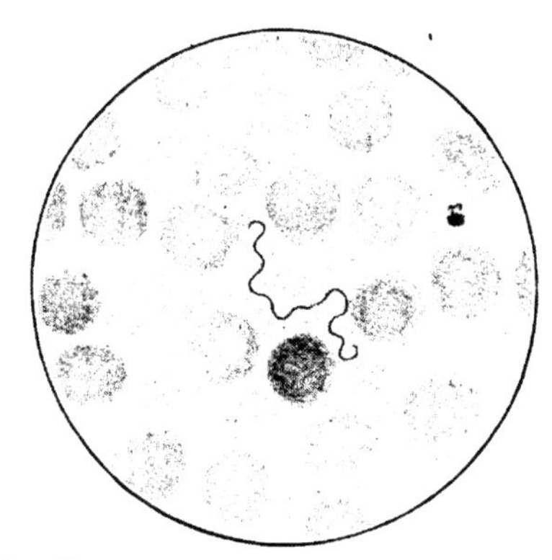

Fig. 76.—Treponema recurrentis. (Kolle and Hetsch.)

Pathogenesis.—The organism was found to be non-pathogenic for rabbits and monkeys.[1]

Relapsing Fever.—The disease known as relapsing fever was described by Obermeier in 1878; he recognized the organism which received his name, Spirocheta obermeieri, now called Treponema recurrentis, in the blood of his patients. Obermeier's observations were made in Europe. Somewhat later the disease was observed in India by Carter, in Africa by Koch, and in America by Norris, Pappenheimer and Fluornoy. In 1896 Novy showed that the organisms found respectively in the relapsing fevers of Europe, India, Africa and America exhibited constant morphological differences which warrant their tentative separation into four distinct types: the European,

[1] Jour. Exp. Med., 1912, xv, 90.

Indian, African and American. Noguchi grew the organisms in pure culture for the first time in 1912. Relapsing fever appears to be transmitted chiefly, if not exclusively, by suctorial insects.

Treponema Recurrentis.—Synonyms.—Spirillum obermeieri, Spirocheta obermeieri, Spirillum recurrentis, Treponema obermeieri, Spirocheta recurrentis.

Relapsing fever is an acute contagious disease which begins abruptly with a chill. The fever which follows immediately after the chill reaches the fastigium (104° to 106°) usually within twenty-four hours, remains high for five to seven days, and falls by crisis. There is an afebrile intermission of five to seven days, then the fever is repeated.[1] Convalescence usually begins at the close of the second paroxysm; it may not occur until the close of the third or even fourth paroxysm. The incubation period is from two to fourteen days. The mortality is low, less than 4 per cent. of all cases. The spleen is enlarged, there is profuse sweating, frequently jaundice, and occasionally diarrhea.

Morphology.—The organism is spiral in outline and of moderate size. Schellack[2] states that the average diameter is 0.4 micron; the length varies from 15 to 20 microns. Other investigators give as measurements, diameter 0.25 micron, length from 7 to 10 microns. The discrepancy appears to be attributable to the fact that younger organisms are about 10 microns in length, the older forms being much longer. There are from twenty to forty spirals in each individual cell, the number depending upon its length. Very frequently the ends are tapered. Fresh preparations viewed by dark-field illumination exhibit three distinct types of motion: a rotation around the long axis, which causes the organism to move rapidly through the medium in which it is suspended, an undulatory movement, and a lateral movement in all planes. The motility is caused by the rhythmic contractions of a terminal flagellum, according to Novy and Knapp.[3] Zettnow[4] believes the organism possesses peritrichic flagella. This has not been confirmed. Reproduction takes place typically by longitudinal fission according to Noguchi.[5] Less commonly he has observed transverse fission.

Isolation and Culture.—The organism appears in the blood stream only during the pyrexia. Novy and Knapp[6] observed multiplication

[1] Obermeier, Centralbl. f. d. med. Wissensch., 1873, xi; Berl. klin. Wchnschr., 1873, x, 152, 378, 391, 455.
[2] Arb. a. d. kais. Gesamte, 1908, xxvii, 364.
[3] Jour. Inf. Dis., 1906, iii, 291.
[4] Deutsch. med. Wchnschr., 1906, xxxi.
[5] Jour. Exper. Med., 1912, xvi, 207.
[6] Jour. Am. Med. Assn., 1906, xlvii, 2152

of Treponema recurrentis in defibrinated rat's blood and succeeded in keeping these organisms alive on blood agar for forty days, at the end of which time they were still infective for rats. No actual multiplication, however, was observed in this medium. Noguchi[1] has grown the organisms in pure culture, using the method described previously (see Treponema pallidum). The organisms develop with considerable rapidity, a distinct clouding of the medium being observed after twenty-four to forty-eight hours' incubation at 37° C. The maximum growth is reached at the end of a week.

Pathogenesis.—*Animal.*—Pure cultures retain their original virulence for rats and mice for several transfers in the agar ascitic fluid tissue medium described by Noguchi. The lesions produced in experimental animals are essentially the same as those observed in man. The disease can be transmitted by inoculation from man to monkeys, from monkey to monkey, and from monkey to mice and rats, which are all susceptible. Rabbits and guinea-pigs appear to be refractory. The disease produced by inoculation of the organisms in monkeys and mice exhibits the characteristic relapses, and it may be fatal.

Human.—There are no characteristic lesions observed in relapsing fever other than a hyperplastic enlargement of the spleen. There may be a catarrhal inflammation of the stomach, bile ducts and liver, which is usually enlarged. All of the organs exhibit parenchymatous degeneration postmortem.

Bacteriological Diagnosis.—The organisms are found in the blood stream only during the paroxysms. During the period of apyrexia they disappear from the blood stream, but are found in the spleen in large numbers, where they are engulfed by leukocytes.

Immunity.—According to Novy,[2] blood drawn from a patient at the beginning of the fever acts as a good culture medium for the organisms; that drawn at the end of a paroxysm or after recovery from the disease appears to possess germicidal properties for the organisms. It is supposed that the organisms are taken up by phagocytes during the afebrile periods, and that they are either weakened or killed at this time. Active immunity follows recovery from the infection. It has been claimed that the blood serum of immunized animals (which exhibit immunity after repeated injections of the organism) or of animals which have recovered from an attack will induce passive immunity and temporarily prevent infection when it is introduced into susceptible animals prior to inoculation of the organisms.

[1] Loc. cit., p. 208.

[2] Jour. Inf. Dis., 1906, iii, 291.

Transmission.—The disease appears to be transmitted by suctorial insects. Mackie[1] believes the human louse, Pediculus vestimenti, is commonly the one involved, but Manteufel[2] has produced evidence suggesting the rat louse, Hematopinus spinosus, is at times a carrier of the organism.

Treponema Novyi.—Norris, Pappenheimer and Fluornoy[3] appear to have been the first to report relapsing fever in America. Several cases were studied; the incubation period averaged from five to seven days, and the mortality varied from 2 to 6 per cent. Novy and Knapp[4] studied the organisms in detail and discovered slight but constant differences which distinguished them from Treponema recurrentis and Treponema duttoni. Schellack[5] named the organism Spirocheta novyi. Mackie[6] was able to differentiate Treponema novyi from Treponema duttoni by agglutination reactions, and Manteufel[7] showed that the serum of patients infected with the organism of American relapsing fever did not agglutinate Treponema recurrentis and *vice versa*, thus confirming Novy and Knapp's observations. Noguchi[8] grew the organism in pure culture.

Treponema Carteri.—The causative organism of the relapsing fever of India. In 1879 Carter[9] observed the organism originally named Spirocheta carteri, but now known as Treponema carteri, in the blood of patients suffering with Indian relapsing fever, and he succeeded in inoculating mice with the organism. Novy and Knapp[10] have shown that this organism differs from those of the European, African and American relapsing fevers.

According to Schellack,[11] Treponema carteri measures from 0.3 to 0.35 micron in diameter and from 15 to 20 microns in length. The organism has not been grown in pure culture.

Treponema carteri is infective for rats and for experimental animals, but it typically causes but one relapse, contrasting in this respect with the organisms of the American, European, and African relapsing fevers respectively. It also differs from the other Treponemata in its agglutination reactions.[12]

[1] Brit. Med. Jour., December 14, 1907. [2] Arb. a. d. kais. Gesamte, xxxix, No. 2.
[3] Jour. Inf. Dis., 1906, iii, 266. [4] Ibid., p. 291.
[5] Arb. a. d. kais. Gesamte, 1908, xxvii, 364.
[6] British Med. Jour., December 14, 1907.
[7] Arb. a. d. kais. Gesamte, 1908, xxvii, 327. [8] Jour. Exp. Med., 1912, xvi, 208.
[9] Deutsch. med. Wchnschr., 1879, v, 189, 351, 386.
[10] Jour. Inf. Dis., 1906, iii, 291.
[11] Arb. a. d. kais. Gesamte, 1908, xxvii, 364.
[12] Manteufel, Arb. a. d. kais. Gesamte, 1908, xxvii, 327.

Treponema Duttoni.—**Synonym.**—Spirocheta duttoni.

Ross and Milne,[1] studying South African tick fever, observed an organism in the blood of their patients which they called Spirocheta duttoni. Dutton and Todd[2] confirmed the discovery. The disease runs a course clinically like European relapsing fever, but the paroxysms usually number four or five with corresponding periods of apyrexia before the onset of convalescence.

Morphology.—Treponema duttoni (Spirocheta duttoni) is somewhat thicker and longer than Treponema recurrentis; it measures about 0.45 to 0.50 micron in diameter and from 24 to 30 microns in length. The motility is similar to that of the organism of European relapsing fever. Noguchi[3] has grown Treponema duttoni in pure culture.

Immunity.—Rats are readily infected with the organism; those which have recovered from infection with Treponema duttoni are easily infected with Treponema recurrentis and *vice versa*. They are refractory to a second injection of the same organism, indicating that the immunity conferred by one Treponema is not protective against infection with Treponemata of another type.

Ross[4] found that the horse tick (Ornithodorus moubata) would transmit the disease from man to monkey, provided the insect bit the man during, or very shortly before, the febrile period. The organism may be demonstrated in the ovaries and eggs of female ticks which have fed upon man. This appears to be a case of true hereditary transmission; the organism is transmissible by the adult and larval insects, and through the eggs as well.

Treponema Pertenue.—**Synonyms.**—Spirillum pertenue. Treponema pallidulum.

Castellani[5] has reported the constant association of an organism which he called Spirillum pertenue, in frambesia tropica (Yaws). Frambesia is a specific infectious tropical disease characterized anatomically by peculiar specific granulomatous eruptions. The disease, like syphilis, presents three stages: (1) a primary lesion, which is a papule situated at the site of infection—this papule becomes indurated and may ulcerate; (2) a generalized eruption, papular in character, which gives rise to characteristic granulomata; this may appear after the primary lesion has healed—the disease frequently ends at

[1] British Med. Jour., 1904, ii, 1453.
[2] Ibid., 1905, ii, 1295.
[3] Jour. Exp. Med., 1912, xvi, 202.
[4] British Med. Jour., February 4, 1905.
[5] Lancet, August, 1905; British Med. Jour., November, 1905.

the second stage; (3) tertiary stage, characterized by gumma-like processes which may undergo deep ulceration.

Morphology.—Treponema pertenue is a very delicate, slender spiral organism, measuring about 0.30 to 0.50 micron in diameter, and from 6 to 18 microns in length. The ends of the organism are frequently pointed, but one or both ends may be rounded, or, rarely, somewhat swollen. There are usually from six to twenty spiral turns in each organism. Blanchard[1] states that the organism possesses an undulatory membrane, but the consensus of opinion is against this view. Very delicate polar flagella, one at each end, have been demonstrated by flagella stains. It will be observed that the size and arrangement of the organism do not differ essentially from that of Treponema pallidum. The organism fails to stain by ordinary methods, but the morphology is well brought out by Giemsa's stain. Treponema pertenue may be demonstrated by the methods applicable for Treponema pallidum. It has never been cultivated in artificial media.

Specificity of Organism.—Paulet[2] inoculated fourteen negroes with the secretion from granulomata and all developed yaws, the initial lesion appearing at the site of inoculation. There is a possibility that these negroes might have been naturally infected, however. Charlouis injected thirty-two Chinese prisoners with scrapings from the granulomata of a case of yaws and twenty-eight developed the disease, the primary lesion again appearing at the site of inoculation. This series is suggestive, but not conclusive, because the possibility of natural infection can not be ruled out.

According to Castellani,[3] yaws and syphilis are distinct diseases, because a native who had been inoculated successfully with yaws was subsequently infected with material from a chancre; this resulted in a typical attack of syphilis superimposed upon the yaws. In Ceylon syphilis is not uncommonly observed in cases of yaws which are in the secondary or tertiary stages.

Pathogenesis.—*Animal.*—The disease may be transferred to monkeys by direct inoculation. The organisms are found in the lesions.

Human.—The distribution of Treponema pertenue in the lesions of yaws is somewhat different from that of Treponema pallidum in syphilis. In the former the organisms are numerous in the spaces between the papillary pegs of the malpighian layer of the epidermis,

[1] Arch. d. Parasit., 1906.

[2] Quoted by Castellani and Chalmers, Manual of Tropical Medicine.

[3] Loc. cit.

not necessarily in intimate association with bloodvessels; in syphilis the organisms are found in considerable numbers around thickened arteries. Treponema pertenue is found constantly in the primary lesion and in unbroken papules of the generalized eruption characteristic of the secondary stage of yaws. In broken down lesions many bacteria, including Treponemata indistinguishable from Treponema refringens, complicate the picture. They are frequently not found in the tertiary stage. At autopsy the spleen, lymph glands and bone marrow contain many Treponemata as a rule; the cerebrospinal fluid is free from them ante- or postmortem.

The disease is transmissible by direct contact, and it is probable that the virus may be transmitted by biting insects as well.

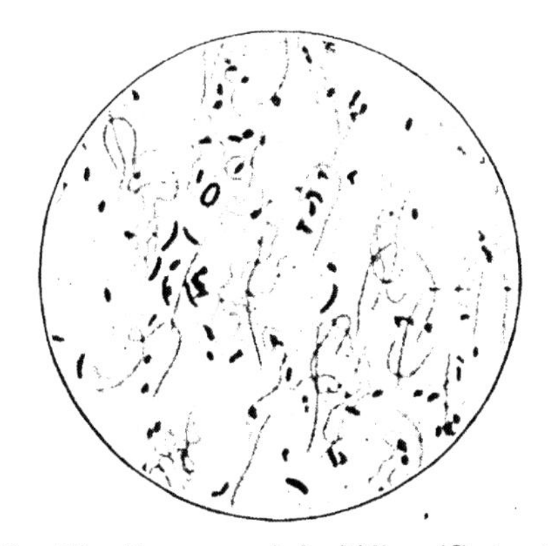

FIG. 77.—Treponema balanitidis. (Corbus.)

Treponema Phagedenis.—**Synonym.**—Spirocheta balanitidis.

Schaudinn and Hoffmann,[1] Mühlens,[2] Hoffmann and Prowazek[3] and others have described spiral organisms resembling Treponema refringens in size, shape and motility in genital and perigenital ulcerations and in phagedenic ulcers. Similar organisms have been observed in noma. Corbus and Harris[4] and Corbus[5] have described a spiral organism resembling Vincent's spiral in several cases of erosive and gangrenous balanitis, and Brault[6] has observed a similar spiral associated with a fusiform bacillus in two cases of noma. The identity of the various organisms is as yet undetermined, and their etiological

[1] Arb. a. d. kais. Gesamte, 1905, xxii, Heft 2.
[2] Centralbl. f. Bakt., Orig., 1907, xlii, 277.
[3] Ibid., 1906, xli, 741, 817.
[4] Jour. Am. Med. Assn., 1909, lii, 1474.
[5] Ibid., 1913, lx, 1769.
[6] Bull. Derm. et Syph., 1908, 2.

relationship to genital ulcerations, phagedenic ulcers, and noma is not satisfactorily established. Noguchi[1] has isolated a spiral organism in pure culture from a phagedenic ulcer, using the technic employed by him for cultivation of Treponema pallidum. This organism is the only member of the group observed in genital ulcerations and phagedenie ulcers which has been satisfactorily studied up to the present time.

Morphology.—The organism measures about 0.75 micron in diameter and about 15 microns in length, although the length varies between the limits of 4 and 30 microns. The number of spirals varies materially in different organisms in the same culture, from two complete turns to as many as eight. The ends of the organisms are found to be distinctly pointed, but not attenuated. In young cultures the organisms were found to be fairly uniform in size, from 10 to 15 microns long. In older growths the length is greater on the average, varying from 20 to 30 microns. The number of spiral turns and the spiral turns themselves are more irregular in the older growths. This organism appears to be devoid of a terminal flagellum or a terminal projection. In very old cultures signs of degeneration appear, and spherical bodies measuring about 0.5 micron in diameter are found either attached to degenerating organisms or free. These spherical bodies do not take the spore stain. In addition various semi-spherical bodies, some exhibiting refractile dots in their substance, are also found in old cultures, but none of these bodies appear to be spores in the ordinary sense.

Treponema phagedenis stains with difficulty by the more penetrating anilin dyes, and it is Gram-negative. It is colored red with the Giemsa stain. The organism is obligately anaërobie, and cultures in artificial media develop an odor suggesting butyric acid.

Pathogenesis.—Noguchi found that pure cultures of Treponema phagedenis produce an acute inflammatory reaction at the site of inoculation (intradermal) both in monkeys and rabbits, but this inflammatory area does not ulcerate. Hoffmann and Prowazek[2] inoculated two monkeys with material from a case of balanitis rich in organisms. They found some erosion had taken place at the site of inoculation after two to three days, with numerous spiral organisms in the lesion. Noguchi did not consider that his observations established the relationship of his organism to the lesion, and the experiments of Hoffmann and Prowazek are not conclusive.

[1] Jour. Exp. Med., 1912, xvi, 261.

[2] Loc. cit., p. 818.

FUSIFORM BACILLI AND SPIRILLUM FUSIFORMIS.

Fusiform bacilli, frequently in association with spiral organisms, have been observed by Plaut[1] and Vincent[2] in diphtheroid angina; by Vincent[3] in cases of hospital gangrene; by Bernheim[4] in stomatitis ulcerosa and angina ulcerosa; in noma[5] and in erosive and gangrenous balanitis by Corbus.[6]

The organism, Bacillus fusiformis, is a long, thin bacillus with distinctly tapering ends measuring from 0.5 to 0.8 micron in diameter at the centre, and varying in length from 3 to 10 microns. The bacilli appear to be rigid and straight as a rule, but occasional rods are observed to be slightly curved. In fluid media there is a tendency for the organisms to develop long tangled filaments in which granules

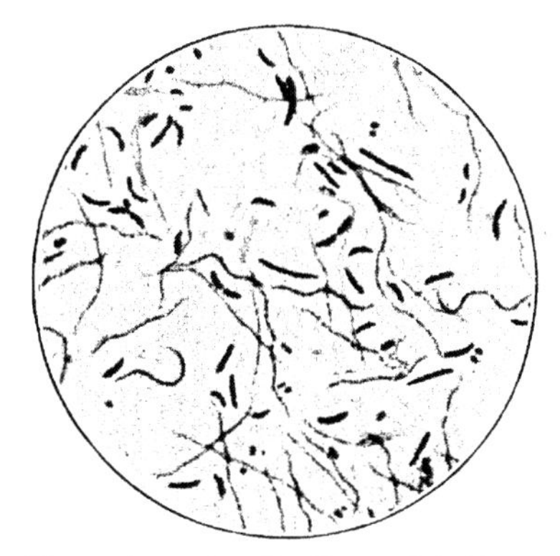

FIG. 78.—Vincent's angina, Bacillus and Spirillum fusiformis.

may be absent. Motility has not been observed, and spores and capsules have never been demonstrated. Ordinary stains color the organisms faintly, but stains containing mordants, as carbolfuchsin and carbolthionin, stain them readily and one or two intensely colored granules are frequently observed in each organism. The organisms are Gram-negative.

Tunnicliff[7] obtained development of fusiform bacilli in ascitic fluid media (anaërobic) at 37° C., but subcultures were usually negative. Krumwiede and Pratt,[8] using an improved anaërobie culture method, obtained pure cultures in anaërobie ascitic agar or serum agar from

[1] Deutsch. med. Wchnschr., 1894, xlix, 922.
[2] Ann. Inst. Past., 1899, 609.
[3] Ibid., 1896, 488.
[4] Centralbl. f. Bakt., 1898, xxiii, 177.
[5] Brault, Bull. Derm. et Syph., 1908, 2.
[6] Jour. Am. Med. Assn., 1913, lx, 1769.
[7] Jour. Inf. Dis., 1906, iii, 148.
[8] Ibid., 1913, xii, 199; xiii, 438.

a variety of lesions of the type mentioned above. The colonies were small, more or less circular in outline with projecting, hair-like growths, which attain a diameter of 1 to 2 mm. In all, fifteen strains were isolated in pure culture, all of which produced indol and possessed a disagreeable odor. Two distinct cultural types were distinguished; all strains produced acid, but no gas, in dextrose, galactose and levulose; one type produced acid in saccharose, the other type was without action upon this sugar. There was no demonstrable relation between the source of the culture and the fermentation of saccharose, which is in harmony with Tunnicliff's observation that the fusiform bacilli obtained from a variety of lesions presented no demonstrable distinctive characters.

No spiral organisms developed in the cultures, although they were present in smears from the original material. This points strongly to the non-identity of the fusiform bacillus and the spiral organism so frequently associated with it, although Tunnicliff[1] claims that the spirilla and the fusiform bacilli are different forms of a single organism.

The relation of the fusiform bacilli to morbid processes is not finally established, although the injection of material rich in these organisms has frequently led to necrosis and suppuration in experimental animals. The most convincing evidence of their pathogenicity is the occasional demonstration of fusiform bacilli in considerable numbers in tissues from cases of noma and similar severe lesions.

[1] Jour. Inf. Dis., 1911, viii, 316.

SECTION III.

HIGHER BACTERIA, MOLDS, YEASTS, FILTERABLE VIRUSES, DISEASES OF UNKNOWN ETIOLOGY.

CHAPTER XXVIII.

TRICHOMYCETES, ACTINOMYCETES, HYPHOMYCETES, SACCHAROMYCETES.

THE PATHOGENIC HIGHER BACTERIA.
- Trichomycetes.
- Leptothrix.
- Cladothrix.
- Nocardia (Streptothrix).
- Actinomyces Bovis.
- Mycetoma (Madura Foot).

HYPHOMYCETES.
- Eumycetes or Molds.

SACCHAROMYCETES.

THE PATHOGENIC HIGHER BACTERIA.

Trichomycetes.—The Trichomycetes occupy a position intermediate between the true bacteria (Schizomycetes) and the molds (Hyphomycetes), in the system of classification. Their method of reproduction is more complex than that of the bacteria, but their cycle of development is simpler than that of the molds. The organisms usually grouped in the Trichomycetes are heterogeneous in their characteristics and there is a decided lack of agreement concerning the limitation of the several subdivisions of these microörganisms. Foulerton[1] places all the members of the higher bacteria in one genus, Streptothrix, including the older genera, Leptothrix, Cladothrix, Streptothrix and Actinomyces. Wright[2] and others have not subscribed to this view and their evidence is impressive. Additional investigations are required before final judgment can be made.[3] For the present the older grouping of the Trichomycetes, Leptothrix, Cladothrix, Nocardia (Streptothrix), and Actinomyces will be adhered to.

[1] Allbutt and Rolleston, System of Medicine, 1906, ii, Part I, 302; British Med. Jour., 1912, i, 300.

[2] Jour. Med. Research, 1905, xiii, 349.

[3] See Musgrave, Clegg, and Polk, Philippine Jour. of Sci., 1908, iii, 447, for very full bibliography and discussion.

Leptothrix.—Leptothrices are frequently found in the mouth, so commonly indeed that Leptothrix buccalis is regarded as a regular inhabitant of the oral cavity. Suppurative processes incited by this organism have been reported by a few observers, but the evidence is by no means conclusive. The organisms are cultured with great difficulty upon artificial media and no cultures were obtained from the cases reported.

Cladothrix.—The important cultural differentiation of the Cladothrices from the Streptothrices rests upon the false branching of the former. The few meager reports of cases of Cladothrix infection cited in the literature are not sufficiently definite to determine the type of organisms involved.

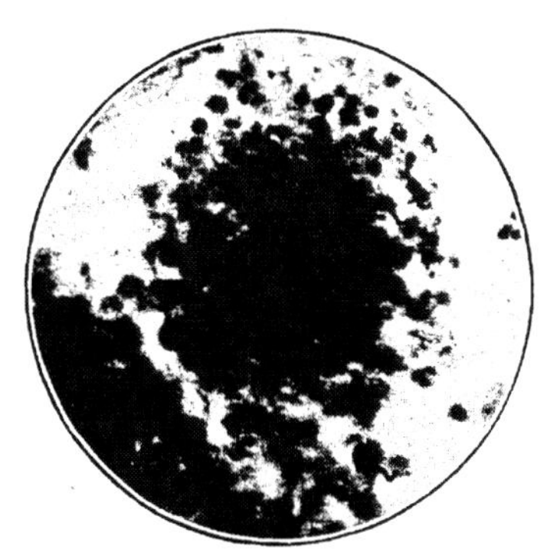

Fig. 79.—Streptothrix hominis.

Nocardia (Streptothrix).—The more common name of the group is Streptothrix, but investigation has shown that the latter term was previously given to a mold; according to rules of botanical nomenclature, it must be replaced by a name hitherto unused. Nocardia appears to be appropriate. The first organism was described by Nocard[1] as the inciting agent of a disease of cattle in Guadaloupe, known as farcin. Since that time many cases have been reported both in animals and in man.

Nocardia mycoses have occasionally been confused with tuberculous infections in the past. Farcin was suspected to be a tuberculous process until Nocard[2] clearly demonstrated that the organism was an acid-fast Nocardia.

In man the disease usually progresses slowly and the lesions are markedly localized, but it may run a rapidly fatal pyemic or pneu-

[1] Ann. Inst. Past., 1888, ii, 293.

[2] Loc. cit.

monic course of one or two weeks' duration. A chronic case may abruptly become generalized and terminate fatally. It is not definitely known if all chronic cases prove fatal or if some eventually recover. The Nocardia appear to be widely distributed in the soil, water, upon foodstuffs and upon plants and it is suggestive that nearly 50 per cent. of all cases reported have been infections of the head and neck.[1] About 20 per cent. of cases are chest infections and the clinical symptoms are very like those of tuberculosis. If repeated sputum examinations are negative although the syndrome suggests tuberculosis, search should be made for Nocardia.

Morphology.—The Nocardia are very pleiomorphic; in purulent material and other discharges the organisms are of varying length, some short and rod shaped, others long-branched filaments (mycelia). The filaments usually segment or fragment, producing the shorter bacillary forms and, in artificial media, forming chains of spores as well. Old cultures in artificial media are composed chiefly of bacilloid forms—long, somewhat curved filaments which may or may not be branched, and spores which occur singly or in small groups and pairs. The organisms may or may not be acid-fast, but they are Gram-positive. The granules or "drusen" so characteristic of actinomycotic infections are not found in Nocardial mycoses.

Cultivation.—Nocardia may frequently be grown upon artificial media—gelatin or agar—directly from pus or other morbid material. The colonies develop slowly and after five to seven days they appear as gray, opaque, shining plaques which may reach 3 to 5 mm. in diameter after prolonged incubation. A densely matted pellicle composed of branched and unbranched filaments forms upon the surface of broth and a flocculent sediment gradually collects at the bottom of the tube. Löffler's blood serum appears to be the most favorable medium for the initial growth of Nocardia directly from the tissues.

The inoculation of cultures into rabbits or guinea-pigs frequently leads to chronic abscesses, bronchopneumonia or a rapidly fatal generalized infection, depending upon the virulence of the organism and the site of inoculation. Recently Claypole[2] has prepared a series of "Streptotrichins;" glycerin bouillon cultures made from non-acid-fast mycelial organisms and the partly acid-fast bacillary forms of Nocardia, which give definite skin reactions on persons with nocardial

[1] Claypole, Jour. Am. Med. Assn., 1914, xiii, 604.
[2] Jour. Am. Med. Assn., 1914, lxiii, 603.

infections. Controls (normal, uninfected individuals), do not react, but a Nocardial mycosis and tuberculosis may exist simultaneously in the same individual, as shown by the appearance of both organisms in the sputum, and both the streptotrichin and tuberculin skin reactions. Claypole also finds that glandular and bone infections with Nocardia may be demonstrated as readily as the lung infections by the skin reaction with streptotrichin.

Actinomyces Bovis.—**Synonyms.**—Discomyces bovis; Nocardia actinomyces; Streptothrix israeli.

The causative organism of the disease of cattle known as "lumpy jaw" or "big jaw," Actinomyces bovis, was first described by Bollinger,[1] although the granules or "drusen," consisting of colonies of the organism, were described by von Langenbeck as early as 1845. The first human cases were reported by Israel.[2]

FIG. 80.—Actinomyces colony showing peripherally arranged clubs.

Considerable confusion has arisen concerning the identity of the organisms found in suppurative lesions which superficially closely resemble those of Actinomycosis.[3] Wright[4] has clearly shown that true actinomycotic infections are characterized not only by suppurative processes and granulation tissue formation, but that the pus from these lesions contains the characteristic granules or "drusen," which are composed of branched filamentous organisms densely packed together, with characteristic club-shaped bodies radially arranged

[1] Centralbl. f. klin. Med. Wissensch., 1877, xv, 481.
[2] Virchows Arch., 1878, lxxiv, 15; 1879, lxxviii, 421.
[3] See Foulerton (Trans. Path. Soc., London, 1902, liii, Part 1, 56), and Neukirch (Ueber Strahlenpilze, Strassburg, 1902), for literature.
[4] Jour. Med. Res., 1905, xiii, 349.

at the periphery of the colony. The pus from so-called pseudo-tuberculosis, streptothrix and cladothrix infections do not exhibit these characteristic "drusen."

Morphology.—Actinomyces bovis is a pleiomorphic organism belonging to that group of microörganisms intermediate between the true bacteria (Schizomycetes) and the molds (Hyphomycetes) known as the Trichomycetes. It is best observed in pus from active lesions, in which it occurs in gray or yellowish colonies or granules (drusen), frequently large enough to be visible to the naked eye. The colonies vary in size but usually measure from 0.5 to 2 mm. in diameter. Such a colony, crushed between two slides or a slide and cover glass, appears as a rosette-shaped aggregation of densely packed filaments

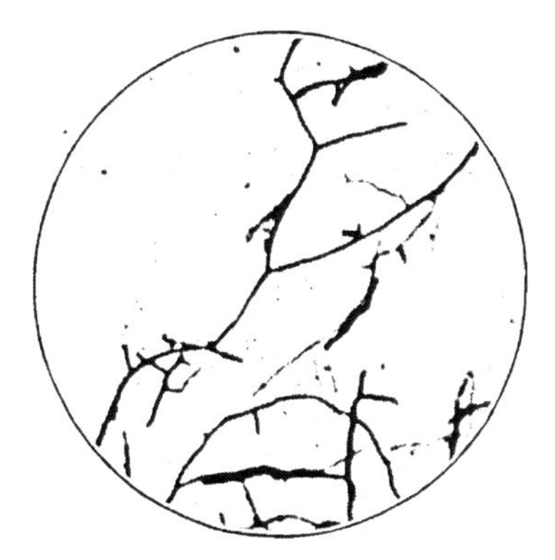

Fig. 81.—Actinomyces, bouillon culture.

which exhibit a radial arrangement. The centre is so crowded with organisms that it appears opaque and usually contains many ovoid bodies measuring from 1 to 1.5 microns in diameter. According to Wright,[1] these ovoid or coccoid bodies are formed by the disintegration of the filaments. The periphery of the colony contains many interlaced branching filaments, many of which exhibit on their distal ends, an enlargement or "club" which is a hyaline layer or sheath about the extremity of a filament. These filaments measure about 10 to 12 microns in length and the clubs 20 to 30 microns in length by 8 to 10 microns in diameter. Grown in artificial media club formation is absent unless blood or blood serum[2] is added, but even in enriched media the formation of clubbed forms is irregular.

[1] Loc. cit.

[2] Wright, loc. cit., p. 336.

Actinomyces bovis stains by Gram's method, but the clubs are not colored. Eosin brings them out clearly. It has been held by Boström[1] that the clubs are degenerative phenomena, but Wright[2] believes their chief function is a protective one, shielding the filaments from the harmful action of the body fluids and cells of the host.

Isolation and Culture.—The organism is anaërobic and appears to grow with moderate luxuriance in deep glucose-agar stab cultures. Material for inoculation is best obtained by crushing a granule between sterile glass slides, or rubbing it on the inside of a sterile test-tube, after two to three preliminary washings in sterile salt solution to remove or diminish surface contamination. The finely macerated colony is distributed evenly in deep dextrose-agar tubes and incubated

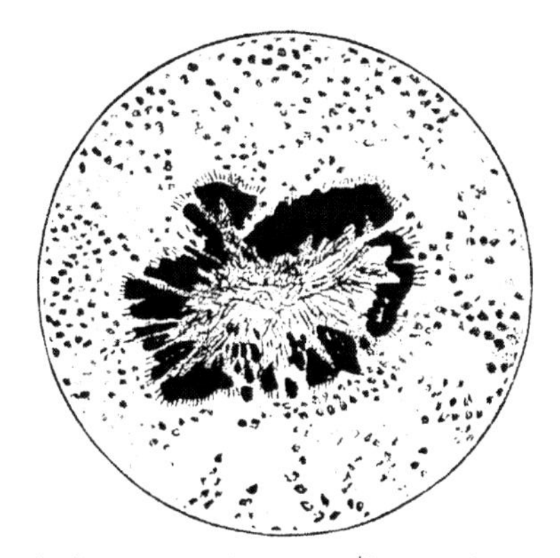

Fig. 82.—Actinomyces—club formation, semi-diagrammatic.

at 37° C. After two to five days colonies appear scattered through the depths of the medium and are generally very numerous in a zone 0.5 to 1 cm. below the surface. They do not ordinarily grow above this level. The deeply lying colonies increase in size until they measure 1 to 3 mm. in diameter at the end of a week's incubation. Microscopically these colonies consist of masses of radially arranged, branching filaments which exhibit a decided tendency to break up into short bacilloid or ovoid segments. A colony at this stage becomes a mass of compact short filaments and bacillary forms. Clubs are not seen under these conditions unless blood or blood serum is added to the medium.

In bouillon the organisms grow in dense white or gray masses of

[1] Beitr. z. path. Anat., u. z. allg. Path., 1890, ix, 1.

[2] Loc. cit., p. 397.

interwoven filaments which develop only at the bottom of the tube. Surface growth is never observed and turbidity practically never occurs. Freshly heated broth, in which the dissolved oxygen has been driven off, appears to afford a somewhat more luxuriant growth, particularly during the first few days' inoculation, but this precaution is by no means absolutely necessary to obtain development. Prolonged cultivation in broth frequently causes the organisms to lose the discrete, mulberry-like colony; the growth becomes somewhat flocculent and viscid. Milk and other artificial media, aside from agar and bouillon, are not favorable for the development of the organisms.

Fig. 83.—Actinomyces—mycelioid development, semi-diagrammatic.

Actinomyces bovis does not grow at temperatures much below 37° C. Development ceases at room temperature. The resistance to drying is considerable, fifty days being about the minimal time required to prevent growth. In artificial media, however, the organism usually becomes non-viable in a shorter period. The thermal death point is about 62° C. for five minutes. Toward ordinary antiseptics, Actinomyces is very resistant, but it is claimed that methylene blue is strongly germicidal to it.

Products of Growth.—Neither toxins nor enzymes have been detected in cultures of Actinomyces bovis. It is believed that toxins are not produced.

Pathogenesis.—*Animal and Human.*—Actinomycosis occurs as a spontaneous infection both in cattle and in man; much more commonly, however, in the former. Other mammals—horses, asses and sheep—are occasionally infected. The lesions belong to the group of the infectious granulomata and the portal of entry of the organism is usually the mouth, although cutaneous infections have been described. The mouth and adnexa and the pharynx are more commonly the site of the initial localization of the organism, but the lungs or the alimentary canal may be first involved. The earliest stage of the infection is a small nodule not unlike a tubercle; microscopically it is made up of small round cells, epithelioid cells and giant cells. This soon softens and sinuses often are formed, through which the pus escapes. The surrounding connective tissue proliferates rapidly, forming a dense encapsulation through which invasion of neighboring tissue takes place; often the disease spreads in one direction while simultaneously the older lesion becomes cicatrized. Death frequently occurs through secondary invasion by adventitious bacteria.

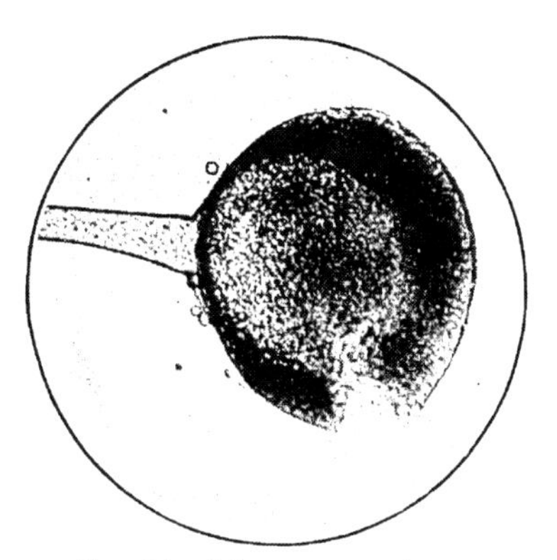

FIG. 84.—Mucor sporangium.

Actinomycosis is not a contagious disease and it is practically impossible to infect experimental animals, as guinea-pigs and rabbits, with the virus. Wright[1] has been unable to produce progressive actinomycosis in these animals, although he succeeded occasionally in inducing a localized purulent nodule formation in guinea-pigs, in which granulation tissues and colonies of Actinomyces appeared, some of which showed poorly defined clubs.

The disease is stated to be transmitted through wounds caused by certain grains, particularly those which possess barbs, but the evidence is not wholly convincing.

The diagnosis of actinomycosis is best made by microscopic examination of sputum, or the pus from the lesions. The demonstration of the characteristic "drusen" with their club-shaped peripheral

[1] Loc. cit.

filaments is conclusive. Sometimes actinomycotic pus does not contain granules; if the sinus be curetted, the organisms will frequently be demonstrable in the scrapings, even though they are absent from the pus.

Mycetoma (Madura Foot).—The term Mycetoma is a generic one, including purulent inflammations of the foot chiefly, but also of the hands and less commonly of other parts of the body. The lesions superficially resemble those of actinomycosis.

Three varieties of the disease have been described, depending upon the color of the granules found in the pus—the melanoid or black type, the ochroid or white, and a red type which has been less thoroughly investigated.

Several organisms have been isolated from the various lesions, including not only an Actinomyces (Actinomyces maduræ), but a mold, Aspergillus bouffardi, as well.

The mutual relations of the organisms and the various types of Madura foot have not been satisfactorily determined.

HYPHOMYCETES.

Eumycetes or Molds.—The molds are a group of organisms which are structurally somewhat more complex than Bacteria for, with a very few exceptions, there is a physiological division of function into vegetative cells which provide the nutrition of the organism and reproductive cells which are concerned in the perpetuation and multiplication of the species. They are widely distributed in nature, the majority living saprophytically upon lifeless organic matter—some are parasitic upon animals and plants; few types, however, incite disease in man, animals, or plants.

In human pathogenesis their activities are usually restricted to the skin and adnexa, but occasionally spreading over mucous membranes and even involving the respiratory tract. Among the hyphomyceal diseases of man are favus, ringworm, thrush, pityriasis versicolor, sporotrichosis and aspergillosis.

The cells of molds are larger than bacteria, as a rule, measuring on the average from 2 to 10 microns in diameter, and they grow into long filaments or threads called *hyphae,* which tend to branch and form intricately interwoven networks called *mycelia.* Like all true plant cells, each hypha exhibits a clearly defined, doubly contoured ectoplasm or limiting membrane within which is confined the cytoplasm, which is usually coarsely or finely granular. In the lower forms, Phycomycetes, each hypha is a unicellular multinuclear cell, which may be branched; in the higher forms, Mycomycetes, the filaments are multicellular, each cell being separated from its fellows by distinct septa. A nucleus is demonstrable in a majority of the molds and it is probable that it is present in all.

Reproduction.—The reproductive cells of the lowest and simplest forms are scarcely differentiated morphologically from the vegetative cells, indeed in some instances the distinction has never been made. The hyphæ break up and the fragments give rise to new colonies. Reproduction in the Phycomycetes, of which the widely distributed genus Mucor is a familiar type, occurs in the following manner—a constriction occurs near the tip of an aërial hypha and the extremity

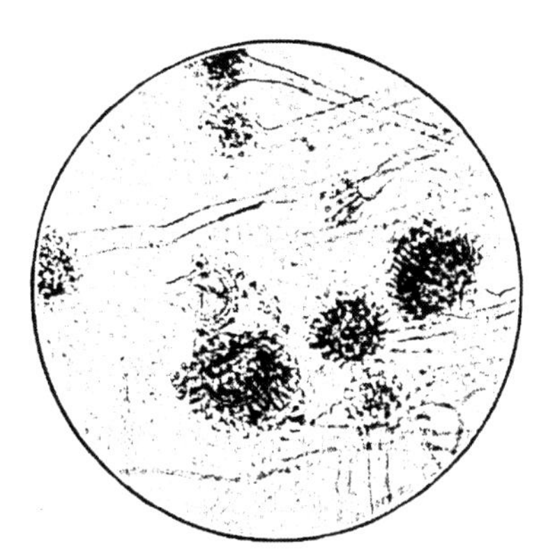

FIG. 85.—Aspergillus sporangia.

then increases in size until a spherical mass, the sporangium, is formed, which divides into a number of spores. These escape with the rupture of the sporangium and, if they reach a favorable medium, form the starting points of new colonies. This is asexual reproduction. Sexual reproduction takes place somewhat differently: lateral branches from two adjacent hyphæ meet and fuse. These branches or gametophores are morphologically indistinguishable but differ in sex. The fused cell enlarges to form a zygospore, separated from the hyphæ by septa, and eventually grows into a sporangium, from which asexual spores escape and start new colonies.

Among the Mycomycetes or higher molds, asexual reproduction alone occurs. The simplest type begins as a thickening of the end of a hypha, which soon constricts at regular intervals to form small spherical or oval spores. The spore-containing cell is known as an ascus and the spore ascospore.

Somewhat more complex is reproduction of the common green mold, Aspergillus. An aërial hypha or conidiophore develops, thicker at the distal than at the proximal end, and from this thickened end radially arranged spherical or oval conidia arise.

Microscopical Examination of Molds.—Molds are usually best examined in water, to which an equal volume of glycerin has been added, and unstained. The general arrangement of mycelium, spores and sexual bodies can be observed with the lower powers of the microscope—the finer details of structure require a greater magnification. Anilin dyes color molds readily and the Weigert fibrin-staining method is very good to demonstrate molds in tissue sections.

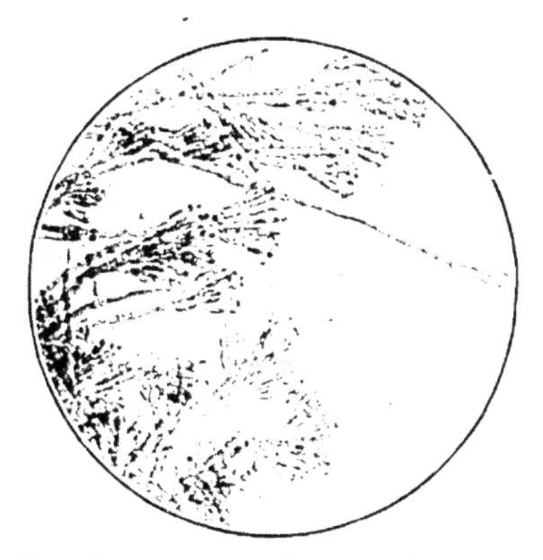

Fig. 86.—*Penicillium*; conidiophores, sterigmata, and conidia.

Growth on Artificial Media.—Molds are almost invariably aërobic and their development in artificial media requires abundant free oxygen. A slightly acid reaction is best for their growth, but media with an alkaline reaction and even a relatively strong acid reaction (organic acids, not mineral acids), will usually permit of their multiplication. Even on very dry media development takes place.

Pathogenic Molds.—*Favus.*—Favus or tinea favosa is a skin disease limited chiefly to the hairy parts of the body; more frequently the head alone is involved, but the disease may spread over the entire surface of the body. It is not limited to man—dogs, cats, mice and rabbits are also susceptible. The disease is contagious and is transmitted from man to man or from animal to man by contact. Uncleanliness is a potent predisposing factor, but individuals with lowered vitality, as poorly nourished children and consumptives, appear to be relatively more readily infected than the more robust. The organism spreads slowly and the disease is a chronic one, difficult to influence by treatment. The initial lesions are small red pimples, which soon enlarge somewhat, forming gray or sulphur-yellow crusts grouped around the base of hairs. These crusts, known as scutella (singular scutellum), slowly increase in size peripherally and tend to coalesce. If a scutellum is removed it is found to be somewhat thicker in the centre and cup shaped. Examined under the microscope it consists of a dense, matted mycelium which in the centre may be so compact as to obscure the individuality of the filaments; at the periphery the growth is less luxuriant and the individual filaments are clearly defined. Spores are very numerous at the centre of the scutellum, but at the periphery they are much fewer in numbers. The hair enclosed by the colony of mold is destroyed.

The organism, Achorion schönleinii, was first observed by Schönlein in 1839. It is readily cultivated at room temperature upon gelatin, or better, upon agar at 30° to 35° C. Media with a neutral or slightly alkaline reaction are more favorable for its development than acid media. In this respect Achorion schönleinii differs culturally from the majority of molds. Material taken directly from the centre of a scutellum, streaked upon agar, usually develops into white or gray colonies in which the mycelia and spores are readily recognizable with the lower powers of the microscope. Frequently adventitious organisms overgrow the more slowly developing favus parasite. If a piece of the scutellum is ground in a sterile mortar with sterile powdered water glass and the powder well distributed upon gelatin-agar or Sabouraud's medium,[1] pure cultures are usually obtained. The yellow-brown colony usually exhibits a central depression resembling somewhat that of the scutellum. The swollen ends of the filaments are quite characteristic. There appear to be several varieties of the mold, but there is only one type of the disease.

Herpes Tonsurans.—Herpes tonsurans, ringworm, Tinea tonsurans or sycosis is a disease chiefly of the hairs of the head or beard, but it often spreads to the skin as well, Tinea circinata. The axillary or pubic hairs are occasionally involved. It occurs in children rather

[1] SABOURAUD'S MEDIUM.

Peptone (Witte)	2.0 grams
Glycerine, redistilled, pure	4.0 grams
Water	100.0 c.c.
Agar	1.2 grams

more frequently than in adults. The disease is characterized clinically by the formation of inflamed scab-areas or patches on the skin immediately surrounding hairs and these patches exhibit a decided tendency to spread. They itch intensely and within them the hairs fall out. Usually the inflammation is not accompanied by exudation, but in very severe cases pustule formation may occur. The disease is contagious and is transmitted by towels, the hands, hairdressers' utensils and very commonly in the tropics through laundry. The initial lesion appears in the outer layers of the skin and extends downward through the hair follicle and then invades the inner layers of the hair itself, through which both the mycelia and spores develop in large numbers.

The organism, Trichophyton tonsurans, was described by Gruby and by Malmsten in 1845. Several subvarieties have been described, but their differential characteristics are imperfectly established. It is readily demonstrated in the hair bulb by adding a few drops of NaOH solution, gently heating and examining under the microscope.[1] The mycelial filaments appear in the bulb and penetrate for some distance along the hair shaft. The spores are usually restricted to the outer layers of the hair.

The mold grows readily upon neutral agar and gelatin, the latter becoming liquefied. After a few days' incubation, multicellular mycelia with their nodal thickenings within which chlamydospores develop appear and frequently the colony becomes pigmented—brownish—after prolonged cultivation. Plant[2] states that there are two varieties of Trichophyton tonsurans, one, less common, producing relatively large spores, the other producing smaller spores. Guinea-pigs may be successfully infected with cultures of the organisms grown on artificial media; a small area on the back is epilated and the culture rubbed in. The lesions are self-limited and usually heal spontaneously after a few weeks.

Pityriasis Versicolor.—Pityriasis versicolor is a disease of the epidermis which differs from favus and ringworm anatomically in that the infecting organisms neither penetrate the deeper layers of the skin, nor do they cause any noteworthy alterations in the skin or hair. Usually the epidermis of the chest, abdomen, joints and axilla are involved, rarely the neck. The disease is observed in the uncleanly

[1] Water should not be added after the addition of the NaOH, else the hair will very quickly crumble.

[2] Plant removes a hair to a sterile moist chamber, seals the cover glass with melted paraffin and incubates for several days. When the spores have germinated the mycelia may be removed and cultivated upon agar or gelatin.

and particularly in those who prespire freely. The tuberculous and diabetics are not infrequently infected. The disease is characterized by the development of light brown or yellow patches which are not noticeably raised above the surrounding surface; these patches are irregular in outline and tend to spread and coalesce.

The inciting organism, Microsporon furfur, was described in 1846 by Eichstedt. The organism resembles the Achorion schönleinii rather closely. It occurs in abundance in the scales where the relatively short, thick, septate hyphæ surrounded by large groups of spores are quite characteristic. The hyphæ measure from 3 to 4 microns in diameter and the spores are frequently observed to be enclosed in a spirally coiled covering. They measure about 3 to 6 microns in diameter.

Cultivation of the organism upon artificial media is accomplished with difficulty and glycerin media are best adapted for this purpose. The colonies are very minute—0.5 to 1 mm. in diameter. They are white or brownish and tend to spread over the medium. The hyphæ are usually definitely curved and the ends are somewhat club shaped. The spores occur in masses very similar in arrangement to those observed in the scale itself. Cultures rubbed into an epilated area on the back of rabbits may induce the characteristic colored patches if the inoculated area is protected with a thick covering to induce hyperemia.

Thrush or Soor.—Thrush is primarily a localized disease of the mouth, occurring chiefly in weakly children. It has also been found in the vagina of pregnant women and in adults suffering from severe nutritional disturbances, diabetes and typhoid fever. The early lesion is a small white plaque which has a velvety appearance, differing in this respect from the pseudomembrane of diphtheria and from the gray throat of scorbutus. The plaque is made up of epithelial cells overgrown with the organism. The lesion may spread to the larynx and esophagus and lead to a generalized fatal infection. Usually, however, the prognosis is favorable.

The organism, Oidium albicans, was described by Langenbeck in 1839, but it was first successfully cultured by Grawitz in 1871. The classification of Oidium albicans is not clear, for the organism grows both as a yeast and produces mycelia and spores. The yeast-like cells are oval or round, measuring about 4 to 6 microns in diameter, and they frequently form buds precisely like true yeasts. They stain mahogany brown with strong Gram's solution. The mycelia are doubly contoured and form chlamydospores. If a bit of the membrane be macerated in a drop of acetic acid the epithelial cells are cleared and the parasite is readily observed. Two distinct types are recognizable in gelatin cultures, one of which liquefies the medium, the other does not. In solid media yeast-cell formation predominates and many of the cells are observed to bud; in fluid media mycelia are produced and spore-formation usually occurs after several days' incubation. The spores—chlamydospores—usually enlarge and develop into filaments when they are transplanted into fresh media. The organism is not uncommon in the air.

The organism does not produce thrush when introduced into experimental animals, but it may cause a generalized thrush mycosis when injected intravenously in rabbits.

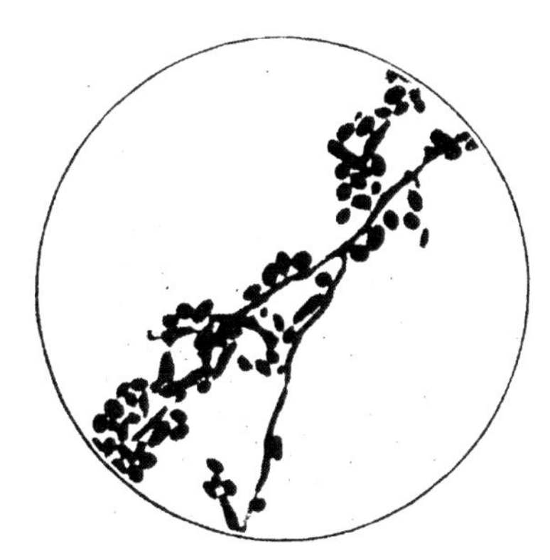

FIG. 87.—Sporothrix.

Aspergillus Mycosis.—Aspergillus fumigatus occasionally incites a disease of the lungs and bronchi in birds and rarely in man. The organism penetrates to the alveoli and the mycelia and spores may be demonstrated in sections of the lungs in fatal cases. It also has been found rarely in middle-ear infections and in the nasopharynx.

The mold grows readily upon ordinary media and the colonies, after several days, become dark green in color. The organism belongs to the genus Aspergillus, which is widely distributed in damp cellars and upon food. Microscopically, aërial hyphæ arise from the filamentous mycelium, whose distal ends are swollen into club-shape masses of undivided sterigmata, from which chains of conidia arise. The conidia are spherical, greenish, and measure about 3 microns in diameter. It is differentiated from many of the aspergilli by its green color, other members of the group exhibiting black, brown and other colored colonies.

Rabbits, guinea-pigs and pigeons are susceptible to infection with Aspergillus fumigatus. The lesions produced resemble tubercles somewhat on superficial examination, but microscopic examination always reveals the mycelium and spores.

Sporotrichosis.—The disease known as sporotrichosis was first described by Schenck[1] and later by Hektoen and Perkins.[2] The latter observers named the causative organism Sporothrix (Sporotrichon) schencki.

Usually sporotrichosis runs a chronic course, characterized by small discrete nodules in the subcutis, which at first are hard and inelastic, indolent and resemble multiple disseminated gummata. The lesions progress slowly and after some time soften, break through the skin and discharge a slimy, serous, yellowish pus. The skin around the nodules is not usually greatly indurated and there is little pain, febrile reaction or constitutional disturbance. Not infrequently regional lymph channels become thickened with a few gumma-like nodules at irregular intervals, which break down and ulcerate. The lesions resemble syphilitic gummata, or, occasionally, tuberculous ulcerations. Rarely the disease may be acute with fever, emaciation and prostration and sporotrichic nodules form on mucous surfaces in the peritoneum, the lungs or kidneys. The Wassermann reaction is negative and neither Treponemata nor tubercle bacilli are found in uncomplicated cases.

The organism develops readily upon ordinary culture media which have an acid reaction. Material for inoculation is best obtained from a softened but unopened nodule. The colonies grow slowly as small plaques which develop into white fluffy masses that become brown after prolonged cultivation. Secondary transfers to artificial media develop much more rapidly. Many strains grow better at room than at body temperature.

The organism as seen in the pus consists almost exclusively of oval spores measuring from 2 to 4 microns in diameter and from 3 to 6 microns in length; they are frequently collected in groups or masses of from 3 to 30 or more, at the ends of the filaments. They are Gram-positive. The mycelia are found in cultures as filaments about 2 microns in diameter and from 20 to 40 microns long.

[1] Johns Hopkins Hosp. Bull., 1898, ix, 286.
[2] Jour. Exp. Med., 1900, v, 77.

Rats are quite susceptible to inoculation with pus from lesions or from cultures. The disease may follow an acute or a chronic course, but the cutaneous nodules are not regularly produced in this animal —otherwise the lesions are fairly typical. In the acute disease the animal usually dies within two weeks, frequently in consequence of a degeneration of the parenchyma of the kidney. The organism may be recovered from the blood stream or the kidneys—a true sporotrichon septicemia. In the chronic type of the disease the mold localizes and results in the formation of multiple abscesses in the internal organs and especially in the testes. Intraperitoneal injections usually lead to the appearance of small nodules in the testes and internal organs which may remain discrete or become confluent, with central necrosis and suppuration. They resemble miliary tubercles superficially. Microscopically the relatively large oval spores, but not the mycelia are found. The disease appears to occur spontaneously in rats, especially the testicular type.

The serum of cases of sporotrichosis frequently agglutinates the spores of the organism (best obtained by grinding cultures to dryness in a sterile mortar, then diluting with salt solution and filtering through filter paper) in dilution from 1 to 200 even 1 to 1000. The sera of normal individuals possesses no agglutinating power for the organism. Actinomycotic serum may agglutinate with the organism in dilutions as great as 1 to 50, suggesting common group agglutinins for both organisms. Complement fixation is apparently not specific.

SACCHAROMYCETES.

The Saccharomycetes or yeasts are especially characterized by their method of multiplication. Unlike the Bacteriaceæ, which reproduce by transverse fission, the resulting cells being of equal size, the yeasts reproduce by budding. A yeast cell about to reproduce sends out an evagination or bud, which is first visible as a minute enlargement on the surface of the parent organism. This gradually increases in size, still maintaining an ovoid shape and remaining adherent by a small isthmus until it reaches approximately the size of the original cell. The isthmus then is broken, continuity between the two cells is interrupted and the fully mature individual reproduces in like manner. It is not uncommon to find budding in the daughter cell before it severs its connection with the mother cell; if the environmental conditions are favorable for rapid growth. Many yeasts form highly refractile bodies—ascopores—within their cytoplasm when invironmental conditions become unfavorable for further development and, unlike the bacteria, each yeast commonly produces more than one spore, usually two, three or four, but rarely or never more than four. The ascospore is outlined by a doubly contoured membrane and usually it remains within the intact maternal cell. At sporulation each ascospore develops into a mature yeast cell, consequently sporulation in this group is, in a sense, a process of reproduction, for each ascospore is potentially equivalent to a bud in that it develops into a complete vegetative cell.

The yeasts are of considerable importance commercially; some varieties are extensively used in the fermentation of malt and others

Fig. 88.—Yeast cells showing budding.

are employed in the manufacture of bread. In either case the organism liberates carbon dioxide from carbohydrates, and alcohol as well. This activity is brought about by an intracellular enzyme, "zymase," which may be obtained in an active state, free from yeast cells, by crushing the latter with hydraulic presses and filtering off residual cells through porcelain filters. Little or no acid is formed and the yeast fermentations are, in general, different in this respect from bacterial fermentations in which acid formation, but not alcohol formation, is the rule.

Structurally, yeasts exhibit greater complexity than the bacteria. The cytoplasm of the yeast cell usually exhibits a granular or vacuolated appearance and nuclear material, or at least structures that color like nuclei have been demonstrated.

The view was formerly held that yeasts had some etiological relationship to cancer. Sanfelice[1] and others have cultivated organisms closely resembling Blastomycetes from cancerous tissue and have attempted to harmonize the appearance of the yeasts with certain inclusion bodies within cancer cells. The consensus of opinion at the present time is wholly against this hypothesis.

Certain varieties of yeast are definitely known to incite disease in man and animals. Busse[2] isolated a yeast which he called Saccharomyces hominis from a fatal infection in a woman which began in a tibial abscess and somewhat later Gilchrist[3] reported a case of blastomycetic dermatitis in man. Since that time numerous similar cases have been recorded, a majority of them around Chicago.[4] The causative organism (Blastomyces), has been variously grouped with the yeasts and with the oidia. It is usually referred to as a yeast.

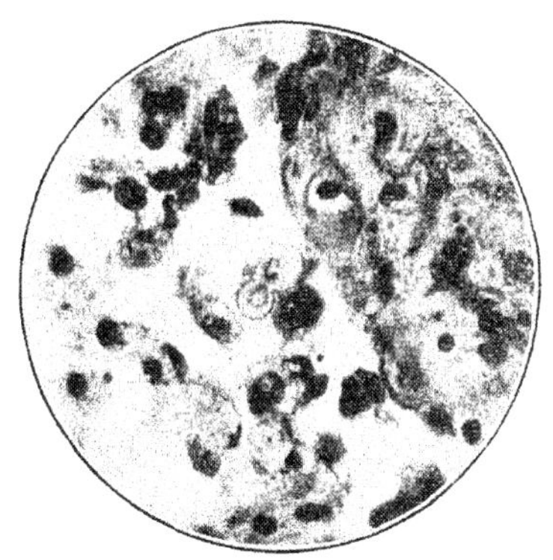

FIG. 89.—Blastomyces—section of lung.

Morphology.—Blastomycetes, as found in the tissues, are ovoid or spherical cells measuring from 3 to 30 microns in diameter, the smaller dimension being the more common. Mycelial and hyphaeal forms are found in cultures, but they are rarely met with in the tissues. The mycelial filaments measure from 5 to 10 microns in diameter. The cells usually occur in groups of fifteen or twenty or even more, but occasionally single organisms are met with. The variation in size within large groups of Blastomyces is usually very considerable. A thick membrane or capsule is frequently found around mature cells within the tissues of the body, but ascospores have not been definitely demonstrated. The Blastomyces stain with ordinary

[1] Centralbl. f. Bakt., Orig., 1902, xxxi, 254.
[2] Ibid., 1894, xvi, 175.
[3] Johns Hopkins Hosp. Rep., 1896, i, 296.
[4] See Arch. Int. Med., 1914, xiii, No. 4, for Case Reports.

anilin dyes and they are Gram-negative. They are best observed unstained in hanging drop preparations previously treated with NaOH, which brings out their outline sharply, also the refractile layers of the cell membrane. Of particular importance is the recognition of budding, which at once distinguishes the organisms. The cytoplasm is granular while the cell as a whole possesses no flagella and is consequently non-motile; the granules frequently exhibit Brownian movement.

Isolation and Culture.—The Blastomycetes grow with moderate luxuriance upon Löffler's blood serum and glucose agar. Initial pure

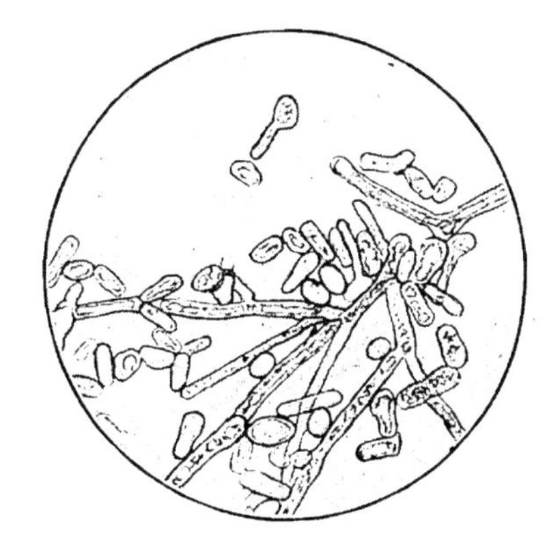

FIG. 90.—Blastomyces, maltose broth culture.

cultures are somewhat difficult to obtain, however, chiefly because adventitious organisms are almost always present, which overgrow the more slowly developing Blastomycetes. It is necessary to dilute material containing the organisms with sterile salt solution or broth and to crush the tissue into minute fragments. Once pure colonies are obtained, their perpetuation by subculturing is readily accomplished. Slightly acid maltose agar, according to Ricketts,[1] is an excellent medium both for isolation and subsequent cultivation. The colonies upon solid media are at first small, white, elevated plaques which later become gray or brownish. After a few days the growth becomes wrinkled and the mycelial threads and aërial hyphæ develop, which gives the culture a moldy appearance. The hyphæ fill the tube around the colony. In fluid media the growth at first is a flocculent mass which collects at the bottom of the tube; a membrane or pellicle usually develops on the surface of the medium, falls to the

[1] Jour. Med. Research, 1901, vi, 377.

bottom, and a new membrane forms. A moderate growth develops in gelatin, but the medium is not liquefied. A slight acidity, but no other visible change, develops in milk cultures.

The organism is strongly aërobic and grows at room or body temperature. At the lower temperature hyphæ are more freely formed; at the higher temperature the typical budding predominates and few or no hyphæ appear until after several days' incubation. Freezing does not kill Blastomycetes, but an exposure to 60° C. for five minutes is fatal to them.

Products of Growth.—The fermentation reactions are variable. Some strains fail to ferment dextrose or maltose, while others produce gas (CO_2) in this medium. On the whole, the fermentative powers of the Blastomycetes are much less than those of the saprophytic yeasts. Toxins and enzymes have not been detected in cultures of the organisms.

Pathogenesis.—*Human.*—The initial lesion, usually cutaneous, is a papule surrounded by an area of hyperemia, which soon becomes a pustule yielding a tenacious pus. The ulceration spreads slowly, discharging small amounts of thick, purulent material and surrounded by a red areola in which numerous papules are frequently detectable. As the lesion spreads the older portions of the lesion tend to become cicatrized and to heal. The progress of the disease is very slow, frequently requiring years to cover an area of a few square inches. It does not often spread to mucous surfaces, but occasionally metastases occur in the lungs. According to Stober,[1] involvement of bones and metastatic foci in the spleen, liver and kidney have been observed in a few cases.

Animal Experimentation.—Attempts to reproduce blastomycetic infections in dogs, rabbits, guinea-pigs, white rats and mice have been unsuccessful when artificially cultivated organisms from human lesions have been inoculated, although Klein[2] isolated a blastomycete from milk, which produced gelatinous, tumor-like swellings and glandular enlargement when injected subcutaneously into guinea-pigs. Intraperitoneal injections resulted in the formation of firm nodules in the liver, lungs, pancreas, testes, ovaries and intestines. The nodules were composed chiefly of masses of the organisms. Tokishige[3] and Tartakowsky[4] have isolated organisms belonging to the

[1] Arch. Int. Med., 1914, xiii, 509.
[2] Brit. Med. Jour., 1901, ii, 1.
[3] Centralbl. f. Bakt., 1896, xix, 105.
[4] Die afrikanische Rotz der Pferde, St. Petersburg, 1897.

Blastomycetes Group from a cutaneous infection of horses, and Sanfelice[1] recovered a similar organism from a lymph gland of an ox which had a generalized carcinoma. This organism was pathogenic for white rats, rabbits, guinea-pigs, sheep and cattle.

The defensive mechanism which tends to limit the spread of the organism in the body is largely phagocytic, together with a proliferation of regional connective tissue which tends to encapsulate and then restrict the progress of the lesion.[2]

The diagnosis of blastomycetic infections is best made by a microscopical examination of the contents of a papule or pustule teased out in diluted NaOH, and unstained.

[1] Centralbl. f. Bakt., 1895, xviii, 521.

[2] Christensen and Hektoen, Jour. Am. Med. Assn., 1906, xlvii, 247. Davis, Jour. Inf. Dis., 1911, viii, 190.

CHAPTER XXIX.

FILTERABLE VIRUSES, DISEASES OF UNKNOWN ETIOLOGY.

FILTERABLE VIRUSES.
- Acute Anterior Poliomyelitis. Epidemic Poliomyelitis.
- Typhus Fever (Tabardillo, Brill's Disease).
- Yellow Fever.
- Foot and Mouth Disease.
- Contagious Pleuropneumonia of Cattle.

DISEASES OF UNKNOWN ETIOLOGY.
- Measles.
- Scarlet Fever.
- Rabies.
- Trachoma.
- Smallpox (Variola) and Vaccinia.
- Dengue.
- Rocky Mountain Spotted Fever.
- Mumps.

FILTERABLE VIRUSES.[1]

THE viruses of certain diseases of plants, animals and of man are fully virulent after they have been passed, suspended in fluid, through filters of unglazed porcelain or diatomaceous earth of definite degrees of fineness. These filters will not permit the passage of organisms as minute as Micrococcus melitensis, but Wherry[2] has shown that the bacillus of guinea-pig pneumonia, an actively motile bacillus 0.3 to 0.5 micron in diameter and 0.7 micron in length, will also pass through such filters unharmed.

The restraining action of filters of unglazed porcelain and diatomaceous earth appears to depend rather upon the tortuous passages in the walls of the filter than upon the ultimate minuteness of these channels. This possibility is suggested by experiments using filters of theoretically equal degrees of fineness of material, but of varying thickness; it has been shown that bacteria may be forced through the thinner walled filter, but not through the thicker. Longer bacteria, Bacillus typhosus for example, will pass through filters, provided time enough for their development is given. The supposition is that the organisms grow around and through tortuous passages which effectually hold the bacteria in the channels when pressure is applied. For this reason filtration must not be prolonged much more than an hour, and too much pressure (or suction) must be avoided.

[1] See Wolbach, Jour. Med. Research, 1913, xxvii, 1, for résumé of literature.
[2] Jour. Med. Research, 1902, viii, 322.

The passage of a virus through a filter of the type mentioned does not necessarily indicate that the virus is too small to be visible with the highest powers of the microscope, although the filtrates of the so-called "ultramicroscopic viruses" are clear and do not contain particles demonstrable with the ultramicroscope. Filters used for the study of filterable viruses should be new, sterile, and tested for permeability with suitable known bacteria. A preliminary test, forcing air under pressure through the submerged filter, will reveal "pin holes." The virus to be tested should be forced through at a

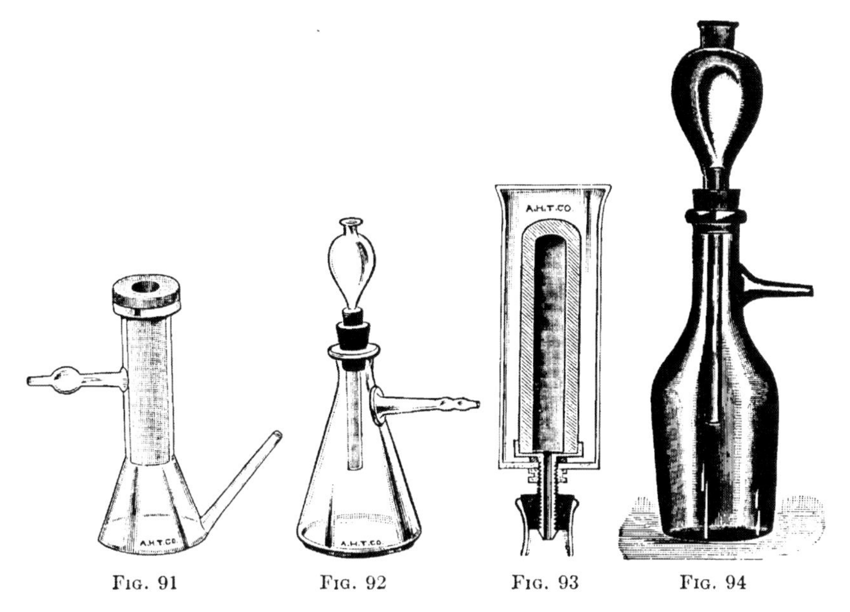

FIG. 91 FIG. 92 FIG. 93 FIG. 94

FIGS. 91 to 94.—Types of unglazed porcelain filters. (*Park.*)

temperature of about 20° C., and the process should be completed within one and a half hours, using as little pressure or suction as possible. The filtrate, proved to be free from visible particles (best by adding a known organism to the fluid to be filtered), should reproduce the disease in susceptible animals; the virus should be recovered, again filtered, and again reproduce the disease. Some of the filterable viruses will pass only the coarser filters, others go through those with finer pores.

Ultramicroscopic viruses with few exceptions are of unknown morphology, and, with the exception of their resistance to desiccation and physical agents, but little is known about them. The viruses of

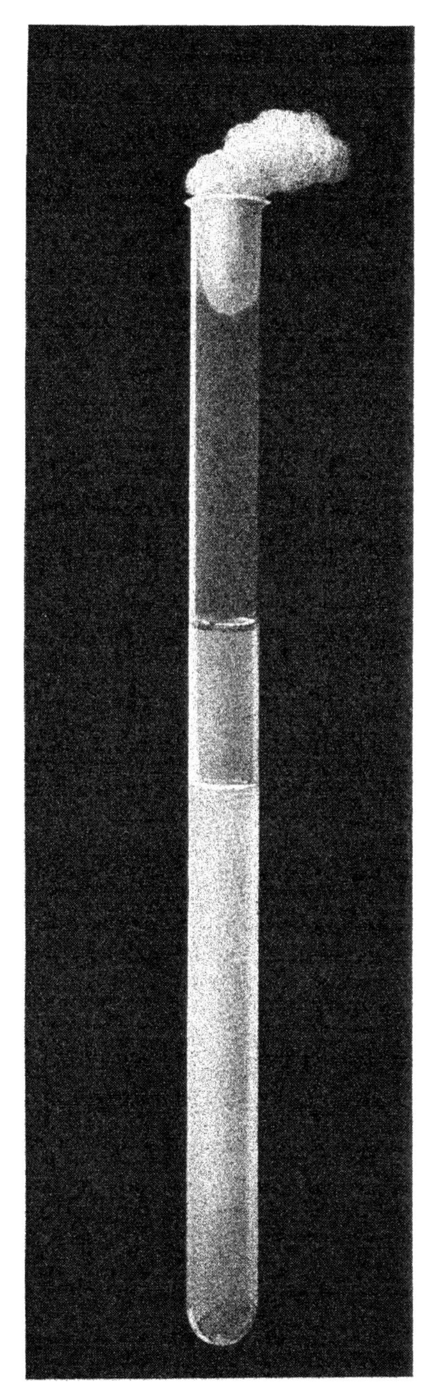

Fig. 1

Fig. 2

icroörganism Causing Epidemic Poliomyelitis. (Flexner and N

Culture in ascitic fluid-tissue medium of No. hi

pleuropneumonia of cattle and of poliomyelitis have been cultivated on artificial media; thus far the remainder have resisted attempts at cultivation.

Acute Anterior Poliomyelitis. Epidemic Poliomyelitis.—Epidemic poliomyelitis is an acute disease observed more frequently in children, although adults are by no means immune. The onset is usually abrupt, although in some cases the earliest symptom is fever, with or without sore throat. The most striking feature is a paralysis of one or more limbs, which may be the first clinical indication of the disease. The principal lesion of the earlier stages is a hyperemia of the vessels of the cord together with thrombosis, and leukocytic infiltration of the perivascular lymph spaces, more commonly in the cervical and lumbar regions, and in the spinal fluid as well. The older lesions are essentially a degeneration of the ganglion cells, particularly of the anterior horn, and eventually their atrophy. The motor nerves appear to suffer most—there are few, if any, indications of sensory disturbance. The relation of the disease to Landry's ascending paralysis, if any, is unknown.

The etiology of acute anterior poliomyelitis was for many years a matter of conjecture. In 1909, however, Landsteiner and Popper[1] transmitted the disease to two monkeys through the injection of a saline emulsion of the spinal cord from an acute case. The animals developed paralysis of their limbs, and were killed and studied bacteriologically and pathologically. The lesions were similar to those found in human cases; the cultures were wholly negative. An attempt to introduce the disease in other monkeys by the injection of material from the two successfully inoculated animals proved futile. They believed the virus belonged to the group of filterable viruses. Flexner and Lewis[2] and Landsteiner and Levaditi[3] soon confirmed the filterable nature of the virus, and Flexner and Lewis succeeded in transmitting the virus through a succession of monkeys. The success of their transmission lies in the choice of inoculation site—intracerebral inoculations are reliable, but intraperitoneal injections are usually barren of results. Of great importance are the observations of Flexner and Clark[4] and Osgood and Lucas[5] that the virus may survive in the mucosa of the nasopharynx of infected monkeys for several weeks.

[1] Ztschr. f. Immunitätsforsch., 1909, ii, 378.
[2] Jour. Am. Med. Assn., 1909, liii, 2095.
[3] Compt. rend. Soc. biol., 1909, lxvii, 592.
[4] Proc. Soc. Exper. Biol. and Med., 1912, 13, x, 1.
[5] Jour. Am. Med. Assn., 1911, lvi, 495.

Landsteiner, Levaditi and Pastia[1] have established the presence of the virus in the tonsils and pharyngeal mucosa of an acute fatal case of infantile paralysis. Flexner, Clark and Fraser[2] have shown definitely that the virus was carried in the upper respiratory mucous membranes of healthy human adults, the parents of a child suffering from an acute attack of the disease. Kling, Wernstedt and Patterson[3] claim, on the basis of experimental evidence, that the nasal secretion may also harbor the virus. Neustaedter and Thro[4] have found that the virus may remain viable in dust. The transmission of the virus, therefore, would appear to be largely through the upper respiratory tract. Flexner and Amoss[5] have brought forth experimental evidence to show that the atrium of infection is the upper respiratory mucous membrane, and that the virus travels to the meninges by way of the lymphatics; not, as a rule, through the blood. Available evidence would indicate that insects play no part, or at best, a very minor role in the transmission of the virus.[6] The observations of Flexner and Amoss[7] and of Clark, Fraser and Amoss[8] would indicate that the amount of virus circulating in the blood stream is usually very small, thus suggesting the improbability of insect transmission except in unusual instances.

A very important advance in the study of the etiology of epidemic poliomyelitis was that of Flexner and Noguchi.[9] Using the technic of Noguchi[10] for the cultivation of Treponemata (unheated ascitic fluid and fragments of sterile rabbit tissue under strictly anaërobic conditions) they obtained minute, slowly growing colonies composed of "globular and globoid" bodies occurring singly, in pairs, masses, and in short chains. The elements measure from 0.15 to 0.3 micron in diameter. Bizarre forms are prone to appear in older cultures. The organisms stain feebly with the Giemsa stain and by Gram's method—they stain variably with the latter. The organism has also been demonstrated in tissues by a modified Giemsa technic. The first cultivations upon artificial media are difficult to obtain, but subcultures grow more readily. No action was observed on the

[1] Semaine Médicale, 1911, 296.
[2] Jour. Am. Med. Assn., 1913, lx, 201.
[3] New York Med. Jour., 1911, xciv, 813.
[4] Ztschr. f. Immunitätsforsch., 1911, xii, 316, 357; 1912, xiv, 303.
[5] Jour. Exp. Med., 1914, xx, 249.
[6] Howard and Clark, Jour. Exp. Med., 1912, xvi, 850. Sawyer and Herms, Jour. Am. Med. Assn., 1913, lxi, 461. Clark, Fraser, and Amoss, Jour. Exp. Med., 1914, xix, 223
[7] Jour. Exp. Med., 1914, xix, 411.
[8] Loc. cit.
[9] Jour. Am. Med. Assn., 1913, lx, 362; Jour. Exp. Med., 1913, xviii, 461.
[10] Jour. Exp. Med., 1911, xiv, 99; 1912, xv, 90; xvi, 199, 211.

PLATE VII

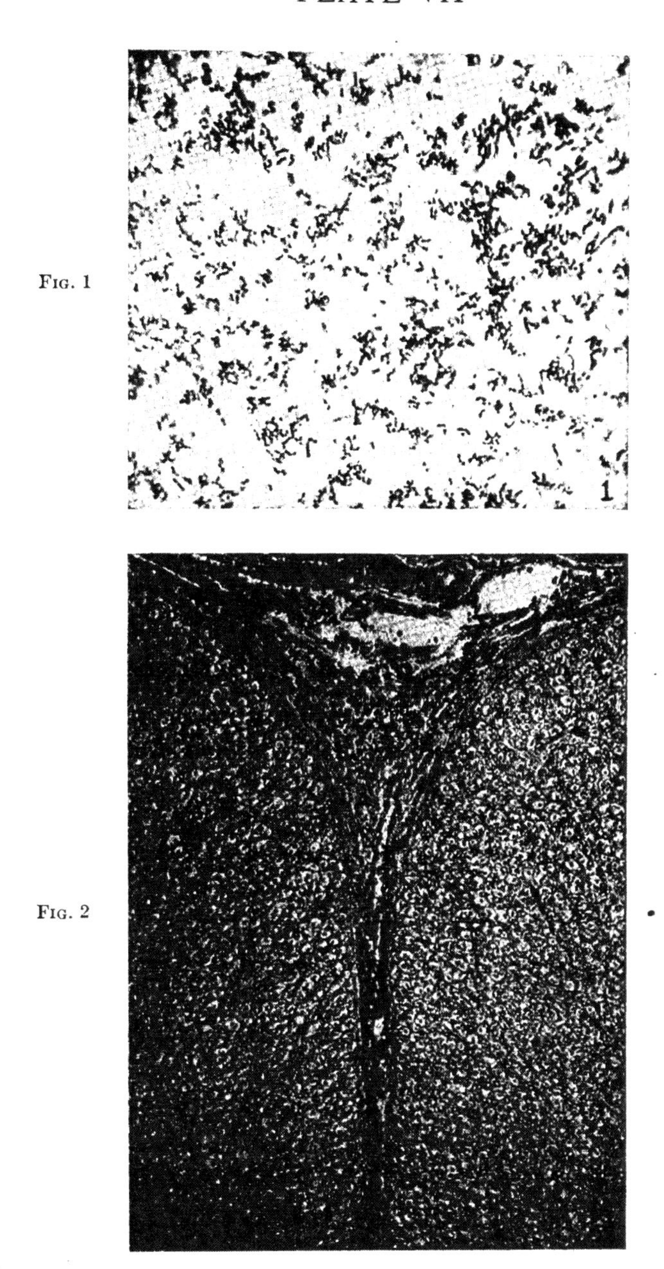

Survival and Virulence of Poliomyelitic Microörganism.
(Flexner, Noguchi, and Amoss.)

FIG. 1.—Sediment showing the minute microörganisms after three days' growth in mixed ascitic fluid and bouillon in a flask employed for mass cultivation. Giemsa stain. × 1000.

FIG. 2.—Spinal cord showing meningeal cellular infiltration extending into the anterior median fissure. × 1000.

PLATE VIII

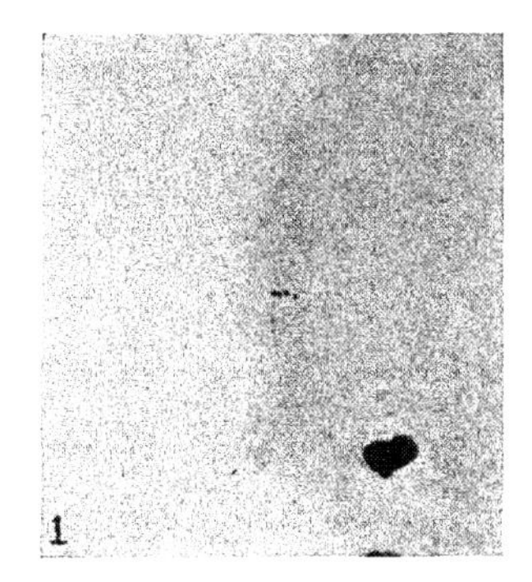

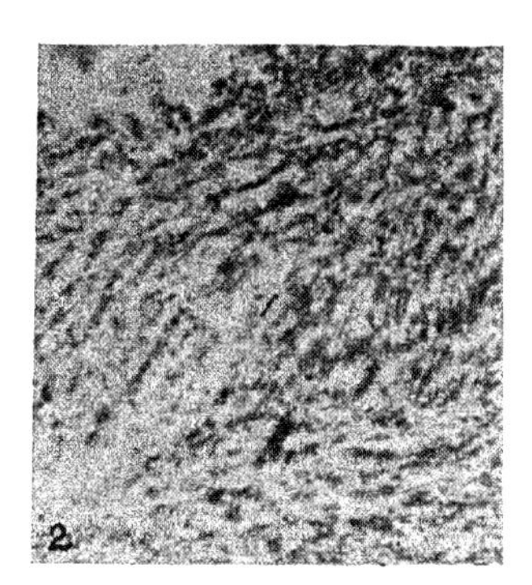

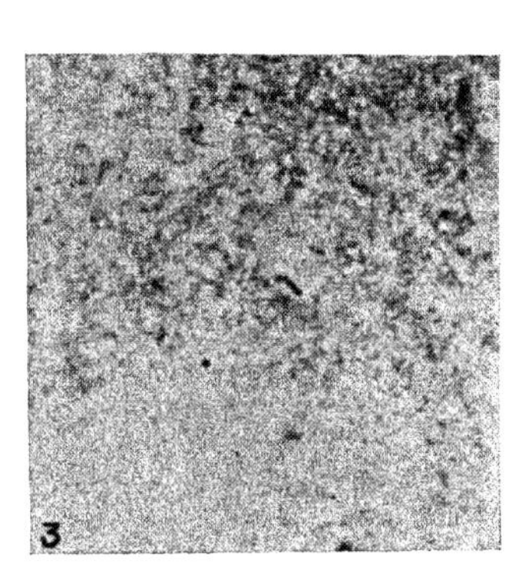

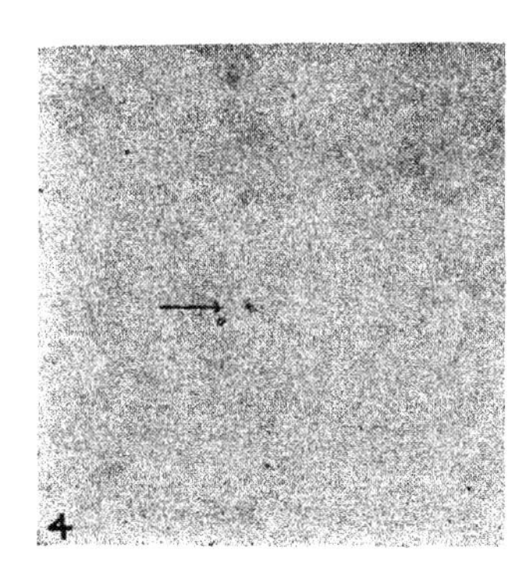

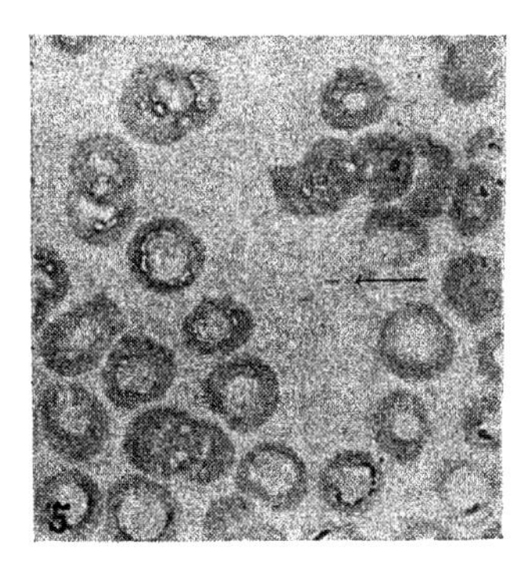

Etiology of Epidemic Poliomyelitis. (Amoss.)

IG. 1.—Globoid bodies in chain formation in brain tissue after six days' incubation. × 1000
IG. 2.—Globoid bodies in chain formation in brain tissue after eight days' incubation. × 1000
IG. 3.—Globoid bodies in chain formation in brain tissue after ten days' incubation. × 1000
IG. 4.—Globoid bodies in mass formation in brain tissue after thirty days' incubation. × 1000
IG. 5.—Globoid bodies in chain formation in heart's blood. × 1000.

ordinary sugars and alcohols. Growth appears to take place in litmus milk reënforced with bits of sterile tissue, but no visible change in the medium can be detected. Cultivations can be readily made from Berkefeld filtrates of ascitic fluid growths, thus showing that the organisms, or at least some of them, are filterable. Cultures of the organism were shown to cause the typical disease with characteristic lesions in monkeys.

Flexner, Noguchi and Amoss[1] have shown that cultures of the organism may retain their virulence for monkeys at least a year, and Flexner, Clark and Amoss[2] have shown that the virus retains its pathogenicity in 50 per cent. glycerin for eleven months; in 0.5 phenol for five days, and frozen at – 2° to – 4° C. for at least six weeks. Amoss[3] has improved the technic for cultivating the virus of epidemic poliomyelitis.

Pieces of brain from infected animals are incubated in the kidney-ascitic fluid of Noguchi for about two weeks, then crushed carefully and reincubated for three days longer. The globoid bodies appear to multiply in the brain tissue and their subsequent recognition and cultivation is rendered more certain. Stained sections of such brain tissues show increased numbers of organisms.

Immunity.—One attack appears to confer immunity in man, but the evidence is not conclusive. Flexner and Lewis[4] have been unsuccessful in reinfecting monkeys which have recovered from a typical infection and they lean toward the view that one attack confers immunity in these animals.

The characteristic disease has not been produced in experimental animals other than primates.

Typhus Fever (Tabardillo, Brill's Disease).—Typhus fever is an acute, febrile disease of man characterized by an incubation period varying from four to five days to twelve days, an acutely developing febrile reaction which persists for about two weeks, falling by crisis, or rapid lysis, and an extensive erythematous eruption, maculopapular in character, which appears usually within three to four days after the onset, and persists for about ten to fourteen days. The disease pathologically is to be regarded as a hemorrhagic septicemia; the lesions postmortem are not distinctive, and the changes in the organs are those produced by an intense febrile reaction. The

[1] Jour. Exp. Med., 1915, xxi, 91.
[2] Ibid., 1914, xix, 207.
[3] Ibid., 1914, xix, 212.
[4] Loc. cit.

mortality varies greatly; in the eastern part of the United States the disease is mild in character,[1] so mild in fact that the malady was spoken of as Brill's disease in honor of Brill who described the clinical features of it in great detail. The mortality in Europe, where the disease is very prevalent in certain areas, especially the more southern lands, is usually high.

The first definite communication relating to the mechanism of infection was that of Nicolle[2] and of Nicolle, Comte and Conseil.[3] They succeeded in infecting an anthropoid ape with the blood of a typhus patient, and very shortly afterward, and independently, Anderson and Goldberger[4] infected two monkeys, a Macacus rhesus and a capuchin, in the same manner. These results have been confirmed by Ricketts and Wilder,[5] and others. Anderson[6] states that guinea-pigs may also be infected with the blood of typhus patients.

It has been shown by animal experimentation that one attack of typhus confers immunity, and this method has been taken advantage of to show that typhus, Brill's disease and tabardillo mutually confer immunity on monkeys; that is, an animal recovered from either of the three clinical types is immune to infection with the other two.

The filterability of the virus of typhus has been a subject of discussion; the concensus of opinion appears to be that blood serum filtered through stone filters has not been definitely shown to be infective for monkeys, although Nicolle, Anderson and Goldberger, and Ricketts and Wilder have noticed that the injection of filtered serum appears to render monkeys refractory to subsequent inoculation with the virus.

Recently Plotz[7] has isolated a small, anaërobic, Gram-positive bacillus from the blood of a series of cases of Brill's disease and of typhus which when used as an antigen caused fixation of complement with the sera of these cases. The bacillus measures from 0.2 to 0.6 micron in diameter and from 0.9 to 2 microns in length.[8] It is non-acid-fast, possesses no capsule, and exhibits bipolar staining. In the latter respect it suggests the organism seen but not cultivated

[1] Brill, Am. Jour. Med. Sc., April, 1910, 484; August, 1911, 196.
[2] Compt. rend. Acad. sci., 1909, cxlix, 157.
[3] Ibid., p. 486.
[4] Public Health Rep., 1909, 1861; ibid., p. 1941; 1910, 177.
[5] Jour. Am. Med. Assn., 1910, liv, 463; ibid., 1304, 1373.
[6] Public Health Rep., 1915, xxx, 1303.
[7] Jour. Am. Med. Assn., 1914, lxii, 1556.
[8] Plotz, Jour. Am. Med. Assn., 1914, lxii, 1556.

by Ricketts and Wilder,[1] both in the blood of patients and in the intestinal contents of lice which had been permitted to bite these patients.

The injection of cultures of the Plotz bacillus into guinea-pigs resulted after an incubation period of from twenty-four to forty-eight hours in a febrile reaction which dropped by lysis after four to five days. This organism, Bacillus typhi-exanthematicus, as it has been named, must be regarded tentatively as the etiological factor of typhus fever.

Typhus is transmitted by the body louse Pediculus vestimenti, as was shown by Nicolle, Anderson and Goldberger, and Ricketts and Wilder.

Yellow Fever.—Yellow fever is an acute fever of tropical and subtropical countries, characterized by jaundice, albuminuria, and a tendency to hemorrhage from mucous membranes; the latter is especially marked in the stomach and the "black vomit" which occurs frequently is a regurgitation of altered blood which has collected in the stomach.

For many years the etiology and mode of transmission of yellow fever were wholly unknown, although many and divers organisms were reported as the inciting factor. Finlay,[2] as early as 1882, believed that mosquitoes played an important part in the transmission of the disease and he actually attempted to infect non-immunes by mosquitoes which had previously bitten yellow fever patients. His experiments were wholly negative, partly because the extrinsic cycle of development in the insect was unknown. Carter[3] made the very important observation that a latent period of about two weeks elapses between primary and secondary cases of yellow fever. This discovery explained some of Finlay's negative results and paved the way for the success of the American Yellow Fever Commission. Finally Reed, Carroll, Agramonte and Lazear,[4] a commission appointed from the Medical Corps of the United States Army, carried out a series of experiments never excelled from a scientific standpoint, which showed conclusively:

1. The virus of yellow fever circulates in the blood stream of a patient at least three days after the initial chill. An injection of blood

[1] Jour. Am. Med. Assn., 1910, liv, 1373.
[2] Ibid., 1901, xxxvii, 1387.
[3] Public Health Rep., 1905, xx, 1350; New York Med. Rec., 1906, lxix, 683.
[4] Jour. Exp. Med., 1900, v, 215; Am. Public Health Assn., 1900, xxvi, 37; Boston Med. and Surg. Jour., 1901, No. 14; Jour. Am. Med. Assn., 1901, xxxvi, 413.

from a patient at this stage of the disease will reproduce the disease in a non-immune.

2. The virus will pass through a Berkefeld filter; it belongs, therefore, to the group of filterable viruses. Berkefeld filtrates of the blood will establish the disease through a series of cases, thus indicating that a living virus is being perpetuated.

3. The disease is transmitted ordinarily by the bite of a female mosquito belonging to the genus Aedes. The insect is now known as Aedes calopus.[1]

4. A patient is infective for a mosquito only during the first seventy-two hours after the initial chill and onset of the disease.

5. A latent period, during which the insect is non-infectious, must elapse before the disease may be transmitted to a non-immune subject through the bite of the yellow fever mosquito.

6. One attack appears to confer lasting immunity, provided the individual resides continuously in the tropics.

The two cardinal features of the transmission of yellow fever—infectivity of the patient during the first three days of the disease, and the part played in its transmission by the mosquito, Aedes calopus, were immediately put to the acid test of practical sanitation by Gorgas,[2] first in Havana and later in Panama, where he organized and directed the sanitation of these pestilential cities along lines which soon freed them from yellow fever and other diseases of endemic origin as well.

The importance of the work of the American Yellow Fever Commission and of Gorgas cannot be overestimated; the completion of the Panama Canal and the liberation of the tropics from the dreaded yellow fever mark a new era in Epidemiology and Preventive Medicine.

Foot and Mouth Disease.[3]—Foot and Mouth disease is an acute, highly infectious exanthematous disease which attacks cloven-footed animals chiefly. The characteristic eruptions, which are vesicular at first and filled with a clear fluid, soon become grayish, and the epidermis sloughs off, leaving a raw reddened surface. The eruption usually appears at three distinct sites—the mucous membrane of the mouth, the teats, and interdigital spaces. The incubation period is from one to six days, and little or no immunity to subsequent attacks is conferred on an animal by successful recovery.

[1] The original name of the insect was *Culex fasciatus*; it has been changed successively to Stegomyia fasciata, Steogomyia calopus, and finally to Aedes calopus.

[2] See Jour. Am. Med. Assn., 1906, xlvi, 322, for brief summary.

[3] For an excellent discussion of various aspects of the disease, see the *Cornell Vet.*, February, 1915, Foot and Mouth Disease Number.

The milk of infected cows contains the virus, and the disease is transmissible to man, particularly young children, through raw or imperfectly pasteurized milk, and possibly from butter and cheese made from infected milk. The disease is mild, as a rule, in older children, but it may be severe or fatal for infants.

The virus belongs to the group of filterable viruses and, in its purest state, is found in the contents of the vesicles. Early in the disease the virus also circulates in the blood stream. Löffler and Frosch,[1] who discovered the filterable nature of the virus, found that the vesicular fluid, filtered through unglazed porcelain filters, retained its infectiousness for some time, provided the fluid be kept cool and in the dark.

Contagious Pleuropneumonia of Cattle.—This disease was the first to be described in which the virus passes through unglazed porcelain filters, although the filtration of the virus was not attempted at that time. Nocard and Roux[2] examined the exudate from the lungs of diseased cattle microscopically with negative results. They suspended it in broth, enclosed in collodion capsules, in the peritoneal cavities of guinea-pigs. After two to four weeks the medium became turbid, while controls remained clear. Examination of the fluid under a magnification of 2000 diameters revealed very minute, highly refractile spots which exhibited Brownian movement. They claim to have cultivated the virus in a peptone-serum medium and to have obtained minute colonies (0.5 mm. diameter) on peptone-serum agar. Later the virus was shown to pass through Berkefeld filters and the coarser grades of porcelain filters, but not the finer grades.

The disease is confined to cattle; man is immune so far as is known.

DISEASES OF UNKNOWN ETIOLOGY.

Measles.—The etiology of measles is unknown, but Hektoen[3] produced the disease in two susceptible individuals by injecting blood from a patient exhibiting typical symptoms. The blood was removed about thirty hours after the appearance of the eruption, and the disease induced was clinically perfectly typical. Anderson and Goldberger[4] report a successful inoculation of several monkeys with blood from human cases; four out of a total of nine animals developed a febrile reaction and a limited eruption. The virus was carried through three monkey generations in one experiment. Growth was

[1] Centralbl. f. Bakt., I Abt., 1898, xxiii, 371. [2] Ann. Inst. Past., 1898, xii, 240.
[3] Jour. Inf. Dis., 1905, ii, 238. [4] Public Health Rep., 1911, xxvi, No. 24.

not obtained in artificial media heavily inoculated with blood from patients, shown by experiment to contain the virus.

Buccal and nasal secretions contain virus of measles which passes a Berkefeld filter.[1]

Scarlet Fever.—The etiology of scarlet fever is unknown. The very common occurrence of streptococci in this disease has led many observers to attribute to the streptococcus an etiological relationship. No satisfactory evidence in support of the view that any type of streptococcus is the causative agent has been brought forward.

Döhle[2] described small oval, round and rod-shaped bodies measuring about 1 micron in diameter, lying within the cytoplasm of polymorphonuclear leukocytes in a series of cases of scarlet fever. It was assumed at first that these inclusion bodies were fragments of a spirochete (the hypothetical inciting agent of scarlet fever) which had been phagocytized and disintegrated by the polymorphonuclear leukocytes. This view is now discredited. Numerous investigations, especially that of Hill,[3] indicated that the Döhle bodies are fragments of the nucleus of the leukocyte, presumably a reaction to injury by bacterial toxins. They are present, however, in a majority of cases of scarlet fever up to the tenth day and especially numerous during the first four days of the clinical disease, as the following table by Hill shows. The Poppenheim stain (two parts of a saturated aqueous solution of pyrosin and four parts of a saturated aqueous solution of methyl green) is especially recommended for the demonstration of the inclusion bodies of Döhle. The nuclei of the cell are colored greenish blue, the Döhle bodies bright red.

	Positive.	Negative.	Total.
Scarlet fever	43[4]	29[5]	72
Erysipelas	5	0	5
Pneumonia	4	1	5
Syphilis	0	2	2
Empyemia	0	1	1
Secondary anemia	0	1	1
Serum rash	0	1	1
Normal	0	13	13

Hill concludes that the Döhle inclusion bodies are present in a majority of cases of scarlet fever up to the tenth day, but they are not

[1] Goldberger and Anderson, Jour. Am. Med. Assn., 1911, lvii, 476, 971.
[2] Centralbl. f. Bakt., Orig., 1911, lxi, 63.
[3] L. W. Hill, Boston Med. and Surg. Jour., 1914, clxx, 792; excellent summary of literature.
[4] 25 cases examined before tenth day; 18 after tenth day; latest case forty-fifth day.
[5] All except 6 cases after tenth day; remaining 6 cases had normal temperature and very slight rash.

specific for the disease; they are found in other infections, especially erysipelas, sepsis, pneumonia and tonsillitis. They are more likely to be found in disease with which the streptococcus is associated. Diagnostically they possess some value. If they are not found in a doubtful case which has a rash and a marked fever, the case is probably not one of scarlet fever.[1]

Rabies.—Rabies is a disease primarily observed among the carnivora —dogs, wolves and cats—but it is transmissible to horses and to man. Laboratory animals are readily infected with the virus. The saliva of rabid animals is infectious and the natural mode of inoculation is through bites of infected animals. The disease is also readily transmissible in an experimental way through the injection of emulsions of the cord or brain of rabid animals directly into the central nervous system of other animals. The infectious nature of rabies was first clearly shown by Pasteur, Chamberland and Roux.[2]

The incubation period for "street rabies" is, on the average, from one to two months, but it may be considerably longer. The incidence of the disease among those bitten by rabid dogs depends largely upon the location of the bite—if upon the body protected with several layers of clothing, infection may fail to develop; the virus is held back by the clothing and fails to enter the wound. In general the inoculation period is shortest when the hands or face are attacked, because the virus acts upon the central nervous system and reaches it through the peripheral nerves.

The disease in man is practically always acute and death usually terminates the infection within three to six days after the onset of the symptoms. The initial symptoms are premonitory and consist typically of slight irritation at the site of inoculation, together with psychic depression. The characteristic symptoms are paralysis of the muscles of deglutition—which leads to extreme difficulty in swallowing—hyperesthesia, extreme restlessness and irritability, and violent reflex spasms. Even so slight an effort as that required to swallow water frequently causes such violent paroxysms that the mere sight of water is distressing—hence the name hydrophobia—the dread of water. It is important to remember that the hydrophobic phenomena are much less commonly seen in rabid dogs than in man; indeed rabid dogs frequently swim across streams that they happen to encounter. The final stage of rabies is a progressive paralysis,

[1] Hill, loc. cit.
[2] *Compt. rend. Acad. Sc.*, 1881, xcii, 159.

which usually first becomes manifest in the limbs and arms; it ascends gradually and death occurs when the higher centres are reached.

The disease occurs in every country except England, and possibly Australia. The elimination of rabies from England dates from the law of 1889 which required all dogs to be muzzled and all imported animals to be quarantined for several months. The law was allowed to lapse for a time, the disease reappeared, but a new and rigid enforcement of the muzzling and quarantine laws has completely eliminated rabies from the British Isles. No cases have been reported since 1903.

The first definite lesions characteristic of rabies were described by Negri,[1] who found characteristic cell inclusion bodies in the ganglion cells, in the cells of Purkinje, and other large nerve cells. These minute granular pleiomorphic bodies are now recognized as specific, or nearly so, for hydrophobia, but there is discussion of their nature. Williams[2] regards them as protozoa and conferred upon them the name Neurorrhyctes hydrophobiæ; in collaboration with Lowden[3] she has made a careful study of the occurrence of Negri bodies and considers them the true etiological agent of rabies. Remlinger,[4] Poor and Steinhardt,[5] and others have found that the virus is filterable, and Noguchi[6] has cultivated an organism from "street" virus and from the central nervous system of animals infected with "street" virus, "fixed" virus and with "passage" virus, which resemble Negri bodies observed in lesions in many particulars. The smallest of these bodies are just visible with the highest magnifications obtainable; larger nucleated or oval bodies occasionally appear in older cultures. Inoculation of dogs, rabbits and guinea-pigs with cultures containing the granular pleiomorphic or nucleated bodies was followed by typical symptoms of rabies. The relation of the organisms grown by Noguchi to Negri bodies is not definitely determined as yet, but the organism has been kept alive for over three months in artificial cultures and found to be virulent after the twenty-first transfer in artificial media. This would suggest strongly that Noguchi's organism was the etiological agent of rabies. The possibility that a filterable virus was growing in these cultures cannot be overlooked, as Noguchi has pointed out, but there is no evidence that such is the case.

The most important rapid laboratory method for the diagnosis of

[1] Ztschr. f. Hyg., 1903, xliii, 507; 1909, lxiii, 421.
[2] Proc. New York Path. Soc., 1906, vi, 77.
[3] Williams and Lowden, Jour. Inf. Dis., 1906, iii, 452.
[4] Ann. Inst. Past., 1903, xvii, 834; 1904, xviii, 150.
[5] Jour. Inf. Dis., 1913, xii, 202.
[6] Jour. Exp. Med., 1913, xviii, 314.

rabies is a demonstration of Negri bodies. If they are found the diagnosis is complete. Failure to find them does not necessarily exclude a diagnosis of rabies, and an emulsion prepared from the central nervous system, using the gray substance as far as possible, is injected subdurally into an experimental animal for a final diagnosis. The method of animal inoculation, while slower than the microscopic examination of the brain, is the final test in doubtful cases. Of course, treatment should not await the results of animal inoculation if there is suspicion that a patient has been bitten by a rabid dog, especially if the hands, face or other unprotected surface be the site of the wounds.

Staining Negri Bodies.—Williams and Lowden[1] have developed a technic for the rapid demonstration of Negri bodies, which is widely followed at the present time. A small piece of the gray substance from the region of the hippocampus major and from the cerebellum of the animal is placed upon a clean glass slide and covered with a clean coverglass. Pressure is applied to the latter until the tissue is flattened and spread uniformly. The pressure is now shifted to one edge of the coverglass and the flattened tissue is forced along the slide, leaving a thin film as it passes. Fixation with neutral absolute methyl alcohol (Merck reagent) containing about 0.1 per cent. picric acid (about ten minutes are required) is followed by removal of the fixing agent with filter paper.

A small amount of a freshly prepared staining mixture, made in the proportions of 30 c.c. of distilled water, 10 c.c. of a saturated alcoholic solution of methylene blue and 0.5 c.c. of a saturated alcoholic solution of basic fuchsin is poured over the slide, warmed till steam arises, then poured off. The excess stain is removed in running water and the preparation is carefully dried with filter paper. The preparation is examined with an oil immersion lens.

Negri bodies, which vary in size from about 1 micron to 25 microns in diameter, are stained magenta with blue granules by this process; the cytoplasm of the nerve cells is pale blue; the nuclei of the nerve cells are colored a darker blue.

The Pasteur Treatment for Rabies.—Pasteur[2] made the very important observation that the virus of rabies as it exists in rabid dogs (street virus) could be so attenuated by repeated passages through rabbits that it lost much of its original virulence for the dog. This change in

[1] Loc. cit. [2] Loc. cit.

virulence was fully established when passage of the virus from rabbit to rabbit caused each successive animal to sicken in about six or seven days, and to die regularly on the ninth day. No further increase in pathogenicity for the rabbit could be induced, and the virus at this level of virulence was called "virus fixé" by Pasteur. The spinal cord of such a rabbit, dried for two weeks over caustic soda at room temperature, lost its virulence for rabbits, although cords dried for a week or ten days killed the animal when injected subdurally; the period of incubation was, of course, increased when the partly dried cords were used.

The original Pasteur treatment consisted in grinding a piece of dried cord half a centimeter in length in 5 c.c. of sterile salt solution, and injecting the emulsion subcutaneously, preferably on the abdomen of the patient. Daily injections, using fresher and fresher cords were used, until finally a cord from a rabbit dead but twenty-four hours furnished the material for inoculation. The entire treatment required about three weeks, at the end of which time a very decided degree of immunity was induced. The incubation period of the naturally acquired disease is usually not less than six weeks; the advantage of instituting treatment at the earliest possible moment is obvious.

The mortality from rabies among those treated by the Pasteur method of immunization is less than 0.5 per cent.; the average mortality of untreated cases is about 16 per cent.

Modifications in the original Pasteur treatment, principally along the lines of injecting more virulent material, have been made from time to time, and the tendency at present is to administer a shorter treatment to mild cases (judged according to the location of the bite and the extent of local injury) on the one hand, and to administer a much more intense treatment in the severe cases. The present routine followed in the Pasteur Institute of Paris is shown in the accompanying table[1] (see page 569).

Statistics indicate that a considerable degree of immunity is developed by the end of the second week of the treatment. The duration of the immunity has not been definitely established, but it appears to last for several years. Exposure to extreme cold and excesses of various kinds, especially alcoholism, are said to be dangerous immediately after the treatment is completed; they may reduce the

[1] Kraus and Levaditi, Handbuch der Technik und Methodik der Immunitätsforschung, 1908, i, 713.

PLATE IX

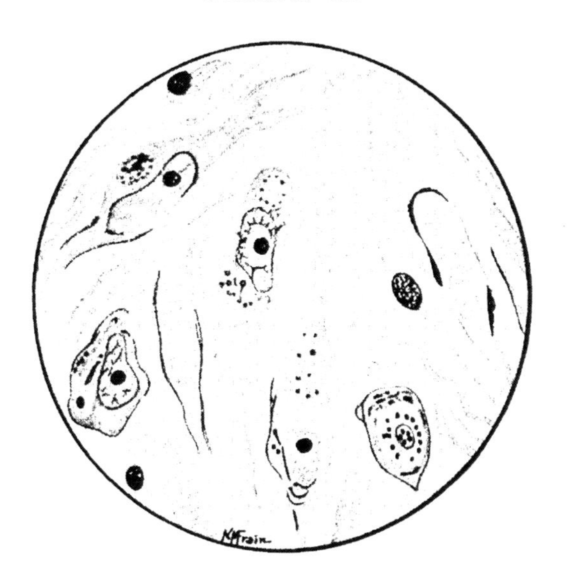

Negri Bodies.

Redrawn from Kolle and Hetsch. (Lentz stain.)

acquired resistance to the virus to such a degree that the patient will succumb to a latent infection.

The dangers attending the treatment are slight; in a moderate number of cases the sites of earlier injections may become inflamed after the treatment has been continued for ten days or two weeks, but this reaction is regarded as a modified Arthus phenomenon depending upon local sensitization. By far the most serious complication

PASTEUR INSTITUTE IMMUNIZATION FOR RABIES.
(KRAUS AND LEVADITI.)

Days.	Mild cases.		Moderate cases.		Severe cases.		
	Dried cord.[1]	Amount	Dried cord.[1]	Amount	Dried cord.[1]		Amount
					A.M.	P.M.	
1	14 + 13 day	3 c.c.	14 + 13 day	3 c.c.	14 + 13 day	12 + 11 day	3 c.c.
2	12 + 11 day	3 c.c.	12 + 11 day	3 c.c.	10 + 9 day	8 + 7 day	3 c.c.
3	10 + 9 day	3 c.c.	10 + 9 day	3 c.c.	6 day	6 day	2 c.c.
4	8 + 7 day	3 c.c.	8 + 7 day	3 c.c.	5 day		2 c.c.
5	6 + 6 day	2 c.c.	6 + 6 day	2 c.c.	5 day		2 c.c.
6	5 day	1 c.c.	5 day	2 c.c.	4 day		2 c.c.
7	5 day	1 c.c.	5 day	2 c.c.	3 day		1 c.c.
8	4 day	1 c.c.	4 day	2 c.c.	4 day		2 c.c.
9	3 day	1 c.c.	3 day	1 c.c.	3 day		1 c.c.
10	5 day	2 c.c.	5 day	2 c.c.	5 day		2 c.c.
11	5 day	2 c.c.	5 day	2 c.c.	5 day		2 c.c.
12	4 day	2 c.c.	4 day	2 c.c.	4 day		2 c.c.
13	4 day	2 c.c.	4 day	2 c.c.	4 day		2 c.c.
14	3 day	2 c.c.	3 day	2 c.c.	3 day		2 c.c.
15	3 day	2 c.c.	3 day	2 c.c.	3 day		2 c.c.
16	....	..	5 day	2 c.c.	5 day		2 c.c.
17	....	..	4 day	2 c.c.	4 day		2 c.c.
18	....	..	3 day	2 c.c.	3 day		2 c.c.
19	....	..	....	..	5 day		2 c.c.
20	....	..	....	..	4 day		2 c.c.
21	....	..	....	..	3 day		2 c.c.

Injections daily for two to three weeks.

of the treatment is a paralysis which, in rare instances, appears during the progress of the treatment, or shortly afterward. This usually results fatally. The cause of this paralysis is not definitely known, but it is assumed that it is a modified form of the disease.

Trachoma.—The etiology of trachoma—contagious granular conjunctivitis characterized by the formation of small granular elevations of the eyelids that atrophy and lead to scar formation—is not definitely settled.

[1] One centimeter of cord of the age indicated, ground in 5 c.c. of sterile salt solution, and injected as per schedule.

Halberstädter and Prowazek[1] have described endocellular bodies lying within the conjunctival epithelium and usually near the cell nuclei, which are minute oval or round granules frequently occurring in pairs, of somewhat variable size, but smaller than ordinary cocci. They are typically enclosed in a somewhat indefinitely defined homogeneous matrix which is regarded as a reaction product. The earlier lesions contain moderate-sized oval or round bodies which stain a faint bluish color with Giemsa's stain; later very minute oval or spherical bodies appear, which color reddish with the same stain. These observations were soon confirmed. Somewhat later the same investigators described inclusions in the conjunctival epithelium of uncomplicated cases of blennorrhea neonatorum which were practically identical histologically with those described. This observation naturally led to new investigation of the subject.

Berterelli and Cecchetto[2] claimed to have reproduced trachoma in a Macacus monkey with a filtrate (Berkefeld) prepared from a human case. Nicolle, Guénod and Blaisot[3] were unable to infect monkeys, but stated that anthropoid apes were susceptible to the trachoma virus.

Herzog[4] believed that the "trachoma bodies" were involution forms of the gonococcus which, under certain unknown conditions, develops into very small forms that are indistinguishable from the trachoma bodies when they are within the epithelial cells. Herzog claims to have developed these very minute forms (microgonococci) in artificial media through a series of rapid transplantations, and he states that this minute state in the development of the organism is the one which leads to trachoma. Williams[5] has studied trachoma extensively and believes that the cellular inclusions characteristic of trachoma are degenerated hemoglobinophilic bacilli. Noguchi and Cohen[6] have cultivated an organism from cases of conjunctivitis in which the inclusion bodies were present, and from an older case in which no inclusion bodies were found, which repeats in culture many of the important morphological appearances of the trachoma bodies. It is certainly neither a gonococcus nor a member of the group of hemoglobinophilic bacteria, but its identity with the trachoma bodies is

[1] Deutsch. med. Wchnschr., 1907, xxxiii, 1285; Arb. a. d. Kais. Gesundheitsamte, 1907, xxvi, 44.

[2] Centralbl. f. Bakt., Orig., 1908, xlvii, 432.

[3] Compt. rend. Acad. sc., 1911, clii, 1504.

[4] Centralbl. f. Bakt., Ref., 1910, xlviii, 276; Arch. f. Ophth., 1910, lxxiv, 520; Ueber die Natur und Herkunft d. Trachomaerregers, Berlin and Wien, 1910.

[5] Arch. Ophth., 1913, xlii, 506; Jour. Inf. Dis., 1914, xiv, 261.

[6] Jour. Exp. Med., 1913, xviii, 572; 1915, xxii, 304.

not yet determined by its discoverers. Noguchi and Cohen have made the important observation that the conjunctivæ of certain monkeys are susceptible to infection with material containing the von Prowazek inclusion bodies, but not to the hemoglobinophilic bacilli isolated from cases of epidemic conjunctivitis; on the other hand, pure cultures of hemoglobinophilic bacilli cause an acute inflammation in the testes of rabbits; at certain stages of the infection numerous clumps of the organisms occur, which stimulate the von Prowazek cell inclusions. Injection of conjunctival scrapings containing the cell inclusion bodies alone is without effect in the rabbit.

These observations have led Noguchi and Cohen to conclude that a group of cases exists in which epithelial cell inclusions alone may be demonstrated in smears; pneumococci and hemoglobinophilic organisms are absent. The conjunctiva may become infected both with the inclusion bodies and hemoglobinophilic organisms. The susceptibility of the conjunctiva of certain monkeys to infection with the hemoglobinophilic bacilli would appear to be an important method for diagnosis of the von Prowazek inclusion bodies.

Smallpox (Variola) and Vaccinia.—Smallpox (variola) and vaccinia, now generally regarded as an infection produced by the virus of smallpox modified by successive passages through the cow, are of unknown etiology. Guarnieri[1] has observed and described cell inclusions in the epithelia of both smallpox and vaccinia lesions and in experimental lesions in the cornea of rabbits as well, which he regards as protozoa, and to which he gave the name Cytoryctes variolæ. Councilman, Magrath and Brinkerhoff[2] have studied these vaccine bodies in detail and incline to the view that they are parasites specific for the disease. Calkins[3] has construed the various forms of the cell inclusions to be distinct stages in the life history of a protozoal parasite. The protozoal nature of the "vaccine bodies" is not universally conceded, and the conservative statement of Ewing[4] that they may be regarded as degenerative phenomena characteristic for the disease is widely accepted at the present time.

The close relationship between smallpox and vaccinia (cowpox) has been recognized since Jenner's[5] classical researches published in

[1] Centralbl. f. Bakt., 1894, xvi, 299.
[2] Jour. Med. Research, 1904, xi, 12.
[3] Ibid., p. 136.
[4] Jour. Med. Research, xiii, 233.
[5] An Inquiry Into the Causes and Effects of the Variolæ Vaccinæ, a disease discovered in some of the Western Counties of England, particularly Gloucestershire, and Known by the Name of the Cow Pox, London, Sampson Low, 1789. (See Epoch-making Contributions to Medicine, Surgery, and Allied Sciences, Carmac, Saunders and Co.)

1789; he showed experimentally that a successful inoculation of man with cowpox virus protected the individual against infection with the virus of smallpox.

The change which the smallpox virus undergoes during passage through calves is not definitely known, but Councilman, Magrath, Brinkeroff and others are of the opinion that the smallpox virus is somewhat widely distributed in the viscera and different organs of the body (in man); passage of the virus through calves so modifies its activities that it localizes rather specifically in pavement epithelium. The relatively insignificant local lesions of vaccinia in contrast to the general distribution of the eruption and lesions of smallpox are in harmony with this view.

FIG. 95.—Guarnieri cell inclusion bodies.

Jenner's remarkable studies upon the immunity to smallpox that follows vaccination with cowpox virus have been amply confirmed by the observations of Brinkerhoff and Tyzzer,[1] who showed that vaccination of monkeys protects them from subsequent infection with the smallpox virus.

Originally vaccine virus was perpetuated by arm to arm inoculation, but the danger of transmitting syphilis or other disease as well as the uncertainty of the method have led to the use of calves as a source of vaccine virus.

The source of the virus is threefold:[2]

1. Virus descended from spontaneous cowpox and continued through an indefinite series of animals—the true animal vaccine.

[1] Jour. Med. Research, 1905, xiv, 209.
[2] Theobald Smith, Med. Soc. Proc., June 10, 1903.

2. Virus obtained from animals which have been inoculated with lymph from human vaccine pustules, either directly or indirectly, through a series of calves—this is known as retrovaccine.

3. Vaccine obtained by passing smallpox virus through the cow—the so-called variola vaccine.

Preparation of Vaccine Virus.—Healthy female calves about three months of age are selected. After thorough cleansing the animal is fastened upon an operating table of special design and the abdomen and inner aspect of the thighs are shaved. If disinfectants have been used they are removed with sterile water. Shallow parallel incisions about half an inch apart and just deep enough to become slightly reddened are made, and the vaccine is thoroughly rubbed into the scarified area. The quarters in which inoculated calves are kept are scrupulously clean; the animals are preferably fed an exclusive milk diet. Dust is reduced to a minimum and excreta are promptly removed by flushing with a stream of water.

Four to six days after inoculation, depending upon the rate of development of the vaccine vesicles, the calf is again placed upon the table, the vaccinated area washed with sterile water and then rubbed gently with sterile absorbent cotton; any crusts or scabs are removed. The slightly elevated confluent eruption is curetted away and appears as a pulpy mass, which is thoroughly ground in a mill of special design with three or four times its volume of 60 per cent. glycerin.[1] The ground and comminuted glycerized virus thus prepared contains variable numbers of bacteria;[2] as many as 700,000 per c.c. have been found.[3] Of the more common microörganisms, various molds, yeasts and members of the coccal group are usually present. Very rarely cases of tetanus have been reported following vaccination.[4] The extreme rarity of these cases and the possibility of infection from uncleanly conditions after the vaccination was made make it doubtful that vaccine may be a vehicle for the transmission of tetanus.[5]

The addition of the glycerin to the pulp obtained from vaccinated calves plays an important part in reducing the number of bacteria

[1] Carbolic acid (1 per cent.) is frequently added to the glycerin before mixing it with the pulp; experience indicates that the carbolized vaccine virus loses its potency more rapidly than when glycerin alone is used.

[2] See Rosenau, Am. Med., 1902, iii, 637, for Bacteriology.

[3] Theobald Smith, loc. cit.

[4] Wilson, Jour. Am. Med. Assn., 1902, xxxviii, 1147, 1222. McFarland, Jour. Med. Research, 1902, vii, 474.

[5] See Francis, Bull. No. 95, U. S. P. H. and Marine Hosp. Service, 1914, for results of implanting tetanus spores directly into vaccine.

which are invariably present in "green vaccine"—it does not seriously impair the activity of the virus itself. After one to two months' storage, which is generally practiced to reduce the number of bacteria, the vaccine is relatively free from microörganisms, although it is practically never sterile.

The ripened vaccine is subjected to a bacteriological examination to determine the number of bacteria per cubic centimeter, the absence of tetanus bacilli and streptococci, and a guinea-pig inoculation is is made with about a cubic centimeter of it to guard against an accidental excess of carbolic acid, before it is tested clinically for its potency. The potency test is made upon several children (previously unvaccinated) in the usual manner. Generally at least a dozen cases are vaccinated and a high percentage of "takes" must be obtained before the product is finally marketed.

Recently Noguchi[1] has cultivated an absolutely sterile vaccine virus of high potency in the testes of rabbits and bulls. The entire freedom of the preparation from alien microörganisms not only eliminates the necessity of a ripening process to reduce bacterial contamination; it also makes it possible to reduce the cost of production materially. The vaccinal eruption induced in the cornea, skin and testes of rabbits and the skin eruptions in calves were identical with those induced by the virus perpetuated in the ordinary manner. The eruptions induced in man also were perfectly typical. Finally, the sterile testicular vaccine induced immunity reactions in experimental animals identical with those obtained with the ordinary "skin" vaccine.

Phenomena of Vaccination.—1. *Technic.*—The site of vaccination, preferably the outer aspect of the arm about the deltoid muscle, is cleansed thoroughly with soap and water, and finally with alcohol if possible. When the surface is dry a light scratch about an inch long is made with a sterile needle,[2] deep enough so that the bottom of the incision is slightly reddened, but not deep enough to draw blood. The virus is then spread over the area and brought into intimate contact with the epidermal layer by gentle rubbing with the side of the needle. The safest method of vaccination is by puncture either with a charged needle, or through a shallow abrasion made with a von Pirquet tuberculin chisel. The chances of successful vaccination by the puncture method are much less than by the linear incision, however. The older method of vaccination was through a scarified

[1] Jour. Exp. Med., 1915, xxi, 539.
[2] An ordinary sewing needle is excellent for the purpose.

area, varying from a square centimeter to nearly twice that size. The crust that forms over such a wound furnishes excellent anaërobic conditions for the growth of bacteria, and the thickness of the crust offers mechanical opposition to the formation of the vesicles, which are prone to appear around the area in consequence. Vaccination by scarification is forbidden by law in Germany.

2. *The Course of the Disease, Vaccinia.*—The initial reddened site of inoculation soon disappears, leaving only a small scratch or puncture; about the third or fourth day, however, one or several small bright red papules appear, which become vesicular by the end of seven days and surrounded with a bright red areola. The contents of the vesicle become yellowish, usually from the eighth to the tenth day, and discharge a yellowish fluid if they are opened. The contents then become dessicated, and a crust forms which drops off in about two weeks.

From the third to the fifth day after the vaccination a febrile reaction of one or two degrees is usually experienced, and the site of the vaccination itches intensely and is painful. There is frequently loss of appetite and general symptoms of malaise quite out of proportion to the amount of local reaction. By the end of the second week the symptoms have disappeared and the sunken multilocular scar is the principal residual evidence of a successful vaccination. It is generally believed that already by the ninth to the eleventh day after inoculation the patient is relatively refractory to infection with smallpox virus.

3. *Immunity.*—The duration of immunity is not definitely known, but it is stated to be from seven to ten years on the average. In Germany, where vaccination has been enforced by law for five decades, a child is required to be vaccinated by the end of the first year, again about the time it enters school, and a third time at the age of sixteen or thereabouts.

Occasionally a first vaccination is unsuccessful. Frequently old or inactive vaccine, poor technic, or a deliberate sterilization of the vaccined area with disinfectants are responsible, because man does not, as a rule, exhibit immunity to natural vaccinia. Several successive negative results should be obtained before the individual is pronounced refractory.

4. *Revaccination.*—Revaccination frequently does not lead to a "take," but in a fair proportion of individuals a typical reaction may take place; this may be an accelerated reaction. The accelerated reaction runs a more rapid course than the ordinary reaction and

reaches maturity usually within four to six days in place of seven to ten days. Less commonly an "immediate" reaction is met with; the site of inoculation becomes reddened and the lesion is greatest within twenty-four hours after the inoculation. The reddened area fades rapidly and the entire process heals almost as quickly as the simple reaction of trauma excited by the scratch in the epidermis. The accelerated and immediate reactions are usually regarded as potentially equivalent to a typical reaction, provided they are induced by revaccination.

Dengue.—The etiology of dengue has not been definitely established, but it appears to belong to the group of filterable viruses and to be transmitted by Culex fatigans, a mosquito very common in the tropics. Graham[1] claimed to have transmitted the disease to non-immune individuals not only through the bite of infected female Culices, but also by injecting the ground up salivary glands of a mosquito that had previously bitten a patient. Ashburn and Craig[2] state that the virus will pass a Berkefeld filter and that both whole blood and serum filtered through Berkefeld filters will reproduce the disease in non-immune individuals. The incubation period in these cases was about four days. Ashburn and Craig believe with Graham that the virus of dengue is ordinarily transmitted by Culex fatigans.

Rocky Mountain Spotted Fever.—Rocky Mountain Spotted Fever is an acute fever characterized by a purpuric eruption of the skin. The disease is rather strictly limited to the Northern Rocky Mountain States, Montana, Wyoming and Idaho.

The etiological agent is not definitely known. Wilson and Chowning[3] believed the causative agent to be a Babesia transmissible by a tick, Dermacentor reticularis (now known as Dermacentor occidentalis). This view was not supported by later observers. Ricketts[4] in numerous investigations has shown that the virus circulates in the blood stream, and that infected ticks may transmit the disease. He was also successful in infecting monkeys (Macacus rhesus) and guinea-pigs with the virus. One attack conferred immunity to subsequent infection in experimental animals, and the serum of an immune animal protected a susceptible animal from infection. As a curative agent the serum was of little value. A minute diplococcoid or bipolar

[1] Jour. Trop. Med., 1903, vi, 209.
[2] Philippine Jour. Sci., 1907, ii, 93.
[3] Jour. Inf. Dis., 1904, i, 31.
[4] Jour. Am. Med. Assn., 1906, xlvii, 33, 358; 1907, xlix, 24, 1278; Trans. Chicago Path. Soc., 1907; Jour. Inf. Dis., 1908, v, 221; Jour. Am. Med. Assn., 1909, lii, 379.

staining structure was observed in great numbers in the blood of infected men and in the eggs of infected ticks. These were not successfully cultivated, but agglutinated with the serum of an immune animal. Their relation to the disease has not been established.

Mumps.—Mumps or epidemic parotitis is a specific infectious disease which is more commonly observed among children from four to fifteen years of age, although younger children and adults are by no means immune. The incubation period averages from seventeen to twenty-eight days. It is probable that the infectious period begins a few days—about four—before the characteristic syndrome appears, and the disease is probably transmitted directly from person to person through infected material from the nasopharynx. The mortality is very low and cases that terminate fatally are generally very young children and infants.

The causative agent is not definitely known: a diplococcus has been isolated from inflamed parotid glands by Laveran and Catrin[1] in sixty-seven out of a total of ninety-two cases. Mecray and Walsh,[2] Michaelis and Bienn,[3] Busquet and Feri[4] have made similar isolations. Teissier and Esmein[5] report the successful culture of a similar organism from a case of suppurative parotitis. Herb[6] has also isolated a diplococcus from a case of suppurative parotitis which ended fatally. Animal experiments with these cultures have not been convincingly positive.

Nicolle and Conseille[7] and Gordon,[8] working independently, state that fluid separated from the parotid glands of patients having mumps, injected into the parotid glands of monkeys, reproduced a syndrome strikingly like that of mumps in these animals. Gordon also found that the virus retained its virulence after passage through a Berkefeld filter. It is destroyed by a brief exposure to 55° C. It would appear from his observations that the virus of mumps belongs to the group of filterable viruses.

[1] Compt. rend., Soc. biol., 1893, 9 sér., v, 528.
[2] Medical Record, 1896, i, 440.
[3] Verhandl. XV Kongress f. inn. Med., 1897, xv, 441.
[4] Rev. d. Med., 1896, xvi, 744.
[5] Compt. rend. Soc. biol., 1906, lx, 803, 853.
[6] Arch. Int. Med., 1909, iv, 201.
[7] Compt. rend. Acad. sc., 1913, clvii, 340.
[8] Lancet, 1913, ii, 275.

SECTION IV.

GASTRO-INTESTINAL BACTERIOLOGY.

CHAPTER XXX.

GASTRO-INTESTINAL BACTERIOLOGY.

General Considerations.—An examination of the feces[1] of a healthy adult with the higher objectives of the microscope will show that a large portion of the fecal mass is made up of bacterial cells. An average-sized bacterial cell is very small indeed, measuring about 1 micron in diameter and 2 microns in length, hence it is not surprising that various investigators have estimated the daily excretion of bacteria by a healthy adult on a mixed diet at one hundred to thirty-three hundred billions. The bacteria when dried would weigh more than 5 grams and would contain about 0.6 grams of nitrogen. A very considerable proportion of the total nitrogen of the feces is contained in these bacteria.

It is apparent that the ingested food does not contain this prodigious number of bacteria, consequently it must be assumed that there is a rapid development of the organisms in the intestinal tract. The theoretical progeny of a single bacterial cell of the more rapidly developing types may number millions in twenty-four hours, so that the mechanical possibility of a very great daily proliferation of bacteria is well established. It is obvious, therefore, that the alimentary canal, from the viewpoint of bacteriology, is a most efficient incubator and cultural medium combined, in which bacterial growth exceeds both in intensity and complexity, that of any known medium. The range of reaction and composition of nutritive substances at different levels of the intestinal tract are such that theoretically a great variety of bacteria capable of developing at body temperature may find conditions favorable for their growth there.[2] The prominent types of

[1] Average weight 100 to 200 grams per diem.

[2] Kendall, Jour. Biol. Chem., 1909, vi, 499; Wisconsin Med. Jour., 1913, xii, No. 1.

bacteria that appear in the intestinal flora of a normal person are fairly constant in their occurrence, but there may be well-marked seasonal and even annual variations in the relative proportions of the individual groups of organisms which comprise this flora. This suggests that the normal bacterial flora is acclimatized to the intestinal environmental conditions of temperature, reaction and composition of food, and of intestinal secretions at different levels. It also indicates that the activities of the organisms which comprise the normal intestinal flora are not in active opposition to those of the host.[1]

Adventitious bacteria, frequently in considerable numbers, undoubtedly reach the intestinal tract from time to time. The fate of these organisms depends upon a number of factors, some of which are little understood. If their activities are greatly at variance with those of the normal types they usually fail to gain a foothold; either they are unable to develop in competition with the well-acclimatized normal flora, or they cannot accommodate themselves to the physiological and chemical conditions which prevail there. If, on the contrary, these organisms can adapt themselves readily to the prevailing conditions at some level of the alimentary canal they may continue to develop either in association with preëxisting types, or gradually replace the latter.[2] It is doubtless through this process that the seasonal prevalence of some types of intestinal bacteria has its origin. It is not unlikely, furthermore, that the occasional unusual type of organism characteristic for an individual or a group of individuals gains entrance to and develops in the intestinal tract in this manner.

The nature of the process whereby progressively pathogenic bacteria (usually of exogenous origin) replace or modify the normal intestinal flora is as yet little understood. There is evidence in favor of the view that exogenous bacteria which invade the body through the intestinal

[1] The general phenomena governing the parasitism of bacteria in the alimentary canal are not unlike those leading to bacterial parasitism upon the skin, the conjunctiva, or other surfaces of the body which are in communication with the exterior. One important phase of intestinal parasitism is not manifested in other parts of the body, however. The bacteria of the intestinal flora change along rather definite lines from infancy to adult life, as the diet of the host changes from the monotonous pabulum of infancy to the varied regimen of the adult. The organisms parasitic upon the skin and other surfaces of the body do not exhibit this change in type, and it is reasonable to attribute the relative stability of the skin flora to the relative constancy of environmental conditions there, while the succession of types of intestinal bacteria from infancy to adult life is rather definitely associated with corresponding changes in the diet of the host.

[2] Undoubtedly repeated inoculation of the alimentary canal with adventitious strains of bacteria plays an important part in determining their acclimatization in the intestines; possibly a simultaneous absence of the preëxisting intestinal types in the environment, leading to a reduction or even absence of these normal inhabitants in the food of the host may materially affect the outcome of the "replacement" process.

tract may become somewhat widely disseminated in restricted areas and appear in the intestinal contents of many individuals without inciting noteworthy symptoms, prior to the appearance of disease in epidemic proportions[1] and with characteristic symptoms.

THE GASTRO-INTESTINAL FLORA OF NORMAL INFANTS, ADOLESCENTS AND ADULTS.

The fetal intestinal contents, the meconium, are sterile at birth; the first bacteria appear in the meconium from eighteen to twenty-four hours postpartum. This is a period of adventitious infection during which a variety of bacterial types, largely determined by the environment of the infant, gain entrance to the alimentary canal by way of the mouth or anus and are excreted in the residual embryonic feces. This initial non-characteristic intestinal flora is usually more varied in summer than in winter and more luxuriant when the infant is exposed to relatively uncleanly surroundings than when the reverse is the case. Escherich[2] and others have called attention to the occurrence of a rather large bacillus in the meconium, possessing a terminal spore closely resembling Bacillus tetani. This organism, known as the Köpfchen bacillus, has been identified by some observers as Bacillus putrificus of Bienstock;[3] it has not been studied culturally, however, and this identification cannot be regarded as final. Other spore-forming bacteria, both aërobic and anaërobic, are also usually present in the meconium at this period. Of these Bacillus aërogenes capsulatus and members of the Bacillus Mesentericus Group are the best known. Bacillus coli, Bacillus proteus, Bacillus lactis aërogenes and Micrococcus ovalis[4] also occur commonly.

The initial period of adventitious bacterial infection of the intestinal contents merges more or less imperceptibly through a transitional stage to the period of dominance of the characteristic infantile intestinal flora, which becomes settled usually about the third day postpartum. At this time the breast milk diet of the nursling is well established and the intestinal tract is permeated with it. The bacteria throughout the alimentary canal become more numerous, the spore-

[1] Kendall, Boston Med. and Surg. Jour., 1915, clxxii, 851.
[2] Escherich, Darmbakterien des Saüglings, Stuttgart, 1886, p. 9.
[3] Arch. f. Hyg., 1899, xxxvi, 335; ibid., 1900, xxix, 390.
[4] Micrococcus ovalis (Escherich, loc. cit., p. 89) appears to be identical with the enterocoque of the French writers, with Streptococcus lacticus of Kruse (Centralbl. f. Bakt., Orig., 1903, xxxiv, 737) and Streptococcus enteriditis of Hirsch (ibid., 1897, xxii, 369), and Libman (ibid., 1897, xxii, 376).

forming types disappear for the most part and rather abruptly, and the coccal forms and Gram-negative bacilli of the colon aërogenes type diminish relatively, but never quite disappear. Simultaneously rather long, thin bacilli, occurring singly, in pairs, or in groups with their axes parallel, become strikingly prominent. These bacilli are frequently slightly curved and occasionally their ends are somewhat attenuated. Typically they are Gram-positive and stain uniformly, but in many instances they exhibit a central Gram-positive granule in an otherwise Gram-negative rod, presenting the "punctate" appearance described by Escherich.[1] Occasionally the cytoplasm of these organisms is collected into small, round or oval granules which stain intensely; the remainder of the rod stains faintly or not at all. At

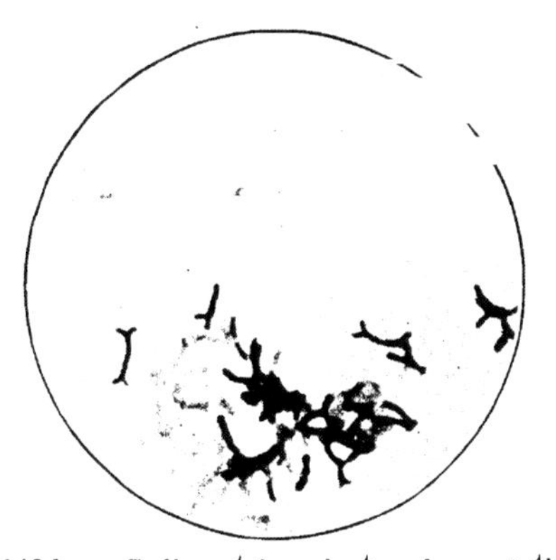

FIG. 96.—Bacillus bifidus. Sediment from lactose fermentation tube. × 1000.

first sight these granules resemble chains of cocci. This somewhat pleiomorphic organism is Bacillus bifidus, first observed by Escherich, but isolated in pure culture and studied in detail by Tissier.[2] It is an obligate anaërobe,[3] fermentative in character, which typically forms considerable amounts of acid from lactose and other sugars, but no gas. The organism received the name "bifidus" from its remarkable property of developing well-defined bifid ends when it is grown in artificial media; it does not ordinarily exhibit bifid ends in the intestinal tract. Moro[4] and, independently, Finkelstein[5] have isolated and described an organism very similar in morphology to Bacillus bifidus

[1] Loc. cit.

[2] Recherches sur la Flore Intestinale des Nourrissons, etc., Thèse de Paris, 1900, p. 85.

[3] Noguchi (Jour. Exp. Med., 1910, xii, 182) appears to have shown that Bacillus bifidus, under laboratory conditions, may become aërobic and form spores.

[4] Wien. klin. Wchnschr., 1900, xiii, 114.

[5] Deutsch. med. Wchnschr., 1900, xxii, 263.

as it occurs in the intestinal contents, but which differs materially from the latter both in its aërobiosis and in its inability to develop bifid ends in artificial media. This organism, Bacillus acidophilus, is more commonly found in the intestinal contents of artificially fed babies than in nurslings, and it is more tolerant of organic acids than Bacillus bifidus. It belongs to the group of Aciduric Bacteria.[1]

In addition to Bacillus bifidus[2] and Bacillus acidophilus, which typically comprise a majority of the characteristic intestinal bacteria, smaller numbers of Micrococcus ovalis, Bacillus coli, Bacillus lactis aërogenes and other bacteria are found in the feces of nurslings.

Escherich[3] has emphasized the very significant fact that putrefactive (proteolytic) bacteria are uncommon in the dejecta of normal nurslings; there is little or no evidence of the development of these organisms in the intestinal tract during this stage. The putrefactive bacteria, as a rule, do not develop in an acid medium in competition with organisms like Bacillus bifidus and other acidogenic types which dominate the alimentary canal of the normal nursling.

Distribution of the Intestinal Flora of the Normal Nursling.—The principal portal of entry of the intestinal bacteria is the mouth. There is no doubt that a great variety of organisms may from time to time enter this atrium, including not only the ordinary organisms of the nurslings' environment, but pathogenic bacteria as well. A majority of these pass to the stomach, and they may pass to the intestinal tract. The flora of the mouth and stomach are not well known, but they appear to be of relatively slight importance as a rule. Those adventitious organisms which pass from the stomach to the duodenum rarely appear to gain a foothold there, or at lower levels of the intestines.

The duodenal flora, which in health is composed chiefly of coccal forms of the Micrococcus ovalis type, is most numerous during those periods when the food is passing through; during interdigestive periods there appear to be relatively few bacteria at this level. From the jejunum to the ileocecal valve, members of the Bacillus lactis aërogenes group occur more commonly. Bacillus coli and other members of the colon group are most numerous at the ileocecal valve and the cecum,

[1] Kendall, Jour. Med. Research, 1910, xxii, 153; Rahe, Jour. Inf. Dis., 1914, xv, 141.

[2] Madame Tsiklinsky (Ann. Inst. Past., 1903, xvii, 317) has been unable to demonstrate B. bifidus in normal nurslings' feces as frequently as has been reported elsewhere; the consensus of opinion appears to be, however, that bifidi are the most characteristic bacilli of the normal nursling intestinal flora.

[3] Loc. cit.

and Bacillus bifidus or similar organisms dominate the large intestines from this level to the sigmoid flexure. The remainder of the large intestine to the rectum is somewhat sparsely populated with living bacteria, partly because the fecal mass is relatively desiccated by the absorption of water, partly because of the accumulation of waste products of bacterial activity—principally acids resulting from fermentation of lactose, formed higher up in the tract—which inhibit the development of bacteria in the lower levels.[1]

It must be remembered that while the greatest number of important bacteria mentioned above occur at the levels indicated, there is a mechanical transportation of all intestinal bacteria from the higher to the lower levels, so that some organisms of all types are found in the dejecta. It is particularly important to realize that the types of bacteria outlined are those which can be identified by staining methods as numerically prominent at the various intestinal levels; these observations can be corroborated by appropriate cultural methods. Nevertheless, there is a wide disproportion between the numbers of each of the respective bacteria seen in stained preparations and the numbers of each type which develop in artificial media. Thus, Escherich[2] observed that a preponderance of bacteria of normal nurslings' feces were Gram-positive bacilli, yet he never succeeded in growing these bacilli in artificial media; the principal types which developed in his cultures were Bacillus coli and Bacillus lactis aërogenes, organisms which are numerically in the minority in the intestines, but which grow luxuriantly outside the body. It is now realized that he did not employ suitable conditions of culture to isolate the most prominent types of organisms. Undoubtedly much of the confusion which has attended the study of intestinal bacteriology in the past is attributable to the lack of appreciation of the cultural peculiarities of the intestinal organisms.

Distribution of the Intestinal Flora of Artificially Fed Infants.—Escherich[3] directed attention to the striking dissimilarity between the intestinal flora of the breast-fed and the artificially fed infant; culturally, morphologically and chemically the former is more homogeneous than the latter. The most distinctive features of the dejecta of artificially fed infants are: the relative increase of Gram-negative bacteria of the coli-aërogenes type, and of coccal forms of the Micrococcus ovalis type, together with a diminution of Bacillus bifidus.

[1] Kendall, Jour. Med. Research, 1911, xxv, 117, et seq.
[2] Loc. cit.
[3] Loc. cit.

Bacillus acidophilus is relatively more numerous, as a rule, in the artificially fed infant than in the nursling. Proteolytic bacteria of several types are also of frequent occurrence,[1] but they are not commonly found in the dejecta of normal nurslings. These organisms are frequently spore-forming bacilli, of which two principal groups are recognized—members of the aërobic group, of which Bacillus mesentericus is a prominent type, and anaërobic bacteria; of the latter, Bacillus aërogenes capsulatus is most widely known; it frequently occurs in small numbers in the feces of artificially fed infants.[2] The reaction of normal feces of artificially fed babies is usually alkaline; culturally and chemically, the evidence of intestinal proteolysis of bacterial causation is more marked in these infants than in normal nurslings.

The general distribution of types of bacteria at the different levels of the intestinal tract is similar to that observed in normal nurslings; the principal differences are found in the cecum and large intestine, where the obligately fermentative bacteria of the bifidus type are replaced to a considerable degree by an extension of the habitat of the colon bacillus, of Bacillus acidophilus, and the appearance of moderate numbers of proteolytic bacteria, both aërobic and anaërobic; many of the latter are sporogenic.

The prevailing bacteria of the artificially fed infant may be changed along fairly definite lines by varying the proportion of protein to carbohydrate in the diet, and by substituting one carbohydrate for another. Thus, a continued preponderance of protein leads to a partial or even practically complete suppression of the activity of the bifidus-acidophilus group, and a noteworthy increase in the activity of proteolytic organisms;[3] of the latter, aërogenic bacteria of the colon-proteus group and spore-forming bacteria of the mesentericus group appear to be the more prominent. A relative increase in carbohydrate leads to a diminution or suppression of proteolytic activity in the intestinal tract, and an increase in the fermentative activities of the intestinal organisms.[4] Those bacteria—as Bacillus coli—which

[1] Escherich, loc. cit.

[2] See Hibler (Untersuchungen über die pathogenen Anaëroben, Jena, 1908), Jungano and Distaso (Les Anaërobies, Paris, 1910) for description of various intestinal anaërobes. Unfortunately, so little is definitely known about a majority of these organisms, culturally, chemically and numerically, that almost nothing can be said of their importance.

[3] Kendall, Jour. Biol. Chem., 1909, vi, 268; Herter and Kendall, ibid., 1910, vii, 203.

[4] Provided, of course, the digestion of the infant remains normal. It is obvious that a disturbance of the digestive function of the alimentary canal may lead to new factors which may play an important part in determining the prevalence of one or several types of intestinal bacteria.

can accommodate their metabolism to either a protein or carbohydrate regimen become fermentative and produce lactic acid and other products of the fermentation of carbohydrate in place of H_2S and NH_3, indol, and other putrefactive products which characterize their development in protein media[1] under these conditions. The obligately proteolytic organisms tend to decrease in number because they are unable to thrive in the presence of active fermentation, and the carbohydrophilic bacteria increase both in numbers and in activity; the type of carbohydrophilic organisms which develops depends upon the carbohydrate fed and upon the length of time the diet is continued; Bacillus bifidus tends to increase in numbers[2] when lactose is the sugar, Bacillus acidophilus if maltose is substituted for lactose, provided the regimen is maintained for several days.[3]

The changes in the intestinal flora from the bottle-fed infant to adolescence and adult life depend somewhat upon the diet of the individual. The general tendency in individuals on an average mixed diet is for Bacillus coli to become the dominating organism; usually about 75 per cent. of the viable bacteria of the feces are colon bacilli. Of the remaining organisms, spore-forming organisms of the mesentericus group are usually numerous, and gas bacilli may be found relatively frequently, but in small numbers. Bacillus coli and Bacillus mesentericus are among the most persistent of the intestinal bacteria of adults. Those two organisms and no others were found in the lower part of the large intestine of a man who abstained from all food for thirty-one days.[4] The characteristic feature of the normal adult fecal flora as compared with the infantile nursling flora is the very heterogeneous variety of types of bacteria in the former, in sharp contrast to the homogeneity of types of bacteria in the latter.

Distribution of the Intestinal Flora in the Adolescent and Adult.—The stomach in health is quite free from bacteria as a rule. It has been assumed in the past that the hydrochloric acidity may be a factor in the destruction of organisms, but it should be remembered that protein undergoing gastric digestion binds hydrochloric acid. Nevertheless, bacterial activity is very limited in the stomach under normal conditions.

[1] Kendall, Boston Med. and Surg. Jour., 1910, clxiii, 322; *Pediatrics*, 1910, xxii, No. 9.

[2] It is apparent that this change cannot take place unless there is a residuum of bifidi in the intestinal tract to develop from. The same is true for Bacillus acidophilus. In the absence of these types the dominant fermenting organisms will vary with the flora of the individual.

[3] Kendall, Boston Med. and Surg. Jour., 1910, clxiii, 322.

[4] Kendall, Publication 203 of the Carnegie Institution of Washington, 1915, p. 232.

The duodenum of adults is relatively poorly populated with bacteria in interdigestive periods, and Cushing and Livingston[1] have called attention to the relative innocuousness of gunshot wounds at this level as contrasted with those at lower levels, where peritonitis practically invariably follows perforation of the gut. This phenomenon is not wholly attributable to the comparative paucity of bacteria in the duodenum as contrasted to lower levels; a final explanation is lacking at the present time. According to Gessner,[2] staphylococci and streptococci are numerous in the duodenum, and Tavel and Lanz[3] have made similar observations. Recently Hess, using a duodenal catheter,[4] has studied the duodenal flora in normal individuals. He finds the bacterial content very low in interdigestive periods; staphylococci and a few Gram-positive and Gram-negative bacteria were the prevailing types. These Gram-negative bacteria were not Bacillus coli. Breast-fed infants showed fewer bacteria in the duodenal region than did bottle-fed babies.

The lower levels of the small intestines become progressively richer in bacteria. The relative slowness with which food passes through the intestines at the lower levels probably is a potent factor in creating conditions favorable for continual bacterial growth. As a rule cocci still predominate in the lower jejunum and upper ileum, but Gram-negative bacilli of the colon group appear in moderate numbers.

The cecum and ascending colon are the regions of most intense bacterial proliferation in health, but the number of living bacteria in the intestinal contents diminishes rather abruptly from the sigmoid to the rectum. It has been stated that at least 90 per cent. of the bacteria of the feces are dead, or so attenuated in vitality that they are incapable of growing in artificial media. For various reasons the accuracy of this statement may be questioned, but there is little doubt that the numbers of viable bacteria in the relatively desiccated feces are less than those in the more fluid intestinal contents at the level of the cecum.

The bacteria commonly present in the ileocecal region are undoubtedly of many and varied types, but in general aërogenic bacilli of the colon type[5] (including probably members of the proteus group as well)

[1] Contributions to the Science of Medicine by the pupils of William Welch, 1900, 543.

[2] Arch. f. Hyg., 1889, ix, 128.

[3] Mitt. a. klin. d. Schweiz, i.

[4] Ergebnisse der inn. Med. u. Kinderheilk., 1914, xiii, 530.

[5] Ford, Classification and Distribution of the Intestinal Bacteria in Man, Studies from the Royal Victoria Hospital, 1903, i, No. 5; MacConkey, Jour. Hyg., 1905, v, 333, have described the common types of aërobic bacilli in the intestinal tract. The cultural characters of the various aërogenic lactose-fermenting organisms, grouped for convenience as the colon group, are clearly set forth in these monographs.

and aërobic spore-forming bacteria of the mesentericus group are the most readily recognized. The important feature of the intestinal flora at the lower levels of the intestinal tract of adolescents, and more especially of adults, is the presence of facultative fermentative bacteria which appear to thrive equally well when the intestinal contents at this level contain protein and carbohydrate as when the carbohydrate is absent. Members of the colon-proteus group, particularly the former, various aërobic liquefying bacilli—both spore-forming, and non-spore-forming—and, to a limited extent, anaërobic bacteria as well are characteristic of the bacterial flora of the large intestines of adults. This is in striking contrast to the distinctive monotonous fermentative flora of the normal nursling, whose diet contains a sufficient amount of carbohydrate (lactose) to bathe the entire alimentary canal. It contrasts also, to a somewhat lesser degree, with the lower intestinal flora of young children on a cow's milk diet, where the proportion of carbohydrate to that of protein, although decidedly less than that of the nursling, is usually still sufficient to restrain an excessive development of proteolytic bacteria.

It will be seen that the carbohydrate of the infant diet is lactose, which is utilizable as such by the dominant bacteria of the infantile intestinal and fecal flora. A not inconsiderable portion of the carbohydrate of the adult, on the contrary, is starch, which is not readily utilizable as such by a great majority of the intestinal or fecal bacteria; it is very probable that a very considerable proportion of the assimilable products of hydrolysis of the starch are absorbed rapidly from the intestinal contents and therefore there is normally but little utilizable sugar available for the intestinal flora of adults. This is especially the case in the lower levels of the intestinal tract, where the stasis of the intestinal contents results in a differential accumulation of the more slowly hydrolyzed and absorbed protein. It would appear from these considerations that the relative absence of utilizable carbohydrate in the large intestine of adults would naturally be associated with a diminution of the obligate fermentative or carbohydrophilic organisms, and available evidence indicates that such is the case.

Significance of Intestinal Bacteria.—The striking differences in morphology, chemistry and in cultural characters between the intestinal floras characteristic respectively of nurslings, artificially fed infants and adults suggest at once that nutritional stimuli may be an important factor in determining the dominance of types of bacteria. An intestinal flora does not appear to be essential for the well-being of mammals in the Arctic regions; Levin[1] has found that the feces of polar bears are practically sterile. It must be remembered, however, that similar animals kept in captivity in more temperate climates exhibit a very definite intestinal and fecal flora. Attempts to rear chicks,[2] turtles,[3] tadpoles[4] and guinea-pigs[5] in a sterile environment have not added materially to available knowledge of the physiological significance of the intestinal flora, partly because the rigorous conditions under which such observations must be made interfere greatly with the normality of the animals' environment. It is probable that the significance of the intestinal flora lies rather in its potential antagonism to alien bacteria which certainly gain entrance to the alimentary canal from time to time, than in any specific participation in the normal digestive process of the host.[6]

The normal intestinal flora may be regarded as intestinal parasites just as the various bacteria which occur commonly on the skin are regarded as cutaneous parasites. It is important to realize that the normal intestinal organisms, like the cutaneous organisms, are "opportunists," potentially capable of becoming invasive whenever the barriers which ordinarily suffice to limit their development to the lumen of the alimentary canal become impaired, giving rise to endogenous infections.

Unlike the cutaneous parasitic flora or that of other surfaces of the body which does not appear to vary materially from infant to adult life, the intestinal flora changes in a most definite and striking manner as the individual develops from infancy to senescence. This change does not appear to depend fundamentally upon bacteria ingested with the food, for Escherich[7] and many others have shown that sterilization of the food does not cause a noteworthy reduction in the number of types of fecal bacteria in young children.

The most important normal factor in determining the intestinal

[1] Ann. Inst. Past., 1899, xiii, 558; Skandinavisches Arch. f. Physiol., 1904, xvi, 249.

[2] Schottelius, Arch. f. Hyg., 1902, xlii, 48.

[3] Moro, Jahrb. f. Kinderheilk., 1905, xii, 467.

[4] Metchnikoff, Ann. Inst. Past., 1901, xv, 361.

[5] Nuttall and Thierfelder, Ztschr. f. physiol. Chem., 1895, xxi, 109; 1896, xxii, 62; 1897, xxiii, 231.

[6] Hilgermann (Arch. f. Hyg., 1905, liv, 335) and others have produced experimental evidence in favor of the view that the immature intestinal tract of the young infant is more permeable to bacteria than that of adolescents and adults. It may be inferred from these observations that the normal nursling intestinal flora is somewhat protective in its relation to the host, in that the normal fermentative activities of the organisms comprising the intestinal flora create conditions throughout the alimentary canal which are inimical to the development of alien proteolytic and fermentative bacteria.

[7] Centralbl. f. Bakt., 1887, ii, 633; also, Jahrb. f. Kinderheilk., 1900, lii, 1.

flora in health is the chemical composition of the ingested food.[1] Escherich,[2] as far back as 1887, clearly showed that a very characteristic change in the intestinal flora of dogs could be brought about by feeding protein, during which bacteria that liquefy gelatin become abundant in the feces.

Assuming that food is an important factor in determining the more common types of bacteria found respectively in the intestinal tracts of nurslings, artificially fed children and adults, it would be reasonable to expect that the same or similar bacteria should develop in the intestinal tracts of experimental animals, provided they were fed upon the same foods as nurslings or adults. A prolonged series of experiments upon monkeys,[3] dogs and cats have shown that alternations in diet do influence the prevailing types of bacteria in the intestinal tract to a marked degree. The essential features of these experiments were that monkeys, dogs and cats fed upon cow's milk containing sufficient lactose solution to bring the percentage of protein and carbohydrate approximately to that of human breast milk excreted feces which, in appearance and in bacterial content, approached very closely those of the normal human nursling. The acid reaction, practical absence of obligately proteolytic bacteria, the dominance of Bacillus bifidus and acidophilus and the appearance of Micrococcus ovalis in numbers similar to corresponding types in normal nurslings' feces were in striking contrast to the feces of the same animal after a prolonged feeding with a purely protein diet. In the latter event large numbers of proteolytic bacteria were present in the feces, which were alkaline in reaction and rich in indol, phenols, hydrogen sulphide, ammonia and other products indicative of intense proteolytic decomposition. Obligately fermentative bacteria of the bifidus-acidophilus type were few in number, or practically absent. Recently Rettger[4] has made somewhat parallel observations in mice and rats.[5]

The nature of the dominant organisms which develop in diets rich in carbohydrate varies with the carbohydrate itself. Bacillus bifidus is more commonly predominant when lactose is the sugar fed, without

[1] See Kendall, Jour. Med. Research, 1911, xxv, 136, for résumé.
[2] Darmbakterien, etc., p. 111.
[3] Kendall, Jour. Biol. Chem., 1909, vi, 499. Herter and Kendall, ibid., 1910, vii, 203. Kendall, Jour. Med. Research, 1910, xxii, 153; ibid., 1911, xxiv, 411; 1911, xxv, 117.
[4] Centralbl. f. Bakt., Orig., 1914, lxxiii, 362.
[5] It is rather more difficult to replace a proteolytic flora in adult animals by a fermentative flora than it is in young animals of the same species; the explanation of this relative refractoriness to substitution of obligately fermentative types of bacteria for the facultative organisms commonly found in the intestinal tracts of the older animal is by no means clear.

an excess of protein; if maltose or dextrose is substituted for lactose under the same conditions, Bacillus acidophilus is very frequently the more prominent. In like manner, the nature of the protein influences the types of proteolytic bacteria to a very marked degree; in general, animal proteins other than casein appear to encourage a somewhat more active proteolytic flora than vegetable proteins. These observations are in harmony, in essential features at least, with those made under like conditions in man. A monotonous diet in which lactose and protein are fed in proportions and amounts similar to breast milk leads to the gradual development of an intestinal flora in experimental animals closely simulating that of nurslings. A preponderance of protein, on the other hand, encourages the development of bacteria which are more proteolytic in nature.

It is a striking fact that the above alternation in intestinal bacteria following changes along definite lines in the diet is elicited only when the feeding is maintained for several days; rapid alternations between a purely protein diet and a diet rich in sugar (as cow's milk diluted with an equal volume of 4 per cent. lactose solution) do not ordinarily lead to such noteworthy changes in the types of bacteria excreted in the feces.[1] The general trend of such rapid alternations between a protein regimen and one in which sugars predominate (starches do not necessarily react in this manner) is to establish a flora which is relatively heterogeneous, in which there is neither a decided predominance of obligately carbohydrophilic bacteria, as B. bifidus or acidophilus, nor of obligately proteolytic bacteria.

A most striking and important influence of diet upon bacterial activity in the intestinal tract does not manifest itself in a study confined exclusively to the changes in bacterial types of the intestinal flora. The monotony of the typical nursling flora depends in a large measure on the continual presence of lactose (a sugar not fermented by a majority of bacteria) throughout the intestinal tract. A substitution of other sugars—as dextrose, saccharose or maltose—leads to a replacement of Bacillus bifidus by other more or less obligately fermentative organisms, provided an excess of the respective carbohydrate be maintained, but the same monotony of types is observed.

The proportion of carbohydrate to protein in the diet of normal adults is far less than in nurslings and, furthermore, a considerable proportion of the carbohydrate is in the form of starches which, as

[1] This probably explains some of the irregularities experienced during brief feeding experiments.

such, are not readily fermented by most bacteria. Again, sugars, if they are present, are largely absorbed from the higher levels of the small intestine, leaving residual unhydrolyzed starches and protein in relatively great concentration in the lower levels of the large intestine. It is not surprising, under these conditions, to find that the more obligate fermentative bacteria—the Cocci—are prominent at the higher levels, as is the case normally in infants; that facultative bacteria, as Bacillus coli, are common in a transitional zone between a medium containing moderate amounts of utilizable carbohydrate and one in which the utilizable carbohydrate is frequently absent,[1] and finally, that proteolytic organisms are most abundant in the large intestines, where carbohydrate in significant amounts is practically absent, but where the protein concentration is still considerable. Practically all the bacteria found in the large intestine of normal adults exhibit a preferential action upon dextrose (a product of the hydrolysis of starches and many bioses as well), but they are, for the most part, unable to utilize lactose.

There are, therefore, two important factors to consider in discussing the influence of diet upon the intestinal flora: The substitution of types of organisms, which frequently follows a monotonous diet; and a change in the metabolism of existing types of intestinal bacteria when dietary conditions are such that the intestinal medium at one or another level fluctuates in its content of utilizable carbohydrate and other nutrient substances.[2]

From time to time modifications or changes in the types of bacteria in the intestinal flora and of their activities takes place. The nature and extent of these modifications and their effects upon the host vary very much, not only qualitatively, but quantitatively as well. An invasion of the intestinal tract by exogenous bacteria, as the dysentery bacillus or the cholera vibrio, may lead to a more or less pronounced replacement of some of the normal intestinal types by these alien organisms, and to the production of disease. Normal intestinal organisms or types indistinguishable from them by ordinary methods of study also may multiply with abnormal luxuriance through unusual

[1] Bacillus coli and various closely related bacilli are among the most labile of intestinal bacteria in adapting their metabolism to the composition of the intestinal contents. In a medium containing both utilizable carbohydrate and utilizable protein these organisms act principally upon the carbohydrate, forming lactic and smaller amounts of other acids. In a protein medium the products of metabolism are indol, phenols, and other products of proteolysis.

[2] For a brief general discussion of the influence of nutritional factors upon bacterial metabolism, see Section on Bacterial Metabolism.

conditions, extend their normal habitat, and crowd out some of the existing organisms, eventually leading to abnormal reactions in the alimentary canal which may be detrimental to the host.

There are many intestinal disturbances of unknown causation, presumably unrelated to bacterial activity, which naturally are not of interest in this connection. There is a second group of conditions in which bacteria may conceivably play a secondary part; in some instances abnormal physiological conditions in the alimentary canal may be justly regarded as the antecedent factors. The boundaries of these two groups are poorly circumscribed and they merge through imperceptible or poorly defined limits into a third group of cases in which the activities of endogenous or exogenous bacteria in the alimentary canal may be the causative factor in morbid processes of the gastro-intestinal tract.

For convenience of discussion this last group may be divided into three types: (*a*) Those cases in which products resulting from the action of bacteria upon proteins or their derivatives appear to be the prominent factors in the production of the morbid process; (*b*) those cases in which products resulting from the fermentation of carbohydrates by the action of bacteria are the prominent substances concerned in the morbid process. A third group, practically unstudied at the present time, would include those cases in which symbiotic activities of proteolytic and fermentative bacteria would result in the production of substances derived both from proteins and from carbohydrates.[1]

The action of bacteria on fats is little understood at present and no statement can be made covering this type of abnormality. It is expressly understood that products of the nature of endotoxins resulting from the dissolution of bacteria are not considered in this connection, which relates exclusively to a discussion of the activities of living organisms.

The symptomatology induced from the products arising from the decomposition of proteins or protein derivatives by the action of bacteria in the intestinal tract depends largely upon the organism or organisms concerned; it varies from the somewhat insidious, slowly progressing, so-called auto-intoxication, in which a marked increase

[1] Thus, in occasional severe diarrheas of children strains of Bacillus coli and Bacillus mesentericus are occasionally isolated, which grow symbiotically in milk, causing a deep-seated change both in the protein and carbohydrate content of the medium. The result of their mutual development is much greater than the sum of their separate activities. Ordinary strains of these organisms frequently do not exhibit this symbiotism. It is by no means improbable that similar symbiotic activity in the intestines, if unrestrained, may lead to conditions incompatible with the well-being of the host.

of urinary ethereal sulphates may be a suggestive index, to the acute toxemias characteristic of bacillary dysentery, typhoid, paratyphoid or cholera. Of course, a variety of other bacteria than the few mentioned specifically may be concerned, either alone or in symbiosis. Thus streptococci alone and streptococci in association with dysentery bacilli may be justly regarded as the etiological agents in their respective syndromes. The important factor, from the viewpoint of this discussion, is to realize that the formation of nitrogenous products from proteins or protein derivatives which are being utilized by various types of intestinal bacteria for energy may be injurious to the host. These substances are of unknown composition for the most part, but beyond doubt they are nitrogenous. Some, as phenols, cresols, or indol are simple in structure and ordinarily harmless, or nearly so, although long-continued absorption may gradually lead to cumulative effects. Others, as beta imidazoleethylamine and other primary amines formed from amino acids may be physiologically active. The unknown poisons of the meat poisoning group and those characteristic of the various bacteria which cause acute infections of intestinal origin are of unknown structure and complexity.

The other prominent type of abnormal bacterial activity in the alimentary canal—the fermentative type—is of entirely different origin; the essential factor is either a decomposition of carbohydrates, with the formation of products abnormal for the intestine, or of excess of normal fermentative products. The abnormality may be a simple hyperacidity, as, for example, that caused by an overgrowth of aciduric bacteria when certain sugars, as maltose, fed in too large amounts, lead to an overdevelopment of the aciduric bacteria; or it may be more complex. This happens frequently when there is an overgrowth of Bacillus aërogenes capsulatus, or of members of the Mucosus Capsulatus Group. In the latter event the exact nature of the irritative substance is as yet unknown, but it is in all probability not a nitrogenous compound. It is formed from carbohydrates, which contain no nitrogen. The factors leading to an overgrowth of these organisms in the intestinal tract appear to be an excess of carbohydrate and a lack of normal lactic-acid-forming bacteria. It is a significant fact that diarrheal cases associated with an overgrowth of the gas bacillus even of several years' duration do not exhibit signs or symptoms of toxemia in spite of the protracted illness.

It is unfortunate that practically none of the bacteria which incite intestinal disturbances or illness produce soluble toxins against which antitoxins can be prepared; sera likewise have been unsatisfactory. There is little, therefore, that can be accomplished serologically with present methods in the treatment of intestinal disturbances of bacterial causation. Attempts to permanently eliminate or destroy undesirable bacteria with cathartics and intestinal antiseptics have not been productive of results in the past[1] and prolonged starvation[2] *per se* does not lead to intestinal sterility or to a significant reduction in the offending bacteria.

Fig 97.—Bacillus bulgaricus. (Photograph by Dr. J. H. Stebbins, Jr., from the Fairchild culture of the Bacillus bulgaricus.

There are two ways, however, in which direct influence may be applied to bacteria in the intestinal tract: By a substitution of harmless types of organisms for abnormal types, and by varying the diet of the host in such a manner that the intestinal contents at the desired level shall contain nutritive substances that may be reasonably expected to shift the metabolism of the offending organism, and therefore radically change the character of the products of its metabolism.

A substitution of bacteria may be accomplished, theoretically at least, either by feeding cultures of organisms whose products of growth

[1] Kendall, Jour. Med. Research, 1911, xxv, 149, for brief résumé.

[2] Even after thirty-one days' starvation, a large number of viable bacteria were found in the lower part of the intestinal tract of the one case studied with this possibility in view.

are harmless to the host and more or less inimical to the bacteria it is desirable to supplant, or by administering a diet which contains appropriate nutritive substances in sufficient amounts to create conditions favoring the development of normal intestinal bacteria whose activities are in opposition to those it is desired to restrict or supplant.

The effects of a monotonous diet maintained for considerable periods of time upon the intestinal flora of a normal individual are clearly shown in the normal nursling, where intestinal organisms are largely carbohydrophilic and fermentative in character. Feeding experiments in normal animals indicate that the development of a nursling intestinal flora follows the prolonged administration of a nursling diet.

If the intestinal flora to be modified does not contain sufficient numbers of the desired types of bacteria, or if these latter organisms are inactive, it may be important to reënforce the weakened or inactive residual types with suitable cultures from without. Herter[1] was the first to recognize the possibility of introducing desirable types of bacteria into the alimentary canal and Metchnikoff[2] has extended and popularized this form of bacteriotherapy through his extensive studies upon the effects of milk soured with the Bulgarian bacillus as a therapeutic measure in excessive intestinal putrefaction. The Bulgarian bacillus[3] is a large Gram-positive organism, which is nonmotile and forms neither spores nor capsules. It develops feebly in ordinary media, but luxuriantly in milk, producing considerable amounts of lactic and other acids, but no gas. It is a milk parasite, having been perpetuated in this medium for many decades by the Bulgarian peasants.

The underlying principles of sour milk therapy as set forth by Metchnikoff are: a restriction of the protein in the diet, to reduce the available putrescible material in the intestinal tract; and the administration of liberal amounts of sour milk to flood the alimentary canal with preformed lactic acid. It was originally believed that the Bulgarian bacillus would become acclimatized in the intestinal tract and continue to produce lactic acid from the ingested carbohydrate, thus maintaining an acidity throughout the intestinal contents; this should create conditions inimical to the development of putrefactive organisms, which are said to be intolerant of acids. It is doubtful if the Bulgarian bacillus does become acclimatized in the large intestines,

[1] British Med. Jour., 1897, ii, 1847.
[2] Prolongation of Life.
[3] See Rahe, Jour. Inf. Dis., 1914, xv, 141, for description and differentiation from other aciduric bacteria.

where putrefactive action is maximal.[1] The theoretical and practical difficulties of acclimatizing a milk parasite in the intestinal tract would suggest that a normal intestinal organism of the lactic-acid type, as Bacillus acidophilus[2] (whose habitat is the large intestine), would be theoretically more efficient in those cases where Bacteriotherapy is indicated.

Bromatherapy.—The very direct and striking relation between the nature of the food of bacteria and the character of their products of metabolism has an important theoretical and practical application in relation to intestinal bacteriology in health and disease. It has been stated in another section that products of bacterial metabolism harmful to the host may be classified as nitrogenous compounds derived from proteins and protein derivatives, and non-nitrogenous compounds derived from carbohydrates and fats. The former are produced by bacteria acting upon proteins and their derivatives in the absence of utilizable carbohydrates; the latter are formed by bacteria which are utilizing carbohydrates or fats. Thus, the diphtheria bacillus forms a powerful toxin in protein media, but does not form toxin when available carbohydrate is added to the medium; Bacillus coli forms indol in protein media, but does not form indol when available carbohydrate is added to the medium. If these bacteria were developing in the intestinal tract at levels where a continuous supply of carbohydrate could reach them it would be theoretically possible to reduce or even prevent the formation of toxin or indol respectively when utilizable carbohydrates are present.

There are a number of intestinal conditions of bacterial causation in which available evidence points strongly to the formation of products arising from the metabolism of protein or protein derivatives by specific organisms as important etiological factors in the morbid process. Thus cholera, bacillary dysentery, typhoid, paratyphoid and many less acute infections are associated definitely with the development of these organisms within the body and, to some degree at least, at the expense of the body tissues.

All of these organisms produce lactic and other acids when suitable carbohydrates are available; the products of fermentation of these bacteria, chiefly lactic and other acids, are almost certainly no more harmful to the host than are those formed by Bacillus bulgaricus,

[1] Herter and Kendall, Jour. Biol. Chem., 1908, v, 293; Rahe, Jour., Inf. Dis., 1915, xvi, 210.
[2] Rotch and Kendall, Am. Jour. Dis. of Children, 1911, ii, 30.

Bacillus coli or Bacillus acidophilus, produced under like conditions. In other words, available evidence points strongly to the view that cholera vibrios, typhoid, dysentery and paratyphoid bacilli and similar organisms produce their characteristic and harmful effects when they are developing in media free from utilizable carbohydrate; when utilizable carbohydrates are added to these media, non-characteristic, harmless products are formed. It is frankly admitted that the chemistry of the products of nitrogenous metabolism of pathogenic bacteria is wholly unknown, and a rigorous proof of a relation between nitrogenous metabolism and disease is yet to be elucidated; the significant fact that the products of fermentation of these organisms are almost certainly innocuous to the host cannot be disregarded.

In the absence of any definite indication to the contrary it would be logical to attempt to maintain a sufficient concentration of carbohydrate within the intestinal canal in these infections as a therapeutic measure. This would be advantageous to the patient as a physiological procedure, as Coleman and Shaffer[1] have shown in their classical studies in typhoid, and it would provide continuously at least a minimal amount of readily utilizable carbohydrate which would shift the metabolism of all the intestinal organisms, pathogenic and non-pathogenic, in such a manner that harmless lactic acid would be formed by them. The bacteria under these conditions would theoretically, and in all probability practically, derive their energy from the readily fermentable carbohydrate and thus not only minimize their action upon the proteins of the intestinal contents,[2] but would tend to create an acid reaction there which in itself would be a potent agent in restricting the activity of the pathogenic organisms in the alimentary tract.

The associated bacteria of the intestinal tract also form acids under these conditions; Bacillus coli does not form indol, and other products of putrefaction are absent. Within a few days, under favorable circumstances, the cumulative effect of a diet liberal in carbohydrate will lead to a considerable development of aciduric bacteria, especially

[1] Arch. Int. Med., 1909, iv, 538.

[2] It is a well-attested fact that typhoid bacilli develop within the tissues of the body, and it might appear that a carbohydrate diet would therefore be ineffective; it is important to remember that the blood normally contains about 0.08 per cent. dextrose, an amount amply sufficient to protect protein from their attack. A liberal carbohydrate diet should tend to maintain the concentration of blood sugar at its physiological level. Recently Simonds (Jour. Inf. Dis., 1915) has shown that the products arising from the autolysis of typhoid bacilli grown in dextrose media are decidedly less toxic for rabbits than those grown in dextrose-free media when acted upon by specific lytic sera. This observation may well have an important bearing upon the case in question.

of the bifidus-acidophilus type if any be present in the alimentary canal to start with.[1] The intestinal contents are acid in reaction at this time and unfavorable for the development of the pathogenic types.

It must be realized that a number of conditions may reduce the theoretical efficiency of a diet rich in carbohydrate in intestinal infections; not infrequently the intestinal mucosa is inflamed and covered with an exudate of mucus and serum, alkaline in reaction and rather impermeable to intestinal medication. Stasis in the large intestine will frequently lead to a residue of protein derivatives there, quite free from carbohydrate, because the latter is readily hydrolyzed and absorbed as dextrose. There may be, and undoubtedly is, in some cases, a deficiency of the more effective lactic-acid-forming bacteria in the intestinal contents; whatever organisms are present, however, almost without exception form acids from carbohydrate, especially dextrose. The possibility of an overgrowth with the gas bacillus must be borne in mind if considerable quantities of sugars are to be administered.

Notwithstanding these difficulties, a diet rich in carbohydrate has been shown to be well tolerated in this type of infection, be it acute or chronic. Coleman and Shaffer,[2] using the high calory diet of the former in typhoid fever, have shown by careful chemical studies that the severe loss of nitrogen and of weight which occurs on a low calory diet can be very largely prevented by a diet comparatively rich in carbohydrate, and the symptoms of toxemia are materially reduced as well. Torrey[3] has shown that the changes in the intestinal flora in typhoid fever with the Coleman diet are, in general, a replacement of the more proteolytic bacteria by greater or lesser numbers of aciduric organisms, a change similar to that observed in bacillary dysentery,[4] in which the same general plan of liberal feeding of lactose was tried. The reduction in symptoms of toxemia in typhoid patients following a high calory diet including several ounces of lactose is significant; it can hardly be explained entirely on the theory of calories; it is very probable that a change in the metabolism of the typhoid bacillus is a potent factor in this phenomenon.

[1] Kendall, Boston Med. and Surg. Jour., 1910, clxiii, 398; 1911, clxiv, 288; Jour. Am. Med. Assn., 1911, lvi, 1084; Jour. Med. Research, 1911, xxiv, 411; 1911, xxv, 117. Kendall and Walker, Boston Med. and Surg. Jour., 1911, clxiv, 301. Kendall and Smith, ibid., 1911, clxv, 306. Kendall, Bagg and Day, ibid., 1913, clxix, 741. Kendall and Day, ibid., 1913, clxix, 753.

[2] Arch. Int. Med., 1909, iv, 538.

[3] Jour. Inf. Dis., 1915, xvi, 72. [4] Kendall, Boston Med. and Surg. Jour., 1911, clxiv.

To summarize, the important effects to be accomplished by a liberal carbohydrate diet in those infections where the decomposition of proteins or protein derivatives by bacterial activity leads to chronic or acute illness of intestinal origin are—a change in the metabolism of the offending organism resulting in the formation of lactic and other acids in them in place of putrefactive products, and a gradual replacement of the proteolytic and pathogenic types by bacteria of the fermentative varieties.

Another type of intestinal disturbance depends upon an unusual or an excessive decomposition of carbohydrate. The excessive formation of acid within the intestinal tract by an overgrowth of aciduric bacteria is well illustrated in young infants, especially those fed upon too much maltose.[1] The dietary treatment of such cases is too obvious to require further remarks. A group of cases which vary in severity from mild, long-continued diarrhea of several years' duration to very severe acute bloody diarrhea with great prostration are apparently caused by an overgrowth of the gas bacillus in the intestinal tract. This organism is relatively intolerant of lactic acid, and a diet practically free from carbohydrate, rich in protein, and reënforced by a liberal consumption of very acid buttermilk usually effects a rapid improvement in the acute cases, and a gradual improvement in those cases which are of months' or years' duration. Members of the Mucosus Capsulatus Group of bacteria may also, by overgrowth, set up a fermentative type of diarrhea which resembles that of the gas bacillus in its general features. The dietary treatment of these cases is like that of gas bacillus diarrheas.

[1] Kendall, Boston Med. and Surg. Jour., 1910, clxiii, 322.

SECTION V.

APPLIED BACTERIOLOGY.

CHAPTER XXXI.

BACTERIOLOGY OF MILK.

A VERITABLE river of milk, collected from many sources, flows daily into the larger cities of the country. Milk is an important food, particularly for infants and children, partly because it is relatively inexpensive and requires little or no preliminary preparation, chiefly because it contains in a small volume, all the essential nutritive elements combined in readily utilizable form. Herein lies its potential danger. It is a good culture medium for bacteria and its opacity precludes the possibility of visually detecting the contamination. Indeed, considerable amounts of dirt and filth may be introduced into milk without visibly changing its normal appearance.

It is inevitable, from existing conditions, that milk from many sources must be mixed before it appears in the open market; there may be an element of danger or a measure of safety in this homogenizing process. If milk from a single dairy happens to be infected with pathogenic bacteria, the degree of infection may be sufficient to effectively seed the entire volume with which it is mingled, or the degree of dilution may reduce the numbers of bacteria per volume below the danger point of infection for man.

The various manipulations to which milk is necessarily subjected before it reaches the consumer afford ample opportunity for bacterial contamination and the time which necessarily elapses between production and consumption furnishes one of the additional elements necessary for the development of adventitious bacteria. The temperature at which the milk is maintained is another important physical element which determines the extent of bacterial growth in it.

A moderate number of bacteria pathogenic for man may lead to infection of those who drink milk containing them, even if no development of these organisms has taken place. On the other hand, the growth of bacteria ordinarily not regarded as pathogenic may induce changes in this medium which render it unfit or even harmful for human use. If these changes are not of sufficient magnitude to alter the physical appearance of the milk, or if they are not perceptible to the senses, they may easily escape detection and yet lead to illness of the consumer. It is obvious, therefore, that those very elements which make milk a valuable food create conditions, themselves innocuous, through which it may become actively or passively a vehicle for the transmission of disease to man.

One of the great hygienic problems of the present time is that of maintaining and safeguarding the milk supply.

Sources of Bacterial Contamination of Milk.—Milk freshly drawn from the udder of a healthy cow, although practically never sterile, rarely contains many bacteria. The greatest contamination of milk probably takes place from unsterile utensils, although undoubtedly unclean animals, filthy surroundings and dusty air contribute many bacteria to it. Organisms introduced into milk from the hands of the milker and from his respiratory tract may be far more formidable to the consumer than mere numbers of saprophytic bacteria.

The ever-increasing application of complicated machinery for handling and bottling milk, while reducing to a large degree the possibility of contamination from human sources, provides a fruitful source of contamination with saprophytic organisms. The sterilization of machinery of this type is difficult to accomplish and not infrequently incomplete cleansing between periods of actual use leaves a residuum of fluid sufficiently rich in nutritive substances to permit of extensive bacterial development. The first portion of milk run through a machine in this condition must inevitably be grossly seeded with microörganisms.

The development of bacteria which have gained entrance to milk depends to a very considerable degree upon the temperature at which the milk is kept and the time which elapses between production and consumption.

Estimation of the Bacterial Content of Milk.—It is obvious from the preceding observations that adventitious milk bacteria may be harmful to man either because they are pathogenic or because they produce changes in the composition of milk which make it unfit for human consumption. From an hygienic point of view, therefore, milk offered for sale should be free from pathogenic microörganisms and of low bacterial content.

The numbers of bacteria in milk are determined in practice by two distinct methods:

(*a*) The numbers of organisms which will grow upon ordinary laboratory media, as nutrient agar (cultural count), and:

(*b*) By direct microscopic count.

(*a*) Cultural Count. Method: 1 c.c. of a well-mixed sample of milk is diluted $\frac{1}{100}$, $\frac{1}{1000}$, $\frac{1}{10000}$ or even $\frac{1}{100000}$ with sterile water, depending upon the grade of the sample, and plated on nutrient agar. The number of colonies which develop after forty-eight hours' incubation at 37° C. multiplied by the dilution is taken as the bacterial count of the milk. It is customary in some laboratories to make a parallel count at 20° C., after four days' incubation. The numbers of colonies developing on agar at the lower temperature may be much greater than those incubated at body temperature. The difference between the counts is usually more marked in samples of milk which have been maintained for some time at a relatively low temperature, and in ice-cream. In such cases bacteria whose minimal temperature of growth is relatively low—4° to 12° C.—may multiply with considerable rapidity. These organisms frequently fail to develop at 37° C.

The cultural count possesses advantages and disadvantages. The principal advantages are: the simplicity of the method, comparative accuracy of results provided uniform conditions are maintained, and some differentiation of the types of organisms present in the milk. The disadvantages are: the time required to obtain results—milk is perishable and cannot be held pending examination by this procedure. Furthermore, by no means all the bacteria which may theoretically gain access to the milk will grow upon plain agar; this is particularly true of pathogenic microörganisms. Bacteria which remain adherent in groups or chains are frequently not separated during the shaking of the sample and a single colony may originate from such a clump or chain. This naturally introduces an error which may be very considerable if, for example, a long chain of streptococci develops as a single colony.

(*b*) Direct microscopic count. Milk hygienists have long recognized the advantages of a direct estimation of the bacterial count of milk and numerous methods have been proposed, from time to time, to accomplish this object. The most practical method thus far prescribed appears to be that of Prescott and Breed.[1] The theory involved is to

[1] Centralb. f. Bakt., 1911, I, 246.

spread a definite volume of milk upon a definite area on a glass slide, evaporate the fluid, fix the sediment (which contains all the bacteria in the sample), and stain it in such a manner that the microörganisms are distinctly colored. The organisms of a definite area are counted under the microscope. The number in the original sample are readily computed, knowing the volume of milk examined, the area over which it is spread and the size of the microscopic field.

In practice 0.01 c.c. of a well-mixed sample of milk is spread uniformly over an area of 1 square centimeter on a glass slide. (This area is readily outlined with a wax pencil, using a pattern previously ruled on a piece of paper as a guide and following the outline on the glass slide; the wax pencil mark tends to limit the spread of the milk beyond the limits of the square.) The film of milk is then air-dried or dried at 40° C., immersed in absolute methyl alcohol for a few minutes to fix the sediment to the slide and to remove some of the milk lipoids and fats which interfere somewhat with the staining, and stained (after drying), with aqueous methylene blue. Alkaline methylene blue should not be used because the alkali tends to loosen the film of casein.

The bacteria are counted with an oil immersion lens. It is necessary to adjust the optical combination of lens and eye-piece so that the diameter of the microscopic field is exactly 0.0016 cm., corresponding to an area of 0.005 sq. cm. This can be readily accomplished with a stage micrometer.

Each organism in a microscopic field corresponds to one-five-hundred-thousandth the number in a cubic centimeter of the original sample of milk ($.005 \times 0.01 = 0.00005$), because 0.01 c.c. of milk was spread on an area of 1 sq. cm. and $\frac{1}{500}$ of the volume is viewed in the microscopic field. In other words, the microscopic field contains the bacteria of $\frac{1}{500000}$ c.c. of the original sample of milk and it is potentially equivalent to an agar plate culture of the milk in a dilution of $\frac{1}{500000}$.

If the bacteria were uniformly distributed, the number of bacteria observed in one field multiplied by 500,000 would give directly the number of bacteria per cubic centimeter in the milk; usually, however, the organisms are somewhat irregularly distributed and in practice several fields are counted and the average number of organisms per field is multiplied by 500,000. Duplicate determinations should always be made. The results obtained are fairly uniform when the exact details of the method are closely followed.

The advantage of the direct microscopical count are: a very material reduction in the time necessary to obtain results; milk which conforms to the standard may be quickly passed. Badly contaminated milk can be detected by simple inspection without even the formality of a count. There are also certain disadvantages. All bacteria which are stainable with methylene blue are visible by this method and dead organisms as well as those which are viable appear in the count. This is a decided source of error in pasteurized milk, where a relatively large proportion of bacteria are killed by heat; the method also does not distinguish sharply between different types of organisms.

On the whole, the advantages very materially outweigh the disadvantages and employed judiciously the method is of great practical value in the bacterial control of dairies and milk supplies.

The information obtained by the bacterial count is of importance chiefly from the viewpoint of the past history of the milk. Milk produced in cleanly surroundings, handled carefully in sterile utensils, kept cool and delivered promptly, should contain relatively few bacteria. If the milk is handled properly but not kept cool the numbers of organisms usually increase greatly, but as a rule the variety of organisms present will be limited. Improperly handled milk kept cool will frequently exhibit several types of bacteria, but not necessarily a high total count. A consistent low count with but few types of bacteria usually indicates a satisfactory milk supply.

Identification of Bacteria in Milk.—The bacterial types found in milk may be very varied; the opportunity for contamination does not cease when the milk is drawn from the cow—every step in the handling of the milk from the producer to the consumer offers new avenues for infection. A catalogue of all the bacteria which have been isolated from milk would be very extensive, but of little practical value. Of vastly greater importance is the recognition of the pathogenic organisms which may be transmitted to man and the chemical changes which ordinary saprophytic milk-bacteria induce in it. There are relatively few bacteria which are pathogenic both for the cow and for man. Of these, the bovine tubercle bacillus, the unknown virus of foot and mouth disease and the virus of the disease known as trembles of cattle are transmissible to man, the latter causing a well-defined symptom complex known as milk sickness. Goats, particularly Maltese goats, infected with the specific organism Micrococcus melitensis, transmit the disease Malta fever to man through their milk.[1]

[1] The detection of tubercle bacilli in milk has been discussed in the chapters on tuberculosis and bacillus abortus. Malta fever has been discussed in the chapter on Micrococcus melitensis and foot and mouth disease in the section relating to filterable viruses.

In addition, the viruses of certain infections specific for man may be transmitted in milk. These organisms gain entrance to the milk directly from human sources, incidental to the various handlings which it undergoes, and they may persist in unheated milk in sufficient numbers to infect the consumers. Typhoid, diphtheria, scarlet-fever, epidemic sore throat and pseudodiphtheria infection, dysentery (bacillary), various types of epidemic diarrhea and even Asiatic cholera are the more important diseases thus transmitted.

Except in very rare instances, specific pathogenic bacteria other than the bovine tubercle bacillus and Micrococcus melitensis have not been isolated directly from milk. The evidence of the transmission of pathogenic bacteria through infected milk rests largely upon statistical data. It is very conclusive, however, and many severe epidemics of typhoid fever and other infections have been satisfactorily traced to carriers or mild cases of the same disease among those who have undoubtedly handled the milk.

Conradi,[1] however, appears to have isolated the typhoid bacillus from infected milk which was shown to be responsible for a small outbreak of typhoid fever, and Bruck[2] and others have shown that typhoid bacilli and similar pathogenic bacteria may persist and even multiply in the presence of the various microörganisms commonly present in ordinarily good grades of milk.

The virus of foot and mouth disease and the bovine tubercle bacillus have been detected in butter and cheese prepared from milk containing these viruses.

The origin and relation of streptococci to milk-borne epidemics of septic sore throat and tonsillitis have been subjects of controversy. There appear to be two theories: one theory maintains that the streptococci are of bovine origin and presumably derived from the udders of cows which are suffering from mastitis or garget. The other theory assumes that these streptococci are usually of human origin and have gained entrance to the milk at some stage of its post-bovine history. Theobald Smith[3] and Brown have made an extensive study of this subject and their conclusions are of particular interest in this connection. They state that "there is at present no satisfactory evidence that bovine streptococci associated with mastitis or garget are the agent of tonsillitis in man. Whenever cases of

[1] Centralbl. f. Bakt., Orig., 1906, xl, 31.
[2] Deutsch. med. Wchnschr., 1903, xxix, 460.
[3] Jour. Med. Research, 1911, xxxi, 501.

garget are suspected as sources of infection in man, both human and bovine types should be looked for."

The most numerous of the saprophytic bacteria commonly found in raw milk belong to the group of organisms which form lactic acid, but no gas, from lactose. They are frequently referred to as lactic acid bacteria, but this name is not wholly appropriate nor is it distinctive; many unlike organisms possess this property in common. The best known and most widely distributed of these lactic acid bacteria is a streptococcus, Streptococcus lacticus,[1] an organism which is present not only in moderate numbers in the feces of the cow, but also upon the udder and flanks of the animal as well if cleanliness is not strictly observed. The initial infection of milk with Streptococcus lacticus is usually not extensive, but milk appears to be a particularly favorable medium for its development and even after a few hours the organism may have increased greatly in numbers if the temperature conditions are favorable. The most noteworthy chemical change associated with the growth of Streptococcus lacticus is a rapid accumulation of acid, principally lactic acid, which soon results in an acid coagulation of the casein. The degree of acidity is usually sufficient to inhibit the development of proteolytic bacteria and also a majority of pathogenic bacteria as well. Occasionally other types of fecal bacteria may be isolated from milk. Of these Bacillus coli has received much attention, chiefly through its constant association with human as well as with bovine excrement. Papasotirin and Prescott[2] have isolated bacteria indistinguishable from Bacillus coli by cultural methods from hay and dried grains and the organism is very frequently present in flour, consequently the identification of it in milk does not furnish conclusive evidence of contamination either from human or bovine sources. Bacillus coli does not produce more than minimal amounts of gas in milk, although its aërogenic activity in dextrose and lactose broth is one of its noteworthy cultural characters. It does, however, form sufficient acid from lactose to cause an acid coagulation of the casein. In this respect it does not differ markedly from other lactic acid bacteria. Occasionally, in association with a strongly proteolytic bacterium, as certain strains of Bacillus mesentericus, a deep-seated change is brought about in milk by the combined action of the two organisms. Bacillus mesentericus acting alone liquefies

[1] Kruse, Centralbl. f. Bakt., 1903, I. Abt., xxxiv, 737; Heinemann, Jour. Inf. Dis., 1906, iii, 173.
[2] Centralbl. f. Bakt., Ref., 1903, xxxiii, 279.

the casein; in symbiosis with Bacillus coli not only are the protein constituents of the milk thoroughly decomposed—a large volume of gas is formed as well and the milk-sugar is converted into carbon dioxide, hydrogen and lactic acid.[1] The alkaline products of putrefaction formed by Bacillus mesentericus neutralize, to a large degree, the acid products formed by Bacillus coli and the net change in the chemical composition of the milk is much greater than the sum of their separate activities.

Abnormal bacterial fermentations of milk are occasionally sources of great trouble to dairymen. One of the more common of these is known as ropy or shiny milk, in consequence of the viscidity which develops. Several kinds of bacteria cause ropiness, but of these Bacillus lactis viscosus appears to be more frequently concerned. A bitter flavor may be imparted to milk either from the feed of the cow or by the growth of bacteria. The latter is usually due to the partial digestion of the milk proteins resulting in an accumulation of peptones. The gas bacillus—Bacillus aërogenes capsulatus—produces an energetic fermentation of milk-sugar and eventually a rather deep-seated digestion of the casein if its activity is not restricted. The spores of the organism are very resistant to physical agents and are often found in commercial grades of lactose, which is prepared from milk. There is evidence that this organism, transmitted through milk, may incite mild or severe diarrhea in children, less frequently in adults. Pasteurized milk, particularly that originating in unclean dairies, occasionally contains considerable numbers of gas bacilli and the absence of lactic-acid-forming bacteria in such milk (which normally restrain their activity) may be a factor in its ability to develop rapidly.

Proteolytic bacteria, particularly spore-forming varieties of the Subtilis-Mesentericus Group, decompose milk proteins with the formation of casein peptones or even polypeptids. They occasionally multiply rapidly in pasteurized milk, when the degree of heat applied has been sufficient to kill the lactic-acid-producing bacteria; ordinarily lactic acid restrains the growth of proteolytic bacteria.

Pathogenic bacteria, as a rule, produce very little change in the appearance of milk and the chemical composition also is not greatly altered during their development.[2] Ordinarily it is impracticable to search for pathogenic bacteria in this medium, for the chances of success are minimal.

[1] Kendall, Boston Med. and Surg. Jour., 1910, clxiii, 322.
[2] Kendall, Day, and Walker, Jour. Am. Chem. Soc., 1914, xxxvi, 1937–1966.

Milk and Its Relation to the Public Health.—The importance of milk as a medium for the transmission of pathogenic bacteria is shown in the following list transcribed from the compilation of Trask.[1] Statistics of 317 epidemics of typhoid fever, 125 epidemics of scarlet fever, 51 epidemics of diphtheria and 7 of septic sore throat are set forth therein. This list is by no means regarded as complete; it includes only those epidemics of recent years in which satisfactory evidence of the origin and spread of disease is available.

Milk that is free from frankly pathogenic microörganisms is not necessarily a suitable food for man; it may be deadly for young children and infants. In the past little was definitely known of the relation of market milk to the high death rate among children, although a very direct connection was suspected. Park and Holt, however, made an extensive study of this very important question and their results are illuminating. Their plan was to feed ten groups of children with milk of known origin; this milk was mixed to secure uniformity and divided into ten portions. One-half, containing about 1,200,000 bacteria per cubic centimeter at the time of feeding, was distributed to one group; the other half was pasteurized before delivery. It contained, on the average, about 50,000 viable bacteria per cubic centimeter. The observations were carried on during the three warmest months of the year. Within a week nearly two-thirds of the infants fed with raw milk developed mild or severe diarrhea; about 25 per cent. remained well. Of those receiving pasteurized milk about 25 per cent. developed diarrhea and 75 per cent. remained well. A similar experiment was made the following summer. Their conclusions were:[2]

"1· During cool weather, neither the mortality nor the health of the infants observed in the investigation was appreciably affected by the quality of the market milk or by the number of bacteria which it contained. The different grades of milk varied much less in the amount of bacterial contamination in winter than in summer, the store milk averaging only about 750,000 bacteria per cubic centimeter.

"2· During hot weather, when the resistance of the children was lowered, the kind of milk taken influenced both the amount of illness and the mortality; those who took condensed milk and cheap store milk did the worst and those who received breast milk, pure bottled milk and modified milk did the best. The effect of bacterial contam-

[1] Bulletin 41 of the Hygienic Laboratory, Washington, D. C., January, 1908.
[2] Park and Holt, Arch. Pediat., December, 1903, 881.

39

ination was very marked when the milk was taken without previous heating; but unless the contamination was very excessive, only slight when heating was employed shortly before feeding.

"3· The number of bacteria which may accumulate before milk becomes noticeably harmful to the average infant in summer differs with the nature of the bacteria present, the age of the milk and the temperature at which it has been kept. When the milk is taken raw, the fewer the bacteria present the better are the results. Of the usual varieties, over 1,000,000 bacteria per cubic centimeter are certainly deleterious to the average infant. However, many infants take such milk without apparently harmful results. Heat above 170° F. (77° C.) not only destroys most of the bacteria present, but, apparently, some of their poisonous products. No harm from the bacteria previously existing in recently heated milk was noticed in these observations unless they had amounted to many millions, but in such numbers they were decidedly deleterious.

"4· When milk of average quality was fed, sterilized and raw, those infants who received milk previously heated did, on the average, much better in warm weather than those who received it raw. The difference was so quickly manifest and so marked that there could be no mistaking the meaning of the results.

"5· No special varieties of bacteria were found in unheated milk, which seemed to have any special importance in relation to the summer diarrhea of children. A few cases of acute indigestion were seen immediately following the use of pasteurized milk more than thirty-six hours old. Samples of such milk were found to contain more than 100,000,000 bacteria per cubic centimeter, mostly spore-bearing varieties. The deleterious effects, though striking, were neither serious nor lasting.

"6· After the first twelve months of life, infants are less and less affected by the bacteria in milk derived from healthy cattle. According to these observations, when the milk had been kept cool, the bacteria did not appear to injure the children over three years of age at any season of the year, unless in very great excess.

"7· Since a large part of the tenement population must purchase its milk from small dealers, at a low price, everything possible should be done by health boards to improve the character of the general milk supply of cities by enforcing proper legal restrictions regarding its transportation, delivery and sale. Sufficient improvements in this respect are entirely feasible in every large city, to secure to all a milk which will be wholesome after heating. The general practice of heating milk, which has now become a custom among the tenement population of New York, is undoubtedly a large factor in the lessened infant mortality during the hot months.

"8. Of the methods of feeding now in vogue, that by milk from central distributing stations unquestionably possesses the most advantages, in that it secures some constant oversight of the child and, since it furnishes the milk in such a form that it leaves the mother least to do, it gives her the smallest opportunity of going wrong. This method of feeding is one which deserves to be much more extensively employed and might, in the absence of private philanthropy, wisely be undertaken by municipalities and continued for the four months from May 15 to September 15.

"9. The use, for infants, of milk delivered in sealed bottles, should be encouraged whenever this is possible, and its advantage duly explained. Only the purest milk should be taken raw, especially in summer.

"10. Since what is needed most is intelligent care, all possible means should be employed to educate mothers and those caring for infants, in proper methods. This, it is believed, can most effectively be done by the visits of properly qualified trained nurses or women physicians to the homes, supplemented by the use of printed directions.

"11. Bad surroundings, though contributing to bad results in feeding, are not the chief factors. It is not, therefore, merely by better housing of the poor in large cities that we will see a great reduction in infant mortality.

"12. While it is true that even in tenements the results with the best bottle feeding are nearly as good as average breast feeding, it is also true that most of the bottle feeding is at present very badly done; so that, as a rule, the immense superiority of breast feeding obtains. This should, therefore, be encouraged by every means and not discontinued without good and sufficient reasons. The time and money required for artificial feeding, if expended by the tenement mother to secure better food and more rest for herself, would often enable her to continue nursing with advantage to her child.

"13. The injurious effects of table food to infants under a year old, and of fruits to all infants and young children in cities, in hot weather, should be much more generally appreciated."

These observations do not correlate the incidence of diarrhea with specific microörganisms, but they do furnish strong presumptive evidence of the relative salubriety of milk containing small numbers of bacteria. The importance of a consistently low bacterial content in milk designed for human consumption has been generally recognized by city, state and national health bureaus, and the grading and control of public milk supplies has been one of the great hygienic questions of the last decade. The older conception of a chemical standard to safeguard the financial interest of the consumer has been broadened to include a bacteriological standard which aims to exclude milk containing an excessive number of bacteria from the public market. The bacterial standard adopted varies somewhat in different cities, but in general it is so defined that all milk which meets its requirements must of necessity be produced in clean dairies, handled carefully and consistently maintained at a low temperature. The bacterial standard is based upon the number of bacteria per cubic centimeter of milk and it is rapidly becoming a custom to recognize grades of milk, each of which must conform to certain regulations regarding production, handling and bacterial count.

Certified milk is the hygienic grade milk. It is usually the product of a single dairy; the cows must be free from tuberculosis or other disease and stringent regulations for the condition of the entire plant are laid down. The milk as delivered must contain less than the maximum number of bacteria per cubic centimeter, as set forth in the standard. Usually the standard specifies 10,000 to 30,000 bacteria per cubic centimeter. Certified milk is usually safe milk, but contamination of it with human pathogenic organisms is not at all impossible. Ordinary market milk is produced under less rigorous conditions and the bacterial content is usually much greater; from 100,000 to 500,000 bacteria per cubic centimeter, or even 1,000,000 bacteria represent the usual standards enforced.

Pasteurization of milk is rapidly becoming obligatory in many cities, particularly for the ordinary grades of milk. Pasteurization is carried out by heating milk to about 145° F. (the degree of heat varies in different places), and maintaining it at that temperature for thirty minutes. This degree and duration of heat is deemed sufficient to weaken or destroy pathogenic organisms without altering the nutritive value. The ideal method of pasteurization is to heat the milk to the required temperature for the required time in the bottle which goes to the consumer, thus entirely eliminating the danger of human contamination subsequent to the process.

The pasteurizing process does not kill many of the milk bacteria;

thus Ayers has shown that an exposure of thirty minutes at a temperature of 145° C. fails to kill all colon bacilli.[1] The bacteria which survive pasteurization at this temperature are chiefly acid formers.[2]

Cellular Elements of Milk.—It has long been known that milk drawn from healthy cows contains variable numbers of cellular elements; these elements have been variously referred to as leukocytes, milk leukocytes, pus cells or gland cells. They may be either mononuclear or polymorphonuclear, and there is little unanimity in interpreting their significance. Harris[3] believes they have little sanitary significance as a general rule. Attempts have been made to correlate the numbers of cellular elements in milk with the leukocyte and erythrocyte count of the blood of the homologous animal, but without avail.[4]

It is a fact, however, that an inflammation of the udder of the cow is frequently associated with an unusually large number of cells in the milk, indistinguishable from polymorphonuclear leukocytes, and at times these cells are phagocytic. The increase in cellular content may be definitely restricted to one quarter of the udder.

An examination of the milk freshly drawn from 168 normal cows was made quantitatively for cellular elements and over 80 per cent. of the animals (composite sample from four quarters of the udder) showed less than 400,000 cells per cubic centimeter of milk. The period of lactation appeared to exercise little influence upon the cellular content, provided the samples were collected at least two weeks after parturition.[5]

[1] Jour. Agr. Res., 1915, iii, No. 5.

[2] Ayers and Johnson, Bull. 126, Bureau Animal Industry, 1910; ibid, Bull. 161, 1913.

[3] Jour. Inf. Dis., 1907, Supp. III, p. 50.

[4] At present comparatively little attention is directed to the cellular content of milk, and inasmuch as it is usually impossible to trace the milk to its source after it is bottled in the city, the method is not of much practical importance. A careful histological study of the cellular elements of milk by a competent cytologist might reasonably be expected to throw at least some light upon the origin and significance of milk leukocytes.

[5] Kendall, Collected Studies from the Research Laboratory, New York City, iii, 169.

CHAPTER XXXII.

BACTERIOLOGY OF THE SOIL, WATER, AND AIR.

SOIL.

THE upper layers of the soil in arable regions of the Torrid and Temperate Zones are densely populated with bacteria, many of which occur with such regularity that they are properly regarded as the normal bacterial flora of the soil. Others are of transitory or accidental occurrence, reaching the soil from the air, from water, from excrement and other waste products of man and animals, and from the dead bodies of man, animals, and plants.

The very uppermost layer of the soil, the first two or three centimeters, which is exposed to sunlight and frequent desiccation, usually contains fewer bacteria than the next layer, from 15 to 20 cm. in depth. Here the bacterial population is enormous, frequently reaching several millions of organisms per gram earth. Below this level the number of microörganisms diminishes rapidly, as Fraenkel[1] showed many years ago. At a depth of from one to two inches in undisturbed soil the bacterial flora is relatively insignificant in numbers and frequently no microörganisms are found.

The character of the soil and its state of cultivation are reflected in the bacterial population which will develop upon ordinary media. Thus sandy soil may contain but a few hundred thousand organisms.[2] Actively cultivated soils frequently contain one to several millions of bacteria.[3] Soil permanently covered with grass is usually relatively poor in bacteria.[4] The dust of streets may contain from one to ten million bacteria per gram,[5] and soil intimately contaminated with manure may exhibit as many as 78,000,000 bacteria per gram.[6] It is not surprising, from these figures, to find that the fertility of the soil is closely related to its bacterial population. Normal fertile soils

1 Ztschr. f. Hyg., 1887, ii, 521.
2 Adametz, Untersuch. ü. niederen Pilze der Akerkrume, 1886.
3 Chester, Delaware Agr. College Expt. Station Report, 1900-1901.
4 Chester, Bacteria of the Soil, etc., Bull. No. 98, U. S. Dept. of Agriculture.
5 Manfredi, Atti della R. Acad. della Science di Napoli, 1891, ii.
6 Maggiora, Roy. Accad. di Medicina, 1897, No. 3.

contain large numbers of microörganisms, and sand, which is notoriously infertile, contains relatively few.

The normal bacterial flora of fertile soil consists essentially of at least two distinct types of organisms; they may be classified according to their chemical activity into those which effect a rapid deep-seated decomposition of dead organic matter into simple combinations of the elements which enter into its composition—ammonia, carbon dioxide, hydrogen sulphide, and so on—and those which transform these simple compounds, especially ammonium salts, into nitrites and eventually into fully-oxidized (mineralized) nitrates. In the latter

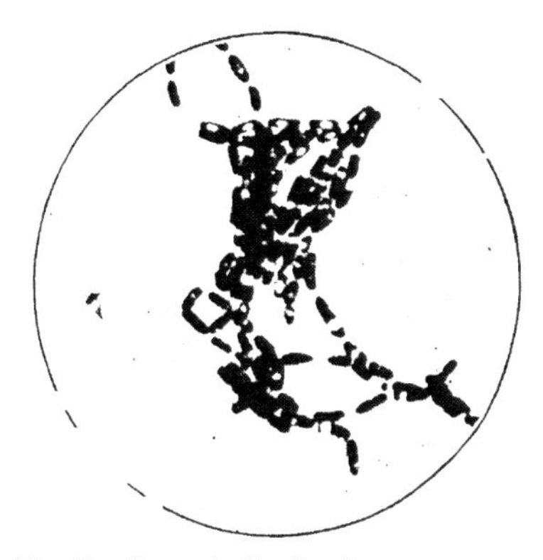

FIG. 98.—Bacillus subtilis showing spores. × 1000.

form the nitrogen originally present in organic matter is available for plant synthesis into protein through the action of sunlight upon the chlorophyll of the vegetable kingdom, thus completing the cycle.

The initial phase in the degradation of dead organic matter to ammonium salts and simple compounds of the other elements which comprise the protein molecule appears to be accomplished largely through the activity of bacteria of the Subtilis-Mesentericus and Proteus Groups. These organisms elaborate powerful active soluble proteolytic enzymes which liquefy protein, and eventually the intracellular digestion of the hydrolytic cleavage products of protein by these microörganisms results in ammonia formation.[1]

The Proteus Group has been discussed elsewhere.[2] The cultural characters of the Subtilis-Mesentericus Group are as follows:

[1] Kendall, Day and Walker, Jour. Am. Chem. Soc., 1913, xxxv, 1243; ibid., 1914, xxxvi, 1966; Jour. Inf. Dis., November, 1915.

[2] Page 359.

Morphology.—Rod-shaped organisms with rounded ends, occurring usually in chains of greater or lesser length. The individual cells measure from 0.7 to 1.2 microns in diameter, and vary in length from 2.5 to 9 microns. The members of the group are actively motile prior to sporulation and possess numerous peritrichic flagella. No capsules are formed, but spore formation is a characteristic feature of the group. The morphological details of spore formation and spore germination are relied upon largely to distinguish the various members of the group, but these details are of no practical significance in this discussion.[1]

Isolation and Culture.—The organisms of the Subtilis-Mesentericus Group grow with great luxuriance upon ordinary cultural media. The colonies on agar are irregular in shape, opaque, and spread rapidly. Gelatin colonies are similar in appearance and the medium is rapidly liquefied. Blood serum and casein are also liquefied. Milk is coagulated and the coagulum dissolves; the reaction, at first slightly acid, soon becomes alkaline as a rule. Indol, ammonia in considerable amounts,[2] hydrogen sulphide and other products of protein decomposition are formed in dextrose-free media and cultures of the organisms contain very powerful soluble proteases. The addition of dextrose to such media definitely prevents the formation of such proteases, however.[3]

As a rule the Subtilis-Mesentericus bacilli are non-pathogenic, but Silberschmidt[4] and others have described a type of ophthalmia in Switzerland, apparently incited by Bacillus subtilis, and Spiegelberg,[5] Flügge,[6] Ardoin[7] and more recently Vincent[8] have presented evidence in favor of the view that the organisms may become temporarily localized in the intestinal tract and incite severe gastro-intestinal disturbances.

It is stated that Bacillus subtilis differs from Bacillus mesentericus and other members of the group in its inability to ferment dextrose. The other varieties form acid but no gas from this sugar.

The foregoing observations have shown that the normal bacterial flora of the soil plays a prominent part in agriculture; it transforms dead unavailable organic matter and certain minerals as well into

[1] Gottheil, Centralbl. f. Bakt., 1901, vii, II Abt. Arthur Meyer, Practicum d. botanischen Bakterienkunde, Jena, 1903. Chester, Delaware College Agricultural Expt. Station, Ann. Rept., 1902–1903.

[2] Kendall, Day and Walker, loc. cit.

[3] Kendall and Walker, loc. cit.

[4] Ann. Inst. Past., 1903, xvii, 268.

[5] Jahrb. f. Kinderheilk., 1899, xlix, 194.

[6] Ztschr. f. Hyg., 1894, xvii, 272.

[7] Thèse de Paris, 1898, p. 78.

[8] Intestinal Toxemia in Infants, 1911.

compounds suitable for plant food. It is essential to relate in some detail the manner in which these transformations are accomplished.

The amount of nitrogen available at the present time for synthesis by plants exists chiefly in an organized state, and as nitrates in the soil. Nitrates are very soluble and it is obvious that large amounts of available nitrogen are yearly carried in solution to the ocean where they are practically lost. Brandt[1] estimates this loss to be about 40,000,000 kilograms annually. It is obvious that this loss must be compensated for.

It is a matter of common observation that soil left uncultivated gains in fertility from year to year and in 1875 Barthelot, and Nobbe and Hiltner[2] made the important discovery that nitrogen from the air is fixed in the soil. It was found that soil heated to 100° C. lost its power of fixation of nitrogen, suggesting that microörganisms played a part in the process. In 1888 Beijerinck[3] made the very important discovery that nodules[4] upon the roots of leguminous plants contain pleimorphic organisms, Bacillus radicicola, which were able to fix atmospheric nitrogen. Mazé[5] and others have confirmed this observation. Somewhat later Winogradsky[6] isolated an anaërobic spore-forming bacillus, Clostridium pasteurianum, not depending upon plants for its sustenance, but free living, which accomplished the same transformation, and in 1901 Beijerinck[7] isolated and described the very important group of Azobacteria, which are widely distributed in the soil and are able to fix atmospheric nitrogen. These organisms are most active when associated with other soil bacteria, but are fully able to fix nitrogen when grown in pure culture in artificial media free from nitrogenous compounds.

The oxidation of ammonia salts to nitrites and then to nitrates is effected through the activities of nitrifying bacteria, first isolated and described by Warrington and Winogradsky. Two organisms are concerned, a coccus, Nitrosococcus, which transforms ammonium salts to nitrites, and a small bacillus, Nitrobacter, which oxidizes nitrites to nitrates. These organisms do not thrive in the presence of complex organic matter and appear to derive their nutritive require-

[1] Report Kommission zur Untersuch. d. deutsche Meere, 1899–1901.

[2] Landwirthsch. Versuchsstat, xlv.

[3] Bot. Zeitung, 1888, 725.

[4] These nodules were first described by Hellriegel (Tageblatt Naturforsch. Vers., Berl., 1886, 290) and Willforth (ibid., 1887, 362).

[5] Ann. Inst. Past., 1897, xi; 1898, xii.

[6] Compt. rend. Soc. biol., 1893, cxvi, 1385; 1894, cxviii, 353.

[7] Centralbl. f. Bakt., 1901, vii, 562, II Abt.

ments chiefly from inorganic salts. The nitrates are taken up by chlorophyll-bearing plants and, with the energy of sunlight transform them, together with carbon dioxide, water, phosphates and various salts, into the complex vegetable proteins upon which the animal kingdom primarily subsists.

It is obvious, therefore, that there is a well-defined nitrogen cycle—an intricate series of changes which proteins and their derivatives undergo, through which complex, lifeless nitrogenous compounds are reduced through bacterial activity to simple, stable mineralized inorganic combinations of their elements. These elements are restored, chiefly through the synthetic activity of plant life, to the animal kingdom. The nitrogen cycle is, in a sense, a measure of the metabolism of the living earth, in which the anabolic or synthetic processes occur in plants and indirectly in animals; the catabolic or analytic process is brought about chiefly by bacteria.

In addition to the normal bacterial flora of the soil and adventitious saprophytic organisms, pathogenic bacteria are occasionally found; Bacillus typhosus, dysentery and cholera organisms and other excrementitious bacteria are occasionally deposited on the ground with human excrement. These microörganisms do not, as a rule, survive prolonged exposure to air, sunlight and other unfavorable environmental vicissitudes, however. Certain spore-forming bacteria—Bacillus tetani, anthrax, symptomatic anthrax, malignant edema and gas bacilli are very common in certain places. These bacteria, except anthrax, appear to multiply in the intestinal tracts of the herbivora.

The natural or biological degradation and mineralization of dead organic matter by bacterial activity in the upper layers of the soil, so essential to promote fertility, is of paramount importance in the purification of water and sewage. Indeed, the essential features of the nitrogen cycle are involved in both instances.

WATER AND SEWAGE.

The very general distribution of bacteria in the superficial layers of the soil makes it almost inevitable that waters which wash the surface of the earth shall receive some bacteria, consequently rivers and smaller streams, lakes and other surface waters always contain bacteria and other microörganisms. The number of bacteria per unit volume, however, is far less in water than upon the land, unless floods carry large amounts of soil with adherent organisms directly into water courses. Then the bacterial content of the water is greatly increased.

The bacterial flora of surface waters is normally considerably reduced by the action of sunlight—which is germicidal at a depth of several feet in quiet, clear water—by dilution, sedimentation, oxidation, and by the activities of predatory aquatic animals. The average soil pollution of water by surface contamination in sparsely populated drainage areas is not harmful to man, and such waters would ordinarily be suitable for domestic use.

Unfortunately water courses are convenient channels for the removal of human waste, including excreta, and such waste is potentially dangerous because it may contain pathogenic bacteria. Extensive epidemics of water-borne excrementitious disease—as typhoid and cholera—have focused attention upon the potential dangers attending the use of unpurified surface water for domestic purposes, and the statistical evidence of a reduction in the incidence of intestinal diseases when water supplies have been purified by filtration or by other methods is conclusive proof of the occasional transmission of excrementitious diseases through polluted water.

Ground water—from deep wells and from springs—is usually relatively free from bacteria unless surface pollution occurs. The water which feeds these sources is filtered free from bacteria during its passage through the deeper layers of the soil. Ground water is not extensively used for municipal supplies at the present time. Surface waters furnish the principal available sources of this commodity for domestic use, and in thickly settled areas it has been found necessary to purify the water before it is safe for human consumption.

The objects of water purification are: To eliminate pathogenic bacteria, and to reduce the dissolved and suspended organic matter to a state of complete oxidization and mineralization. It will be remembered that bacteria of the soil effect a mineralization of organic substances, and the purification of water and of sewage, which is grossly polluted water, is ordinarily accomplished by a direct application of the same natural process.

For convenience in operation, filters are constructed which are essentially water-tight basins (to prevent the entrance of extraneous, unpurified water) containing underdrains covered with a layer of sand of uniform size, from two to four feet in thickness.[1] The under-

[1] The details of structure and operation of filters designed for the purification of water and sewage are beyond the scope of this volume.

drains are designed to remove the purified water and they have little or nothing to do with the actual process or purification. The sand layer *per se* has little action in the purifying process; it does not strain out bacteria, because the spaces between the sand grains are very great compared with the size of the organisms. The sand does support upon its upper surface, however, a thin, delicate continuous layer of microörganisms, the Schmutzdecke, through which the water (or sewage) passes. This layer is so compact and so closely matted together that all suspended matter (including both pathogenic and non-pathogenic bacteria) in the supernatant water is strained out, and the dissolved organic substances pass with the raw or unfiltered water through the bodies of the microörganisms which collectively comprise the Schmutzdecke. During this passage the dissolved organic matter undergoes the same general degradation to nitrates and other fully-mineralized products of microbic digestion that organic substances in the upper layers of the soil undergo; the purification of water by sand filtration is, therefore, a catabolic phase in the nitrogen cycle, brought about by bacterial activity precisely as the mineralization of organic substances in the upper layers of the soil is a catabolic phase of the nitrogen cycle. The final products in each case are normally nitrates and other inorganic salts.

The efficiency of the purification of water or of sewage by the method of sand filtration is therefore to be measured chemically and bacteriologically. Chemically a complete transformation of complex organic compounds (ordinarily determined as albuminoid and "free ammonia") to nitrates is an indication that the digestive power of the filter is at par. Bacteriologically a disappearance of all bacteria derived from human or animal excrement and a great reduction of the total numbers of bacteria in the filtered water as compared with the unfiltered water is evidence of the bacterial efficiency of the filter.

The chief source of danger in potable waters is bacterial contamination from human sources. A simple inspection of water frequently fails to detect contamination, and even a chemical examination may not suffice to reveal pollution. Millions of typhoid bacilli may be introduced into a liter of water without inducing changes that could be detected visually or chemically. The bacteriological examination of water, therefore, is from ten to one hundred times more delicate than the chemical examination as a means of detecting contamination of water with human or animal waste.

Bacteriological Examination of Water.—A bacteriological examination of water requires relentless attention to details, from the collection of the sample to its final analysis and interpretation.

Collection of Sample.—It must be borne in mind that a small volume of water—100 c.c. or less—is ordinarily collected as a sample representing thousands or millions of gallons, consequently sampling is an important detail in the bacteriological analysis of water. The colleeting bottle must be clean and sterile, and the site at which the sample is taken must be representative.

It is customary to obtain a sample of water from brooks, rivers and lakes at a distance from the shore, and preferably samples from different depths should be taken. The bottle must be immersed below the surface before water is allowed to enter it, to avoid surface scums. If water is taken from faucets or pumps the sample should not be collected until a sufficient flow has been established to make certain that the fluid has come directly from the water mains, or from the well itself.

As soon as the sample has been collected it should be examined; frequently this is impracticable, and the bottle should be surrounded with ice at once and shipped to the laboratory. Ice restrains bacterial development for some hours and this maintains the sample at approximately its original bacterial content.

Bacteriological Analysis of Water.—A bacteriological examination of water ordinarily includes a determination of the numbers of bacteria which develop in ordinary nutrient media at 20° C. and 37° C., a search for organisms characteristic of the excrement of man or animals, their approximate enumeration, and other tests which vary according to the source of the sample.

Counting Bacteria.—The counting of bacteria ordinarily signifies the numbers of microörganisms which will grow on gelatin incubated at 20° C., and those that develop on agar at 37° C. Unpolluted waters usually contain relatively few bacteria that will grow at body temperature, consequently the gelatin plate seeded with the same volume of water as the agar plate will show many more colonies than the latter; polluted waters show a more even distribution of types of bacteria that grow respectively at 20° C. and 37° C.

The amount of water to be plated in gelatin and in agar depends upon the source of the sample. Water from deep wells and from springs should contain relatively few organisms, and a cubic centimeter of the sample is usually "planted." Surface waters almost invariably contain more bacteria than ground waters; it may be necessary to dilute a cubic centimeter of the sample with 99 c.c. of sterile water to obtain the requisite distribution of organisms for an accurate estimation, or even higher dilutions may be necessary. Grossly polluted waters are diluted one thousand or even ten thousand times with sterile water before they are plated. In any event, not more than 200 colonies or less than 50 colonies should be present in the final dilution, for experience has shown that greater numbers of organisms materially restrict development, and fewer than fifty colonies upon a plate introduces an error in dilution.

Technic of Plating.—The sample of water, diluted to the required degree if necessary, is shaken vigorously to break up groups and chains of bacteria; a cubic centimeter of water is then removed with a sterile pipette into each of two sterile Petri dishes, being careful to prevent contamination.

A tube of sterile nutrient gelatin (10 c.c.) previously melted and cooled to 42° C., is then carefully poured over the water in one Petri dish, and melted nutrient agar is similarly poured into the other Petri dish. The water and culture fluid are intimately mixed by carefully tilting the plates, and then set aside to harden. The agar plate is inverted after it has hardened to prevent condensation of moisture upon the surface of the medium; this procedure reduces the possibility of confluence of surface colonies. The gelatin plate is not inverted.

Incubation at 20° C. for the gelatin plate and 37° C. for the agar follows. The agar plate is counted after forty-eight hours' incubation, the gelatin plate after four days.

Interpretation of Bacterial Count.—At best the quantitative estimation of bacteria in water and sewage is inexact and relative only. The many factors of error in sampling, lack of uniformity in media, the difficulties of counting colonies when several hundred have grown in one plate—all tend to reduce the accuracy and precision of the method. Again, the normal difference in bacterial content between ground waters, surface waters and polluted waters makes an interpretation of the bacterial count somewhat difficult. For example, 100 bacteria per cubic centimeter in a deep well water might have greater sanitary significance than 500 bacteria per cubic centimeter in a surface water, where bacterial counts are almost invariably higher.

Attempts have been made to establish arbitrary bacterial standards; thus, waters containing less than 100 bacteria per cubic centimeter were formerly regarded as safe waters; those containing from 100 to 500 organisms per cubic centimeter were regarded with suspicion, and those containing 1000 or more organisms were pronounced dangerous for domestic use. In the abstract these standards are fictitious; surface waters even in uninhabited districts may contain many hundreds of bacteria per cubic centimeter after rains, yet the bacterial count would convey but little information of the actual sanitary status of the water. Successive bacterial counts carried out over long periods of time, on the other hand, are frequently of very great value.[1]

A direct examination of water for pathogenic bacteria, as the typhoid bacillus, if it were practicable, would be a most satisfactory method of evaluating domestic water supplies, for it is the presence of these organisms harmful to man which, in the last analysis, makes water containing them dangerous for human consumption. Unfortunately it is not practicable, as numerous observers have amply demonstrated, to isolate pathogenic organisms of this type directly from water, and there are but few authentic records of a successful cultivation of the typhoid bacillus from water supplies known to be infected, in spite of numerous attempts.

The practical impossibility of isolating pathogenic bacteria from water has led to the development of methods for the detection of Bacillus coli and organisms found practically constantly in human and animal excrement. Bacillus coli is somewhat more tolerant of environmental conditions as they exist in water than Bacillus typhosus, and its constant presence in fecal discharges makes it somewhat more effective as an indicator of excrementitious contamination than the frankly pathogenic organisms. The simplest and in many respects the best method for detecting Bacillus coli in water is to add graduated amounts of the sample to be analyzed—beginning with 1 c.c. and decreasing the amount one-tenth in successive cultures—to lactose fermentation tubes.[2] A production of gas within twenty-four or forty-eight hours is suggestive, but not conclusive evidence of the presence of the organism. If gas develops some of the culture should be placed on Endo medium, and red colonies that develop are tested for their ability to produce acid and gas in dextrose and lactose media, for indol production in sugar-free broth, for their action upon milk,

[1] For an excellent résumé of the subject, see the Bacteriology of Surface Waters in the Tropics, Clemesha, London and Calcutta, 1912, and Prescott and Winslow, Elements of Water Bacteriology, New York, 1913.

[2] Theobald Smith, Notes on Bacillus coli communis and Related Forms, Am. Jour. Med. Sci., September, 1895, 283.

and an absence of liquefaction in gelatin. These reactions are regarded as satisfactory to establish the identity of Bacillus coli.

In some laboratories a direct plating of the sample of water in lactose litmus agar or upon Endo medium is practiced, but this procedure is considerably less sensitive than the fermentation enrichment method outlined above.

Colon bacilli may occasionally be isolated from considerable volumes of water—10 or 100 c.c.—when they cannot be detected with regularity in 1 c.c. or less. Very little significance attaches to such results, because experience has shown that even springs in uninhabited regions may occasionally contain a few colon bacilli, derived probably from chance contamination with the feces of wild animals. If, on the contrary, colon bacilli are regularly present in a water supply to such an extent that a cubic centimeter of the water gives a positive culture in a decided majority of attempts, that water is viewed with suspicion. If the organism is regularly present in one-tenth of a cubic centimeter, the water is judged unfit or dangerous for human consumption until it is purified.

Other organisms have from time to time been proposed as indicators of pollution—thus streptococci and gas bacilli have been studied in this connection—but up to the present time they have not been accepted as authoritative criteria for evaluating the potability of water supplies.

BACTERIA OF THE AIR.

Bacteria when dried and attached to dust particles may be wafted into the air and remain suspended there for considerable periods of time. Even the gentlest air currents suffice to prevent their settling out. At high altitudes and over large bodies of water the bacterial population of the air is very small indeed; over large cities and cultivated land the number of organisms in the air is frequently much greater. Heavy rains and snow tend to remove bacteria from the atmosphere, while dry windy weather increases the aerial contamination.

Usually the more hardy organisms alone are found in the air, but in houses and hospitals pathogenic bacteria may be detected occasionally; probably the extrusion of minute droplets of sputum[1] containing these organisms is a most potent factor in air contamination by bacteria.

[1] See Droplet Infection, p. 91.

Several methods have been proposed for the estimation of the number of bacteria in the air; that of Winslow,[1] which consists essentially in aspirating a definite volume of air through two flasks, each of which contains melted nutrient gelatin, is the simplest and most direct. Comparatively little has been accomplished thus far from a quantitative study of the bacterial population of the air; it is possible that an attempt to isolate specific types of pathogenic bacteria from theatres and other places where large numbers of people meet might throw some light upon certain features of the air transmission of bacterial infections which are not well understood at the present time.

[1] Science, 1908, xxviii, 28.

AUTHOR INDEX.

C

D

E

F

G

H

I

J

K

S

T

U

V

W

Y

Z

GENERAL INDEX.

B

C

D

E

Q

R

S

U

V

W

X

Y

Z

9 780265 396346